W9-CRL-668

Metallocene-based Polyolefins

Volume One

Wiley Series in Polymer Science

Series Editor:
John Scheirs
ExcelPlas Australia
PO Box 163
Casula, NSW 2170
AUSTRALIA

Modern Fluoropolymers
High Performance Polymers for Diverse Applications

Polymer Recycling
Science, Technology and Applications

Forthcoming titles:

Dendritic Polymers

Polymer–Clay Nanocomposites
Unique Polymers with a Dispersed Nanophase

Metallocene-based Polyolefins

Preparation, properties and technology

Volume One

Edited by

J. SCHEIRS
ExcelPlas Australia, Casula, NSW, Australia

and

W. KAMINSKY
Institut für Technische und Makromolekulare Chemie,
Universität Hamburg, Germany

WILEY SERIES IN POLYMER SCIENCE

John Wiley & Sons, Ltd
Chichester · New York · Weinheim · Brisbane · Singapore · Toronto

Chemistry Library

Chem

Copyright © 2000 John Wiley & Sons Ltd,
Baffins Lane, Chichester,
West Sussex PO19 1UD, England

National 01243 779777
International (+44) 1243 779777
e-mail (for orders and customer service enquiries): cs-books@wiley.co.uk
Visit our Home Page on http://www.wiley.co.uk
or http://www.wiley.com

All Rights Reserved. No part of this publication may be reproduced, stored in a retrieval system, or transmitted, in any form or by any means, electronic, mechanical, photocopying, recording, scanning or otherwise, except under the terms of the Copyright Designs and Patents Act 1988 or under the terms of a licence issued by the Copyright Licensing Agency, 90 Tottenham Court Road, London W1P 9HE, UK, without the permission in writing of the Publisher.

Other Wiley Editorial Offices

John Wiley & Sons, Inc., 605 Third Avenue,
New York, NY 10158-0012, USA

WILEY-VCH Verlag GmbH, Pappelallee 3,
D-69469 Weinheim, Germany

Jacaranda Wiley Ltd, 33 Park Road Milton,
Queensland 4064, Australia

John Wiley & Sons (Asia) Pte Ltd, Clementi Loop #02-01,
Jin Xing Distripark, Singapore 129809

John Wiley & Sons (Canada) Ltd, 22 Worcester Road,
Rexdale, Ontario M9W 1L1, Canada

Library of Congress Cataloging-in-Publication Data
Metallocene-based polyolefins : preparation, properties, and
    technology / edited by J. Scheirs and W. Kaminsky.
              p.   cm. — (Wiley series in polymer science)
     Includes bibliographical references and index.
     ISBN 0-471-98086-2 (set : alk. paper)
     1. Polyolefins— Synthesis.   2. Metallocenes.   3. Metallocene
catalysts.   I. Scheirs, John.   II. Kaminsky, W. (Walter), 1941–   .
III. Series.
     TP1180.P67M48   1999
     668.4'234—dc21   99-21151
                                                                    CIP

British Library Cataloguing in Publication Data

A catalogue record for this book is available from the British Library

ISBN 0 471 99911 3 (Vol. 1)
ISBN 0 471 99912 1 (Vol. 2)
ISBN 0 471 98086 2 (Set)

Typeset in Times by Techset Composition Ltd, Salisbury, Wiltshire
Printed and bound in Great Britain by Biddles Ltd, Guildford, Surrey
This book is printed on acid-free paper responsibly manufactured from sustainable forestry, in which at least two trees are planted for each one used for paper production. ac

Contents

TP 1180
P67
M 48
2000
v. 1
CHEM

I   NATURE OF METALLOCENE CATALYSTS

III PROPYLENE POLYMERIZATION

Contributors

A. Akimoto
Tosoh Corporation Research
Laboratory
1–8 Kasumi
Mie-ken 510, Yokkaichi-shi
Japan

M. Akiyama
Mitsui Chemicals, Inc.
Performance Materials R & D Center
1190 Kasama-cho
Sakae-ku
Yokoyama 247-8567
Japan

T. Asanuma
Central Research Institute
Mitsui Toatsu Chemicals, Inc.
1190 Kasama-cho, Sakae-ku
Yokohama 247-8567
Japan

M. C. Baird
Department of Chemistry
Queen's University
Kingston, Ontario
Canada K7L 3N6

A. R. Barron
Department of Chemistry
Rice University
6100 Main Street

Houston, TX 77005-1892
USA

J. C. W. Chien
Amherst Polymer Technology, Inc.
15 Coach Lane
Amherst, MA 01002
USA

S. P. Chum
Dow Chemical Co.,
Polyolefins and INSITE Technology
R&D
2301 N. Brazosport Blvd, B1470-D
Freeport, TX 77541-3257
USA

T. Dall'Occo
Montell Italia
G. Natta Research Centre
Piazza le G. Donegani 12
44100 Ferrara
Italy

U. Dietrich
Anorganische Chemie II
Molekülchemie & Katalyse
Universität Ulm
Albert-Einstein-Allee 11
D-89069 Ulm
Germany

S. W. Ewart
Department of Chemistry
Queen's University
Kingston, Ontario
Canada K7L 3N6

J. A. Ewen
Catalyst Research Corporation
14311 Golf View Trail
Houston, TX 77059
USA

D. Fischer
Targor GmbH
Font-Malakoff Park
Rheinstr. 4G
D-58116 Mainz
Germany

I. Fujio
Central Research Institute
Mitsui Toatsu Chemicals
1190 Kasama-cho
Sakae-ku,
Yokohama 247-8567,
Japan

O. Fusco
Montell Italia
G. Natta Research Centre
Piazza le G. Donegani 12
44100 Ferrara
Italy

M. Galimberti
Montell Italia
G. Natta Research Centre
Piazza le G. Donegani 12
44100 Ferrara
Italy

U. Giannini
Montell Italia

G. Natta Research Centre
Piazza le G. Donegani 12
44100 Ferrara
Italy

M. Hackmann
Anorganische Chemie II
Molekülchemie & Katalyse
Universität Ulm
Albert-Einstein-Allee 11
D-89069 Ulm
Germany

S. Harima
Mitsui Chemicals, Inc.
Performance Materials R & D Center
1190 Kasama-cho
Sakae-ku
Yokoyama 247-8567
Japan

G. G. Hlatky
Equistar Chemical Co.
Equistar Research Center
P.O. Box 429566
Cincinnati, OH 45249-9566
USA

N. Inoue
Central Research Institute
Mitsui Toatsu Chemicals Inc.
1190 Kasama-cho, Sakae-ku
Yokohama 247-8567
Japan

S. Jüngling
BASF AG
D-67056 Ludwigshafen
Germany

G. J. Jiang
Department of Chemistry
Chung-Yuan Christian University
Chung-Li
Taiwan 320
Republic of China

L. K. Johnson
DuPont Central Research and
Development
DuPont Experimental Station
Wilmington, DE 19880-0328
USA

E. Kaji
School of Materials Science
Japan Advanced Institute of Science
and Technology
1-1 Asahidai, Tatsunokuchi
Ishikawa 923-1292
Japan

C. I. Kao
Dow Chemical Co.
Bulding 1470D
2301 Brazosport Blvd
Freeport, TX 77541
USA

C. Killian
Eastman Chemical Co.,
Research Laboratories
Polymers Research Division
P.O. Box 1972
Kingsport, TN 37662-5150
USA

I. L. Kim
Department of Chemical Engineering
University of Ulsan
P.O. Box 18
Nam Ulsan

Ulsan 680–749
Korea

S. Kimura
Mitsui Chemicals, Inc.
Performance Materials R & D Center
1190 Kasama-cho
Sakae-ku
Yokoyama 247-8567
Japan

G. W. Knight
400 Cedar Creek Drive
Edmond, OK 73034
USA

M. Kohno
Central Research Institute
Mitsui Toatsu Chemicals Inc.
1190 Kasama-cho, Sakae-ku
Yokohama 247-8567
Japan

D.-H. Lee
Department of Polymer Science
Kyungpook National University
Taegu 702-701
Korea

R. Mülhaupt
Institut für Makromolekulare Chemie
Albert-Ludwigs Universität, Freiburg
Stefan-Meier-Str. 23
D-79104 Freiburg 1
Germany

S. K. Noh
School of Chemical Engineering and
Technology
Yeungnam University
Kyungsan 712-749
Korea

F. Piemontesi
Montell Italia
G. Natta Research Centre
Piazza le G. Donegani 12
44100 Ferrara
Italy

L. Resconi
Montell Italia
G. Natta Research Centre
Piazza le G. Donegani 12
44100 Ferrara
Italy

B. Rieger
Anorganische Chemie II
Molekülchemie & Katalyse
Universität Ulm
Albert-Einstein-Allee 11
D-89069 Ulm
Germany

M. J. Schneider
BASF AG
D-67056 Ludwigshafen
Germany

T. Shiomura
Central Research Institute
Mitsui Toatsu Chemicals Inc.
1190 Kasama-cho, Sakae-ku
Yokohama 247-8567
Japan

K. Soga
School of Materials Science
Japan Advanced Institute of Science
and Technology
1-1 Asahidai, Tatsunokuchi
Ishikawa 923-1292
Japan

M. Sone
Tosoh Corporation Research
Laboratory
1-8 Kasumi
Mie-ken 510, Yokkaichi-shi
Japan

W. Spalek
Hoescht AG
Polyolefin-Forschung, G 832
D-65926 Frankfurt am Main
Germany

R. Sugimoto
1190 Kasama-cho, Sakae-ku
Yokohama 247-8567
Japan

J. Suhm
BASF AG
D-67056 Ludwigshafen
Germany

K. Swogger
Polyolefin Research
Dow Chemical Co.
Building 1607
2301 Brazosport Blvd
Freepost, TX 77541
USA

A. Torres
Dow Europe SA
Bachtobelstr. 3
CH-8810 Horgen
Switzerland

N. Uchikawa
Central Research Institute
Mitsui Toatsu Chemicals Inc.
1190 Kasama-cho, Sakae-ku
Yokohama 247-8567
Japan

T. Uozumi
School of Materials Science
Japan Advanced Institute of Science and
Technology
1–1 Asahidai, Tatsunokuchi
Ishikawa 923-12
Japan

J. Vöegele
Anorganische Chemie II
Molekülchemie & Katalyse

Universität Ulm
Albert-Einstein-Allee 11
D-89081 Ulm
Germany

A. Yano
Tosoh Corporation Research Laboratory
1–8 Kasumi
Mie-ken 510, Yokkaichi-shi
Japan

Series Preface

The Wiley Series in Polymer Science aims to cover topics in polymer science where significant advances have been made over the past decade. Key features of the series will be developing areas and new frontiers in polymer science and technology. Emerging fields with strong growth potential for the twenty-first century such as nanotechnology, photopolymers, electro-optic polymers, etc., will be covered. Additionally, those polymer classes in which important new members have appeared in recent years will be revisited to provide a comprehensive update.

Written by foremost experts in the field from industry and academia, these books place particular emphasis on structure–property relationships of polymers and manufacturing technologies as well as their practical and novel applications. The aim of each book in the series is to provide readers with an in-depth treatment of the state-of-the-art in that field of polymer technology. Collectively, the series will provide a definitive library of the latest advances in the major polymer families as well as significant new fields of development in polymer science.

This approach will lead to a better understanding and improve the cross fertilization of ideas between scientists and engineers of many disciplines. The series will be of interest to all polymer scientists and engineers, providing excellent up-to-date coverage of diverse topics in polymer science, and thus will serve as an invaluable ongoing reference collection for any technical library.

John Scheirs
June 1997

Preface

This book provides an overview of metallocene catalysts and metallocene-based polyolefins. The manufacture of polyolefins by metallocene catalysts represents a revolution in the polymer industry. Polymerization of olefin monomers with single-site metallocene catalysts allows the production of polyolefins (such as polyethylene or polypropylene) with a highly defined structure and superior properties. Furthermore, the structure of these metallocene catalysts can be varied to 'tune' the properties of the polymer. Thus, for the first time it is possible to tailor carefully the properties of large-volume commodity polymers such as polyethylene and polypropylene.

Metallocene polymerization catalysts generally have a constrained transition metal (usually a group 4b metal such as Ti, Zr or Hf) which is sandwiched between cyclopentadienyl ring structures to form a sterically hindered site. Variations on this theme include ring functionalization with various alkyl or aromatic groups, ring bridging with either a Si or C atom and metal coordination to either an alkyl group or halogen atom.

While the term metallocene classically described compounds with π-bound cyclopentadienyl ring structures, today's catalysts are better described as being 'single-site catalysts'. They differ from traditional olefin polymerization catalysts by the fact that the catalytically active metal atom is generally in a constrained environment and thereby only allows single access by monomers. By confining the polymerization reaction to a single site instead of the conventional multiple sites, these catalysts permit close control over comonomer placement, side-chain length and branching. Such single-site catalysts allow precise control over polymer design because the polymer grows by a single mechanism rather than multiple routes. Metallocene catalysts can yield a polymer with almost perfect regularity and tacticity. This enables a range of different polyolefins each with a well-defined structure to be manufactured in one reactor configuration. Such polymers are also characterized by very narrow polydispersities and low extractables (one-fifth that of conventional polyethylenes).

Metallocene-based resins offer improved strength and toughness, enhanced optical and sealing properties and increased elasticity and cling performance. Metallocene-based polyolefins are having most impact in applications such as film packaging for products such as baked goods, meat and poultry, frozen foods, snack foods, cheese wraps and shrink film. Metallocene catalysts allow the manufacture of highly flexible and tenacious films. Traditionally cling and stretch films used to stabilize pallet loads contain a cling additive such as polyisobutylene which limit their recyclability. However new metallocene-catalyzed polyolefin films can achieve remarkable stretch properties without the need of such additives. Flexible plastics based on new metallocene polyolefins can also replace plasticized PVC in many applications without requiring the addition of environmentally damaging phthalate plasticizers. In addition to these areas, metallocene penetration is occurring in applications such as adhesives, disposable medical packaging, flexible foam products and wire and cable insulation.

Metallocene technology has matured significantly in the last 3 years. There has been no text published to date which covers these developments. Polyolefin technology is growing faster than any other polymer technology and metallocenes are at the centre of this frenetic activity. It has been estimated that within 10 years, 50% of polyolefins will be made by the metallocene route. In addition, copolymerization of ethylene with styrene opens the door to a new series of commodity polymers. Metallocenes can also polymerize bulky monomers such as norbornene, so a range of polymers await that hitherto were not commercially viable or possible.

This book gives comprehensive coverage to all areas of metallocene technology–catalyst structure, comonomer incorporation, polymerization mechanisms and conditions, reactor configurations, special properties, comparison with conventional polyolefins, rheological and processing behavior and fields of application.

John Scheirs

December 1998

Brief Historical Perspective

Metallocene catalysts for ethylene polymerization were reported as long ago as 1957 by Breslow and Newburg [1]. These were based on bis(cyclopentadienyl)titanium dichloride activated with an aluminium alkyl (AlEt$_3$). Such catalysts however, had no commercial viability because their polymerization activity was far inferior to comparable Ziegler–Natta catalysts based on titanium tetrachloride and AlEt$_3$.

The field was relatively dormant for almost two decades, then 18 years after his original discovery Breslow [2] reported that a substantial increase in the ethylene polymerization rate of bis(cyclopentadienyl)titanium dichloride–dimethylaluminium chloride could be achieved through the addition of small amounts of water to the catalyst. Shortly thereafter, Sinn, Kaminsky and coworkers [3] showed that a titanium-based metallocene mixed with trimethylaluminium became a highly active polymerization catalyst when water was added to the aluminium alkyl cocatalyst before reaction with the metallocene. They attributed the enormous increase in catalytic activity accompanying the addition of water to the formation of an alumoxane—a chain of alternating oxygen and aluminium atoms to which methyl groups are bound [now commonly referred to as methylalumoxane (MAO)]. Methylalumoxane is a hydrolysis product of trimethylaluminium. The bulky non-coordinating MAO anion stabilizes the metallocene cation to create the active site. In 1980, Sinn, Kaminsky et al. [4] took methylalumoxane that they synthesized separately and added this to a bent zirconocene (a zirconium atom bound to two chlorine atoms and sandwiched between two cyclopentadienyl rings) to yield a system that gave unprecedented catalytic activity (10^6 g PE/g Zr.h.bar).

Unfortunately the prototype zirconium metallocene-MAO system developed by Kaminsky was unsuitable for commercial production of polyethylene. In addition the catalyst exhibited very poor activity for propylene polymerization and offered no stereoselectivity. In 1984 John Ewen (then at Exxon) first demonstrated activity and molecular weight improvements for polyethylene by modifying the structure of Kaminsky's catalyst. Ewen also made the very first stereoselective metallocene catalysts for producing isotactic polypropylene [5, 6]. This stereoselective poly-propylene catalyst was a combination of Kaminsky's MAO and a Brintzinger

Figure 1 Professor Walter Kaminsky in his laboratory where he discovered highly active, metallocene polymerization catalysts based on zirconocenes in combination with methylalumoxane

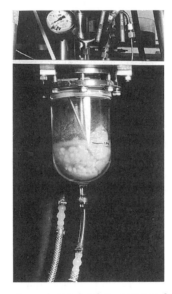

Figure 2 Kaminsky's early polymerization reactor for metallocenes research. While studying zirconocenes in combination with triethyl and methyl aluminium he accidentally discovered that water condensed in a sample tube led to an unexpected yield in the amount of polyethylene produced

Figure 3 John Ewen made the very first stereoselective metallocene catalysts which produced isotactic polypropylene. He is also responsible for the discovery of *ansa* zirconium metallocenes for the production of highly syndiotactic polypropylene

metallocene called *rac*-ethylidenebis(indenyl)titanium dichloride. This so-called *ansa* metallocene was selected because tying of the cyclopentadienyl rings together with a two carbon 'handle' made it more rigid (*ansa* literally means bent handle in Latin). The main feature of these catalysts is that the cyclopentadienyl ligands are connected by a bridge, hence they could not turn against each other and this made the overall molecule more rigid.

Around the same time Kaminsky, Brintzinger and coworkers produced highly isotactic polypropylene by using a zirconium metallocene [7]. A couple of years later Ewen produced yet another *ansa* zirconium metallocene with bilateral symmetry (typical of life: i.e., right and left handed) and predictably it produced highly syndiotactic (i.e. alternating pendant methyl groups), high molecular weight polypropylene when combined with MAO [8]. The syndiotactic polymer was more transparent then isotactic polypropylene but less stiff.

In the late 1980s, James Stevens and coworkers at Dow made the remarkable discovery that titanium catalysts based on certain monocyclopentadienyl (monoCp) metallocenes in which a donor ligand stabilizes the metal center are much better than those based on the corresponding bis-cyclopentadienyl metallocenes [9]. These catalysts have become known as 'constrained geometry catalysts' with one key

Figure 4 Some of the persons responsible for the development of Dow's constrained geometry catalysts for polyolefin polymerization. Back row from left: Kurt Swogger (Vice President of Dow Polyolefins and INSITE® Technology), James Stevens (catalyst development), Brian Kolthammer (process development), Bill Knight (materials science), Che-I Kao (process and polymer fundamentals). Front row from left: Jackie de Groot (applications development), Steve Chum (product development).

advantage being very high activities for ethylene and alpha-olefin copolymerizations. This catalyst technology plus new materials science and process capability were trademarked by the Dow Chemical Company as INSITE® Technology. Such catalysts have a silicon atom bonded to one of the carbons of a cyclopentadienyl ring coordinated to a titanium atom. The silicon atom is also bonded to an amino nitrogen which is coordinated to the titanium. The 'short-leash' silicon–nitrogen bond pulls on the cyclopentadienyl ring and opens up the bond angle between the ring and the other ligands making the metal site more accessible to longer chain monomers. In the 1990s, several new families of polymers were developed based on INSITE® technology.

John Scheirs, April, 1999

REFERENCES

1. Breslow, D. S. and Newburg, N. R., *J. Am. Chem. Soc.*, **79** 5072 (1957).
2. Long, W. and Breslow, D. S., *Liebigs Ann. Chem.*, 463 (1975).
3. Andresen, A., Cordes, H. G., Herwig, J., Kaminsky, W., Merck, A., Mottweiler, R., Pein, J., Sinn, H. And Vollmer, H. J., *Angew, Chem, Int. Ed. Engl*, **15**, 630 (1976).
4. Sinn, H. J., Kaminsky, W., Vollmer, H.-J. and Woldt, R., *Angew. Chem., Int. Ed. Engl.*, **19**, 390 (1980).
5. Ewen, J. A., *J. Am. Chem. Soc.*, **106**, 6355 (1984).
6. Ewen, J. A. and Welborn, H. C., Eur. Patent Appl. 0,129,368 (1984); Ewen, J. A., *Stud. Surf. Sci. Catal.*, **25** 271 (1986).
7. Kaminsky, W., Kulper, K., Brintzinger, H. H. and Wild, F. R., *Angew. Chem., Int. Ed., Engl.*, **24**, 507 (1985).
8. Ewen, J. A., Jones, R. L., Razavi, A. and Ferrara, J. D., *J. Amer. Chem. Soc.*, **110**, 6255 (1988).
9. Stevens, J. C., *Stud. Surf. Sci. Catal.*, **89**, 277 (1994).

About the Editors

John Scheirs

Dr John Scheirs was born 1965 in Melbourne and studied applied chemistry at the University of Melbourne. His PhD thesis was on the mechanism and fate of chromocene catalysts in the gas phase polymerization of high-density polyethylene. This work led to an understanding of how the polymerization catalysts affect polymer morphology and thermooxidative stability. Subsequently he worked with silica-supported polymerization catalysts based on chromocene and silyl chromate for the commercial production of high-density polyethylene at Kemcor Australia (an Exxon and Mobil venture). He has also been involved with metallocene-based packaging films and their characterization by thermal analysis. Dr Scheirs has authored over 50 scientific papers including eight encylopedia chapters, a book on polymer recycling and has given presentations at ACS, IUPAC and ANTEC symposia.

Walter Kaminsky

Professor Walter Kaminsky was born 1941 in Hamburg and studied chemistry at the University of Hamburg. His thesis was in the field of metallocene chemistry. Since 1979 he has occupied the role of Full Professor of Technical and Macromolecular Chemistry at the University of Hamburg.

He is currently supervising a group of 20 students and scientists in the field of metallocene/MAO chemistry and a group in the field of pyrolysis of plastic waste and scrap tyres. His past experience includes discovering a highly active, soluble metallocene catalyst system for the polymerization of olefins and for the copolymerization of ethylene with cyclic olefins. Other research interests are in the fields of chemical engineering, polymer recycling by pyrolysis and macromolecular chemistry.

He has published more than 200 papers and books and holds 20 patents. He has also organized several international symposia in the field of olefin polymerization and metallocene catalysis.

In 1991 he received, together with Hans H. Brintzinger, the Heinz Beckurts Prize for the isotactic polymerization of propylene with metallocene catalysts and in 1995, together with Hans H. Brintzinger and Hansjörg Sinn, the Alwin Mittasch Medal for Metallocene Catalysts. In 1997 he received the Carothers Award of the American Chemical Society (Delaware Section), and in the same year the Walter Ahlström Prize in Helsinki, Finland. Since 1997 he has been an Honorary Member of the Royal Society of Chemistry in London, and Honorary Professor of the Zheijiang University in China since 1998. In 1999 he received the Benjamin Franklin Medal for Chemistry as well as the SPE ANTEC Outstanding Achievement Award.

PART I
Nature of Metallocene Catalysts

1

Metallocene Polymerization Catalysts: Past, Present and Future

JOHN A. EWEN

Catalyst Research Corporation, Houston, TX, USA

1 INTRODUCTION

The science of homogeneous metallocene polymerization catalysis has proved to be very popular because it is both curiosity-driven and application-driven. The turning point from purely academic to applied research came 30 years after their discovery with the realization that changing the structures of the catalysts' cyclopentadienyl (Cp) ligands had favorable and exceptionally powerful effects on the catalyst's activities and, more importantly, on the polymers' structures and chain lengths. This was the 'Big Bang' of metallocene catalysis (Figure 1) [1].

The Cp catalysts rapidly evolved into a mature industry in terms of our understanding the catalyst structural effects on polymerization. The study of ligand effects has become so commonplace now that reviews rarely point out explicitly that this is what has most distinguished the field from traditional heterogeneous Ziegler–Natta catalysis [2]. The lion's share of metallocene research has unquestionably been as empirical as research in any other field of catalysis. However, the structures of both the catalysts and the polymers have provided unprecedented insights into understanding them at a molecular level. An overview is presented in this chapter on the more important breakthroughs in ligand and transition metal effects on polyethylene and polypropylene that resulted in the better known catalyst structures of today and their activators. These breakthroughs led to the syntheses of novel polymer structures, some of which have purely academic but nevertheless very elegant structures and others with desirable properties [2h].

Metallocene-based Polyolefins   Edited by J. Scheirs and W. Kaminsky
© 2000 John Wiley & Sons Ltd

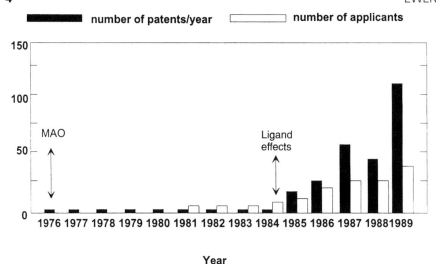

Figure 1 The 1984 'Big Bang' in metallocene catalysis as illustrated by the change in the number of patents issued between 1976 and 1989 [1]

The research described herein has mostly focused on *controlling* what the catalysts make. We now have several different ways of solving the bedeviling low molecular weight problems for LLDPE, EPR and PP found with the bis-Cp Zr prototypes. Ligands have been modified for polypropylenes with rigidity, the right symmetry and the appropriate bulk around the non-Cp coordination sites to give the desirable stereochemistry and stereospecificity. The smaller Ti has been most successful as more open, half-metallocenes in polyethylene and polystyrene where stereochemical control with ligand effects is lost for the most part.

Some of the theories as to how metallocene catalysts actually work are also mentioned because they explain why each catalyst has a particular structure for a given purpose. Catalyst chemists used such mechanisms to make creative guesses as to what structures were needed to make desirable polymerization reactions go and in testing working hypotheses they found an ability to *control* what catalysts with a given symmetry and structure actually make.

In addition, the multiform structures of the catalysts and polymers provided an unparalleled degree of diversity and it has even been said that the degree of a particular catalyst's stereoselectivity can almost be predicted *a priori* [2e]. The ability to anticipate qualitatively and even semi-quantitatively trends in catalyst behavior and the catalysts' high selectivity to desirable and often unique products are what assured the continued high interest that has led to the newly developed and diverse commercial successes with homogeneous catalysts in the polyolefin industry.

2 LIGAND AND POLYMER STRUCTURES

The diversity in possible catalyst structures can be appreciated by inspecting a few of the simplest cyclopentadienyl ligands that can be attached to transition metals (Table 1). Several polymer structures that have been produced using different ligands and simple α-olefins are compiled in Tables 2 and 3.

Table 1   Some Cp anion structures used as 'building blocks' for metallocene catalysts. The hydrogen atoms are routinely replaced with hydrocarbon and hydrosilyl groups. **1**, Cp; **2**, Me$_4$Cp; **3**, Ind; **4**, Flu; **5** and **6** are thiopheno analogs of 2-Me-Ind and Flu respectively

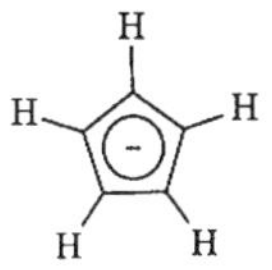

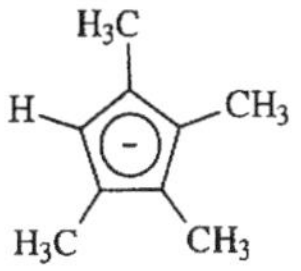

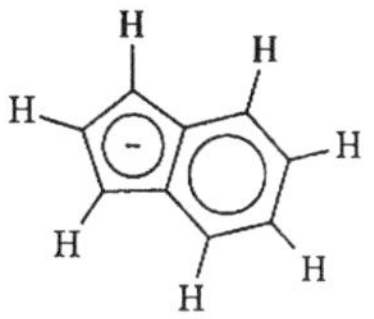

The two five-carbon Cp rings of the prototype Cp$_2$MCl$_2$ complexes (M = Ti, Zr, Hf) offer between them 10 hydrogen atoms that are replaceable with substituents. The 14 and 18 replaceable hydrogen atoms on two indenyl (Ind) and two fluorenyl (Flu) ligands lead to even more possibilities, which increase further in cases where alkyl and silyl groups and bridges on and between the Cps have additional replaceable hydrogen atoms. The number of possible permutations and combinations becomes enormous when the Cp, Ind and Flu ligands are mixed interchangeably. Add to this the recent finding that Montell's heterocenes (heterocyclic analogs) work equally as well or better and the length and complexity of many of the patent claims in the field are accounted for. The number of ligands has in general been constrained so that their steric effects are in the immediate environment of the active site and only limited by the degree of difficulty in preparing them.

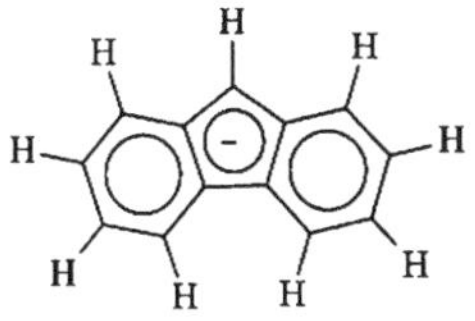

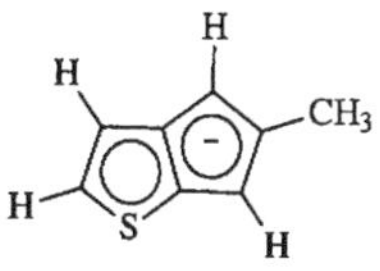

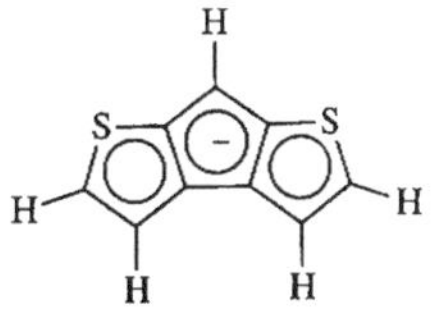

Table 2   Tacticities of polypropylenes made with metallocenes

Entry	Acronym	Polyproplyene structure	Interest
1	APP	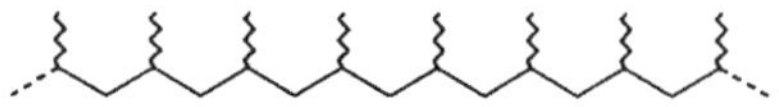	Borderline
2	IPP (chain-end)	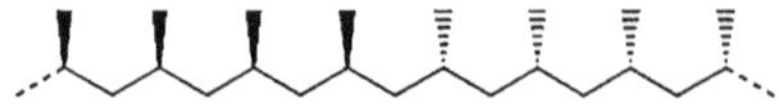	Academic
3	IPP (site)	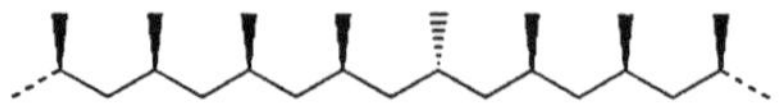	Commercial
4	hit-PP		Academic
5	SPP	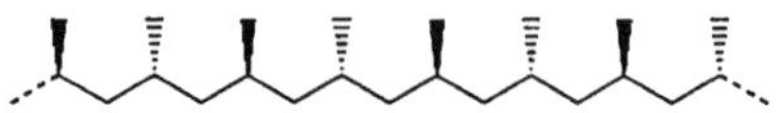	Commercial

Table 3   Selected ethylene and ethylene α-olefin copolymers made with metallocenes

Entry	Acronym	Polymer structure	Interest
1	HDPE		Borderline
2	EPR	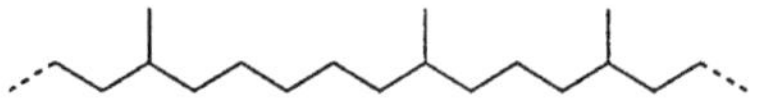	Commercial
3	LLDPE	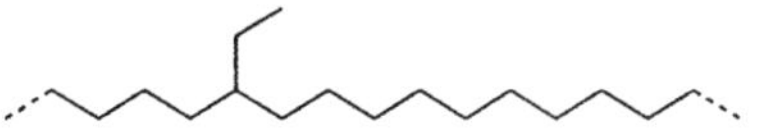	Commercial
4	ES	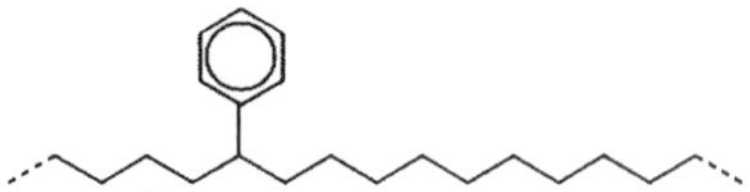	Commercial
5	LCB	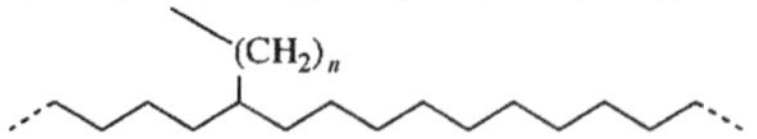	Commercial
6	SPS	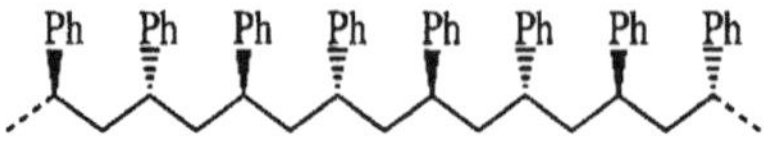	Commercial

3   THE FIRST 30 YEARS (1955–1984)

High density polyethylene (HDPE; Table 3, entry 1) was made in the earliest research on dicyclopentadienyltitanium dichloride activated with aluminum alkyls (e.g. Cp_2TiCl_2/Et_2AlCl) [3]. The *cis*-insertion of ethylene into the Ti—C bond of a pseudo-tetrahedral, monoalkyltitanium cation (Scheme 1) was one of the earliest mechanisms proposed for chain growth [4]. Migration of the growing chain to the ethylene coordination site during chain growth (chain migratory insertion) was

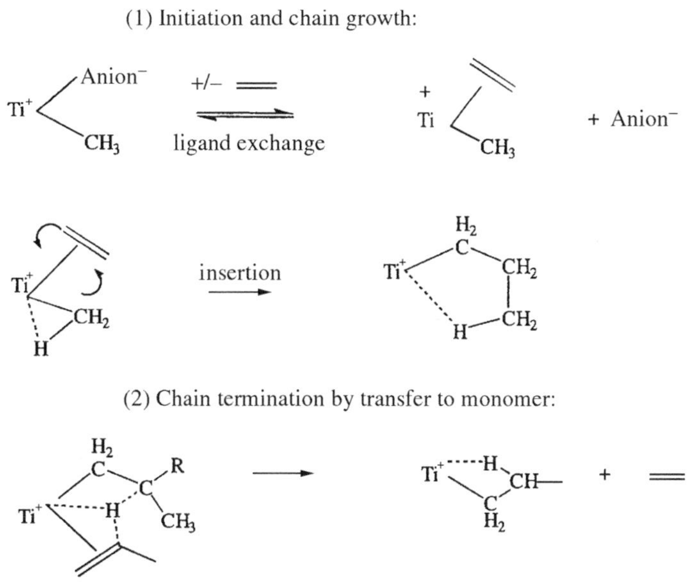

(1) Initiation and chain growth:

Ti$^+$ < Anion$^-$ / CH$_3$ +/− ⇌ ligand exchange   Ti < / CH$_3$ + Anion$^-$

Ti$^+$ / H, CH$_2$ insertion → Ti$^+$ ⟨ C(H$_2$)—CH$_2$ / H—CH$_2$

(2) Chain termination by transfer to monomer:

Ti$^+$ ⟨ C(H$_2$)—H, C—R, CH$_3$ → Ti$^+$ ⟨ H—CH—, C(H$_2$) + =

Scheme 1 The initiation steps in the polymerization of ethylene and the prevalent termination step in ethylene copolymerizations and homopolymerizations of 1-olefins. (1) Ligand exchange between ethylene and a loosely coordinated cocatalyst anion followed by cis-insertion into the TiC bond. (2) Termination by β-H transfer to monomer. The Ti cation forms weak agostic bonds with alkyl and olefin CH groups

proposed on the principal of a reaction path corresponding to the least nuclear motion being the most energetically favorable [5].

Cp_2TiCl_2/Et_2AlCl was active also in copolymerizations made with ethylene and propylene (EPR; Table 3, entry 2) [6] and 1-butene (LLDPE; Table 3, entry 3) [7–9]. The first ligand modifications to the Cp prototypes were on the titanocenes $(CH_3Cp)_2TiCl_2$, $(CH_3CH_2Cp)_2TiCl_2$ [10], $(Ind)_2TiCl_2$ [11] and $CH_2[Cp]_2TiCl_2$ [12] (Table 4). The catalysts with the alkyl-substituted cyclopentadienyls had lower activities than Cp_2TiCl_2 and no comparisons were made with the latter two.

The discoveries by Sinn, Kaminsky and co-workers with the structurally ill-defined, oligomeric methylaluminoxanes [MAO or $Me_2Al(—O—AlMe)_n—O—AlMe_2$; $n \approx 10$] as a cocatalyst (catalyst activator) in place of Et_2AlCl were important breakthroughs [13]. They reported that Cp_2ZrCl_2/MAO had a long catalyst half-life and made atactic polypropylene oil with $MW \approx 500$ (APP; Table 2, entry 1) [2c]. The ethylene turnover numbers for HDPE of 10–100 per second were compared with those of the fastest enzymes [13]. The most important finding was, unlike Cp_2TiCl_2/Et_2AlCl, the Cp_2ZrCl_2/MAO system does not reduce to inactive Zr(III) species rapidly and consequently remains active at polymerization temperatures of 50–70 °C for the 2–4 h residence times required in most commercial

Table 4   The first metallocenes investigated for ligand effects in ethylene polymerizations

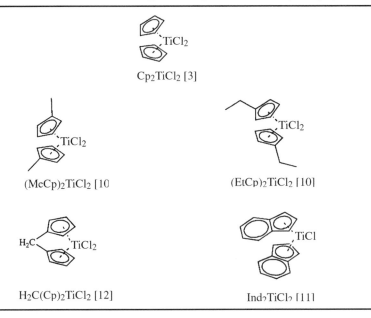

Cp₂TiCl₂ [3]

(MeCp)₂TiCl₂ [10   (EtCp)₂TiCl₂ [10]

H₂C(Cp)₂TiCl₂ [12]   Ind₂TiCl₂ [11]

reactors. In terms of Kaminsky's early studies on ligand effects, (MeCp)$_2$TiCl$_2$/MAO and (Me$_5$Cp)ZrCl$_2$/MAO showed only disadvantages with Cp ligand substitutions [13e,f]. A decisive breakthrough from later studies of ligand effects was that many of the newer catalyst structures work at Al/Zr mole ratios of only a few hundred, as opposed to the impractical ratios in the tens of thousands needed with the Cp$_2$ZrCl$_2$ prototype.

4 CATALYST ACTIVATION

Lewis acid cocatalysts alkylate the transition metal, abstract a non-Cp ligand to form active alkyl ion pairs and scavenge traces of water which would otherwise react with and poison electrophilic catalysts [4]. Structurally well defined anions were developed as MAO substitutes with the chemistry portrayed in Scheme 2.

Following Jordan, a methyl of Cp$_2$Zr(CH$_3$)$_2$ was protonated with tertiary ammonium salts [14] and alternatively abstracted with the trityl salt [Ph$_3$C$^+$] [(C$_6$F$_5$)$_4$B$^-$] and with the strong Lewis acid (C$_6$F$_5$)$_3$B [15]. Me$_3$Al, Et$_3$Al, TIBAL or MAO serve as scavengers. Simple aluminium alkyls can also poison the cations by forming Zr—(Pl')$^+$ ··· RAlR$_2$ adducts when they are present in excess [15,16]. The active ion pairs are presumed to be either associated with very weak Zr ··· F—C coordination for (C$_6$F$_5$)$_4$B$^-$ and with a stronger Zr ··· CH$_3$—B bond for (C$_6$F$_5$)$_3$B(CH$_3$)$^-$ or to be completely dissociated. The catalysts with

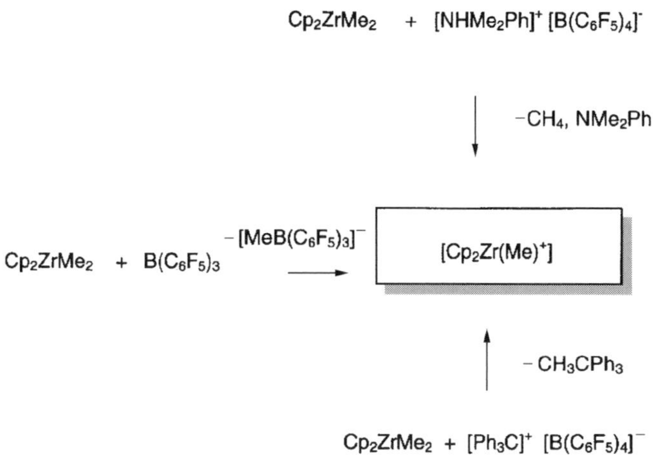

$$Cp_2ZrMe_2 \quad + \quad [NHMe_2Ph]^+ [B(C_6F_5)_4]^-$$

$$\downarrow \quad -CH_4, NMe_2Ph$$

$$Cp_2ZrMe_2 \quad + \quad B(C_6F_5)_3 \quad \xrightarrow{\; -[MeB(C_6F_5)_3]^- \;} \quad \boxed{[Cp_2Zr(Me)^+]}$$

$$\uparrow \quad -CH_3CPh_3$$

$$Cp_2ZrMe_2 \quad + \quad [Ph_3C]^+ [B(C_6F_5)_4]^-$$

Scheme 2 Three ways to generate an alkylzirconium cation loosely associated with weakly or non-coordinating perfluoroborate anions [14,15]

MAO^- and $(C_6F_5)_4B^-$ are generally more active than those paired with $(C_6F_5)_3B(CH_3)^-$.

Structurally well defined higher aluminoxanes from TIBAL have been prepared and structurally characterized [17], but MAO and $(C_6F_5)_3B$ remain the most convenient and economical methods for catalyst activation.

5 POLYETHYLENES

5.1 MOLECULAR WEIGHT CONTROL

Cp_2ZrCl_2/MAO produces HDPE (Table 3, entry 1) with one tenth the molecular weight made with Cp_2TiCl_2/MAO (MW $\approx 150\,000$ vs $1\,500\,000$). This molecular weight problem was further compounded by the propensity of Zr to terminate chain growth prematurely by β-H transfer to monomer both in α-olefin polymerization (MW ≈ 500) and in ethylene/α-olefin copolymerization (Scheme 1) [18]. Experimental evidence for this termination mechanism with α-olefins comes from the independence of polymer MW on monomer concentration with the early simple metallocene structures.

The competing β-H transfer to Zr is a slower termination process. Transfer to monomer has made commercial production of LLDPE impossible with Cp_2ZrCl_2/MAO and is so ubiquitous that the first priority in industrial research programs with Zr catalysts for PP, EPR and LLDPE (Tables 2 and 3) is the polymer molecular weight. A correlation between catalyst activities and HDPE molecular weights is depicted in Figure 2. A steric effect on ethylene polymerization with the

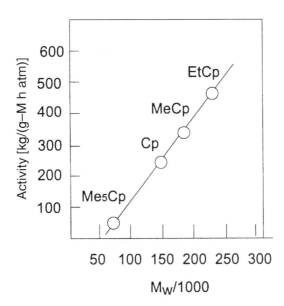

Figure 2   Correlation between catalyst activity and M_w with alkyl-substituted Cp ligands attached to Zr [19d]

bulky Me_5Cp ligand gives a large decrease in the polyethylene propagation rate and a concomitant reduction in HDPE molecular weight relative to Cp_2ZrCl_2.

An electronic effect was the first decisive breakthrough for LLDPE in terms of ligand effects [19]. Replacing one hydrogen atom on each Cp with an alkyl group (R = Me, Et) increased the propagation rates and also increased the polyethylene molecular weights significantly. It has been supposed that increased electron donation to Zr by induction from the alkyl-substituted Cp simultaneously reduces termination rates, R_t (reduced Zr Lewis acidity), and, more importantly, as evidenced by increases the rate of propagation R_p, by weakening the Zr–polymer bond strength. Both of these effects would increase the polymer molecular weight since the degree of polymerization is dictated by the rate of chain growth *vs* chain termination ($P_N = R_p/R_t$).

Electronic effects of the Cp ligands and the transition metals on MW have dominated research on polyethylene (Figure 3). One of today's commercial catalysts for LLDPE production in the Unipol process is simply $(n\text{-BuCp})_2ZrCl_2$ [20]. This outcome is consistent with the organic chemist's dictum: 'methyl, ethyl, propyl, futile', wherein longer side-chains would not be expected to have a further beneficial electronic inductive effect on catalyst performance.

A second benefit of attaching alkyl groups to Cps that was immediately appreciated from an industrial point of view was that the Schultz–Flory molecular weight distribution typical of homogeneous catalysts ($M_w/M_n = 2$) can be broadened to the range of the heterogeneous catalysts (MWD = 5–10) for improved processing [2h]. This was simply done by mixing two or more species with different ligands or transition metals. Indeed, bimodal GPC curves are possible where the peak molecular weights produced by each catalyst differ by an order of magnitude [19d].

The bridged bisfluorenyl catalysts such as $Me_2Si[Flu]_2ZrCl_2$/MAO (APP) and $Ph_2C[CpFlu]ZrCl_2$ (SPP) produce substantially higher molecular weight polypropylenes and ethylene copolymers than Cp_2ZrCl_2/MAO [21,22]. In the case of $Me_2Si[Flu]_2ZrCl_2$/MAO there is steric hindrance of termination by β-H transfer to 1-olefins (see below) and so MW should decrease with increasing ethylene [23].

In the case of $Ph_2C[CpFlu]ZrCl_2$, it was assumed that the higher S-PP MW than with $Me_2C[CpFlu]ZrCl_2$ is due in part to an electronic effect from the bridge, making it the better candidate of the two for LLDPE. The carbon and silicon bridges of these complexes apparently provide a stabilizing chelate effect against ligand-exchange reactions with MAO to produce ill-defined, heterogeneous mixtures of active sites [24].

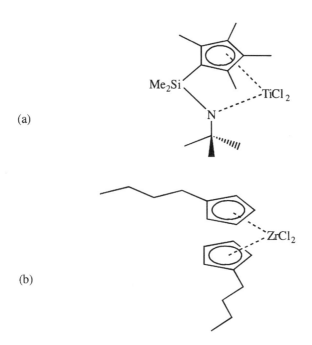

(a)

(b)

Figure 3 Schematics of two metallocenes that exploit electronic effects to produce LLDPE with high molecular weights: (a) [(tetramethylcyclopentadienyl)-(N-*tert*-butylamido) dimethylsilyl] titanium dichloride [25] and (b) bis(n-butylcyclopentadienyl) zirconium dichloride

A serendipitous solution to the Zr MW problem was found on changing back to Ti with the more open, half-titanocenes such as dimethylsilyl-[*tert*-butylamidotetra-methylcyclopentadienyl]titanium dichloride {$Me_2Si[(t\text{-}BuN)(Me_4Cp)]TiCl_2$, Figure 3 in Dow's solution process [25]. This class of catalyst worked beautifully for LLDPE because the short residence times of about 10 min in small solution reactors made catalyst deactivation by reduction to Ti(III) inconsequential.

The open ligand structures and high temperatures gave products with long-chain branches (LCB) suitable for wire and cable coating applications and also unconventional ethylene–styrene copolymers (Table 3, entries 4 and 5). Extraordinarily enough, it has been said that the biscyclopentadienyl metallocenes in gas-phase processes also give rise to LCB for an entirely different reason: the supported catalysts and the vinyl end groups of the polymer chains are both trapped in the amorphous phase. An added benefit of Dow's solution process is that the catalysts do not need be supported on a matrix such as silica or magnesium chloride to obtain a granular product, since LLDPE is miscible with isopar at high temperatures. The only downside was that the loss of one Cp ligand which means that the catalyst symmetry and structural control needed for polypropylene is lost forever. Nevertheless, some electronic effects on the active site are possible in LLDPE using Cp, Flu and Ind in place of Me_4Cp and by exchanging the *t*-Bu amido group for Ph and 4-FC_6H_4 [25].

Credit for the most important industrial breakthroughs for LLDPE with metallocene catalysis must go to the catalysts themselves, since their homogeneity fixed a deficiency in copolymerization by classical heterogeneous catalysts. Fractionation of conventional ethylene–butene random copolymers (LLDPE) shows that 1-butene is preferentially incorporated into very short chains [2h, 26, 27]. This phenomenon is akin to the now historical atactic polypropylene problem because it increases the content of extractable, stickly material making it unsuitable for food packaging applications such as meat-wrap. In contrast, the homogeneous catalysts have comonomer distributions that are independent of chain lengths. Other features of LLDPE produced by metallocenes such as higher transparency and tear strength have also been attributed the random distribution of α-olefin comonomers inherent in homogeneous catalysis [2h].

5.2 *REACTIVITY RATIOS*

The above-mentioned differences between heterogeneous and homogeneous catalysts are represented quantitatively by the product of their rate constant copolymerization parameters (Table 5) with $r_1 r_2 > 1$ for blocky, undesirable materials and $r_1/r_2 < 1$ being consistent with more ideal, random placements.

The zirconocene copolymerization parameters r_1 and r_2 quantify the relative reactivity of each catalyst towards ethylene and propylene. r_1 is the ratio of the rates of L_2Zr—CH_2CH_2—Pl^+ ($L = RCp$, Pl = copolymer) reacting with ethylene *vs* propylene. r_1 is a quantitative measure of how much faster the catalysts react with

Table 5 $\delta TiCl_3$ and metallocene copolymerization parameters

Entry	Procatalyst	r_1	r_1r_2	Ref.
1	$\delta TiCl_3$	7	5.5	28
2	$Cp_2Ti(Ph)_2$	20	0.3	19d, 29
3	$Cp_2Ti=CH_2.DMAC^a$	24	0.2	19d
4	Cp_2ZrCl_2	48	0.7	19d
5	$[Cp_2ZrCl]_2O$	50	0.3	19d
6	$(CH_3Cp)_2ZrCl_2$	60	–	19d
7	$Me_2Si(Cp)_2ZrCl_2$	24	0.7	19d
8	$Me_2C(Cp)(Flu)ZrCl_2$	1	0.3	30
9	$Me_2Si(Ind)_2ZrCl_2$ [b]	25	0.4	31
10	$(Me_5Cp)_2ZrCl_2$	250	0.5	19d

[a] Tebbe reagent with DMAC = dimethylaluminum chloride.
[b] Ethylene–1-hexene copolymer.

the smaller ethylene than with propylene. r_1 is very sensitive to the structures of the Cp ligands attached to the active catalysts and is independent of the non-Cp ligands (Table 5, entry 2 *vs* 3 and entry 4 *vs* 5) which are stripped off the transition metal during in the activation process.

Entry 4 *vs* 7 suggests that $[(CP)_2Zr—CH_2CH_2—Pl]^+$ ($r_1 = 48$) is more crowded than $[Me_2Si(Cp)_2Zr—CH_2CH_2—Pl]^+$ ($r_1 = 24$). Molecular models indicate that the only noteworthy difference is that Cp_2ZrCl_2 can have staggered (s) or eclipsed (e) Cp conformations as compared with the enforced e conformation with $Me_2Si(Cp)_2ZrCl_2$ (Figure 4). On the other hand, the bulky Cp^* ligands with $[(Me_5Cp)_2Zr—CH_2CH_2—Pl]^+$ make this catalyst 5, 10 and 250 times less reactive with the larger propylene monomer than with ethylene than are $[(Cp)_2Zr—CH_2CH_2—Pl]^+$, $[Me_2Si(Cp)_2Zr—CH_2CH_2—Pl]^+$ and $[Me_2C(Cp-9-Flu)_2Zr—CH_2CH_2—Pl]^+$, respectively. The practical consequences of this can be readily appreciated when it is considered that $[(Me_5Cp)_2Zr—CH_2CH_2—Pl]^+$ produces crystalline HDPE under the same polymerization conditions where $[Me_2Si(Cp)_2Zr—CH_2CH_2—Pl]^+$ produces a syrupy copolymer that is predominantly propylene.

$Me_2Si[Cp]_2ZrCl_2$ ($r_1 = 24$, $\angle CNT—M—CNT = 126°$) *vs* $Me_2C[Cp-9-Flu]ZrCl_2$($r_1 = 1.3$, $\angle CNT—M—CNT = 119°$) is consistent with the Cp ligands being peeled further away from the non-Cp coordination sites, much like opening a clam shell, by carbon bridges than by silicon bridges. The CNT—M—CNT angles also have strong influences on the stereospecificities of metallocenes (see below). Perhaps consistent with ligand-accelerated catalysis (LAC) in homogeneous Ziegler–Natta catalyses in general [32] (especially when taken together with the higher reactivities of the more stereoselective catalysts with propylene) is the small r_1 for $[Me_2Si(Ind)_2Zr—CH_2CH_2—Pl]^+$ in ethylene/hexene ($r_1 = 25$: Table 5, entry 9) copolymerization, despite the preferred chain-end conformation locking out reaction with one face of hexene [31].

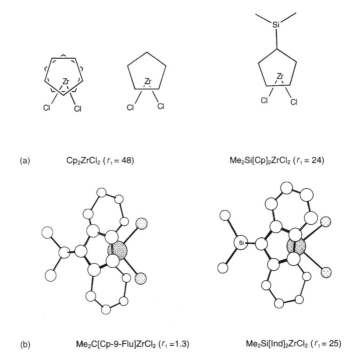

(a) Cp_2ZrCl_2 ($r_1 = 48$) $Me_2Si[Cp]_2ZrCl_2$ ($r_1 = 24$)

(b) $Me_2C[Cp-9-Flu]ZrCl_2$ ($r_1 = 1.3$) $Me_2Si[Ind]_2ZrCl_2$ ($r_1 = 25$)

Figure 4 Subtle differences in the structures of two pairs of metallocenes: (a) s and e ring conformations for Cp_2ZrCl_2 and e conformation for $Me_2Si[Cp]_2ZrCl_2$; (b) dorsal views of $Me_2C[Cp-9-Flu]ZrCl_2$ and $Me_2Si[Ind]_2ZrCl_2$ with the non-Cp coordination sites more exposed for the former

Fabricators manufacturing items with PE often blend polymers having different amounts of comonomer to achieve a final product with the desired properties. For example, manufacturers of shopping bags can mix HDPE and LLDPE to obtain a material with both high tear strength and increased rigidity in the handle. This composite is a mixture of materials with differing comonomer content and therefore, by definition, has a broad composition distribution. It was demonstrated in principle that polyethylene manufacturers could produce reactor blends of such materials by mixing catalysts with differing reactivity ratios [19d]. This technology apparently has yet to be reduced to practice commercially [33]; presumably because a highly active metallocene producing high molecular weight HDPE with r_1 approaching 250 had yet to be discovered.

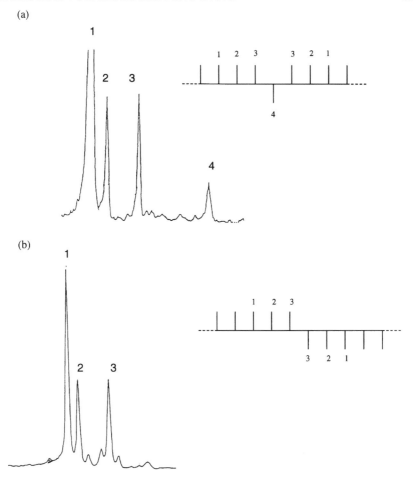

Figure 5 ^{13}C NMR spectra of the methyl pentad regions of IPP: (a) Site-control structure with (Me)$_2$C(3-t-Bu-1-Cp)(9-Flu)ZrCl$_2$/MAO; (b) Chain-end control structure with Cp$_2$TiCl$_2$/MAO

6 POLYPROPYLENE

6.1 INTRODUCTION

A lot of publications and patents on metallocene catalysts deal with polypropylene. The conventional wisdom that isotactic polypropylene could only be made with heterogeneous catalysts [3c] along with the popularity of stereochemistry provided irresistible lures, at least for this author, to stereospecific homogeneous catalysis.

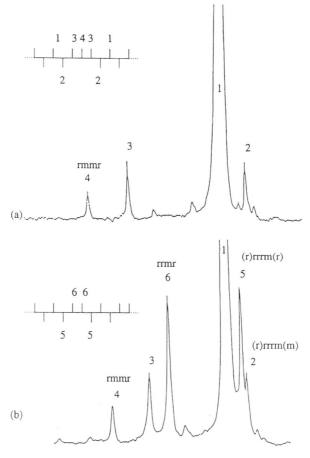

Figure 6 ^{13}C NMR spectra of the methyl pentad regions of SPP: (a) Site-control structure with Me$_2$C(1-Cp)(9-Flu)Zr/Cl$_2$/MAO at $-5\,^{\circ}$C; (b) Skipped-out insertions with Me$_2$C(1-Cp)(9-Flu)Zr/Cl$_2$/MAO at 1.3 M C$_3$H$_6$

Other than modeling heterogeneous catalysts to understand them better, the motivation for studying α-olefins also came from the potential of new materials such as syndiotactic polymers [34,35] and perfectly and partially isotactic polypropylene [36]. Polypropylene research has proved to be particularly interesting because the tactiospecificities of all stereospecific metallocenes have been 'predicted' with hindsight by molecular mechanics and the same reaction mechanisms proposed for catalysts with unknown structures 30 years ago (see below) as outlined in Schemes 1–3, including Cossee's 'back-skip' reaction [5b,37].

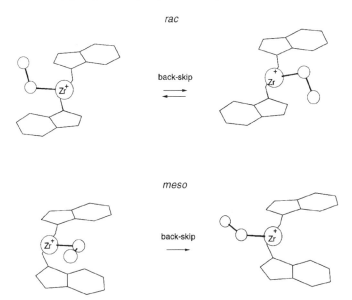

Scheme 3 Illustration of Cossee's back-skip reaction with *rac*- and *meso*-Et[Ind]$_2$ZrCH$_2$CH$_3$]$^+$. Isotactic enchainments are favored with the *rac*-isomer since the preferred conformation of the surrogate chain-end is indistinguishable on either side of the catalyst. For the *meso*-isomer the chain-end preferentially back-skips to the uncrowded side leading to regiospecific but non-stereospecific insertions [5b,37e]

6.2 ^{13}C NMR SPECTROSCOPY

An understanding of polypropylene ^{13}C NMR spectroscopy is essential in research on polypropylene catalysis because this is the only way to determine polymer microstructures. The ^{13}C chemical shifts of the polymers' methyl groups are determined by the relative configurations of the four adjacent methine units around it. Bovey's stereochemical notation for the 10 unique configurational arrangements of five adjacent units are shown as Fischer projection formulae in Scheme 4 [38].

An isotactic or *meso* (m) dyad has both methine units with the same relative configuration and hence a mirror plane between them. Adjacent syndiotactic or racemic (r) dyads have opposite configurations and are mirror images of each other. The isotactic (mm), heterotactic (mr) and syndiotactic (rr) centered pentads are grouped together and numbered in order of decreasing chemical shifts for the central methyl carbon atom. Only nine bands are observed in the methyl region with pentad resolution since the xmrx pentads (mmrm and rmrr) have the same chemical shift.

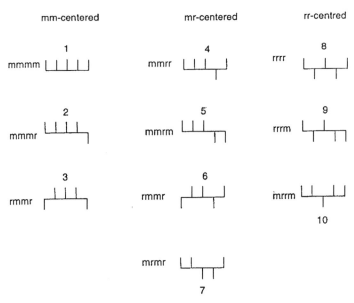

Scheme 4 The 10 possible stereochemical pentads of polypropylene

The ^{13}C NMR spectra of the methyl pentad region is therefore a kind of bar graph with the intensities of the pentads informing us of the polymer microstructure.

6.3 POLYPROPYLENE MICROSTRUCTURES AND THEIR ORIGINS

Perfectly isotactic and syndiotactic polypropylenes would have the uninteresting result of only one mmmm and only one rrrr pentad, respectively. Examples of the ^{13}C NMR spectra for four polymers containing the only possible steric imperfections for isotactic and syndiotactic polypropylene are displayed in Figures 5 and 6 along with the polymer Fischer projection formulae [15c,39,40]. The numbers above the methyls in each formula correspond to the number above the signals they give rise to in the spectra. The relative pentad intensities of the peaks of the four polymers are structural requirements and therefore the microstructures of these polymers unambiguous.

Stereoregular polymers obtained with the chiral complexes [Figures 5(a) and 6(a)] contain the steric imperfection of an isolated unit enchained out of sequence with its neighbors leading to the same 2:2:1 pentad intensity pattern for IPP (mmmr:mmrr:mrrm) and for SPP (rrrm:rrmm:rmmr). In the polymerization at $-30°$C with achiral Cp_2TiPh_2/MAO [Figure 5(b)], the chain-end controls the polymer stereochemistry provided that it rotates more slowly than the rate of insertion. In this situation steric inversions propagate themselves leading to the IPP stereo-block structure with mmmr:mmrm in a 1:1 ratio. The spectrum for SPP in Figure 6(b) shows a steric defect consistent with an insertion that has been skipped out leading to rrrm:rrmr in a 1:1 intensity ratio. This microstructure has been attributed to occasional catalyst isomerization in between insertions *via* the back-skip process [40a] and is identical with the steric defect obtained with chain-end control syndiospecific catalysis [41].

Probability models (Kinetic models) are only useful in analyzing relatively atactic materials such as the hemi-atactic or hemi-isotactic polypropylene [HIT-PP, Figure 7(a)] and random or atactic polypropylene [APP, Figure 7(b)]. The mrmr pentad is forbidden for HIT-PP and the intensities of the eight allowable hit-PP pentads are reproducible with Farina's equations with the probability α equal to the fraction of m dyads [40d,42]. In the case of APP all nine bands are visible in intensity ratios of 1:2:1 (mm-centered pentads), 2:4:2 (mr-centered) and 1:2:1 (rr-centered). The pattern is a consequence of the degeneracy of the xmrx pentads and of the different number of equally possible ways of generating each pentad. The intensities of the pentads have been reproduced with Bovey's Bernoullian statistical equations with the probability P equal to the fraction of $m = 0.5$ [2c,38,39].

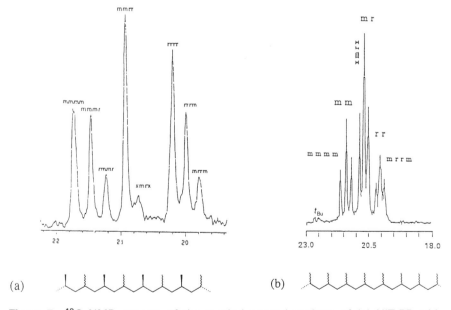

Figure 7 ^{13}C NMR spectra of the methyl pentad regions of (a) HIT-PP with Me$_2$C(3-Me-1-Cp)(9-Flu)Zr/Cl$_2$/MAO and (b) APP with (Me$_5$Cp)(Cp)ZrCl$_2$/MAO

6.4 ISOTACTIC POLYPROPYLENE

The historical sequence of events leading to today's more stereospecific metallocenes began with low-temperature polymerization for chain-end control {Cp_2TiPh_2/MAO, Figure 5(b) [39]}. The material produced was interesting only because IPP with this microstructure had long been speculated on but had never been prepared [2a]. The low stereoregularity (83 % m) meant elastomeric properties and therefore only academic interest. The use of Brintzinger's compounds with C_2 or playing card symmetry, rac-Et[Ind]$_2$TiCl$_2$/MAO and rac-Et[H$_4$Ind]$_2$TiCl$_2$/MAO, showed that with site-control stereoregulation it was possible to get to obtain crystalline IPP with metallocenes [19d, 39].

The homotopic C_2-symmetric catalysts were selected so that the back-skip reaction (Scheme 3) could not influence the successive propylene enantiofacial orientations. It was determined through both ^{13}C isotopic labeling of the chain-ends [43] and through molecular mechanics calculations that the preferred chain-end conformation places it in the most open sector away from the β-Cp substituents (β to the bridgehead carbon) and that it thus acts like a molecular lever in directing the olefin enantiofacial approach (Scheme 5) [37].

An important discovery was the Kaminsky–Brintzinger observation that ligand-accelerated catalysis (LAC) resulted in a higher activity and about a 10–15-fold increase in MW for the rac-Et[Ind]$_2$ZrCl$_2$/MAO relative to Cp$_2$ZrCl$_2$/MAO at 50–70 °C [44]. The task from here on became one of designing better Zr catalysts with new bridges and hydrocarbons replacing the hydrogen atoms on indene because IPP molecular weights were still 5–10 times lower than for commercial polymers and the polymer melting points about 40 °C too low. For the zirconocenes to be competitive under commercial polymerization conditions the stereochemical inversions (mrrm $\approx$ 2.5 %) and especially the regiochemical 2–1 and 1–3 propylene addition errors ($\approx$ 1 %) (Scheme 6 [45]) had to be reduced to so that mrrm < 1 % (mmmm >95 %) and zero, respectively.

The three key ligand effects that led to isospecific metallocenes approaching the performance of the heterogeneous catalysts are portrayed in Table 6. A 10 °C

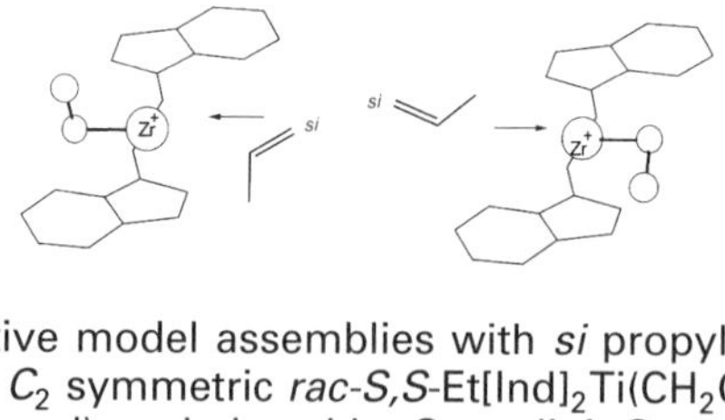

Scheme 5 Isoselective model assemblies with si propylene face coordination for preinsertion with C_2 symmetric rac-S,S-Et[Ind]$_2$Ti(CH$_2$CH$_3$)$^+$ (the ethyl group as a surrogate chain-end) as deduced by Corradini, Guerra and coworkers [37]. The preferred conformation of the 'chain-end' places it in this most open sector. Propylene is enchained with its methyl group trans to the β-carbon of the 'chain-end'. C_2-symmetric complexes do not isomerize with migratory insertion (IPP)

(a) 2-1 insertions

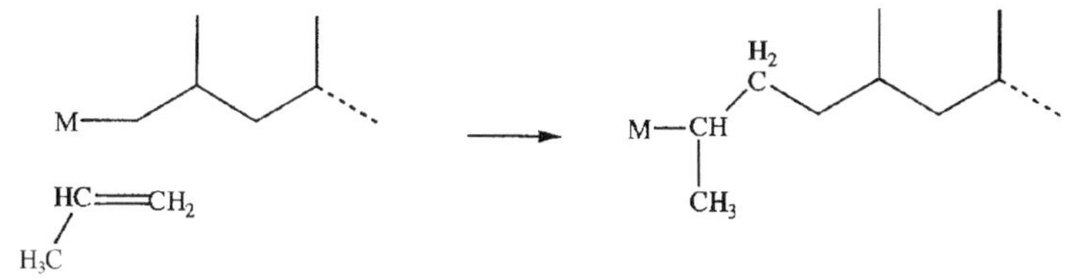

(b) 1-3 additions

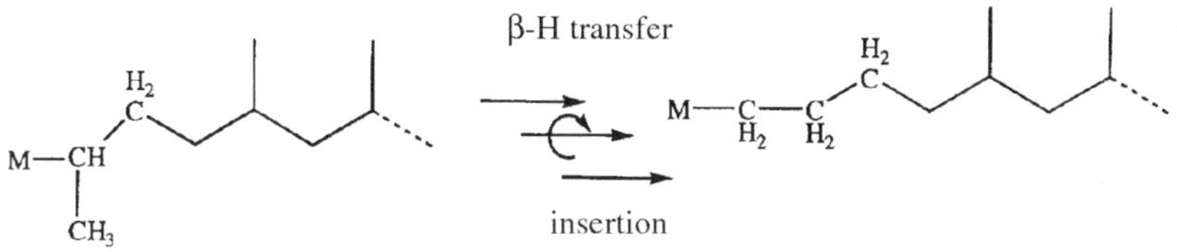

Scheme 6   Regiochemical 'errors'

improvement in IPP m.p. with rac-Me$_2$Si[Ind]$_2$ZrCl$_2$/MAO over rac-Et[Ind]$_2$ZrCl$_2$/MAO was attributed to higher catalyst rigidity [46] and an Me$_2$Si bridge has since been designed into most IPP specific catalysts ever since. The IPP molecular weight (up ~20%) and catalyst activity were also improved. An important practical advantage was the replacement of 1,2-dibromoethane with Me$_2$SiCl$_2$ in the ligand synthesis, which also led to higher yields of the ligand when the indenyl anions are added to dichlorosilanes.

A more important breakthrough was the subsequent discovery by Miya and co-workers [47] that α-methyl substituents on Me$_2$Si(2,4-dimethyl-1-Cp)$_2$ZrCl$_2$ (α to the bridgehead carbon atom) relative to Me$_2$Si(3-methyl-1-Cp)$_2$ZrCl$_2$ causes the polypropylene molecular weights to increase about 10-fold and the polymer melting-points to increase to values approaching those of commercial polymers, albeit at 20–40 °C below commercial polymerization temperatures. Spaleck and co-workers [48] quickly adapted both the Me$_2$Si bridge and α-methyl findings to bis(indenyl) catalysts. Brinztinger and co-workers' analysis of M_n vs [C$_3$H$_6$] is consistent with the increased molecular weight being due to a decrease in rate of β-H transfer to monomer [49]. This was confirmed by molecular mechanics calculations on steric effects in the transition states for termination [50]. Molecular models likewise indicated that the significant increases in stereo- and regio-regularities with α-methyl

Table 6 Three important ligand effects that increased catalyst activity, stereospecificity and IPP MW transforming the C_2 symmetric zirconocenes into today's best catalysts

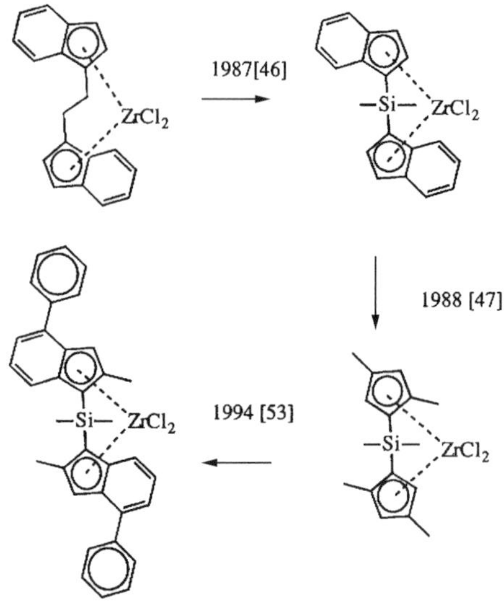

substituents are due to steric effects conferring more rigidity to the catalysts and inhibiting monomer rotation for 2–1 insertions [51].

Experimental evidence anticipated the final breakthrough leading to the most highly stereospecific C_2-symmetric catalysts with large substituents at the position β to the bridgehead Cp atom suspended over the non-Cp coordination sites (Figure 8). In definitive works, Collins and Brintzinger varied the size of β-groups methodically and clearly showed increased stereospecificity with increasing size of the β-groups [52]. Further, APP was produced by rac-Et[3-Me-1-Ind]$_2$ZrCl$_2$/MAO because the non-Cp coordination sites are flanked by a CH$_3$ and an aromatic C—H whose equivalent steric requirements are opposed to each other [46]. Decreasing the effective steric effects of β-substituents by decreasing the CNT—M—CNT angle by 7° on going from rac-Me$_2$Si[Ind]$_2$MCl$_2$/MAO to rac-H$_2$C[Ind]$_2$MCl$_2$/MAO(M = Zr, Hf) significantly decreased the catalyst stereopsecificity [40b]. It was also learned very early on that the more sterically demanding β-substituents for rac-Et[H$_4$Ind]$_2$HfCl$_2$/MAO vs rac-Et[Ind]$_2$HfCl$_2$/MAO make the former more stereospecific [46].

In retrospect, it is only logical that catalysts with larger β-groups such as Me$_2$Si[2-Me-4-Ph-1-Ind]$_2$ZrCl$_2$, Me$_2$Si[2-Me-4,5-Bz-1-Ind]$_2$ZrCl$_2$ and Me$_2$Si(1-Ph-2,5-Me$_2$-4-Cp[1.2-b]Pyr)$_2$ZrCl$_2$ (Figure 8) should more closely emulate their heterogeneous counterparts [49,53]. The real surprise with these structurally

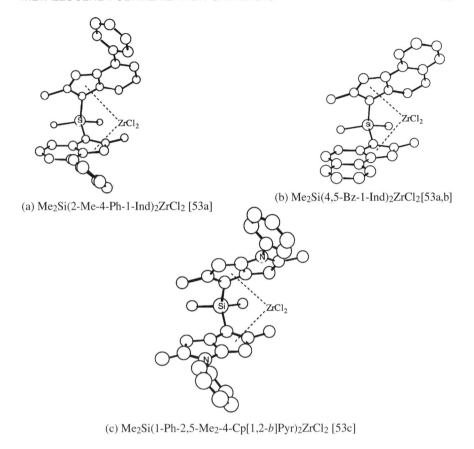

(a) Me₂Si(2-Me-4-Ph-1-Ind)₂ZrCl₂ [53a]

(b) Me₂Si(4,5-Bz-1-Ind)₂ZrCl₂[53a,b]

(c) Me₂Si(1-Ph-2,5-Me₂-4-Cp[1,2-*b*]Pyr)₂ZrCl₂ [53c]

Figure 8 Ligand structures for selected metallocenes having ultra-high activities and stereo-and regio-specificities approaching those of heterogeneous catalysts

more congested catalysts came from the approximately fivefold increase in catalyst activities. This has been speculated to be due LAC (see above) and steric inhibition of bimolecular metallocene interactions [2e]. These latter-day homogeneous catalysts are remarkably improved relative to the prototype *rac*-Et[Ind]₂ZrCl₂/MAO. They can make IPP with mmmm ≈ 0.95 and less than half the earlier regioirregularities. Their syntheses are typically tedious with 50 % of the final product being the *meso* isomer. The remaining problem is therefore that there is as yet no efficient way to join a 2-MeCp moiety to aromatic rings having the appropriate substitution patterns. Until this synthetic difficulty is resolved, it could well be considered the 'Big Crunch' for metallocene IPP catalysis.

6.5 NARROW MWD, ELASTOMERIC, ISOTACTIC POLYPROPYLENE

Chien *et al.* [54] similarly found that an elastomeric IPP (anisotactic PP) with a high level of atactic placements could be made with the weakly stereoregulating, C_1-symmetric HMeC[(Me$_4$Cp)(Ind)]TiCl$_2$/MAO. These lopsided catalysts tend to back-skip the chain to the least crowded side where monomer coordination leads to stereoregular insertions and aspecific insertions occur with the chain on the crowded side. The short isotactic runs presumably crystallize in both these and chain-end control IPP, cross-linking the chains and conferring stress induced crystallization and elastic properties. Elastomeric materials have also been obtained by Reiger and co-workers [55] with C_1-symmetric zirconocenes. However, the high glass transition temperatures for PP ($\sim 0\,^\circ$C) make these materials interesting only for theoretical and highly specialized applications.

6.6 SYNDIOTACTIC POLYPROPYLENE

It was anticipated for many years on the basis of theoretical arguments that the melting-point of syndiotactic poypropylene would be higher than found for isotactic polypropylene. Highly syndiotactic SPP was felt to be potentially important commercially because it could penetrate many high temperature applications closed to IPP [2a].

The pro-catalyst Me$_2$C[Cp-9-Flu]ZrCl$_2$ has C_s symmetry (bilateral symmetry) that results in the active monoalkyl cation isomerizing between left- and right-handed configurations with each monomer addition by chain-migratory insertion (Scheme 1) [40a,56]. The two coordination sites are enantiotopic (mirror images) resulting in alternating enantiofacial orientations of the successive olefin insertions (*R, re* and *S, si*). According to molecular mechanics calculations, stereoregulation is by a site control mechanism that operates at a molecular level in the same manner as the isospecific metallocenes (Scheme 7) with the chain-end lever in the most open quadrant [37].

Fortunately, the insertion reactions (Scheme 1) were faster than catalyst isomerization by Cossee's back-skip reaction (Scheme 3). Otherwise, it would have resulted in APP. The first SPP produced by metallocenes did not have the regio-irregularity and molecular weight problems that bedeviled IPP, apparently because of steric crowding in the transition states for β-H transfer to monomer (Scheme 1) and in propylene coordination for 2–1 insertions [23]. The high M_w of SPP may have had a contribution from the higher pK_a of fluorene than cyclopentadiene and indene but the M_w dependence on propylene concentration showed that transfer to monomer was also inhibited [40].

The Me$_2$C[Cp-9-Flu]ZrCl$_2$-based catalyst produced SPP with 2 % mm defects from the site control mechanism [Figure 6(a)] and an additional 3–4 % m defects caused by skipped-out insertions [Figure 6(b)], consistent with Cossee's back-skip reaction at polymerization temperatures between 50 and 70 °C [40a]. The m and mm

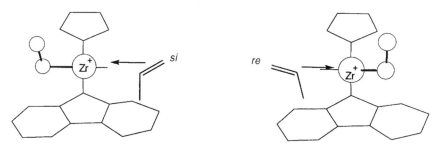

Scheme 7 Syndioselective model assemblies with propylene coordination for preinsertion with C_s symmetric $Me_2C(1\text{-}Cp)(9\text{-}Flu)Zr(CH_2CH_3)^+$. The preferred conformation of the 'chain-end' places it in this most open sector. Propylene is enchained with its methyl group *trans* to the β-carbon of the 'chain-end'. C_s symmetric complexes isomerize with migratory insertion producing SPP with alternating *S*, *si* and *R*, *re* as the preferred arrangements [37]

placements result in SPP with low m.pts. and slow crystallization rates. Scores of catalysts with bilateral symmetry have since been made with the as yet unrealized hope of eliminating or at least improving on the 5% isolated isotactic placements in metallocene produced SPP [57]. Every synthetic trick (such as increasing the CNT—M—CNT angle and bulking up the β-substituents [40c]) has either backfired to make the polymers worse or resulted in no significant improvement over the original SPP specific catalyst under commercial polymerization conditions. However, the polymer has a low solubility, high clarity, high impact strength and high γ-ray resistance, making it interesting in certain medical applications.

6.7 C_1-SYMMETRIC CATALYSTS

More interesting results came on changing the catalyst from C_s to C_1 symmetry (Figure 9, Scheme 8). $Me_2C[3\text{-}MeCpFlu]ZrCl_2/MAO$ [40d] produces hemi-isotactic polypropylene [Figure 7(a)] and both $Me_2C[3\text{-}t\text{-}BuCpFlu]ZrCl_2/MAO$ and $Me_2Si[3\text{-}t\text{-}BuCpFlu]ZrCl_2/MAO$ produced IPP [Figure 5(a)] [40b]. Molecular mechanics are consistent with the preferred conformation of the chain-end at the less crowded sites of giving isotactic placements as usual [37].

As for *rac*-Et[3-Me-1-Ind]ZrCl$_2$ [40d, 46], the chain-end has no preferential conformation at the site sandwiched between the aromatic CH and CH$_3$ of $Me_2C[3\text{-}MeCpFlu]ZrPl^+$ and therefore every other insertion has a random configuration as in Farina's elegantly structured HIT-PP. The chain-end apparently back-skips from the crowded sites of $Me_2C[3\text{-}t\text{BuCpFlu}]ZrPl^+$ and $Me_2Si[3\text{-}t\text{-}BuCpFlu]ZrPl^+$ to give IPP with most insertions taking place with the chain in the more open position and propylene coordinated at the more crowded side. It turned out to be much easier to convert a syndiospecific catalyst to more isospecific ones than to convert it into a more syndiospecific structure.

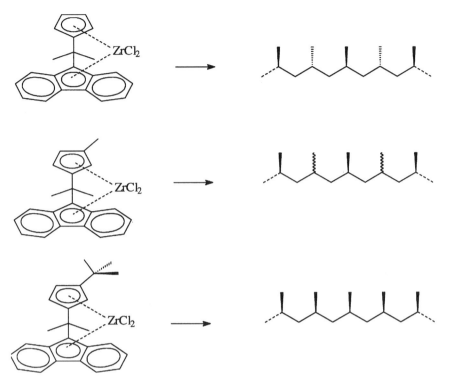

Figure 9 Relationships between ligand structures and stereoselectivities. SPP is produced by Me$_2$C(1-Cp)(9-Flu)Zr-Pl$^+$, HIT-PP by Me$_2$C(3-Me-1-Cp)(9-Flu)Zr-Pl$^+$ and IPP by Me$_2$C(3-t-Bu-1-Cp)(9-Flu)Zr-Pl$^+$. The catalyst selectivities provided the first experimental evidence for Cossee's chain migratory insertion and polmer 'back-skip' mechanisms

7 FUTURE CATALYSTS

It has been said that homogeneous metallocene catalysis is a mature industry and that the revolutionary era of designing new polymers by designing new catalyst structures is over. Nowadays publications and patents are predominantly centered around making improvements or trying to get around one set or another of earlier patent claims. While these observations are certainly true, it must be remarked that very little has been done to investigate the electronic effects of the Pauling electronegativities of elements other than carbon and silicon in the ligand framework. This leaves catalyst chemists with an almost limitless number of opportunities for new permutations and combinations of metallocenes with the conditional stipulation that the synthetic chemistry remains simple. Having determined the optimum transition elements for given processes, we are left with the rest of the

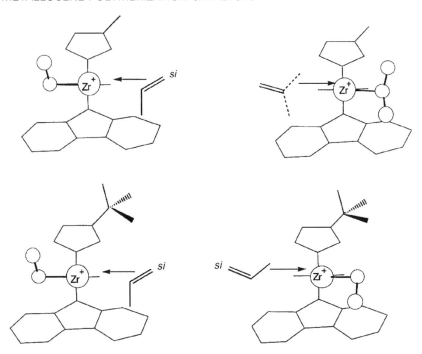

Scheme 8 Stereoselective and unselective model assemblies with propylene coordination for preinsertion with C_1 symmetric $Me_2C(3\text{-}Me\text{-}1\text{-}Cp)(9\text{-}Flu)$ $Zr(CH_2CH_3)^+$ and C_1 symmetric $Me_2C(3\text{-}t\text{-}Bu\text{-}1\text{-}Cp)(9\text{-}Flu)Zr(CH_2CH_3)^+$. $Me_2C(3\text{-}Me\text{-}1\text{-}Cp)(9\text{-}Flu)Zr(CH_2CH_3)^+$ gives hit-PP as the chain has no preferred conformation on the crowded side. $Me_2C(3\text{-}t\text{-}Bu\text{-}1\text{-}Cp)(9\text{-}Flu)Zr(CH_2CH_3)^+$ yields IPP with a non-bonded driving force for the chain to return to the least crowded side after each insertion [37]

Periodic Table to play with for the Cp ligand substituents. Even the discovery of one new, inexpensive ligand such as the heterocene ligand **5** in Table 1 opens up the possibility for scores of interesting new catalyst structures. This approach seems to offer more opportunities for controlled polymerizations than other alternatives because of the two advantages of metallocenes over other types of ligands: their refusal to come off the transition metal when mixed with activators and the large number of structures possible with 10 replaceable Cp hydrogens.

8 REFERENCES

1. Guyot, A., Böhm, L., Sasaki, T., Zucchini, U., Karol, F. and Hattori, I., *Makromol. Chem., Macromol. Symp.*, **66**, 311 (1993).

2. Reviews: (a) Boor, J., *Ziegler–Natta Catalysts and Polymerizations*. Academic Press, New York, 1979; (b) Sinn, H. and Kaminsky, W., *Adv. Organomet. Chem.*, **18**, 99 (1980); (c) Pino, P. and Mülhaupt, R., *Angew. Chem.*, *Int. Ed. Engl.*, **19**, 857 (1980); (d) Kaminsky, W., *Catal. Today*, **20**, 257 (1994); (e) Brintzinger, H. H., Fischer, D., Mülhaupt, R., Rieger, B. and Waymouth, R. M., *Angew. Chem.*, *Int. Ed. Engl.*, **34**, 1143 (1995); (f) Ewen, J. A., *Scientific American*, **276**, 60 (1997); (g) Jordan, R. F. (Ed.), *J. Mol. Catal, A: Chem.*, **128**, (1998); (h) Kashiwa, N., *Stud. Surf. Sci. Catal.*, **89**, 381 (1994).

3. (a) Breslow, D. S., *US Pat. Appl.*, 537 039 (1955); (b) Breslow, D. S. and Newburg, N. R., *J. Am. Chem. Soc.*, **79**, 5072 (1957); (c) Breslow, D. S. and Newburg, N. R., *J. Am. Chem. Soc.*, **81**, 81 (1959).

4. (a) Long, W. P., *J. Am. Chem. Soc.*, **81**, 5312 (1959); (b) Zefirova, A. K. and Shilov, A. E., *Proc. Acad. Sci. USSR, Chem. Sect., Engl. Transl.*, **136**, 77 (1961).

5. (a) Cossee, P., *J. Catal.*, **3**, 80 (1964); (b) Arlman, E. J. and Cossee, P., *J. Catal.*, **3**, 99 (1964).

6. (a) Karrol, F. J. and Carrick, W. L., *J. Am. Chem. Soc.*, **83**, 2654 (1961); (b) Philips, G. W. and Carrick, W. L., *J. Am. Chem. Soc.*, **84**, 920 (1962); (c) Philips, G. W. and Carrick, W. L., *J. Pol. Sci.*, **59**, 401 (1962).

7. Wiman, R. E. and Rubin, I. D., *Makromol. Chem.*, **94**, 160 (1966).

8. Matkovski, P. Ye., Belov, G. P., Russiyan, L. N., Lisitskaya, A. P., Kissin, Y. V., Solov'eva, A. A. Brikenshtein, A. A. and Chirkov, N. M., *Polym. Sci. USSR (Engl. Transl.)*, **12**, 2590 (1971).

9. Montecatini, *Br. Pat.*, 875 078 (1961).

10. Höcker, H. and Saiki, K., *Makromol. Chem.*, **148**, 107 (1971).

11. Marconi, W., Santostasi, M. L. and De Malde, M., *Chim. Ind. (Milan)*, **44**, 235 (1962).

12. Acton, N. and Katz, T. J., *Tetrahedron Lett.*, **28**, 2497 (1970).

13. (a) Sinn, H. and Kaminsky, W., *Adv. Organomet. Chem.*, **18**, 99 (1980); (b) Andresen, A., Cordes, H. G., Herwig, J., Kaminsky, W., Merck, A., Mottweiler, R., Pein, J., Sinn, H. and Vollmer, H. J., *Angew. Chem. Int. Ed. Engl.*, **15**, 630 (1976); (c) Kaminsky, W., Kopf, J., Sinn, H. and Vollmer, J., *Angew. Chem., Int. Ed. Engl.* **15**, 629 (1976); (d) Sinn, H., Kaminsky, W., Vollmer, H. J. and Woldt, R., *Angew. Chem. Int. Ed. Engl.*, **19**, 396 (1980); (e) Herwig, J., Dissertation, Universität Hamburg (1979); (f) Kulper, K., Dissertation, Universität Hamburg (1985).

14. (a) Jordan, R. F., *Adv. Organomet. Chem.*, **32**, 325 (1991); (b) Hlatky, G. G., Stern, C. L. and Turner, H. W., *US Pat. Appl.*, 4 459 921 (1990); *Chem. Abstr.*, **115**, 25689v (1991).

15. (a) Ewen, J. A. and Elder, M. J., *US Pat. Appl.*, 419 017 (1989); *Chem. Abstr.*, **115**, 136998g (1991); (b) Ewen, J. A. and Elder, M. J., *Eur. Pat. Appl.*, 426 637, 426 638 (1991); *Chem. Abstr.*, **115**, 136987c, 136988d; (c) Ewen, J. A. and Elder M. J., *Makromol. Chem., Macromol. Symp.*, **66**, 179 (1993).

16. Bochmann, M. and Lancaster, S. J., *Angew. Chem, Int. Ed. Engl.*, **33**, 1634 (1994).

17. (a) Mason, M. R., Smith, J. M., Bott, S. G. and Barron, A. R., *J. Am. Chem. Soc.*, **115**, 4971, (1993); (b) Harlan, C. J., Mason, M. R. and Barron, A. R., *Organometallics*, **13**, 2957 (1994).

18. Tsutsui, T., Mizuno, A. and Kashiwa, N., *Polymer*, **30**, 428 (1989).

19. (a) Ewen J. A. and Welborn, H. C., *Eur. Pat. Appl.*, 0 129 368 (1984); *US Pat. Appl.*, 5 324 800 (1994); (b) Ewen J. A. and Welborn H. C., *Eur. Pat. Appl.*, 0 128 046 (1984); *US Pat Appl.*, 4 530 914 (1985), 4 935 474 (1990); (c) Ewen J. A. and Welborn H. C., *Eur. Pat. Appl.*, 0 128 046 (1986); *US Pat Appl.*, 4 937 299 (1990) (d) Ewen, J. A., *Stud. Surf. Sci. Catal.*, **25**, 271 (1986).

20. Cheruvu, S., *US Pat.*, 5 608 019, to Mobil Oil Co. (1997).

21. Alt, H. G., Palackal, S. J., Welch, M. B., Rohlfing D. C. and Janzen, J., *US Pat.*, 5 710 224, to Phillips (1998).

22. Resconi, L., Jones R. L., Rheingold, A. and Yap, G., *Organometallics* **15**, 998 (1996).
23. Ewen, J. A., *J. Mol. Catal. A: Chem.*, **128**, 103 (1998).
24. (a) Chien, J. C. W. and He D., *J. Pol. Sci. A: Pol. Chem.*, **29**, 1585 (1991); (b) Razavi, A. and Atwood, L., *J. Am. Chem. Soc.*, **115**, 7529 (1993). Note: the copolymerization parameters in ref. 24a must be incorrect because r_1 was reported to be the same for Cp_2TiCl_2 and Cp_2ZrCl_2 although Ti is in fact twice as reactive with propylene *vs* ethylene as Zr [19d]. However, the broad MWD with Ind_2ZrCl_2/MAO and the APP/IPP mixture with $(1-CH_3Flu)_2ZrCl_2/MAO$ suggests some unbridged catalysts with Cps having either two or four of their π-electron sextet coming from other aromatic rings may become heterogeneous mixtures during polymerization.
25. Stevens, J. C., *Stud. Surf. Sci. Catal.*, **89**, 277 (1994).
26. (a) Wild, T. and Blutz, C., *Polym. Mater. Sci. Eng.* **67**, 153 (1992); (b) Pigeon, M.G. and Rudin, A., *J. Appl. Polym. Sci.*, **51**, 303 (1994).
27. (a) Usami, T., Gotoh, Y. and Takayama, S., *Macromolecules*, **19**, 2722 (1986); (b) Defoor, F., Groeninckx, G., Schouterden, P. and van der Heijden, B., *Polymer*, **33**, 3878 (1992); (c) Montagna, A. A. and Floyd, J. C., *Proc. MetCon Houston*, 173 (1993).
28. Mark, H. F., Bikales, N. B., Overberger, C. G., Menges, G. and Kroschwitz, J. I. (Eds) Encyclopedia of Polymer Science and Engineering Wiley, New York, 1988, Vol. 13, p. 500.
29. Busico, V., Mevo, L., Palumbo, G., Zambelli, A. and Tancredi, *Macromol. Chem.*, **184**, 2193 (1983).
30. Zambelli, A., Grassi, A., Galimberti, R., Mazzocchi, R. and Piemontesi, F., *Makromol. Chem., Rapid Commun.*, **12**, 523 (1991).
31. Denger, Ch., Haase, U. and Fink, G., *Makromol. Chem., Rapid Commun.*, **14**, 697 (1991).
32. Berrisford, D. J., Bolm, C. and Sharpless, K. B., *Angew. Chem., Int. Ed. Engl.*, **34**, 1059 (1995).
33. Montagna, A. A., Burkhart, R. M. and Dekmezian, A. H., *Chemtech.*, December, 26 (1997).
34. Ewen, J. A., Jones, R. L., Razavi, A. and Ferrara, J. D., *J. Am. Chem. Soc.*, **110**, 6255 (1988).
35. Ishihara, N., Seimiya, T., Kuramoto, M. and Uoi, M., *Macromolecules*, **19**, 2464 (1986).
36. Ewen, J. A., Zambelli, A. and Longo, P., *Makromol. Chem. Rapid Commun.*, **19**, 71 (1998).
37. (a) Corradini, P., Guerra, G., Vacatello, M. and Villani, V., *Gazz. Chim. Ital.*, **118**, 173 (1988); (b) Cavallo, L., Guerra, G., Oliva, L., Vacatello, M. and Corradini, P., *Polym. Commun.*, **30**, 16 (1990); (c) Corradini, P., presented at Workshop on Present State and Trends in Olefin and Diolefin Polymerization, May 23–25, 1989, Como, Italy; (d) Cavallo, L., Guerra, G., Vacatello, M. and Corradini, P., *Polymer*, **31**, 530 (1990); (e) Corrdadini, P., Guerra, G., Cavallo, L., Moscardi, G., Vacatello, M., in Fink G. Mulhaupt, R. and Brintzinger, H. H. (Eds.) Ziegler Catalysts; Springer, Berlin, 1995, pp. 237–249; (f) Guerra, G., Corradini, P., Cavallo, L., Vacatello, M., *Macromol. Symp.*, **89**, 307 (1995).
38. Bovey, F. A., *High Resolution NMR of Macromolecules*, Academic Press, New York, 1972.
39. Ewen, J. A., *J. Am. Chem. Soc.*, **106**, 6355, (1984).
40. (a) Ewen, J. A., Elder, M. J., Jones, R. L., Curtis, S. and Cheng, H. P., *Stud. Surf. Sci.*, **56**, 271, 1990; (b) Ewen, J. A. and Elder, M. J., in G., Fink, Mulhaupt, R. and H. H. Brintzinger, (Eds) Ziegler Catalysts, Fink, G., Springer, Berlin, 1995, p. 99; (c) Ewen, J. A., Elder, M. J., Harlan, C. J., Jones, R. L., Atwood, J. L., Bott, S. G. and Robinson, K., *ACS, Polym. Prepri.*, **32**, 469 (1991); (d) Ewen, J. A., Elder, M. J., Jones, R. L., Haspeslagh, L., Atwood, J. L., Bott, S. G. and Robinson, K., *Makromol. Chem., Macromol. Symp.*, **48/49**, 253 (1991).

41. Zambelli, A., Locatelii, P., Zannoni, G. and Bovey, F. A., *Macromolecules*, **11**, 923 (1978).
42. Farina, M., DiSilvestero, G. and Sozzari, P., *Macromolecules*, **15**, 1451 (1982).
43. (a) Longo, P., Grassi, A., Pellecchia, C. and Zambelli, A., *Macromolecules*, **20**, 1015 (1987); (b) Zambelli, A., Pellecchia, L. and Oliva, L., *Makromol. Chem., Macromol. Symp.*, **48/49**, 297 (1991); (c) Zambelli, A. and Pellechia, C., *Makromol. Chem., Macromol. Symp.*, **66**, 1 (1993)
44. Kaminsky, W., Külper, K., Brintzinger, H. H. and Wild, F. R. W. P., *Angew. Chem., Int. Ed., Engl.*, **24**, 507 (1985).
45. (a) Soga, K., Shiono, T., Takemura, S. and Kaminsky, W., *Makromol. Chem., Rapid Commun.*, **8**, 305 (1987); (b) Cheng H. N. and Ewen, J. A. *Makromol. Chem.*, **190**, 1931 (1989).
46. Ewen, J. A., Haspeslagh, L., Elder, M. J., Atwood, J. L., Zhang, H. and Cheng, H. N., in Kaminsky, W. and Sinn, H. (Eds); *Transition Metals and Organometallics as Catalysts for Olefin Polymerization*, Springer New York, 1988, pp. 281–289.
47. (a) Miya, S., Yoshimura, T., Mise, T. and Yamazaki, H., *Polym. Prepr. Jpn.*, **37**, 285 (1988); (b) Miya, S., Mise, T. and Yamazaki, H., *Chem. Lett.*, 1953 (1989); (c) Miya, S., Mise, T. and Yamazaki, H., *Stud. Surf. Sci. Catal.*, **56**, 531, (1990).
48. (a) Herrmann, W. A., Rohrmann, J., Herdtweck, E., Spaleck, W. and Winter, A., *Angew. Chem., Int. Ed. Engl.*, **28**, 1511 (1989); (b) Spaleck, W., Antberg, M., Rohrman, J., Winter, A., Bachmann, B., Kiprof, P., Behm, J. and Herrmann, W. A., *Angew. Chem., Int. Ed. Engl.*, **31**, 1347 (1992).
49. Stehling, U., Diebold, J., Kirsten, R., Röll, W., Brintzinger, H. H., Jüngling, S., Mülhaupt, R. and Langhauser, F., *Organometallics*, **13**, 964 (1994).
50. Cavallo, L. and Guerra, G., *Macromolecules*, **29**, 2729 (1996).
51. Hortmann, K. and Brintzinger, H. H., *New J. Chem.*, **16**, 51 (1992).
52. (a) Röll, W., Brintzinger, H. H., Rieger, R. and Zolk, R., *Angew. Chem., Int. Ed. Engl.*, **29**, 279 (1990); (b) Collins, S., Gauthier, W. J., Holden, D. A., Kuntz, B. A., Taylor, N. J. and Ward, D. G., *Organometallics*, **10**, 2061 (1991); (c) Lee, I. M., Gauthier, W. J., Ball, J. M., Iyengar, B. and Collins, S., *Organometallics*, **11**, 2115 (1992).
53. (a) Spaleck, W., Küber, F., Winter, A., Rohrmann, J., Bachmann, B., Antberg, M., Dolle, V. and Paulus, E. F., *Organometallics*, **13**, 954 (1994); (b) Stehling, U., Diebold, J., Kirsten, R., Röll, W., Brintzinger, H. H., Jünglin, S., Mülhaupt, R. and Langhauser, F., *Organometallics*, **13**, 964 (1994); (c) Ewen, J.A., Elder, M.J. and Dubitsky, Y.A., *Int. Pat. Appl.*, WO98/22486, to Montell (1998).
54. Chien, J. C. W., Rieger, B. and Sugimoto, R., *Stud. Surf. Sci. Catal.*, **56**, 535 (1990).
55. Dietrich, U., Hild, S., Imhof C. and Rieger, S., Abstract in Program for "International Symposium on Organometallic Catalysts for Synthesis and Polymerization" (ed. Kaminsky, W.) Hamburg, September 13–17, 1998.
56. Ewen, J. A., Jones, R. L., Razavi, A. and Ferrara, J. D., *J. Am. Chem. Soc.*, **110**, 6255 (1988).
57. Ewen, J. A., *Makromol. Chem., Macromol. Symp.*, **89**, 182 (1995); (b) Shiomura, T., Kohno, M., Inoue, N., Yokote, Y., Akiyama, M., Asanuma, T., Sugimoto, R., Kimura, S. and Abe, M., *Stud. Surf. Sci. Catal.*, **89**, 327, 1994; (b) Shiomura, T., Kohno, M., Inoue, Asanuma, T., Sugimoto, R., Iwatani, T., Uchida, O., Kimura, S., Harima, S., Zenkoh, H. and Tanaka, E., *Makromol. Symp.*, **101**, 289 (1996); (c) Herzog, T. A., Zubris, D. L. and Bercaw, J. E., *J. Am. Chem. Soc.*, **118**, 11988 (1996); (d) Alt H. G., Milius, W. and Palackal, S. J., *J. Organomet. Chem.*, **472**, 113 (1984); (e) Schmidt, M. A., Alt, H. G. and Milius, W., *J. Organomet. Chem.*, **501**, 101 (1995); (f) Alt, H. G. and Zenk, R., *J. Organomet. Chem.*, **512**, 51 (1996); (g) Alt, H. G., Zenk and R., W. M Milius, *J. Organomet. Chem.*, **514**, 257 (1996); (h) Alt, H. G. and Zenk, R., *J. Organomet. Chem.*, **518**, 7 (1996); (i) Alt, H. G. and Zenk, R., *J. Organomet. Chem.*, **522**, 39 (1996); (j) Alt, H.

G. and Zenk, R., *J. Organomet. Chem.*, **522**, 177 (1996); (k) Schmidt, M. A., Alt, H. G. and Zenk, R., *J. Organomet. Chem.*, **525**, 9 (1996); (l) Schmidt, M. A., Alt, H. G. and Milius, W., *J. Organomet. Chem.*, **525**, 15 (1996); (m) Alt, H. G. and Zenk, R., *J. Organomet. Chem.*, **526**, 295 (1996).

2

Alkylalumoxanes: Synthesis, Structure and Reactivity

ANDREW R. BARRON
Rice University, Houston, TX, USA

1 INTRODUCTION

Although it had been known since the 1950s that compounds of aluminum react with water to give compounds containing aluminum–oxygen bonds, commonly termed alumoxanes, it was not until the work of Manyik *et al.* [1] that their application to olefin catalysis was fully appreciated. These workers showed that alkyl-substituted alumoxanes (alkylalumoxanes) were highly active cocatalysts for olefin polymerization in combination with compounds of the Group 4, 5 and 6 elements, including metallocenes. The discovery, in 1952, by Geoffrey Wilkinson of metallocenes, especially those of titanium and zirconium [2], had a major impact in olefin polymerization catalysis since for the first time truly homogeneous Ziegler–Natta catalysts were possible. Subsequent to the work of Manyik *et al.*, the groups of Meyer [3], Breslow [4] and Kaminsky [5] all showed that the addition of water to the soluble metallocene/alkylaluminum catalyst systems resulted in a large increase in catalyst activity. The *in situ* formation of alumoxanes in all of these systems was recognized [3–5], but the high catalytic activity of a metallocene and preformed methylalumoxanes system was shown by the work of Kaminsky and co-workers [6,7]. Kaminsky and co-workers also demonstrated that zirconium metallocenes were more active than the titanium metallocenes [8,9].

Since 1980 there has been an enormous amount of research activity involving metallocene/alkylalumoxane catalytic systems, as is evident by the publication of this book. These efforts have, however, been largely aimed at the development of substituted metallocenes in order to control the stability and solubility of the catalyst and the molecular weight and stereochemistry of the polymer. Relatively little work

Metallocene-based Polyolefins Edited by J. Scheirs and W. Kaminsky
© 2000 John Wiley & Sons Ltd

has been reported on the alkylalumoxanes, although several significant studies were performed throughout the 1970s, especially by Pasynkiewicz and co-workers [10]. With regard to olefin polymerization, most researchers were content to use alkylalumoxanes as a 'black box' chemical. This view was based on the assumption that the alkylalumoxane 'cocatalyst' was not involved in the catalytic cycle. The lack of any definitive characterization of alkylalumoxanes compounded this state of affairs. The isolation and structural characterization of the first alkylalumoxanes [11,12] has led to a renewed effort for the development of a cohesive picture of the mode of cocatalytic activity of alkylalumoxanes.

The goal of this chapter is to present the current understanding of the synthesis, structure and reactivity of alkylalumoxanes. However, owing to the rapidly evolving nature of our knowledge of alkylalumoxanes, this cannot be the conclusive story. Although there have been no reports of any cocatalytic activity of the galloxanes (the gallium analogs of alumoxanes), the obvious homologous relationship of gallium and aluminum compounds offers the use of galloxanes as structural models of alumoxanes, and so these are also discussed where appropriate.

2 WHAT IS AN ALKYLALUMOXANE?

The term 'alumoxane' is used to describe a molecular species containing at least one oxo group (O^{2-}) bridging (at least) two aluminum atoms, i.e. a compound containing an Al—O—Al subunit. The simplest alumoxane compounds are those containing two aluminum atoms bridged by a single oxygen with n additional ligands (X) bonded to aluminum, e.g. **1** [13–15].

$$X_nAl\!\!-\!\!O\!\!-\!\!AlX_n$$

(1)

Alumoxane does not, however, refer to bridging alkoxides, siloxides or related compounds in which the oxygen is bonded to anything other than aluminum, although the aluminum itself may have a variety of pendant groups (X). Within the overall definition, the term 'alumoxane' is most commonly used (especially with regard to metallocene catalysis) to denote compounds in which the pendant groups on aluminum are organic radical substituents, i.e. alkylalumoxanes.

Alkylalumoxanes are, therefore, oligomeric aluminum compounds which can be represented by the general formulae $[(R)Al(O)]_n$ and $R[(R)Al(O)]_nAlR_2$. In these formulae R is an alkyl group, such as methyl (CH_3), ethyl (C_2H_5), propyl (C_3H_7), butyl (C_4H_9) and pentyl (C_5H_{11}), and n is an integer from 1 to about 20. Such compounds are usually derived from the hydrolysis of alkylaluminum compounds, AlR_3.

3 SOME NOTES ON NOMENCLATURE AND ACRONYMS

It should be noted that while "alkylalumoxane" is generally accepted, alternative terms are found in the literature. These include alkylaluminoxane, poly(alkylalumoxane), poly(alkylaluminum oxide) and poly(hydrocarbylaluminum oxide). Furthermore, although alkylalumoxanes are also simply referred to as alumoxanes, this class of compounds actually includes such materials as sol–gels and antiperspirants with no cocatalytic activity. For clarity, alkylalumoxane will be used throughout this chapter.

It is common practice for both scientists and commercial vendors to use acronyms for various specific alkylalumoxanes. Thus, methylalumoxane formed via the hydrolysis of trimethylaluminum is ordinarily given the acronym MAO. However, PMAO from poly(methylalumoxane) is also used. Table 1 lists the trialkylaluminum compounds and the derived alkylalumoxane, along with their respective acronyms.

Where the structure of an individual alumoxane (or related molecule) is definitively known, μ- and μ_3- are used to indicate structural features consistent with common inorganic nomenclature. Thus, μ indicates that the group or moiety following bridges two aluminum centers. For example, the formula $[(R)_2Al(\mu\text{-OH})]_n$ indicates that the hydroxide group is bridging two aluminum atoms (**2**). Similarly, the μ_3- in $[(R)Al(\mu_3\text{-O})]_n$ shows that the oxo ion is capping three aluminum atoms (**3** or **4**).

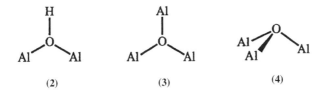

(2)   (3)   (4)

4 SYNTHESIS OF ALKYLALUMOXANES

Alkylalumoxanes may be prepared in a variety of ways. Preferably, they may be synthesized by contacting water with a solution of trialkylaluminum, AlR_3, in a suitable organic solvent such as aromatic or an aliphatic hydrocarbon. In an

Table 1 Acronyms of trialkylaluminum compounds and the derived alkylalumoxane

Trialkylaluminum		Alkylalumoxane	
Trimethylaluminum	TMA	Methylalumoxane	MAO
Triethylaluminum	TEA	Ethylalumoxane	EAO
Triisobutylaluminum	TIBA	Isobutylalumoxane	IBAO
Tri-*tert*-butylaluminum	TBA	*tert*-Butylalumoxane	TBAO

alternative method, the trialkylaluminum can be reacted with a hydrated salt such as hydrated aluminum sulfate. In both cases the reaction is evidenced by the evolution of the appropriate hydrocarbon, e.g. methane (CH_4) during the hydrolysis of trimethylaluminum ($AlMe_3$). While these two routes are by far the most common, and are discussed below in detail, several 'non hydrolysis' routes have been developed to obviate the dangers inherent in performing such a violent highly exothermic reaction ($\Delta H = 1090\,kJ/mol$) [16,17].

4.1 HYDROLYSIS OF ALKYLALUMINUM COMPOUNDS

Conceptually, but not experimentally, the simplest route to alkylalumoxanes involves the reaction of water with a trialkylaluminum compound. Simply reacting water (or ice) [18] with an aromatic or aliphatic hydrocarbon solution of a trialkylaluminum will yield an alkylalumoxane. However, it is important to control the temperature of this highly exothermic reaction both as a safety precaution and in order to maximize the yield and ensure the solubility of the products [19]. In an effort to control the rate at which the water reacts with the trialkylaluminum, several researchers have employed hydrated salts, such as $Al_2(SO_4)_3 \cdot 14(H_2O)$ or $CuSO_4 \cdot 5(H_2O)$, as 'indirect hydrolysis' sources [20], since the water of crystallization in a hydrated salt reacts at a vastly slow rate than dissolved 'free' water.

Irrespective of the source of water, the overall reaction of water with an trialkylaluminum compound to give an alkylalumoxane may be described by

$$AlR_3 + H_2O \longrightarrow \frac{1}{n}[(R)Al(O)]_n + 2RH \tag{1}$$

$$2AlR_3 + H_2O \longrightarrow \frac{1}{n}[R_2AlOAlR_2]_n + 2RH \tag{2}$$

Unfortunately, these equations are a gross oversimplification of the series of reaction steps that occur during the hydrolysis reaction. It is the complex nature of the hydrolysis of trialkylaluminum compound that is most responsible for hampering studies on alkylalumoxanes. In spite of this, a number of the individual reaction sequences have been demonstrated.

The hydrolysis of $AlR_3 (R = Me, Et, iBu)$ has been shown to proceed via the formation of an alkylaluminum water complex [equation (3)] [21], which subsequently eliminates alkane to form a dialkylaluminum hydroxide complex. This rapidly associates to give dimers or larger oligomers in solution [equation (4)].

$$AlR_3 + H_2O \longrightarrow AlR_3(H_2O) \tag{3}$$

$$AlR_3(H_2O) \longrightarrow \frac{1}{n}[R_2Al(OH)]_n + RH \tag{4}$$

Although the existence of alkylaluminum hydroxide intermediates has long been supported by solution NMR data [21] it is only recently that these species have been

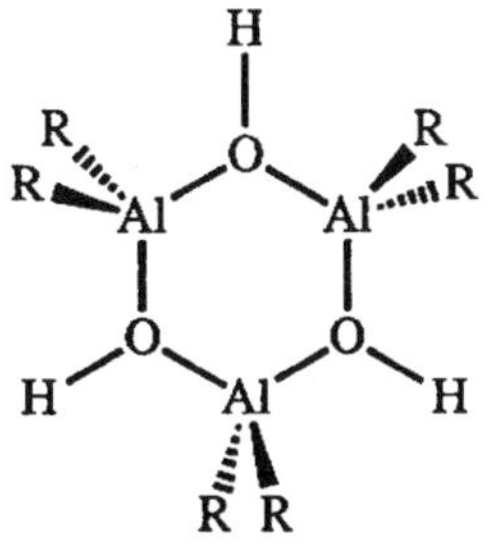

(5) 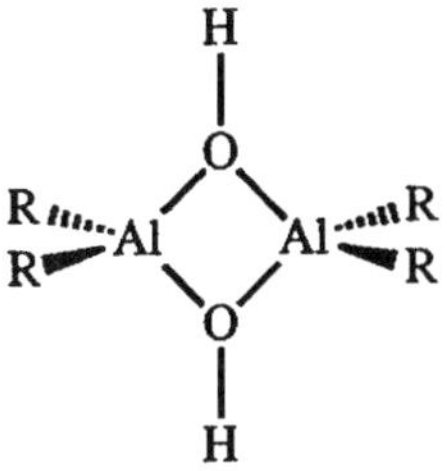   (6)

isolated and structurally characterized, e.g. $[R_2Al(\mu\text{-}OH)]_3$ (R = ^tBu [11], $CEtMe_2$ [22]) and $[R_2Al(\mu\text{-}OH)]_2$ (R = ^tBu [12], $C_6H_2Me_3$ [23]), 5 and 6, respectively.

Further alkane elimination is observed with the formation of alkylalumoxanes as the unstable hydroxide compounds are warmed, e.g.

$$[R_2Al(OH)]_n \longrightarrow [(R)Al(O)]_n + nRH \qquad (5)$$

Alternatively, the dialkylaluminum hydroxide can react with further trialkylaluminum to yield an alumoxane with an aluminum to oxygen ratio of 2:1;

$$[R_2Al(OH)]_n + nAlR_3 \longrightarrow [R_2AlOAlR_2]_n + nRH \qquad (6)$$

Clearly, the ratio of alkylalumoxanes of the type $[(R)Al(O)]_n$ relative to those of the general formula $[R_2AlOAlR_2]_n$ will depend on the ratio of aluminum to water in the reaction mixture. The exact pathway (relative rates of individual reaction steps) and hence product distribution is also dependent on the conditions under which the hydrolysis reaction is performed: direct hydrolysis versus salt hydrolysis and low versus ambient temperature. Furthermore, each of these reactions [equations (4)–(6)] is further complicated by permutations in the extent of oligomerization (i.e. the value of n).

The reaction of an aluminum alkyl with a Brønsted acid (HX) resulting in alkane elimination is ubiquitous in the organometallic chemistry of aluminum [24]. It has been commonly assumed that this reaction occurs via the prior formation of a Lewis acid–base adduct from which the elimination reaction occurs [i.e. equation (7)]. A concerted intramolecular elimination possibly via a planar four-centered transition state (7) was therefore proposed [25,26].

$$AlR_3 + HX \longrightarrow AlR_3(XH) \longrightarrow \frac{1}{n}[R_2Al(\mu\text{-}X)]_n + RH \qquad (7)$$

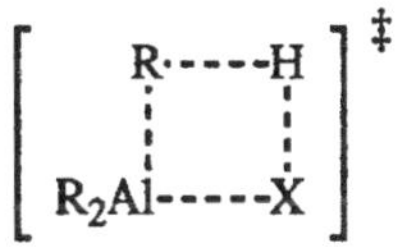

(7)

In apparent contradiction of this proposal, *ab initio* calculations indicate [27] that an intramolecular elimination of methane from $AlMeH_2(H_2O)$ is energetically disfavored ($\Delta H^\ddagger > 98$ kJ/mol) versus dissociation of the water (BDE $\approx 84-95$ kJ/mol). In accordance with the preference for dissociation, Beachley and Tessier-Youngs [28,29] demonstrated (for the reaction of aluminum alkyls and amines) that while a Lewis acid–base adduct is formed, the important step for the elimination–condensation reaction is a prior dissociation of the this adduct [equation (8), where $X = NR_2$]. Recombination of the monomeric aluminum compound and the amine with the appropriate orientation was proposed to result in elimination via a four-centered $S_E i$ (substitution, electrophilic, internal) mechanism (**8**) [30].

$$AlR_3(XH) \rightleftharpoons AlR_3 + HX \longrightarrow \frac{1}{n}[R_2Al(\mu\text{-}X)]_n + RH \qquad (8)$$

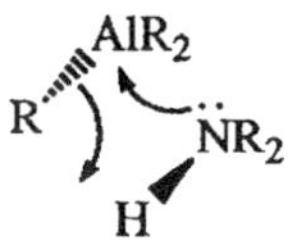

(8)

It is widely observed that complexation of water to a transition metal increases its Brønsted acidity [31,32] and the isolation of the alkoxide$^-$/ammonium$^+$ zwitterionic complex, **9**, suggests a decrease in pK_a of the Brønsted acid's proton upon complexation to aluminum by at least 7 units [33]. It is not obvious, therefore, why the Brønsted acid should react once uncomplexed from aluminum since it is at its most active (acidic) when complexed.

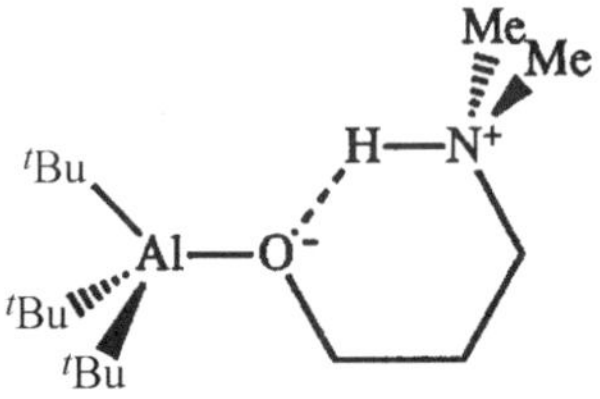

(9)

The contradictory nature of these mechanisms has led to an alternative proposal. The formation of a Lewis acid–base adduct [equation (3)] activates the α-proton (on the water) by increasing its acidity. Intramolecular elimination does not occur, instead adduct dissociation yields 'free', but relatively unreactive, Brønsted acid (H_2O) and 'free', but more reactive AlR_3:

$$AlR_3(H_2O) \rightleftharpoons AlR_3 + H_2O \qquad (9)$$

This uncomplexed AlR_3 reacts directly with the coordinated water on another molecule of the activated complex, $AlR_3(H_2O)$, resulting in an intermolecular elimination–condensation reaction:

$$Al^*R_3(H_2O) + AlR_3 \longrightarrow \frac{1}{n}[R_2Al(\mu\text{-OH})]_n + RH + Al^*R_3 \qquad (10)$$

Hence the rate-determining step involves adduct dissociation, consistent with Beachley and Tessier-Young's experiments [28,29], but the reactive species is the activated complex, $AlR_3(H_2O)$, consistent with ligand activation by the aluminum [33]. Where the trialkylaluminum is dimeric the elimination will be intra-dimer in nature and probably occurs via six-membered transition state (Scheme 1).

The reaction of alkylaluminum compounds (e.g. trimethylaluminum, TMA) with water is a violent highly exothermic reaction and is best carried out at low temperature in an inert solvent; however, syntheses have been reported at temperatures from -78 to $110\,^{\circ}\text{C}$. Those employing the direct addition of water (or ice) are carried out below room temperature, while the salt hydrolysis method generally requires heating. [*CAUTION: the hydrolysis of trimethylaluminum at high temperatures has irresponsibly been proposed to be the preferred synthesis of methylalumoxane* [34]. *This is a highly dangerous reaction and should not be attempted under any circumstances.*] Empirical observation has shown that the temperature of reaction is important in defining the activity and solubility of the resultant alkylalumoxane. This difference appears to be related to the oligomerization of the hydroxide intermediates, i.e. the value of n for $[R_2Al(\mu\text{-OH})]_n$.

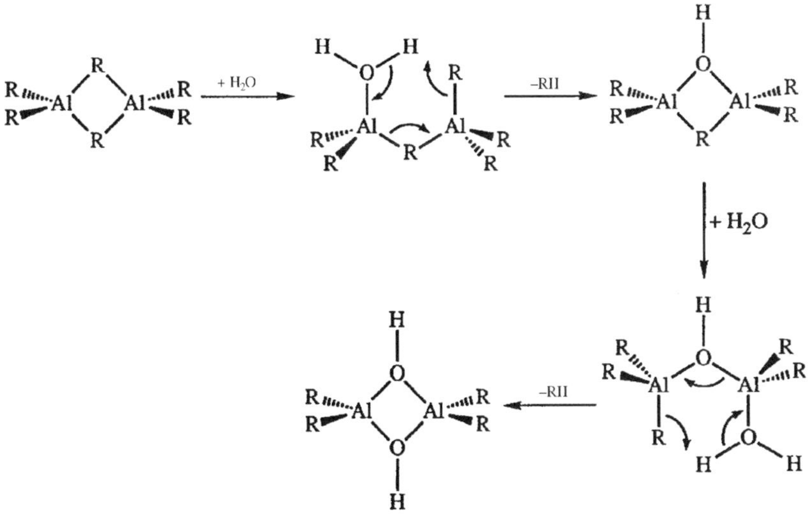

Scheme 1   Proposed pathway for the hydrolysis of AlR_3

The reaction of $Al(^tBu)_3$ with one molar equivalent of water results in the liberation of isobutane and the formation of the trimeric aluminum hydroxide $[(^tBu)_2Al(\mu\text{-}OH)]_3$ [10]:

$$Al(^tBu)_3 + H_2O \longrightarrow \frac{1}{3}[(^tBu)_2Al(\mu\text{-}OH)]_3 + {}^tBuH \qquad (11)$$

In contrast, the addition of $Al(^tBu)_3$ to a suspension of $Al_2(SO_4)_3 \cdot 14(H_2O)$ in toluene followed by heating to reflux yields the dimeric aluminum hydroxide, $[(^tBu)_2Al(\mu\text{-}OH)]_2$:

$$Al(^tBu)_3 \xrightarrow{\frac{1}{14}Al_2(SO_4)_3 \cdot 14(H_2O)} \frac{1}{2}[(^tBu)_2Al(\mu\text{-}OH)]_2 + {}^tBuH \qquad (12)$$

The dimeric hydroxide is also formed from the addition of H_2O to a toluene solution of $Al(^tBu)_3$ heated to reflux at $110\,°C$ [*CAUTION: this reaction is extremely violent.*] Thus, low-temperature hydrolysis of $Al(^tBu)_3$ yields the trimer whereas the high-temperature synthesis results in the formation of the dimer [10,11].

The temperature dependence of oligomerization is common in aluminum chemistry. Aluminum alkoxides, $[Me_2Al(\mu\text{-}OR)]_n$ (R = Me, Et, Pr, etc.) as prepared from the reaction of $AlMe_3$ and an alcohol are trimers and dimers in equilibrium [35] and the relative quantities of dimer and trimer present in a reaction mixture are dependent on the reaction temperature: decreasing the temperature of reaction increases the relative quantity of trimer produced. From this it is concluded that the trimeric alkoxides, $[Me_2Al(\mu\text{-}OR)]_3$, are the thermodynamic products whereas the dimers, $[Me_2Al(\mu\text{-}OR)]_2$, are the entropically favored products at high temperatures [35]. Clearly the same principles appears to apply to the formation of $[(^tBu)_2Al(\mu\text{-}OH)]_n$.

Although the hydroxide compounds are only intermediates to alkylalumoxanes, the reactivity of the dimeric and trimeric forms is sufficiently different to warrant comment. The thermolysis of both $[(^tBu)_2Al(\mu\text{-}OH)]_2$ and $[(^tBu)_2Al(\mu\text{-}OH)]_3$ results in the formation of *tert*-butylalumoxane, *cf.* equation (5). However, whereas $[(^tBu)_2Al(\mu\text{-}OH)]_3$ reacts at room temperature, $[(^tBu)_2Al(\mu\text{-}OH)]_2$ requires heating above $100\,°C$, and the distribution of alumoxane products is very different, i.e. the relative quantities of each oligomer, $[(^tBu)Al(\mu_3\text{-}OH)]_n$. In addition, where mixtures of dimeric and trimeric hydroxide are found, condensation results in the formation of a pentameric alumumoxane, $[Al_5(^tBu)_7(\mu_3\text{-}O)_3(\mu\text{-}OH)_2]$ (Figure 1), thermolysis of which subsequently results in gel formation from which a white powder may be obtained [11]. X-ray powder diffraction of this powder showed broad rings at d-spacings consistent with the two most intense reflections observed for boehmite, and comparable to literature values for gelatinous alumina [36].

It is commonly observed that during the reaction of AlR_3 (especially when R = Me) with water-soluble alumoxanes are only formed at low temperatures and low water:aluminum ratios. As the relative amount of water is increased, or as the reaction is permitted to proceed at higher temperatures, the reaction mixture often

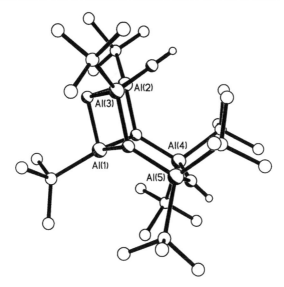

Figure 1   Molecular structure of $[Al_5(^tBu)_7(\mu_3\text{-}O)_3(\mu\text{-}OH)_2]$. The hydrogen atoms of the *tert*-butyl groups are omitted for clarity

contains gels usually assigned to the formation of hydrated alumina from the reaction of AlR_3 with a local excess of H_2O:

$$AlR_3 \xrightarrow[-2RH]{+H_2O} \frac{1}{n}[(R)Al(O)]_n \xrightarrow[-2RH]{+H_2O} Al(O)(OH) \tag{13}$$

Based upon structural similarity between the core of $[Al_5(^tBu)_7(\mu_3\text{-}O)_3(\mu\text{-}OH)_2]$ and the structure of boehmite, it is proposed that the gel accumulation previously observed during alkylalumoxane synthesis is due to the formation of alkylalumoxanes with structures based not on open cage structures (see below), but on a boehmite core.

During the hydrolysis of AlR_3 the formation of the hydroxide trimer is preferred at lower reaction temperatures; thermolysis of the trimer yields soluble alumoxanes. However, if the temperature of the reaction mixture is sufficiently high, some fraction of the trimeric hydroxide will rearrange to the dimer, and/or the dimer will form directly. Condensation of the trimeric hydroxide occurs to yield soluble cage-like alumoxanes. With significant concentrations of dimer, however, the formation of compounds similar to $[Al_5(^tBu)_7(\mu_3\text{-}O)_3(\mu\text{-}OH)_2]$ will result. Further condensation yields the alumina gels containing the boehmite core structure. These reactions are summarized in Scheme 2.

The isolation of $[(^tBu)_2Al(\mu\text{-}OH)]_n$ has allowed the investigation of the condensation reaction that results in the formation of alkylalumoxanes [equation (5)]. Whereas $[(^tBu)_2Al(\mu\text{-}OH)]_3$ eliminates alkane even at room temperature,

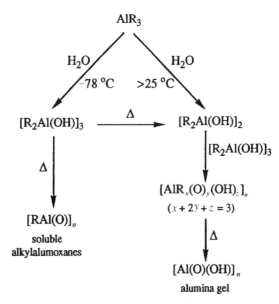

Scheme 2 Proposed reactions responsible for the formation of soluble alkyl-alumoxanes versus alumina gels during the hydrolysis of aluminum alkyls, AlR_3

$[(^tBu)_2Al(\mu\text{-OH})]_2$ shows no reaction until heated in refluxing toluene (110 °C) for 15 h. In this regard $[(^tBu)_2Al(\mu\text{-OH})]_2$ is similar to $[(^tBu)_2Ga(\mu\text{-OH})]_3$, which requires refluxing in xylene (143–145 °C) to permit conversion into the galloxane [37]. The structural similarity but disparate reactivity/stability of $[(^tBu)_2Al(\mu\text{-OH})]_3$ and $[(^tBu)_2Ga(\mu\text{-OH})]_3$ suggests that it is the acidity of the hydroxide proton that is important in determining the reactivity of the aluminum hydroxide compounds. In fact, the acidity of the hydroxides in $[(^tBu)_2M(\mu\text{-OH})]_n$, as determined by ^1H NMR chemical shift measurements [38], follows the same order as their reactivity, i.e. $[(^tBu)_2Al(\mu\text{-OH})]_3 > [(^tBu)_2Al(\mu\text{-OH})]_2 > [(^tBu)_2Ga(\mu\text{-OH})]_3$.

Irrespective of the proton acidity of the hydroxides, the condensation reaction [i.e., equation (5)] may occur via either intra-or intermolecular elimination of alkane. The formation of $[Al_6(^tBu)_8(\mu_3\text{-O})_4(\mu\text{-OH})_2]$ (Figure 2) from the controlled low-temperature thermolysis of $[(^tBu)_2Al(\mu\text{-OH})]_3$ [10] and its quantitative conversion to $[(^tBu)Al(\mu_3\text{-O})]_6$ suggests that the hydroxide condensation reaction is intermolecular. Consistent with this is the observation that the condensation reaction is also concentration dependent.

Although alkylalumoxanes are ordinarily formed via the hydrolysis of tri-alkylaluminum compounds, with the concomitant liberation of the corresponding alkane, hydrolysis of dialkylaluminum alkoxides or amides also results in the formation of alkylalumoxanes. Hydrolysis of the intramolecularly stabilized amine $(^tBu)_2Al[N(Me)CH_2CH_2\overline{N}Me_2]$ yields the base-stabilized alkylalumoxane

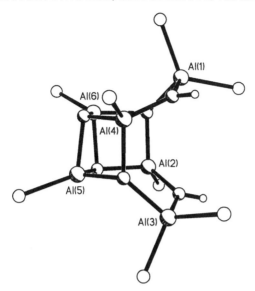

Figure 2 Molecular structure of $[Al_6(^tBu)_8(\mu_3\text{-}O)_4(\mu\text{-}OH)_2]$. The methyl groups are omitted for clarity

$[(^tBu)_2Al\{\overline{NH(Me)CH_2CH_2NMe_2}\}]_2(\mu\text{-}O)$ [39]. Similarly, hydrolysis of $[(^tBu)_2Al$ $(\mu\text{-}OCH_2CH_2OMe)]_2$ in the presence of 'wet' pyridine yields $[(^tBu)_2Al(py)]_2(\mu\text{-}O)$, which is also formed from the reaction of $[(^tBu)_2Al\{(\mu\text{-}OAl(^tBu)_2\}]_2$ with pyridine. In both cases the heteroatom donor–Al bond is hydrolyzed before to the remaining Al–alkyl bonds. Similarly, the galloxane $[Ga_{12}(^tBu)_{12}(\mu_3\text{-}O)_{10}(\mu_3\text{-}OH)_4]$ is formed via the hydrolysis of $(^tBu)_2Ga(Spy)$ (H-Spy) = 2-mercaptopyridine) [40].

4.2 ALTERNATIVE SYNTHESES OF ALKYLALUMOXANES

Although not commercially viable, there is a wide range of non-hydrolysis reactions that allow the formation of alkylalumoxanes.

Ziegler first reported the formation of an alumoxane from the reaction of $AlEt_3$ with CO_2 [41,42]. Presumably, this reaction occurs via the formation of an aluminum acetate [equation (14)], since a similar product is formed from the reaction of aluminum alkyls with carboxylates and amides [43,44].

$$AlEt_3 \xrightarrow{+CO_2} Et_2Al(O_2CEt) \xrightarrow[-Et_2C=O]{+AlEt_3} Et_2AlOAlEt_2 \qquad (14)$$

An interesting synthesis of alkylalumoxanes was developed by Pasynkiewicz and co-workers [45,46] involving the reaction of aluminum alkyls with lead(II) oxide:

$$2AlR_3 + PbO \longrightarrow R_2AlOAlR_2 + PbR_2 \qquad (15)$$

$$8Me_2AlCl + 4PbO \xrightarrow{+Et_2O} 4Me_2AlOAl(Cl)Me(Et_2O) + PbMe_4 + 2PbCl_2 + Pb$$

$$(16)$$

A variation of this oxygen-transfer reaction has been recently reported [equation (17)] [47] as an extension to the well known reaction of aluminum alkyls and hydrides with siloxanes [48].

$$[(Mes^*)AlH_2]_2 + \tfrac{2}{3}(Me_2SiO)_3 \longrightarrow \tfrac{1}{2}[(Mes^*)Al(\mu\text{-}O)]_4 + 2Me_2SiH_2 \qquad (17)$$
$$(Mes^* = 2,4,6\text{-}^tBu_3C_6H_2)$$

Oxygen abstraction is also the basis of the synthesis of sterically hindered alkylalumoxanes via the insertion of oxygen into the Al—Al bond of low-valent aluminum(II) compounds [49]:

$$[(Me_3Si)_2HC]_2Al—Al[CH(SiMe_3)_2]_2 + Me_2SO \longrightarrow$$
$$[(Me_3Si)_2HC]_2Al—O—Al[CH(SiMe_3)_2]_2 + Me_2S \qquad (18)$$

As an alternative to hydrolysis or oxygen abstraction, Araki and co-workers [50] reported that alkali metal aluminates formed from the reaction of trialkylaluminum with alkali metal hydroxides [equation (19)] react with aluminum chlorides to yield alkylalumoxanes [equation (20)]. This route has been studied in detail [51] and several of the intermediates have been isolated [52].

$$AlR_3 + LiOH \longrightarrow R_2Al(OLi) + RH \qquad (19)$$

$$R_2Al(OLi) + R_2AlCl \longrightarrow R_2AlOAlR_2 + LiCl \qquad (20)$$

5 STRUCTURE OF ALKYLALUMOXANES

Ever since the first reported alumoxane synthesis in the 1950s [53], most of the literature has shown alumoxanes as linear chains (**10**) or cyclic rings (**11**) consisting of alternating aluminum and oxygen atoms, based on the known structure of dialkylsiloxane polymers (**12**).

While such structures were originally intended as a pictorial representation, several workers have incorrectly suggested that these precise structures are present in methylalumoxane and similar non-sterically hindered alkylalumoxanes. Clearly, structures **10** and **11** require the aluminum to have a coordination number of three, which is rare, existing only in compounds in which oligomerization is sterically hindered by bulky ligands [54], In fact, a recent report of the structure of the 2,4,6-

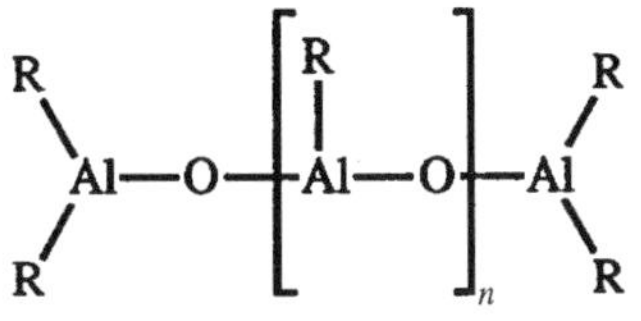

(10) 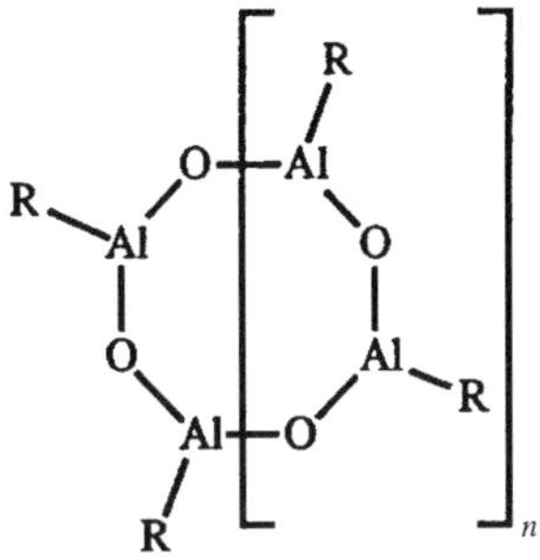   (11) 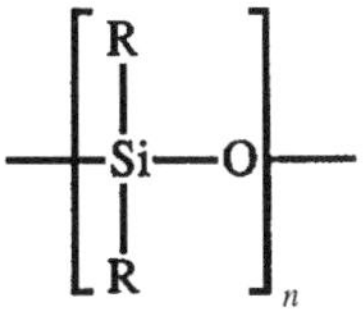   (12)

tri-*tert*-butylphenylalumoxane, [(Mes*)Al(μ-O)]$_4$, has shown that the presence of such a sterically hindered ligand does allow for the isolation of a simple cyclic alumoxane (*cf.* 11) for the sterically hindered arylalumoxane [(Mes*)Al(μ-O)]$_3$ (Mes* = 2,4,6-tri-*tert*-butylphenyl) [47]. However, it should be noted that no catalytic activity is observed for this compound.

5.1 STRUCTURES OF [(R)Al(O)]$_n$

Since it is common for aluminum to maximize its coordination number through the formation of dimers and trimers *via* bridging ligands [55], it is reasonable to expect that for catalytically active alkylalumoxanes (i.e. those without overwhelming steric hindrance), the aluminum must have a coordination number of four. With this in mind, several proposals have been made of two dimensional 'ladder' structures, e.g. 13 and 14.

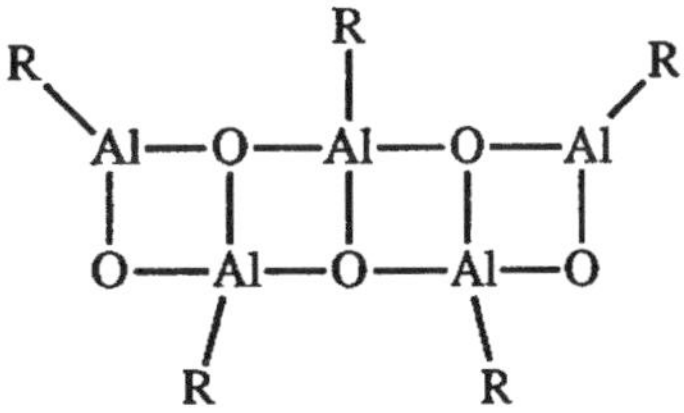

(13) 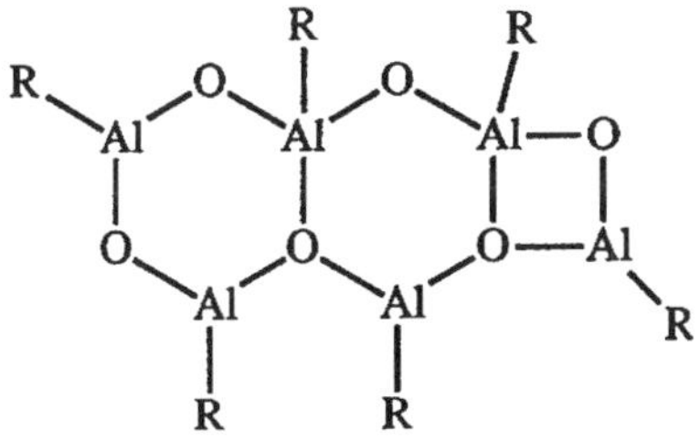   (14)

Although the presence of four-coordinate aluminum in MAO and other alkyl-alumoxanes is confirmed by ^{27}Al NMR, the postulated two-dimensional structures all retained some three-coordinate aluminum, which would make them unstable to further oligomerization. In this regard, the first molecular structures of an alkylalu-moxane of the formula [(R)Al(O)]$_n$ to be structurally characterized were those of the *tert*-butyl derivatives, [(^{t}Bu)Al(μ_3-O)]$_n$ [10,11]. The molecular structures of [(^{t}Bu)Al(μ_3-O)]$_6$, [(^{t}Bu)Al(μ_3-O)]$_8$ and [(^{t}Bu)Al(μ_3-O)]$_9$ were determined by X-ray crystallography to be three-dimensional electron precise cage structures in which the aluminum and oxygen have octet electron counts.

The n...ecular structure of $[(^tBu)Al(\mu_3\text{-}O)]_6$ is shown in Figure 3. The Al_6O_6 core can be described as a hexagonal prism with alternating Al and O atoms, such that every aluminum is four-coordinate and each oxygen has a coordination number of three. The octameric alkylalumoxane, $[(^tBu)Al(\mu_3\text{-}O)]_8$, can be described as being derived from the fusing of a hexameric cage and a square, Al_2O_2, ring (Figure 4). The cage structure of $[(^tBu)Al(\mu_3\text{-}O)]_9$, shown in Figure 5, consists of two parallel six-membered Al_3O_3 rings that are connected by three oxygen atoms and three aluminum atoms, and is isostructural to both $[Na(\mu_3\text{-}O^tBu)]_9$ [56] and $[(^tBu)Ga(\mu_3\text{-}O)]_9$ [37]. Although not fully crystallographically characterized, $[(^tBu)Al(\mu_3\text{-}O)]_7$ (**15**) and $[(^tBu)Al(\mu_3\text{-}O)]_{12}$ (**16**) have also been shown to have cage structures.

tert-butyl groups omitted for clarity

(15)   (16)

Although unexpected at the time, the presence of cage structures for alkylalumoxanes is consistent with the structures of their heavier gallium–sulfur analogs [57–59] and the iminoalanes [60,61]. On the basis of the structures observed for iminoalanes, Smith [62] proposed that in stable cages the number of four-membered rings is constant (six) whereas the number of six-membered rings is dependent on the extent of oligomerization (i.e. n) and must be equal to $n-4$. As would be expected, the alkylalumoxanes follow this rule; thus the octameric compound $[(^tBu)Al(\mu_3\text{-}O)]_8$ ($n = 8$) should consist of a cage with six four-membered Al_2O_2 faces and four ($n-4$) six-membered Al_3O_3 faces. This is indeed the structure observed (Figure 4). Although this rule does not predict the structures of the cages, a consideration of the relationship between each of the known cages [63] allows for a growth relationship between all possible oligomers.

The smallest cage observed for Group 13–16 compounds is the cubane, i.e. $[(^tBu)Ga(\mu_3\text{-}S)]_4$ [57]. The simplest route to cage extension, from the structural rules laid down by Smith, would be addition of an M_2E_2 (M = Group 13 metal, E = Group 16 element) moiety to give the hexagonal prismane $[(^tBu)M(\mu_3\text{-}E)]_6$. Subsequent addition would lead to the octamer and decamer (Scheme 3). Further

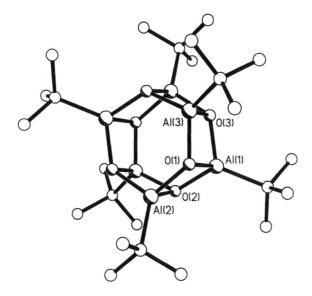

Figure 3   Molecular structure of $[(^tBu)Al(\mu_3\text{-}O)]_6$. The hydrogen atoms are omitted for clarity

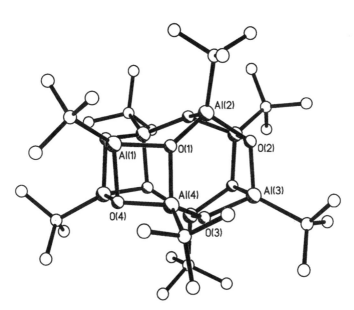

Figure 4   Molecular structure of $[(^tBu)Al(\mu_3\text{-}O)]_8$. The hydrogen atoms are omitted for clarity

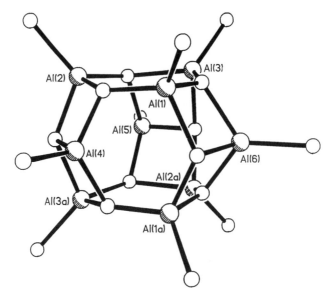

Figure 5 Molecular structure of $[(^tBu)Al(\mu_3\text{-}O)]_9$. The methyl groups are omitted for clarity

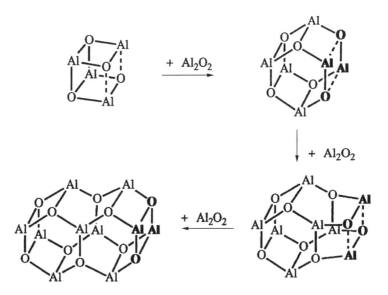

Scheme 3 Proposed relationship between $[(R)Al(\mu_3\text{-}O)]_n$, $n = 4, 6, 8, 10$. Alkyl groups are omitted for clarity. The bold groups and dashed lines represent, respectively, the Al_2O_2 units added and the AlO bonds broken during the conceptual cage expansion

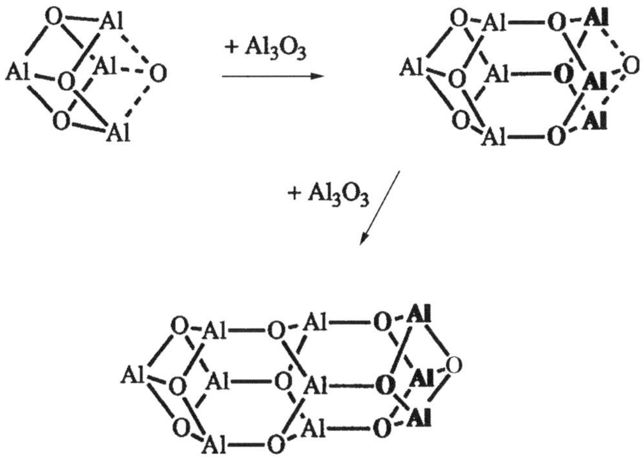

$+ Al_3O_3$

$+ Al_3O_3$

Scheme 4 Proposed relationship between $[(R)Al(\mu_3\text{-}O)]_n$, $n = 4, 7, 10$, etc. Alkyl groups are omitted for clarity. The bold groups and dashed lines represent, respectively, the Al_3O_3 units added and the AlO bonds broken during the conceptual cage expansion

cage expansion via the addition of two M–E units leads to the dodecamer structure observed for $[(^tBu)Al(\mu_3\text{-}O)]_{12}$. The second type of cage expansion from the cubane consists of insertion of an M_3E_3 cycle between one of the apexes and the remainder of the cube to give the heptamer and, subsequently, a second isomer of the decamer (Scheme 4). These simple relationships allow the prediction of the structure of oligomers of alkylalumoxanes higher than the dodecamer.

5.2　STRUCTURAL RELATIONSHIP OF $[(^tBu)Al(\mu_3-O)]_n$ TO METHYLALUMOXANE

As outlined above, a number of *tert*-butylalumoxanes and other alkylalumoxanes with sterically hindered substituents have been crystallographically characterized [10,11]. While the isolation and structural elucidation of these *tert*-butylalumoxanes was an important advance in the chemistry of alkylalumoxanes *per se*, and they have been shown to be catalytically active [64,65], their relevance to the active species present in MAO merits some discussion.

Before discussing similarities between the *tert*-butylalumoxanes and commercial MAO, it is worth noting the important differences between the ligating ability of the *tert*-butyl *versus* methyl alkyl groups, which may possibly alter structure and reactivity. These differences include the relative steric bulk of *tert*-butyl *versus* methyl substituents as defined by their cone angles (126° *versus* 90°) [66] and the ability of the methyl group to act as a stable bridging ligand. Changes in steric bulk of substituents are often cited as the controlling factor in the magnitude of oligomerization. However, based upon mass spectrometry [67] and cryoscopy [68], it has been reported that MAO consists of a range of hydrocarbon soluble species containing 5–12 and possibly as many as 30 aluminum atoms [69]. *tert*-Butylalumoxanes containing 4–12 aluminums have been isolated or spectroscopically characterized [10,11]. Hence concerns that the *tert*-butyl group would, as a consequence of steric bulk, preclude oligomeric forms that are components of MAO appears unfounded. The second dissimilarity between methyl and *tert*-butyl substituents, the ability of the former to act as a bridging ligand, open up the possibility that some structural types present in MAO will not be formed with *tert*-butyl ligation. Although there are no examples of isolable Group 13 compounds with *tert*-butyl bridging ligation, the observation of ligand exchange reactions implies their existence, albeit as transition states [70]. However, through a series of elegant NMR experiments Mole and co-workers [71,72] demonstrated that the three-center, four-electron bridge formed by Group 15, 16 or 17 donor atoms (**17**) are always preferred to the possible formation of a three-center, two-electron methyl bridge (**18**). Thus, in the present case an oxo bridge will always be preferred over an alkyl bridge.

$$
\begin{array}{cc}
\text{R} \cdots \text{Al} \underset{X}{\overset{X}{<}} \text{Al} \cdots \text{R} & \text{R} \cdots \text{Al} \underset{R}{\overset{R}{<}} \text{Al} \cdots \text{X} \\
(\mathbf{17}) & (\mathbf{18})
\end{array}
$$

The solution ^{1}H and ^{13}C NMR spectra of MAO show broad resonances consistent with a fluxional species. Similarly, the solid-state ^{13}C CP/MAS NMR spectrum of MAO exhibits a single resonance sufficiently broad ($\delta - 6.5$, $W_{1/2} \approx 750\,\text{Hz}$ [73]) to preclude assignment of specific Al—Me environments. Apart from confirming the presence of aluminum-bound methyl groups, these ^{1}H and ^{13}C NMR spectra do not provide any additional structural data. A further complication in the ^{1}H and ^{13}C NMR spectral characterization of MAO is the presence of AlMe$_3$ in commercial solutions. In contrast, ^{27}Al and ^{17}O NMR spectroscopic data are useful in defining possible (and eliminating unlikely) coordination geometries of aluminum and oxygen in structurally unknown compounds. In this regard the heteroatom NMR chemical shifts of MAO, the *tert*-butylalumoxanes and related compounds are given in Table 2.

The ^{27}Al NMR spectra of MAO and isobutylalumoxane consist of single broad resonances (δ 149–152). Thus, based upon ^{27}Al NMR spectroscopy, it is difficult to preclude the presence of aluminum centers in three-coordinate R$_2$AlO or four-coordinate R$_2$AlO$_2$ coordination environments in addition to the RAlO$_3$ environments observed for [(^{t}Bu)Al(μ_3-O)]$_n$. However, the three-coordinate RAlO$_2$ environment present in the cyclic structures previously proposed can be reasonably eliminated as a major contributor, since this should appear at approximately 100 p.p.m. [74].

The ^{17}O NMR spectrum of MAO consists of a single broad ($W_{1/2} = 800\,\text{Hz}$) resonance at 55 p.p.m. This resonance is within the range observed for either a three-

Table 2 Heteroatom NMR chemical shifts (δ, ppm) of alkylalumoxanes and related compounds

Compound	^{17}O	^{27}Al
MAO	55	149–152
IBAO	70	149
$[(^tBu)Al(\mu_3\text{-}O)]_n$	55–75	112–150
$[(R)Al(\mu_3\text{-}NR')]_n$		120–160
$[Et_2Al(\mu\text{-}OAlEt_2)]_2$	59	156
$[(^tBu)_2Al\{\mu\text{-}OAl(^tBu)_2\}]_2$	75	142 (4-coordinate)
		200 (3-coordinate)
$[(^tBu)_2Al(py)]_2(\mu\text{-}O)$	17	126
$[(^tBu)_2Al(\mu\text{-}OH)]_3$	0–2^a	139–145^a

^{a} Solvent-dependent shift.

coordinate oxo ligand (δ 55–75) or three-coordinate hydroxide moiety (δ 54) [75]. Since the hydroxide groups may be rejected as contributing significantly to the structure of MAO based on IR and ^1H NMR spectroscopy, the oxo ligands in MAO are clearly trifurcated, i.e. **3** or **4**. At present insufficient data are available to determine the geometry, planar (**3**) *versus* tetrahedral (**4**), of the oxide due to overlapping of the regions. The absence of two-coordinate oxo ligands (**2**) is suggested by the absence of any resonance at approximately δ 17 in the ^{17}O NMR spectrum of MAO.

5.3 STRUCTURES OF $R[(R)Al(O)]_nAlR_2$

The simplest alkylalumoxane with the formula $R[(R)Al(O)]_nAlR_2$ is for $n = 1$, i.e. R_2Al—O—AlR_2. It is oligomers of this general formula (i.e. $[R_2Al(\mu\text{-}OAlR_2)]_n$) which make up the second largest group of alkylalumoxanes after those discussed above. An example of $R[(R)Al(O)]_nAlR_2$ where $n = 2$ has recently been reported [47].

A monomeric alkylalumoxane of the type R_2Al—O—AlR_2 is only stable when the alkyl substituents are of sufficient steric bulk to preclude oligomerization, e.g. the bis(trimethylsilyl)methyl group. The structural characterization of $[(Me_3Si)_2HC]_2Al$—O—$Al[CH(SiMe_3)_2]_2$ indicates that the Al—O—Al unit is linear, as expected (**19**) [49]. Reduction of the steric bulk of the alkyl substituents results in oligomerization to a dimer. For example, the molecular structure of $[(^tBu)_2Al\{\mu\text{-}OAl(^tBu)_2\}]_2$ is shown in Figure 6.

(**19**)

52   BARRON

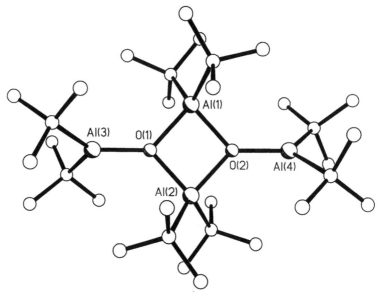

Figure 6 Molecular structure of [(^{t}Bu)$_2$Al{μ-OAl(^{t}Bu)}]$_2$. The hydrogen atoms are omitted for clarity

Given the structural characterization of [(^{t}Bu)$_2$Al{μ-OAl(^{t}Bu)$_2$}]$_2$, it is reasonable to believe, therefore, that the structure proposed by Boleslawski and Serwatowski [76] for the isobutyl analog (**20**) is indeed correct. The presence of three-coordinate aluminum in [R$_2$AlOAlR$_2$]$_n$ ($n = 1, 2$) offers the possibility of the formation of Lewis acid–base complexes. Atwood and Zaworotko reported [77] the [Me$_2$Al(μ-OAlMe$_3$)]$_2^{2-}$ anion (**21**), which may be considered a Lewis acid–base complex of [Me$_2$Al(μ-OAlMe$_2$)]$_2$ and two methyl anions.

(20)   (21)

A similar structure is found for the chloride analog, $[Cl_2Al(\mu\text{-}OAlCl_3)]_2^{2-}$ [78]. Reaction of $[({}^tBu)_2Al\{\mu\text{-}OAl({}^tBu)_2\}]_n$ with nitrogen Lewis bases (L) results in the cleavage of the dimer and formation of $[({}^tBu)_2Al(L)]_2(\mu\text{-}O)$; L = pyridine, $NH(Me)CH_2CH_2NMe_2$.

5.4 STRUCTURES OF HYBRID ALUMOXANES

All the alkylalumoxanes discussed up to this point may be considered as fully condensed, i.e. all the water/hydroxide hydrogens have reacted with aluminum alkyl groups. However, there are several hydroxide-containing alumoxanes that have been structurally characterized. These are either intermediate to fully condensed alkylalumoxanes [11] or formed as a result of the hydrolysis of stoichiometric alkylalumoxanes [40].

The molecular structure of the tetraaluminum compound $[Al_4({}^tBu)_7(\mu_3\text{-}O)_2(\mu\text{-}OH)]$, formed from the partial hydrolysis of $[({}^tBu)_2Al\{\mu\text{-}OAl({}^tBu)_2\}]_2$, is shown in Figure 7 [11]. This compound is thought to be an intermediate in the formation of $[({}^tBu)Al(\mu_3\text{-}O)]_8$ [11], in a similar manner to the intermediacy of $[Al_6({}^tBu)_8(\mu_3\text{-}O)_4(\mu\text{-}OH)_2]$ (Figure 2) in the formation of $[({}^tBu)Al(\mu_3\text{-}O)]_6$ (see above). In line with its formation, the core structure of $[Al_5({}^tBu)_7(\mu_3\text{-}O)_3(\mu\text{-}OH)_2]$ (Figure 1) is best considered to consist of the fusion of a six-membered Al_3O_3 ring with a four-membered Al_2O_2 ring.

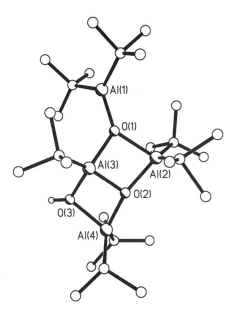

Figure 7   Molecular structure of $[Al_4({}^tBu)_7(\mu_3\text{-}O)_2(\mu\text{-}OH)]$. The hydrogen atoms of the *tert*-butyl groups are omitted for clarity

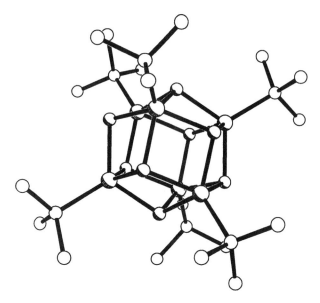

Figure 8 Molecular structure of $[Al_6(^tBu)_6(\mu_3\text{-}O)_4(\mu_3\text{-}OH)_4]$. The hydrogen atoms are omitted for clarity

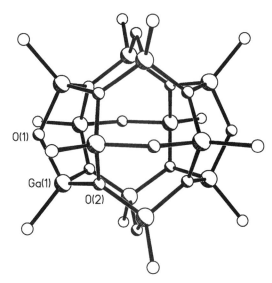

Figure 9 Molecular structure of $[Ga_{12}(^tBu)_{12}(\mu_3\text{-}O)_8(\mu\text{-}O)_2(\mu\text{-}OH)_4]$. The *tert*-butyl groups and hydroxide hydrogen atoms are omitted for clarity

Hydrolysis of $[({}^{t}Bu)Al(\mu_3\text{-}O)]_6$ results in the formation of $[Al_6({}^{t}Bu)_6(\mu_3\text{-}O)_4$ $(\mu_3\text{-}OH)_4]$ (Figure 8) and is iso-structural to $[Ga_6(Mes)_6O_4(OH)_4]$ reported by Roesky and co-workers [79]. The oxide and hydroxide sites in $[Al_6({}^{t}Bu)_6(\mu_3\text{-}O)_4$ $(\mu_3\text{-}OH)_4]$ are disordered as a result of the high symmetry of the $[Al({}^{t}Bu)]_6$ core. A similar disorder between oxide and hydroxide groups is observed in the structure of $[Ga_{12}({}^{t}Bu)_{12}(\mu_3\text{-}O)_8(\mu\text{-}O)_2(\mu\text{-}OH)_4]$. (Figure 9). [40]

The anionic hybrid alkoxide–alkylalumoxane, $[Al_7O_6Me_{16}]^-$, reported by Atwood *et al.* [80], consists of an Al_6O_6 ring capped by a seventh aluminum atom which is bonded to three alternate oxygen atoms in the ring (**22**).

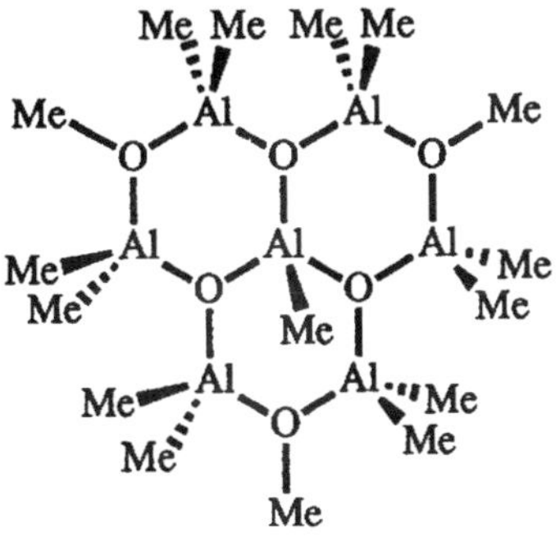

(22)

6 REACTIONS OF ALKYLALUMOXANES

The reactivity of alkylalumoxanes has been extensively studied, in particular reactions with electron donors, e.g. ethers [81,82], amines [83,84], ketones [85], nitriles [86], epoxides [87] and acetylacetone [88]. However, with regard to their activity as cocatalysts for olefin polymerization, two specific reactions are especially important: their reactions with trialkylaluminum compounds and metallocenes.

6.1 THE PRESENCE AND ROLE OF AlMe₃ IN METHYLALUMOXANE

Commercially the hydrolysis of $AlMe_3$ (TMA) is performed with a controlled excess of $AlMe_3$ such that the resulting product often called an 'MAO solution' contains $AlMe_3$ [89]. The amount of $AlMe_3$ present in commercial MAO solutions is variable, depending on vendor and batch. However, usually 5–20 % of the aluminum is in the form of $AlMe_3$. The function of this $AlMe_3$ is, among others, to provide solubility to the alkylalumoxanes formed in hydrocarbon solvents and act as a reactive source of alkyl radicals (see below). The presence of $AlMe_3$ in solutions of MAO is now well understood and a number of methods have been developed to determine accurately the amount of $AlMe_3$ [90–92]. However, it should be noted that by definition alkylalumoxanes, *per se*, do not contain trialkylaluminum and a distinction should be (and is) made about AlMe$_3$-free and AlMe$_3$-containing MAO. The function of the TMA has received some discussion; however, it is generally accepted that its function includes removal of impurities from the catalyst system, alkylation of metallocene dihalides and solubilization of the alkylalumoxanes. Unfortunately, the AlMe$_3$ also appears to inhibit the catalytic activity of the metallocene/alkylalumoxane catalyst system through complexation to the metallocene [93].

As is common with all aluminum alkyls, AlMe$_3$ reacts rapidly with oxygen [equation (21)], water (see above) and acids such as HCl [equation (22)] [94].

$$AlMe_3 \xrightarrow{+O_2} \text{'}Me_2Al(OOMe)\text{'} \xrightarrow{+AlMe_3} \frac{1}{n}[Me_2Al(OMe)]_n \quad (21)$$

$$AlMe_3 + HCl \longrightarrow \tfrac{1}{2}[Me_2Al(\mu\text{-}Cl)]_2 + MeH \quad (22)$$

Such species are commonly present as impurities in commercial sources of olefins and/or may be introduced to the catalyst system during polymerization. The reaction of AlMe$_3$ with oxygen is faster than that of the MAO, although the highly reactive cationic metallocene catalysts probably react at a comparable rate: the use of an excess of AlMe$_3$ with respect to the metallocene is sufficient to 'protect' the catalyst.

Although it has been generally assumed that MAO transfers a methyl group to the zirconocene metal fragments by ligand exchange, there is no direct evidence of the MAO acting as the alkylation agent since AlMe$_3$ is known to alkylate metallocenes such as Cp$_2$ZrCl$_2$(Cp = cyclopentadienyl anion, C$_5$H$_5$).

While the *tert*-butylalumoxanes are readily soluble in hydrocarbon solutions [10,11], attempts to prepare trialkylaluminum-free alkylalumoxanes with sterically less demanding alkyl substituents result in relatively insoluble materials. It has been commonly assumed this was due to the formation of higher oligomers; however, recent work with gallium and indium chalcogenides has suggested that this is not true. The indium selenide cubane [(^{n}Bu)In(μ_3-Se)]$_4$ is essentially insoluble in non-polar solvents despite the presence of long-chain aliphatic alkyl substituents [95]. In contrast, the isostructural *tert*-butyl analog, [(^{t}Bu)In(μ_3-Se)]$_4$, is soluble in toluene. These results suggest that other intermolecular forces must be responsible for the low solubility of Group 13–16 cage compounds with sterically small substituents. In the case of methylalumoxanes, solubility in hydrocarbon solvents is possible through the use of an excess of AlMe$_3$ during synthesis. If this excess AlMe$_3$ is completely removed, the resulting methylalumoxane is insoluble.

Trialkylaluminum compounds, such as AlMe$_3$, are strong Lewis acids which can complex with transition metal halides and alkyls, competing with the alkylalumoxanes, as has been demonstrated by the model system, Cp$_2$ZrX$_2$/Al(^{t}Bu)$_3$. The addition of one molar equivalent of Al(^{t}Bu)$_3$ to Cp$_2$ZrX$_2$(X = Me, Cl) yields the Lewis acid–base complex Cp$_2$Zr(X)[μ-XAl(^{t}Bu)$_3$] (23) [96].

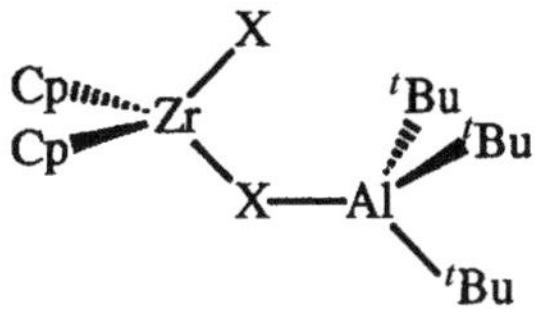

X = Cl, Me

(23)

The relatively strong nature of this type of interaction is sufficient to inhibit the catalytic polymerization of olefins at high trialkylaluminum concentrations. There is, therefore, a subtle balance between the desire to lower the $AlMe_3$ concentration (to enhance the catalysis activity of the metallocene/MAO catalyst system) and the need to use $AlMe_3$ (to solubilize the MAO and getter impurities from the feedstocks).

6.2 REACTION OF AlMe₃ WITH ALKYLALUMOXANES

Based on the presence of a single, broad ^1H NMR signal for the aluminum methyl groups in a solution of methylalumoxane and trimethylaluminum, it has been proposed that a ligand redistribution reaction occurs. Instead of a simple methyl exchange, as is observed for $AlMe_3$, it has been proposed that $AlMe_3$ coordinates to the oxo ligands in methylalumoxane, $[(Me)AlO]_n$. Subsequent alkyl transfer results in the formation of a new alkylalumoxane in which each aluminum has two methyl substituents. An example of this proposal is shown in Equation (23) [10].

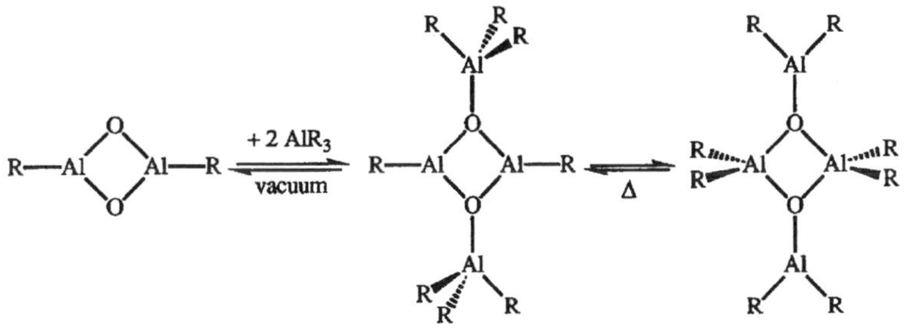

(23)

The removal of the $AlMe_3$ is proposed to be accomplished through heat or the addition of an electron donor such as triphenylphosphine:

$$n(R_2Al—O—AlR_2) + PPh_3 \longrightarrow [(R)AlO]_n + nR_3Al(PPh_3) \quad (24)$$

This type of alkyl transfer has been observed for the reaction of $AlMe_3$ with $[(Mes^*)Al(\mu\text{-}O)]_4$ [equation (25)] [47] but a slightly different reaction is observed for the cage-type structures.

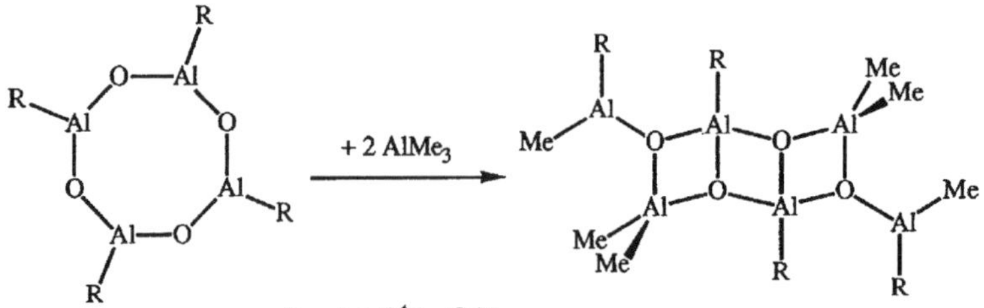

$$R = 2,4,6\text{-}^{t}Bu_3C_6H_2$$

(25)

The reaction of trimethylaluminum with the hexameric *tert*-butylalumoxane $[(^{t}Bu)Al(\mu_3\text{-}O)]_6$ results in the formation of two isomers with the general formula $[Al_7(\mu_3\text{-}O)_6(^{t}Bu)_6Me_3]$ (**24** and **25**) [97], whose structures consist of $[Al_6(\mu_3\text{-}O)_6(^{t}Bu)_5Me]$ alumoxane cages:

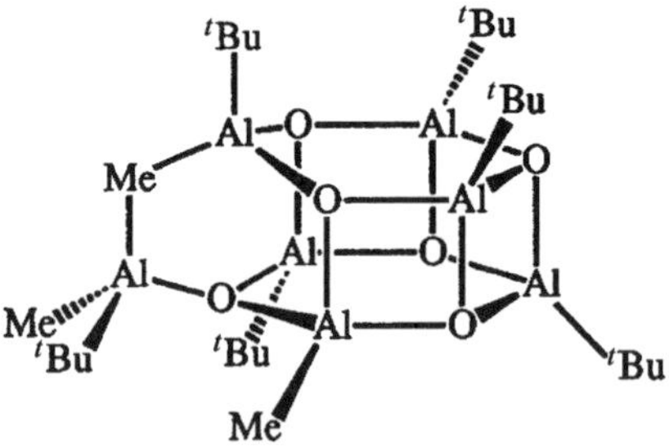

(24) 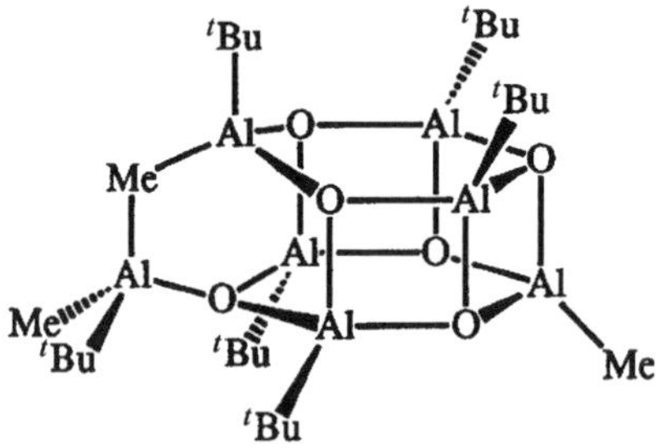   (25)

These are formed via *tert*-butyl–methyl exchange, in which one of the edges of the Al_6O_6 cage is complexed to the $(^{t}Bu)_2AlMe$ formed during alkyl exchange. The difference between the isomers results from the geometric relationship of the cage Al—Me group and the opened edge. A similar product has been proposed by Beared *et al.* [98] (**26**).

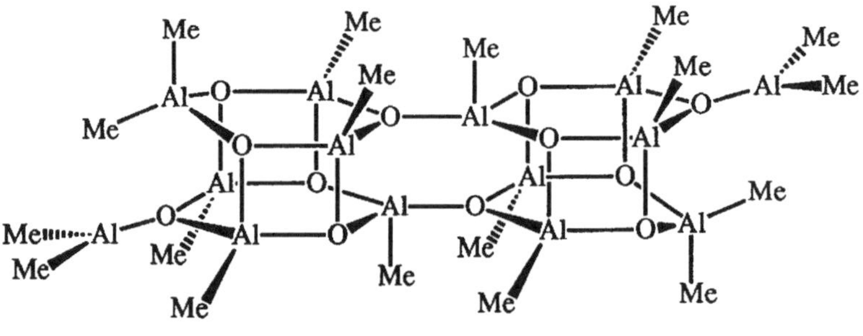

(26)

6.3 REACTION OF ALKYLALUMOXANES WITH METALLOCENES

The accepted function of methylalumoxanes as a cocatalyst to Cp_2ZrX_2 ($Cp = \eta^5$-C_5H_5, X = Me, Cl) in Kaminsky-type olefin polymerization is twofold [99]. First, MAO is assumed to alkylate the zirconocene metal fragments by ligand exchange [equation (26)]. Second, and most importantly, the methylalumoxane abstracts the second chloride to generate a formally cationic zirconocene moiety [equation (27)]. Alternatively, with use of a dialkylzirconocene, the methylalumoxane acts to abstract the alkide [equation (28)] [100–102].

$$Cp_2ZrCl_2 \xrightarrow{\text{+MAO}} Cp_2Zr(Me)Cl \tag{26}$$

$$Cp_2Zr(Me)Cl \xrightarrow{\text{+MAO}} [Cp_2Zr(Me)]^+ \tag{27}$$

$$Cp_2ZrMe_2 \xrightarrow{\text{+MAO}} [Cp_2Zr(Me)]^+ \tag{28}$$

Although the methyl/chloride metathesis [equation (26)] has not been demonstrated for alkylalumoxanes, the isoelectronic iminoalane [HAl(μ_3-N^{t}Bu)]$_4$ has been shown to react with Cp_2ZrMe_2 to yield the methyl–hydride transfer products, [Al(Me)(H)$_3$(μ_3-N^{t}Bu)$_4$] and [Cp$_2$ZrMe(μ-H)]$_2$:

$$[HAl(\mu_3\text{-N}^t\text{Bu})]_4 + Cp_2ZrMe_2 \longrightarrow$$
$$[Cp_2ZrMe(\mu\text{-H})]_2 + [Al(Me)(H)_3(\mu_3\text{-N}^t\text{Bu})_4] \tag{29}$$

Based on kinetics measurements, the metathesis reaction is proposed to occur via an intermediate or transition state containing both five-coordinate aluminum and five-coordinate zirconium (i.e. **27** [103].

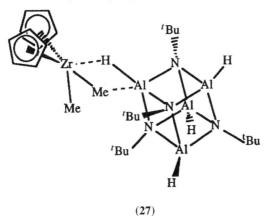

(**27**)

This model reaction suggests that cage alkylalumoxanes, such as MAO, may be able to perform the Me/Cl metathesis assumed to occur [*cf.* equation (26)] in the absence of $AlMe_3$.

Whether alkylalumoxanes can act as alkyl transfer agents to metallocenes is less important than understanding the manner in which they abstract a halide or alkide to form the active catalytic species; the metallocene cation [equations (27) and (28)]. The isolation and structural characterization of the *tert*-butylalumoxanes offer an opportunity to investigate the reactions of individual alkylalumoxanes with metallocenes.

Based upon conventional wisdom, the cage alkylalumoxanes, $[(^tBu)Al(\mu_3\text{-}O)]_n$, should be inactive as cocatalysts with Cp_2ZrMe_2, whereas $[(^tBu)_2Al\{\mu\text{-}OAl(^tBu)_2\}]_2$ would be expected to be an active cocatalyst. However, although $[(^tBu)_2Al\{\mu\text{-}OAl(^tBu)_2\}]_2$ contains two coordinatively unsaturated three-coordinate aluminum centers and reacts as a Lewis acid, it is not sufficiently Lewis acidic either to complex with or to abstract the methyl group from the zirconium. In contrast, the coordinatively saturated alkylalumoxane $[(^tBu)Al(\mu_3\text{-}O)]_6$ reacts with Cp_2ZrMe_2 to yield the methyl transfer product:

$$[(^tBu)Al(\mu_3\text{-}O)]_6 + Cp_2ZrMe_2 \rightleftharpoons [Cp_2ZrMe][(^tBu)_6Al_6O_6Me] \qquad (30)$$

The structure of $[Cp_2ZrMe][(^tBu)_6Al_6O_6Me]$ is proposed to have the structure shown in **28**. The structural relationship between the 'closed' and 'open' alumoxane cages in $[(^tBu)_2Al(O)]_6$ and $[Cp_2ZrMe][(^tBu)_6Al_6O_6Me]$ is shown in Scheme 5.

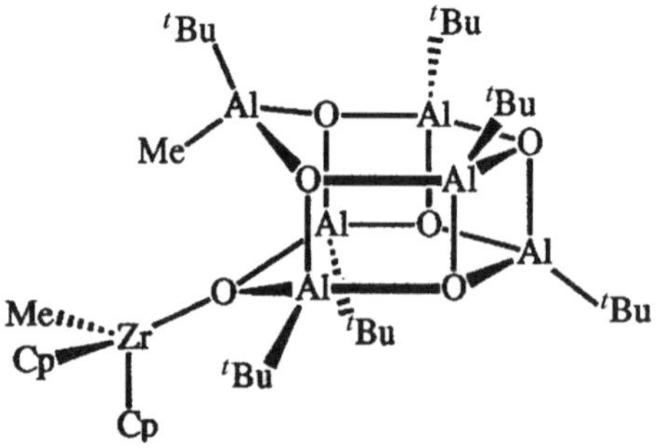

(28)

As was indicated from the lack of reactivity between $[(^tBu)_2Al\{\mu\text{-}OAl(^tBu)_2\}]_2$ and Cp_2ZrMe_2, there is no catalytic activity for this alumoxane, despite it having two three-coordinate aluminum Lewis acidic centers. On the other hand, the electron precise cage compound $[(^tBu)Al(\mu_3\text{-}O)]_6$ is active as a polymerization cocatalyst. Solutions of an equimolar mixture of $[(^tBu)Al(\mu_3\text{-}O)]_6$ and Cp_2ZrMe_2 rapidly polymerize ethylene at room temperature. The $[(^tBu)Al(\mu_3\text{-}O)]_6$ is only about one sixth as active as commercial MAO. Reasons for the decreased activity of $[(^tBu)Al(\mu_3\text{-}O)]_6$ compared with MAO are difficult to determine since the exact speciation of MAO is unknown. However, it is worth noting that at the lowest Al:Zr ratio (*ca* 6), $[(^tBu)Al(\mu_3\text{-}O)]_6$ is active whereas no polymer could be isolated with MAO as the cocatalyst.

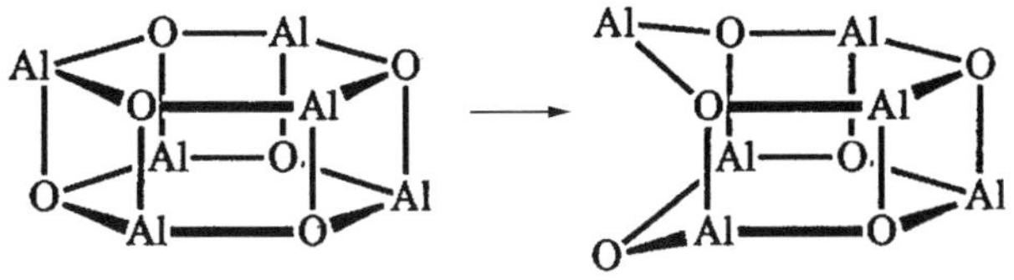

Scheme 5 Structural relationship between the 'closed' and 'open' alumoxane cages in $[(^tBu)_2Al(O)]_6$ and $[Cp_2ZrMe][(^tBu)_6Al_6O_6Me]$, respectively

7 LATENT LEWIS ACIDITY: AN EXPLANATION OF THE COCATALYTIC ACTIVITY OF ALKYLALUMOXANES

Spectroscopic and theoretical data suggest that the primary role of MAO in metallocene catalysis is to abstract an alkide (or halide) from the metallocene, forming a 'cation-like' metal center, i.e. equations (28) and (29) [104–106]. The insistence by many workers that a three-coordinate aluminum center must be present has developed from the idea that compounds with aluminum in a four-coordinate octet environment are not Lewis acidic, whereas compounds with coordinatively unsaturated non-octet three-coordinate aluminum centers are strong Lewis acids. Given this prevalent thinking concerning the activity of alkylalumoxane cocatalysts, the catalytic activity of the cage compounds $[(^tBu)Al(\mu_3\text{-}O)]_n$ is surprising. Equally so is the lack of activity for the three-coordinate aluminum compound $[(^tBu)_2Al\{(\mu\text{-}OAl(^tBu)_2\}]_2$. This result begs the question, why are the coordinatively saturated cage compounds active catalysts? In order to explain the cocatalytic activity, a new concept has been proposed, '*latent Lewis acidity*' [96].

Latent Lewis acidity is defined as the ability of a electron precise molecule, e.g. a cage alkylalumoxane, to undergo cage opening, *via* heterolytic bond cleavage, to generate a Lewis acidic site (along with the concomitant Lewis basic site). For a given bond type (e.g. and Al—O dative bond in alkylalumoxanes), the relative magnitude of the latent Lewis acidity is related to the relative strain present in the cage. Thus, in general, four-membered Al_2O_2 rings are more strained than their six-membered Al_3O_3 homologues, and hence exhibit higher latent Lewis acidity. Based upon the angular distortions of the cage atoms from an ideal geometry, a semi-qualitative value for the latent Lewis acidity may be obtained, allowing a prediction of the relative reactivity of a series of alumoxane cage structures [96].

If the latent Lewis acidity of an Al—O bond in a given alumoxane is a function of the ring strain within a cycle, and as four-coordinate aluminum and three-coordinate oxide prefer tetrahedral and trigonal planar geometries, respectively, then a quali-tative determination of latent Lewis acidity may be made by calculating the sum of the angular distortions of the cage atoms from this ideal. Thus, the distortion about any element (Γ_E) may be estimated according to equation (31), where $\Sigma(X\text{—}E\text{—}X)_{ideal}$ is the sum of the intra-cage angles in an ideal geometry (i.e.

328.5° for tetrahedral, 360° for trigonal planar), and $\Sigma\,(X\!-\!E\!-\!X)_{exp.}$ is the sum of the experimental intra-cage angles in the molecule.

$$\Gamma_E = \Sigma\,(X\!-\!E\!-\!X)_{ideal} - \Sigma\,(X\!-\!E\!-\!X)_{exp}. \tag{31}$$

The total strain (Γ_{E-X}) is therefore defined as the sum of the strain at the two atom centers attached to a particular bond:

$$\Gamma_{E-X} = \Gamma_E + \Gamma_X \tag{32}$$

In $[(^tBu)Al(\mu_3\text{-}O)]_6$, the sum of the intra-cage angles, $\Sigma\,(O\!-\!Al\!-\!O)$, for aluminum is 284°, whereas the ideal value for a tetrahedral geometry is 328.5°; the angular distortion (Γ_{Al}) is therefore ca 45°. Similarly, the sum of the intra-cage angles for oxygen is 312.8° and its $\Gamma_O \approx 48°$ (the ideal value being 360°). The strains at aluminum and oxygen are 45° and 48°, respectively, and the total strain (Γ_{Al-O}) is therefore ca 93°. Similar values may be calculated for each of the Al—O bonds in $[(^tBu)Al(\mu_3\text{-}O)]_n$, $n = 6, 7, 8, 9$ (Figure 10). Hence, based upon the angular distortions of the cage atoms from an ideal geometry a qualitative value for the latent Lewis acidity may be obtained, allowing a prediction of the relative reactivity of a series of alumoxane cage structures [96]:

$$[(^tBu)Al(\mu_3\text{-}O)]_7 > [(^tBu)Al(\mu_3\text{-}O)]_9 > [(^tBu)Al(\mu_3\text{-}O)]_6 \approx [(^tBu)Al(\mu_3\text{-}O)]_8 \tag{33}$$

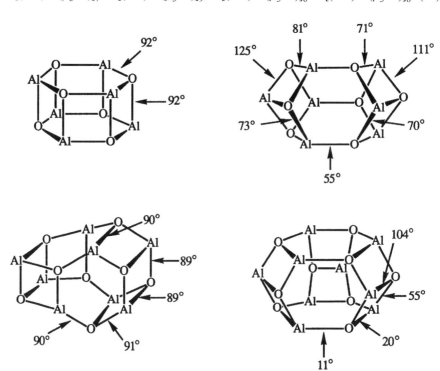

Figure 10 Estimated latent Lewis acidity of individual AlO bonds in cage alumoxanes, $[(R)Al(\mu_3\text{-}O)]_n$. Alkyl groups have been omitted for clarity

Unfortunately, this simplistic approach does not take into account any steric hindrance of the Al—O bond (i.e. the steric bulk of the aluminum alkyl group), or the possible strain in the ring-opened product. A comparison of the relative catalytic rates for a series of cage alumoxanes, equation (34) [107], indicated that a knowledge of the steric effects in the alumoxanes is required in order to develop a reliable predictive measure of latent Lewis acidity.

$$[(^tBu)Al(\mu_3\text{-O})]_7 > [(^tBu)Al(\mu_3\text{-O})]_6 > [(^tBu)Al(\mu_3\text{-O})]_9 \qquad (34)$$

An experimental method for the determination of the steric restriction of the alkylalumoxane's latent Lewis acid sites has been reported [107], involving reaction of the alkylalumoxane with a series of amines with various steric bulk (cone angles [66]):

$$[(^tBu)Al(\mu_3\text{-O})]_6 + 2RNH_2 \longrightarrow [(^tBu)_6Al_6(\mu_3\text{-O})_4(\mu\text{-O})_2(NH_2R)_2] \qquad (35)$$

No reaction indicates the maximum steric bulk of a substrate with which the alkylalumoxane can react. For example, $[(^tBu)Al(\mu_3\text{-O})]_6$ reacts with $^tBuNH_2$ ($\theta = 120°$) but not Et_2NH ($\theta = 140°$), and therefore we propose the steric limitation of the reactivity of $[(^tBu)Al(\mu_3\text{-O})]_6$ to be in between these values.

8 FUTURE DIRECTIONS

As was noted in the Introduction, this chapter is not and should not be considered complete because of the evolving understanding of alkylalumoxanes. There are, however, a few key areas where further research is desirable. First, the synthesis of methylalumoxane is far from controllable. New experimental methods need to be developed to improve the consistency of the product. Second, although great advances have been made in recent years with regard to the structure of alkylalumoxanes, there is no good structural probe available for methylalumoxane. Before a detailed understanding of the reactivity of methylalumoxane may be gained, better information on its composition and structure must be obtained. Third, we are only just appreciating the subtlety of the interaction between alkylalumoxanes, trialkylaluminum and metallocenes. Improved understanding of this interplay will enhance catalyst design.

One topic that has not been covered in this chapter is that of supported and solid metallocene/alkylalumoxane catalysts. Although it is heterogeneous rather than homogeneous catalyst systems that will undoubtedly represent the commercial future of Kaminsky-type metallocene-based polymerization catalysts, there is almost no information available as to the interaction of alumoxanes with typical catalyst supports. Unfortunately, the difficulties that must be overcome in characterizing alkylalumoxanes in solution will only be amplified in the solid state.

9 ACKNOWLEDGMENTS

The following people are acknowledged: my assistant Ms Jane McNeel for creating order from chaos, Dr Simon Bott for collaboration and critically reading the manuscript, longtime colleague and friend Professor David Hoffman for discussions, Dr Tim Westby and Mr David Burgert for bringing me to the party and, last but not least, my research group for their creativity and support.

10 REFERENCES

1. Manyik, R. M., Walker, W. E. and Wilson, T. P., *US Pat.*, 3 242 099 (1966).
2. Wilkinson, G. and Birmingham, J. M., *J. Am. Chem. Soc.*, **76**, 4281 (1954).
3. Reichert, K. H. and Meyer, K. R., *Macromol. Chem.*, **169**, 163 (1973).
4. Long, W. P. and Breslow, D. S., *Leibigs Ann. Chem.*, 463 (1975).
5. Andreson, A., Cordes, H.-G., Herwig, J., Kaminsky, W., Merck, A., Mottweiler, R., Pein, J., Sinn, H. and Vollmer, H.-J., *Angew. Chem., Int. Ed. Engl.*, **15**, 630 (1976).
6. Herwig, J. and Kaminsky, W., *Polym. Bull.*, **9**, 464 (1983).
7. Sinn, H. and Kaminsky, W., *Adv. Organomet. Chem.*, **18**, 99 (1980).
8. Sinn, H., Kaminsky, W., Vollmer, H.-J. and Woldt, R., *Angew. Chem., Int. Ed. Engl.*, **19**, 390 (1980).
9. Kaminsky, W., Miri, M., Sinn, H. and Woldt, R., *Makromol. Chem., Rapid Commun.*, **4**, 417 (1983).
10. Pasynkiewicz, S., *Polyhedron*, **9**, 429 (1990).
11. Mason, M. R. Smith, J. M. Bott, S. G. and Barron, A. R. *J. Am. Chem. Soc.*, **115**, 4971 (1993).
12. Harlan, C. J., Mason, M. R. and Barron, A. R., *Organometallics*, **13**, 2957 (1994).
13. Gurian, P., Cheatham, L.K., Ziller J.W. and Barron, A.R., *J. Chem. Soc., Dalton Trans.*, 1449 (1991).
14. Wynne, K. Y., *Inorg. Chem.*, **24**, 1339 (1985).
15. Kushi, Y. and Frenando, Q., *J. Chem. Soc., Chem. Commun.*, 555 (1969).
16. Storr, A., Jones, K. and Laubengayer, A. W., *J. Am. Chem. Soc.*, **90**, 3173 (1968).
17. Sakharovskaya, G. B., Korneev, N. N., Popov, A. F., Larikov, E. J. and Zhigach, A. F., *Zh. Obshch. Khim.*, **34**, 3435 (1964).
18. Winter, H., Schnuchel, W. and Sinn, H., *Macromol. Symp.*, **97**, 119 (1995).
19. Sakharovskaya, G. B., Korneev, N. N., Popov, A. F., Kissin, Yu. V., Mezhkovskii, S. M. and Kristalanyi, E., *Zh. Obshch. Khim.*, **39**, 788 (1969).
20. Razuvaev, G. A., Sangalov, Yu. A., Nel'kenbaum, Yu. Ya. and Minsker, K. S., *Izv. Akad. Nauk SSSR, Ser. Khim.*, 2547 (1975).
21. Boleslawski, M. and Serwatowski, J., *J. Organomet. Chem.*, **255**, 269 (1983).
22. Harlan, C. J., Gillan, E. G., Bott, S. G. and Barron, A. R., *Organometallics*, **15**, 5479 (1996).
23. Storre, J., Klemp, A., Roesky, H. W., Schmidt, H.-G., Noltemeyer, M., Fleischer, R. and Stalke, D., *J. Am. Chem. Soc.*, **118**, 1380 (1996).
24. Eisch, J. J., in Wilkinson, G., Stone, F. G. A. and Abel, E. W. (Eds), *Comprehensive Organometallic Chemistry*, Pergamon Press, Oxford, 1986, Vol. 1, Chapt. 6.
25. Coates, G. E., Green, M. L. H. and Wade, K., *Organometallic Compounds*, 3rd Ed., Methuen, London, Vol. 1 Chapter 3 (1967).
26. Stone, F. G. A., *Chem. Rev.*, **58**, 101 (1958).

27. Barron, A. R., unpublished results.
28. Beachley, O. T., Jr and Tessier-Youngs, C., *Inorg. Chem.*, **18**, 3188 (1979).
29. Beachley, O. T., Jr, *Inorg. Chem.*, **20**, 2825 (1981).
30. Abraham M. H. and Hill, J. A. *J. Organomet. Chem.*, **7**, 11 (1967).
31. Baes C. F. and Mesmer, R. E., *The Hydrolysis of Cations*, Wiley, New York, 1976.
32. Baran, V., *Coord. Chem. Rev.*, **6**, 65 (1971).
33. McMahon, C. N., Bott, S. G. and Barron, A. R., *J. Chem. Soc., Dalton Trans.*, 3129 (1997).
34. Welborn, Jr., H. C. and Ewen, J. A., *US Pat.*, 5 324 800 (1994).
35. Rogers, J. H., Apblett, A. W., Cleaver, W. M., Tyler, A. N. and Barron, A. R., *J. Chem. Soc., Dalton Trans*, 3179 (1992).
36. Wefers, K. and Misra, C., *Oxides and Hydroxides of Aluminum*, Alcoa Laboratories (1987).
37. Power, M. B., Ziller, J. W. and Barron, A. R., *Organometallics*, 11, 2783 (1992).
38. Apblett, A. W., Warren, A. C. and Barron, A. R., *Chem. Mater.*, **4**, 167 (1992).
39. McMahon, C. N. and Barron, A. R., *J. Chem. Soc., Dalton Trans.*, 3703 (1998).
40. Landry, C. C., Harlan, C. J., Bott, S. G. and Barron, A. R., *Angew. Chem., Int. Ed. Engl.*, **34**, 1202 (1995).
41. Ziegler, K., *Angew. Chem.*, **68**, 721 (1956).
42. Ziegler, K., Krupp, F., Weyer, K. and Larbig, W., *Liebigs Ann. Chem.*, **629**, 251 (1960).
43. Harney, D. W., Meisters, A. and Mole, T., *Aust. J. Chem.*, **27**, 1639 (1974).
44. Zakharkin, L. I. and Khorlina, I. M., *Izv. Akade. Nauk SSSR, Ser. Khim.*, **12**, 146 (1964).
45. Boleslawski, M. and Pasynkiewicz, S., *J. Organomet. Chem.*, **43**, 81 (1972).
46. Boleslawski, M., Pasynkiewicz, S., Jaworski, K. and Sadownik, A., *J. Organomet. Chem.*, **97**, 15 (1975).
47. Wehmschulte, R. J. and Power, P. P., *J. Am. Chem. Soc.*, **119**, 8387 (1997).
48. Apblett A. W. and Barron, A. R., *Organometallics*, **9**, 2137 (1990).
49. Uhl, W., Koch, M., Hiller, W. and Heckel, M., *Angew. Chem., Int. Ed. Engl.*, **34**, 989 (1995).
50. Ueyama, N., Araki, T. and Tani, H., *Inorg. Chem.*, **12**, 2218 (1973).
51. Ishihara, N., DPhil Thesis, University of Oxford (1990).
52. Storre, J., Schnitter, C., Roesky, H. W., Schmidt, H. -G., Noltemeyer, M., Fleischer, R. and Stalke, D., *J. Am. Chem. Soc.*, **119**, 7505 (1997).
53. Andrianov K. A. and Zhadanov, A. A., *J. Polym. Sci.*, **30**, 513 (1958).
54. Healy, M. D. and Barron, A. R., *Angew. Chem., Int. Ed. Engl.*, **31**, 921 (1992).
55. Oliver, J. P. and Kumar, R., *Polyhedron*, **9**, 409 (1990).
56. Greiser, T. and Weiss, E., *Chem. Ber.*, **110**, 3388 (1977).
57. Power M. B. and Barron, A. R., *J. Chem. Soc., Chem. Commun.*, 1315 (1991).
58. Power, M. B., Ziller, J. W. and Barron, A. R., *Organometallics*, **11**, 1055 (1992).
59. Barron, A. R., *Chem. Soc. Rev.*, 93 (1993).
60. Amirkhalili, S., Hitchcock, P. B. and Smith, J. D., *J. Chem. Soc., Dalton Trans.*, 1206 (1979).
61. Del Piero, G., Cesari, M., Perego, G., Cucinella, S. and Cernia, E., *J. Organomet Chem.*, **129**, 289 (1977).
62. Smith, J. D., *Vortrag Anorg. Chem. Colloq. Univ. München*, **25**, 7 (1974).
63. Noth, H. and Wolfgardt, P. *Z. Naturforsch., Teil B*, **31**, 697 (1976).
64. Lenz, R. W., Yang, J., Wu, B., Harlan, C. J. and Barron, A. R., *Can. J. Microbiol.*, **41**, 274 (1995).
65. Wu, B., Harlan, C. J., Lenz, R. W. and Barron, A. R., *Macromolecules*, **30**, 316 (1997).
66. Tolman, C. A., *Chem. Rev.*, **77**, 313 (1977).

67. Kaminsky, W., in Quirk, R. P., Hsieh, H. L., Klingensmith, G. B. and Tait, P. J. T. (Eds), *Transition Metal Catalyzed Polymerization, Alkenes and Dienes, Part A*, MMI Press, New York, 1983, p. 225.
68. Grogorjan, E. A., Dyachkovskii, F. S. and Shilow, A. E., *Vysokomol. Soedin*, **7**, 145 (1965).
69. von Lacroix, K., Heitmann, B. and Sinn, H., *Macromol. Symp.*, **97**, 137 (1995).
70. Cleaver, W. M. and Barron, A. R., *Chemtronics*, **4**, 146 (1989).
71. Jeffery, E. A., Mole, T. and Saunders, J. K., *Can. J. Chem.*, **21**, 137 (1968).
72. Jeffery, E. A., Mole, T. and Saunders, J. K., *Can. J. Chem.*, **21**, 649 (1968).
73. Sishta, C., Hathorn, R. M. and Marks, T. J., *J. Am. Chem. Soc.*, **114**, 1112 (1992).
74. Benn, R., Janssen, E., Lehmkul, H., Rufinska, A., Angermund, K., Betz, P., Goddard, R. and Krüger, C., *J. Organomet. Chem*, **411**, 37 (1991).
75. Apblett, A. W., Warren, A. C. and Barron, A. R., *Chem. Mater.*, **4**, 167 (1992).
76. Boleslawski, M. and Serwatowski, J., *J. Organomet. Chem.*, **254**, 159 (1983).
77. Atwood, J. L. and Zaworotko, M. J., *J. Chem. Soc., Chem. Commun.*, 301 (1983).
78. Thewalt, U. and Stollmaier, F., *Angew. Chem., Suppl.*, 209 (1982).
79. Storre, J., Belgardt, T., Stalke, D. and Roesky, H. W., *Angew. Chem., Int. Ed. Engl.*, **33**, 1244 (1994).
80. Atwood, J. L., Hrncir, D. C., Priester, R. D. and Rogers, R. D., *Organometallics*, **2**, 985 (1983).
81. Sadownik, A., Praca Doktorska, Warsaw (1997).
82. Hagendorf, W., Harder, A. and Sinn, H., *Macromol. Symp.*, **97**, 127 (1995).
83. Sadownik, A. Pasynkiewicz, S., Boleslawski, M. and Szachnowska, H., *J. Organomet. Chem.*, **152**, C49 (1978).
84. Piotrowski, A., Kunicki, A. and Pasynkiewicz, S., *J. Organomet. Chem.*, **186**, 185 (1980).
85. Sadownik, A. Pasynkiewicz, S. and Kunicki, A., *J. Organomet. Chem.*, **141**, 275 (1977).
86. Piotrowski, A., Kunicki, A. and Pasynkiewicz, S., *J. Organomet. Chem.*, **201**, 105 (1980).
87. Harlan, C. J., Bott, S. G., Wu, B., Lenz, R. W. and Barron, A. R. *J. Chem. Soc., Chem. Commun.*, 2183 (1997).
88. Pietrzykowski, A., Pasynkiewicz, S. and Wolinska, A., *J. Organomet. Chem.*, **201**, 89 (1980).
89. Tritto, I, Sacchi, M. C., Locatelli, P. and Li, S. X., *Macromol. Chem. Phys.*, **197**, 1537 (1996).
90. Barron, A. R., *Organometallics*, **14**, 3581 (1995).
91. Jordan, D. E., *Anal. Chem.*, **40**, 2150 (1968).
92. Resconi, L., Bossi, S. and Abis, L., *Macromolecules*, **23**, 4489 (1990).
93. Gianetti, E., Nicoletti, G. M. and Mazzocchi, R., *J. Polym. Sci., Polym. Chem. Ed.*, **23**, 2117 (1985).
94. Eisch, J. J., in Wilkinson, G., Stone, F. G. A. and Abel, E. W. (Eds), *Comprehensive Organometallic Chemistry*, Pergamon press, Oxford, 1982, Vol. 1, Chapt. 6.
95. Stoll, S. L., Bott, S. G. and Barron, A. R., *J. Chem. Soc., Dalton Trans.*, 1315 (1997).
96. Harlan, C. J., Bott, S. G. and A. R. Barron, *J. Am. Chem. Soc.*, **117**, 6465 (1995).
97. Watanabi, M., McMahon, C. N., Harlan, C. J. and Barron, A. R., in press.
98. Beard, W. R., Blevins, D. R., Imhoff, D. W., Kneale, B. and Simeral, L. S., *Polyethylene: New Technology, New Markets*, Institute of Materials, London, 1997.
99. Bochmann, M. *J. Chem. Soc., Dalton Trans.*, 255 (1996).
100. Sishta, C., Hathorn, R. M. and Marks, T. J., *J. Am. Chem. Soc.*, **114**, 1112 (1992).
101. Jolly, C. A. and Marynick, D. S., *J. Am. Chem. Soc.*, **111**, 7968 (1989).
102. Gassman, P. G. and Callstrom, M. R., *J. Am. Chem. Soc.*, **109**, 7875 (1987).

103. Harlan, C. J., Bott, S. G. and Barron, A. R., *J. Chem. Soc., Dalton Trans.*, 637 (1997).
104. Resconi, L., Bossi, S. and Abis, L., *Macromolecules*, **23**, 4489 (1990).
105. Lauher, J. W. and Hoffmann, R., *J. Am. Chem. Soc.*, **98**, 1729 (1976).
106. Dahmen, K. H., Hedden, D., Burwell, R. L., Jr and Marks, T. J., *Langmuir*, **4**, 1212 (1988).
107. Koide, Y., Bott, S. G. and Barron, A. R., *Organometallics*, **15**, 2213 (1996).

3

MAO-free Metallocene Catalysts for Ethylene (Co)Polymerization

LUIGI RESCONI, UMBERTO GIANNINI AND
TIZIANO DALL'OCCO
Montell Italia, G. Natta Research Center, Ferrara, Italy

1 INTRODUCTION

Homogeneous olefin polymerization catalysts based on Group 4 metallocenes and aluminum alkyls have been the subject of great interest since 1957. Natta *et al.* [1] and Breslow and Newburg [2] independently discovered in 1957 that the reaction mixtures of Cp_2TiCl_2 and AlR_3 or AlR_2Cl catalyze the polymerization of ethylene under mild conditions, although these systems were used only in mechanistic studies owing to their low activity.

Reichert and Meyer [3] first reported that small amounts of water, generally considered a strong poison for Ziegler–Natta catalysis, improve the productivity of the $Cp_2TiClEt$–$EtAlCl_2$ system. The activating effect of water was subsequently confirmed by Long and Breslow [4] for the systems Cp_2TiCl_2—Me_2AlCl and Cp_2TiCl_2—$AlMe_3$. These authors postulated the formation of the dimeric alumoxane $(ClMeAl)_2O$, reputed to be a strong Lewis acid and an efficient activator of $Cp_2TiMeCl$ for ethylene polymerization. Probably the low Al:Ti (1.5) and H_2O:Al (0.5) molar ratios used and the low polymerization temperature (30 °C) did not allow these authors to reach the exceptionally high activities subsequently found by Sinn, Kaminsky and co-workers [5,6] by polymerizing olefins with polymethylalumoxane (MAO) and dichloro- or dialkylmetallocenes. The combination of the outstanding

Metallocene-based Polyolefins   Edited by J. Scheirs and W. Kaminsky
© 2000 John Wiley & Sons Ltd

cocatalytic efficiency of MAO towards Group 4 metallocenes, and the high steric and electronic variability of the latter, led to a novel class of olefin polymerization catalysts [7] which is currently revolutionizing the polyolefin industry.

In spite of the industrial interest in replacing the MAO cocatalyst with cheaper and less hazardous Al-alkyl cocatalysts, few data have been reported in the literature concerning the use of alumoxanes from commercially available higher aluminum alkyls. Research in this direction might have been discouraged by the observation that the polymerization activity of Cp_2ZrCl_2 towards ethylene decreases by over two orders of magnitude on replacing MAO cocatalyst with polymeric ethylalumoxane and tetraisobutyldialumoxane (TIBAO) [8]. The low activity of the catalyst system Cp_2ZrCl_2 — TIBAO has recently been confirmed by Reddy [9]. On the other hand, polymeric isobutyl alumoxane (IBAO) was found to be virtually inactive. Ethylalumoxane was confirmed to be a poor activator also for alkyl-substituted zirconocenes of the type $(RCp)_2ZrCl_2$ [10].

At the end of the 1980s it was found that high activity in ethylene polymerization can be achieved also by using TIBAO as cocatalyst by properly choosing the nature of the zirconocene ligands [11–14].

In this paper, we discuss the influence of the kind of cyclopentadienyl ligands on the performance of zirconocene/TIBAO catalysts in ethylene polymerization. In addition, for selected zirconocenes and the model system $Me_2Si(Me_4Cp)_2ZrCl_2/Al(^iBu)_3/H_2O$, we discuss the influence of the type of the Al cocatalyst and the Al:Zr and Al:H_2O molar ratios.

2 THE TIBAO COCATALYST

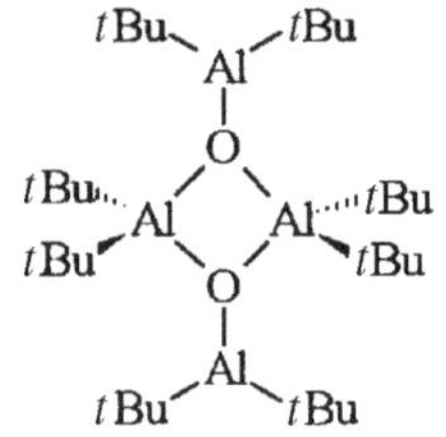

It is well known that MAO is not a structurally well defined compound and it is traditionally described as a mixture of oligomers with an Al—O backbone with linear, cyclic or cage structures [15,16]; generally more or less AlR_3, unreacted or coming from disproportionation reactions between oligomeric components, is present either 'free' or 'associated' in their solution [13,16d,17]. Also, the knowledge of the properties of tetraalkyldialumoxanes, $(R_2Al)_2O$, is still insufficient. Their structure and stability are strictly dependent on the bulkiness of the R group. When R is a *tert*-butyl group, the compound has been isolated in a crystalline state and crystallographically characterized [16] as a dimer with two bridging μ-oxo groups and two three-coordinate and two four-coordinate Al centers.

The compound is stable in solution, does not disproportionate to give $Al(^tBu)_3$ and is not active in ethylene polymerization in the presence of Cp_2ZrMe_2 [16].

When R is a (trimethylsilyl)methyl group, the compound crystallizes as a monomer [18]; nothing is known about its behavior as cocatalyst in olefin polymerization.

On the basis of cryoscopic and ^1H, ^{17}O and ^{27}Al NMR studies, a trimeric structure (1) has been proposed for $(Et_2Al)_2O$, in a dynamic equilibrium, in solution, with structures 2 and 3 [19].

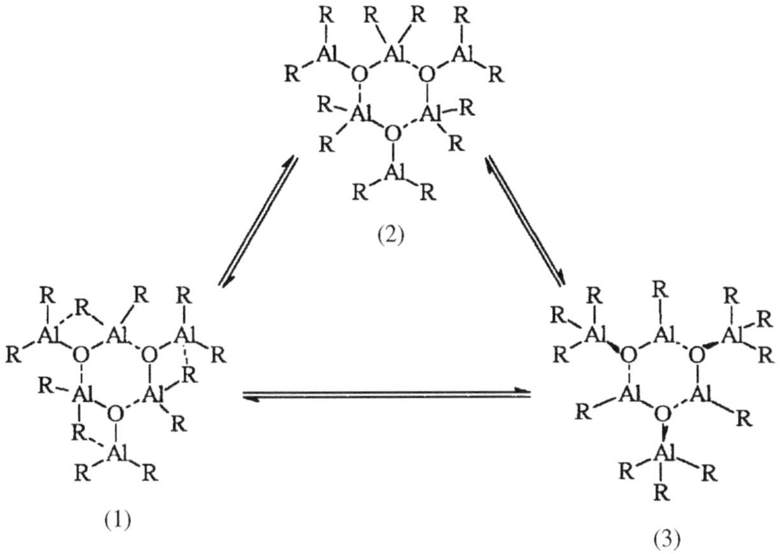

AlEt$_3$ can be abstracted under heating and therefore, in addition to the trimer, AlEt$_3$ and oligoethylalumoxanes are also present in tetraethyldialumoxane solutions [20]. A dimeric structure, analogous to that found for $(^tBu_2Al)_2O$, has been proposed for $(^tBu_2Al)_2O$ prepared by hydrolysis, on the basis of cryoscopic measurements in benzene; however, a maximum degree of association of 1.5 was found in benzene solution for $(^tBu_2Al)_2O$ prepared by reaction between $(^iBu)_2AlCl$ and $(^iBu)_2AlOLi$ [21]. Furthermore, we detected a considerable amount of $Al(^iBu)_3$ in aged solutions of TIBAO at an Al:H$_2$O molar ratio of 2; even higher amounts of AlR$_3$ were found in commercial samples of TIBAO. Comparison of the ^1H NMR spectra of fresh and aged samples revealed that TIBAO is not a stable molecule but undergoes extensive disproportionation to give $Al(^iBu)_3$ and $(^iBuAl)O$ oligomers; therefore, on comparing the results of olefin polymerization with TIBAO cocatalyst, the actual composition of its solutions must be taken into account.

3 ETHYLENE HOMOPOLYMERIZATION WITH ZIRCONOCENE/AI(^{i}Bu)$_3$/H$_2$O SYSTEMS: INFLUENCE OF LIGAND STRUCTURE ON POLYMERIZATION ACTIVITY AND POLYMER MOLECULAR WEIGHT

Our initial studies on metallocene catalysts were aimed at replacing the MAO cocatalyst with cheaper, safer and readily available Al cocatalysts. Some precedents in the literature indicated that replacement of MAO with other Al compounds could be achieved.

A report by Kaminsky *et al.* [8a] indicated that both ethylalumoxane (EAO) and 'tetraisobutyldialumoxane' (TIBAO) showed some ethylene polymerization activity with Cp$_2$ZrX$_2$(X = Cl, Me) at 70 °C, with TIBAO being worse than MAO but better than EAO.

Although in our hands no interesting activity could be obtained with either EAO or IBAO, Cp$_2$ZrCl$_2$ activated by freshly prepared (via the wet nitrogen method [22]) TIBAO gave low activities (25–40 kg$_{PE}$ g$_{Zr}^{-1}$ h^{-1}) for ethylene polymerization (toluene, 80 °C, 3.6 bar ethylene) [23].

We tested methyl-substituted zirconocenes and found that, for the systems Cp$_2$ZrCl$_2$, (Mecp)$_2$ZrCl$_2$ and (Me$_5$Cp)$_2$ZrCl$_2$, polymerization activity is higher in aromatics (toluene) than in aliphatics (hexane or heptane) and indeed increases when methyl groups are added to the Cp ligands, following the order Cp < Mecp < Me$_5$Cp [11]. This is in line with the hypothesis that an increased electron density at the metal center would stabilize the putative cationic active species.

Initially, our results with the TIBAO cocatalyst were plagued by irreproducibility. We later realized that TIBAO is not a stable compound, but undergoes disproportionation upon standing, with release of Al(^{i}Bu)$_3$ and formation of higher [Al(^{i}Bu)O]$_n$ oligomers (see the Introduction). Despite our still largely incomplete understanding of this catalyst system, our early investigation on Al(^{i}Bu)$_3$-based cocatalysts taught us that the only Group 4 transition metal with interesting activity is zirconium, while among the tested AlR$_3$ cocatalysts, Al^{i}Bu$_3$ was the most convenient, although Al(^{i}Hex)$_3$ showed similar activity [11].

After we had realized the activating effect of water [11], two main aspects of ethylene (co)polymerization with Al(^{i}Bu)$_3$-based zirconocene catalysts needed a more thorough investigation: the chemical nature of the cocatalyst (and the mechanism of active species formation) and the influence of the zirconocene structure on polymerization activity and polyethylene molecular weight. In the following we discuss this latter aspect.

We first screened several different zirconocenes, spanning a broad range of symmetries and degrees of substitution, for ethylene polymerization under standard slurry conditions (hexane at 50 °C, 4 bar-a constant ethylene partial pressure for 1 h in a 1 l Büchi autoclave; see Section 8) with a commercial TIBAO [24] and, in some cases, with other cocatalysts for comparison. Polymerization results and polymer characterization are reported in Table 1. Activities were simply calculated from polymerization yields. The choice of hexane as the polymerization solvent is dictated by the impossibility of using toluene as the solvent for large-scale polyolefin manufacture, while a high Al:Zr ratio is used to ensure the highest possible reproducibility.

Concerning catalyst activity for ethylene homopolymerization, the following conclusions can be drawn.

(A) The simplest and readily available zirconocene, Cp_2ZrCl_2, *is virtually inactive with commercial TIBAO*, in hexane at 50 °C. $(MeCp)_2ZrCl_2$, $(^nBuCp)_2ZrCl_2$, Ind_2ZrCl_2 and *rac*-EBIZrCl$_2$ showed activities of 30–150 $kg_{PE}/(g_{Zr} h)$, lower than the activities achieved with MAO with the same zirconocenes. Electron-richer zirconocene/TIBAO catalysts can have ethylene polymerization activities well above $500 kg_{PE}/(g_{Zr} h)$. It is important to stress that polymerization activities for a given zirconocene depend strongly on the solvent, the cocatalyst and the mode of catalyst preparation (aluminum concentration, Al:Zr ratio, aging time of the zirconocene–Al cocatalyst solution). For example, Cp_2ZrCl_2 is one of the most active ethylene polymerization catalysts when activated with MAO in toluene [25]. Earlier experiments had shown that $(Me_5Cp)_2ZrCl_2$ shows much higher activity when the polymerization is carried out in toluene and when the zirconocene and the Al cocatalyst are prereacted in the absence of monomer, and that the $(Me_5Cp)_2ZrCl_2/MAO$ catalyst is *less* active than the $(Me_5Cp)_2ZrCl_2/$ TIBAO catalyst. Hence there are also solvent and cocatalyst effects to be taken into account before a generalization can be made. The polymerization temperature (T_p) is also a crucial parameter, as different catalysts might have very different chemical stabilities and/or different intrinsic activities depending on T_p, so the relative zirconocene activity series might change on changing T_p.

Considering the non-bridged zirconocenes, when simple hydrocarbyl substituents are added to the cyclopentadienyl ligands, using the TIBAO cocatalyst we observe an increase in ethylene polymerization activity as their number increases, reaching a maximum for $(Me_4Cp)_2ZrCl_2$, then decreasing again for the fully methylated $(Me_5Cp)_2ZrCl_2$ (Table 1):

$$Cp \ll MeCp \approx {}^nBuCp < Ind < Me_5Cp < H_4Ind < 4,7-Me_2-Ind < Me_4Cp.$$

There are literature confirmations for this effect in similar systems [10,26–28], indicating that activity increases with increasing the electron density on the metal center but decreases with increasing steric crowding around the metal center. Whether this affects the number of active centers or the value of k_p remains an open question. In particular, Miya *et al.* [26] investigated the complete series of $(Me_xCpH_{5-x})_2ZrCl_2$ complexes ($x = 0-5$) for propylene polymerization at 50 °C, 8 bar, MAO cocatalyst, and the following relative activities were found: $Cp \approx 1, 2, 3-Me_3Cp < MeCp \approx Me_5Cp < 1, 2-Me_2Cp < 1, 3-Me_2Cp < Me_4Cp < 1, 2, 4-Me_3Cp$. This activity/Cp substitution pattern relationship is remarkably similar to that found for ethylene in the above study. These findings suggest that $(Me_4Cp)_2ZrCl_2$ might not be the most active catalyst in the series of unbridged zirconocenes.

Table 1 HDPE synthesis with zirconocene catalysts: effect of ligand structure on zirconocene activity and polymer molecular weight[a]

	Metallocene			Cocatalyst					Yield of HDPE (g)	Activity		$[\eta]^c$ (dl/g)
Entry	Type	mg	μ mol	Type[b]	Al:H$_2$O molar ratio	Al mmol	Al mmol/l	Al:Zr molar ratio		$kg_{PE}/(g_{cat}\,h)$	$kg_{PE}/(g_{Zr}\,h)$	
1	Cp$_2$ZrCl$_2$	0.058	0.20	TIBAO	2	1	2.5	5000	Traces	–	–	
2		1.46	4.99	TIBAO	2	5	12.5	1000	Traces	–	–	
3	(MeCp)$_2$ZrCl$_2$	0.316	0.99	TIBAO	2	1	2.5	1000	2.9	9.2	32	6.4
4	("BuCp)$_2$ZrCl$_2$	0.080	0.20	TIBAO	2	1	2.5	5000	0.7	8.7	38	
5	(Me$_4$Cp)$_2$ZrCl$_2$	0.085	0.21	TIBAO	2	1	2.5	5000	13.2	155.3	689	11.9
6	(Me$_5$Cp)$_2$ZrCl$_2$	0.086	0.20	TIBAO	2	1	2.5	5000	3.4	39.5	187	10.5
7	Ind$_2$ZrCl$_2$	0.082	0.21	TIBAO	2	1	2.5	5000	1.1	13.4	58	
8	(H$_4$Ind)$_2$ZrCl$_2$	0.080	0.20	TIBAO	2	1	2.5	5000	4.4	55.0	242	15.1
9	(4,7-Me$_2$-Ind)$_2$ZrCl$_2$	0.089	0.20	TIBAO	2	1	2.5	5000	9.0	101.1	497	5.8
10		0.215	0.48	MAO	1	2.4	6	5000	10.5 (40 min)	73.2	360	7.2
11	rac-EBIZrCl$_2$	0.085	0.20	TIBAO	2	1	2.5	5000	1.9	22.3	102	
12		0.42	1.00	TIBAO	2	1	2.5	1000	14.0	33.3	153	2.1
13		0.20	0.48	MAO	1	2.4	6	5000	21.5	107.5	493	2.5

Entry	Catalyst	[η]		Activator				Al:H₂O				
14	rac-EBTHIZrCl₂	0.087	0.20	TIBAO	2	1	2.5	5000	3.4	39.1	183	6.4
15		0.432	1.01	TIBAO	2	1	2.5	1000	10.4	24.1	113	4.9
16	rac-EBDMIZrCl₂	0.095	0.20	TIBAO	2	1	2.5	5000	22.0	231.6	1205	3.8
17	meso-EBDMIZrCl₂	0.095	0.20	TIBAO	2	1	2.5	5000	1.3	13.7	71	21.2
18	rac-Me₂SiInd₂ZrCl₂	0.097	0.20	MAO	2	1	2.5	5000	4.2	43.3	225	19.0
19	rac-Me₂Si(H₄Ind)₂ZrCl₂[d]	0.091	0.20	TIBAO	2	1	2.5	5000	0.7	7.7	38	
20	Me₂Si(2-Me-Ind)₂ZrCl₂	0.091	0.20	TIBAO	2	1	2.5	5000	4.0	43.9	220	12.5
21	Me₂Si(4,7-Me₂-Ind)₂ZrCl₂[e]	0.092	0.19	TIBAO	2	1	2.5	5000	11.9	129.3	675	4.0
22		0.099	0.20	TIBAO	2	1	2.5	5000	2.8	28.3	156	6.3
23	Me₂SiFlu₂ZrCl₂[f]	0.109	0.20	TIBAO	2	1	2.5	5000	8.8	80.7	485	9.0
24	Me₂Si(Me₄Cp)₂ZrCl₂	0.050	0.11	TIBAO	2	0.5	1.25	4500	2.9	58.0	292	3.6

[a] Polymerization conditions: 1 l Büchi glass autoclave, 0.4 l n-hexane, 4 bar-a ethylene 50 °C, 1 h, stirring rate 800 rpm; n-hexane contains 2 ppm water. Given Al:H₂O ratios do not include residual water in the solvent. Zirconocene/TIBAO or Zirconocene/MAO aged 5 min at room temperature in toluene.

[b] TIBAO = standard Schering TIBA/H₂O (Al:H₂O = 2), 1.55 M(Al) in cyclohexane. MAO = methylalumoxane, Schering, dried *in vacuo* to a free-flowing powder (residual AlMe₃ ≈ 3–5 mol%).

[c] [η] intrinsic viscosity in THN, 135 °C.

[d] rac 90%

[e] rac 84%

[f] 85% purity (contains LiCl, Et₂O).

At this stage, we could produce too few polymerization results to be able to make a clear dissection of the steric and electronic effects of the different Cp ligands on catalyst performance. Although this knowledge would be of paramount importance in order to tailor catalyst activity, polymer molecular weight and comonomer(s) incorporation, such a task (synthesis of complete series of zirconocenes, polymerization testing under different experimental conditions and with different cocatalysts, polymer analysis) will require a much broader investigation.

However, a simple rule-of-thumb has been identified [10,26–28] which seems to apply also to TIBAO-activated zirconocenes: the more basic (electron-donating) the Cp is, i.e. the higher the electron density on the zirconium atom, the higher is the polymerization activity of the corresponding catalyst. Also, and more intuitively, the lower the steric crowding around the metal center, the higher are the activity and comonomer incorporation.

The adverse steric effect caused by direct methylation of the Cp ligands could be avoided by making use of the electron-donating effect of methyl groups to the Zr atom through the aromatic system of the indenyl ligand, as in *rac*-ethylenebis(4,7-dimethyl-1-indenyl)zirconium dichloride (*rac*-EBDMIZrCl$_2$) [28–30] in which a higher electron density on the zirconium atom is obtained *without* an increase in the steric hindrance around it. Furthermore, the results of a thorough investigation of this zirconocene towards propylene and 1-butene polymerization led us to believe that comonomer incorporation rates should be favorable, at least comparable to those of other racemic metallocene catalysts, such as *rac*-EBIZrCl$_2$ and *rac*-EBTHIZrCl$_2$, and the copolymer molecular weight should be increased owing to a more difficult primary β-hydrogen transfer at *rac*-EBDMIZrCl$_2$ than at *rac*-EBIZrCl$_2$. Indeed, *rac*-EBDMIZrCl$_2$ is the most active zirconocene/TIBAO catalyst found so far and, more important, copolymerization results [31] have shown that it is also an excellent catalyst for the production of ethylene/α-olefin copolymers, increasing their molecular weight with respect to *rac*-EBIZrCl$_2$.

The relevance of steric effects when the TIBAO cocatalyst is used is shown by the lower activity obtained with *meso*-EBDMIZrCl$_2$, about one order of magnitude with respect to the racemic isomer.

The influence of a bridging group between the ligands causes different effects depending on the nature of the bridge and of the π-ligand. *rac*-EBTHIZrCl$_2$ and *rac*-Me$_2$Si(H$_4$Ind)$_2$ZrCl$_2$ show a lower catalytic activity than (H$_4$Ind)$_2$ZrCl$_2$, while activity is higher for *rac*-EBIZrCl$_2$ and slightly reduced for *rac*-Me$_2$Si(Ind)$_2$ZrCl$_2$ in comparison with Ind$_2$ZrCl$_2$. The presence of a methyl group in the 2-position of indene as in *rac*-Me$_2$Si(2-Me-Ind)$_2$ZrCl$_2$ causes a dramatic increase in catalytic activity (about 20-fold); the opposite effect was previously observed in the polymerization of propylene, where *rac*-Me$_2$Si(2-Me-Ind)$_2$ZrCl$_2$ is less active than *rac*-Me$_2$Si(Ind)$_2$ZrCl$_2$. [32]

(B) Different zirconocenes produce very different HDPE molecular weights. In ethylene-based polymers, increasing Cp alkylation increases the molecular weight of the (co)polymer, while there is only a minor dependence on the cocatalyst. For

example, $(Me_5Cp)_2ZrCl_2$ produces HDPE [toluene or hexane, $50\,^\circ C$, 4 atm ethylene, cocatalysts MAO, $Al(^iBu)_3/H_2O$, TIBAO or $AlH(^iBu)_2/H_2O$], with $[\eta] \approx 8-12\,dl\,g^{-1}$; with TIBAO as cocatalyst and under the same experimental conditions, HDPE viscosities (THN, $135\,^\circ C$) range from 2 to 21 $dl\,g^{-1}$ for polymers obtained with different zirconocenes (see Table 1). This means that there is a strong ligand effect also on chain-transfer rates, in addition to the already discussed effect on polymerization rates. This effect is well documented for metallocene/MAO catalysts [10,25–28,33] and it has been used to produce polymers with bimodal molecular weight distributions by using mixtures of metallocenes, either two different transition metals or two zirconocenes with different ligands, or both [33,34]. We note that the same can be achieved with our zirconocene/TIBAO catalysts, by using two different Cp ligands rather than two different metals, being limited by the inactivity of both titanocenes and hafnocenes with TIBAO. Thus, an equimolar mixture of $(Me_4Cp)_2ZrCl_2$ (producing HDPE with $[\eta] = 12$ and $\overline{M}_w/\overline{M}_n = 2.8$) and rac-EBDMIZrCl$_2$ (producing HDPE with $[\eta] = 4$ and $\overline{M}_w/\overline{M}_n = 2.5$) yielded a bimodal distribution with $\overline{M}_w/\overline{M}_n = 3.9$.

It occurred to us that the same effect could be obtained by mixing the racemic and *meso* isomers of the same zirconocene. In the case of propylene polymerization, the *rac*-metallocene isomers produce low to high (depending on the metallocene type) molecular weight iPP, whereas their *meso* isomers usually are less active and produce lower molecular weight atactic PP than their racemic counterparts. Indeed, the use of *rac*–*meso* mixtures of EBIZrCl$_2$ and EBTHIZrCl$_2$ produced HDPE with somewhat broadened MWD (e.g. EBTHIZrCl$_2$, *rac*:*meso* = 40:60, TIBAO cocatalyst, $Al/Zr = 1000$, $\overline{M}_w/\overline{M}_n = 3.8$, vs 2.4 with pure *rac*).

Generally, *rac*–*meso*-zirconocene mixtures are difficult to separate into their pure components, and the *meso* isomers are usually less stable than *rac* isomers. The lucky exception is *meso*-EBDMIZrCl$_2$, which can be efficiently synthesized through a fully diastereoselective synthesis [30]. Thus, *meso*-EBDMIZrCl$_2$ was tested in polymerization and, surprisingly, gave HDPE with $[\eta] = 21$ dl/g, a molecular weight much higher than that obtained with *rac*-EBDMIZrCl$_2$ ($[\eta] = 4$ dl/g), and the highest of all zirconocenes tested so far (Table 1). Note that in propylene polymerization *meso*-EBDMIZrCl$_2$ produces atactic polypropylene with $\overline{M}_n$ as low as 4600 [30]. An $[\eta]$ of 21 dl/g falls into the range of ultra-high molecular weight polyethylene (UHMW-PE), and prompted us to investigate this catalyst further. Polymerization results for two different polymerization temperatures and two cocatalysts are compared in Table 2. As expected, the $\overline{M}_w$ decreases with increasing T_p. We also observe that TIBAO gives slightly higher molecular weights than MAO: as there is more residual $Al(^iBu)_3$ in TIBAO than $AlMe_3$ in MAO, it seems safe to conclude that, at least at $50\,^\circ C$, the main chain transfer with this particular catalyst occurs via transfer to the aluminum cocatalyst, and that $Al(^iBu)_3$ is a poorer transfer agent than $AlMe_3$. In principle, mixtures of *rac*- and *meso*-EBDMIZrCl$_2$ should produce a much broader MWD, given the differences in molecular weights produced by the two catalysts. In practice, we could not demonstrate this effect experimentally,

Table 2 HDPE synthesis with *meso*-EBDMIZrCl$_2$

Zr		Cocatalyst							
mg	μmol	Type	Al:Zr molar ratio	T_p (°C)	P_{tot}[a] (bar-a)	Yield (g)	Activity [kg$_{PE}$/(g$_{cat}$ h)]	$[\eta]$[b] (dl/g)	$\overline{M}_v$[c]
0.1	0.2	TIBAO	5000	50	4	1.3	13.7	21.2	3.5×10^6
0.1	0.2	MAO	5000	50	4	4.2	43.3	19.0	3.0×10^6
0.5	1.0	TIBAO	1000	70	5.5	10.6	22.1	15.1	2.2×10^6
0.5	1.0	MAO	1000	70	5.5	10.0	21.2	12.9	1.8×10^6
0.1	0.2	MAO	5000	70	5.5	7.5	78.1	13.9	2.0×10^6

[a] Same monomer concentration at the two temperatures.
[b] THN, 135 °C.
[c] Calculated from $[\eta] = 3.8 \times 10^{-4} M_v^{0.725}$.

as molecular weights from *meso*-EBDMIZrCl$_2$ are far too high for a correct size-exclusion chromatographic (SEC) determination. However, the use of *rac–meso*-EBDMIZrCl$_2$ enabled us to develop a process for the production of broad molecular weight polyethylenes [35].

From Table 1, it appears that *rac*-EBDMIZrCl$_2$ is the best catalyst found in the present investigation. The comparison of Ind$_2$ZrCl$_2$ with *rac*-EBIZrCl$_2$, where the added ethylene bridge apparently provides only a slight advantage on activity, and of *rac*-EBIZrCl$_2$ with *rac*-EBDMIZrCl$_2$, where there is a remarkable increase in activity and an almost twofold increase in viscosity with the latter, due to the addition of the methyl groups, prompted us to test its non-bridged analog, bis(4,7-dimethyl-1-indenyl)zirconium dichloride, BDMIZrCl$_2$ [36]. The activity of BDMIZrCl$_2$, although not as high as that of *rac*-EBDMIZrCl$_2$, is 3–4 times higher than that of *rac*-EBIZrCl$_2$ at 50 °C.

Note that the activity of BDMIZrCl$_2$ (with the two indenyl ligands freely rotating around the Zr atom) falls between those of *rac*- and *meso*-EBDMIZrCl$_2$, which are stereorigid but have different orientations of the 4,7-dimethylindenyl ligands. Finally, the higher molecular weight obtained with BDMIZrCl$_2$ makes it a good candidate for polymerization at higher temperatures.

4 ETHYLENE HOMOPOLYMERIZATION WITH ZIRCONOCENE AIR$_3$/H$_2$O SYSTEMS: INFLUENCE OF THE COCATALYST TYPE AND Al/H$_2$O RATIO

The influence of different cocatalysts on activity is dramatic. For example, compare in Table 1 the catalysts *rac*-EBIZrCl$_2$/MAO (107 kg/g$_{cat}$) and *rac*-EBIZrCl$_2$/TIBAO (22 kg/g$_{cat}$) and the very high activity of Cp$_2$ZrCl$_2$/MAO (667 kg/g$_{cat}$ in toluene at 30 °C) [25] with the already mentioned inactivity of Cp$_2$ZrCl$_2$/TIBAO at 50 °C.

Given the practical interest of some zirconocene/$Al(^iBu)_3$/H_2O systems, the influence of the most important experimental parameters, such as $Al:H_2O$ and $Al:Zr$ ratios, and also comparison with MAO and other cocatalysts, was studied first on the model system $Me_2Si(Me_4Cp)_2ZrCl_2$ [37,38]. We had originally chosen this zirconocene structure because we predicted that the release of the steric hindrance via reduction of the Cp—Zr—Cp angle (increase in the 'wedge' angle, i.e. the space available for monomer coordination and chain growth between the two Cp's) caused by the presence of a short silyl bridge [37b] would generate a catalyst that would be more active, produce lower molecular weights and have a higher comonomer incorporation rate than $(Me_5Cp)_2ZrCl_2$. These predictions were confirmed by experiment. The results for ethylene homopolymerization are shown in Table 3.

At least in the case of $Me_2Si(Me_4Cp)_2ZrCl_2$/TIBAO, polymerization activities are strongly dependent on the $Al:Zr$ ratio, but relatively insensitive to the $Al:H_2O$ ratio for values in the range 10–1, with the highest activities in the range 10–2. The good activity obtained with the catalyst at $Al:H_2O = 1$ is clearly due to incomplete reaction between water and $Al(^iBu)_3$ when they are contacted at high dilution for a relatively short time, as is the case under our experimental conditions. In fact, when commercial IBAO [apparently $Al(^iBu)_3$-free poly-isobutylalumoxane] is used as the cocatalyst, only traces of polymer are obtained.

We can also observe that this catalyst is activated by very low levels of water in the system: under our experimental conditions, even the residual 2 p.p.m. of H_2O in the solvent (corresponding to an $Al:H_2O$ ratio of ca 40) are sufficient for the activation of about 15 % of the catalyst. This extreme sensitivity of the system zirconocene/$Al(^iBu)_3$ to the presence of very small amounts of water in the polymerization medium led to the erroneous statement that alkyl-substituted zirconocenes can be activated by $Al(^iBu)_3$ [12–14].

As expected, the effect on molecular weight is marginal. Available SEC data are compared in Table 4. Comparing the different cocatalysts, it can be appreciated that TIBAO, which surpasses even MAO, is the best cocatalyst among those tested.

On five selected zirconocenes, we also investigated the influence of different cocatalysts at 80 °C and 10 bar of ethylene partial pressure, conditions similar to the plant operating conditions. The results are shown in Table 5 and Figure 1, and they confirm the trend observed at lower T_p and pressure. Different behaviors are observed depending on the metallocene, structure and type of alumoxane (R in AlR_3, $Al:H_2O$ molar ratio and preparation procedure).

For the simple Cp_2ZrCl_2, only MAO or the reaction product(s) between $AlMe_3$ and water provide reasonable activity in the ethylene polymerization and HDPE with M_w higher for $AlMe_3$/H_2O than MAO is obtained. The other cocatalytic systems are not active [Figure 1(a)].

The introduction of one electron-donating methyl group on each cyclopentadienyl ring provides, with MAO or $AlMe_3$/H_2O as cocatalysts, much higher activities than with the unsubstituted metallocene. Further, the metallocene now shows activity also

Table 3 Ethylene polymerization with $Me_2Si(Me_4Cp)_2ZrCl_2$[a]

Entry	Catalyst		Cocatalyst		Al		Al:Zr molar ratio	Yield of HDPE (g)	Activity		$[\eta]$[c] (dl/g)
	mg	μmol	Al:H₂O molar ratio	Type[b]	mmol	mmol/l			$kg_{PE}/(g_{cat}h)$	$kg_{PE}/(g_{zr}h)$	
24	0.050	0.11	2	TIBAO	0.5	1.25	4500	2.9	58.0	292	3.6
25	0.050	0.11	2	TIBAO	0.9	2.25	9000	8.5	170.0	858	3.5
26	0.048	0.10	2	TIBAO	1.8	4.5	18000	13.5	281.2	1420	3.5
27	0.049	0.11	1	MAO	4.5	11.25	42000	7.2	146.9	742	
28	0.049	0.11	1	MAO	0.89	2.2	8000	4	81.6	412	3.1
29	0.101	0.22	1	IBAO	1.8	4.5	8000	Traces	–	–	
30	0.098	0.21	40.5[d]	Al(^{i}Bu)₃/H₂O	1.8	4.5	8500	3.3	33.7	170	3.6
31	0.102	0.22	8.1[d]	Al(^{i}Bu)₃/H₂O	1.8	4.5	8200	21.1	206.9	1045	4.0
32	0.101	0.22	4.2[d]	Al(^{i}Bu)₃/H₂O	1.8	4.5	8200	22.2	219.8	1110	4.1
33	0.114	0.25	1.9[d]	Al(^{i}Bu)₃/H₂O	1.8	4.5	7300	24.5	214.9	1085	3.6
34	0.098	0.21	1.0[d]	Al(^{i}Bu)₃/H₂O	1.8	4.5	8500	18	183.7	928	3.3

[a] Polymerization conditions: 1 l Büchi glass autoclave, 0.4 l n-hexane, 4 bar-a ethylene, 50 °C, 1 h, stirring rate 800 rpm; n-hexane contains 2 ppm water. Given Al:H₂O ratios do not include residual water in the solvent. Zirconocene/TIBAO or zirconocene/Al(^{i}Bu₃) aged 5 min at room temperature in toluene.
[b] TIBAO = standard Schering Al(^{i}Bu₃)/H₂O (Al:H₂O = 2) 1.55 M (Al) in cyclohexane. MAO = methylalumoxane, Schering, dried *in vacuo* to a free-flowing powder (residual AlMe₃ ≈ 3–5 mol%). IBAO = polyisobutylalumoxane, Schering, heptane solution, AlH₂O = 1. Al(^{i}Bu₃)/H₂O = isobutylalumoxanes prepared *in situ*. H₂O added to the hexane into the autoclave prior to addition of the catalyst/cocatalyst solution.
[c] $[\eta]$: Intrinsic viscosity in THN, 135 °C.
[d] Includes 2 p.p.m. H₂O present in the solvent.

Table 4 Available SEC data for HDPE samples in Tables 1 and 3

Entry	Metallocene type	$[\eta]$ (dl/g) Exp.	Calc.	$\bar{M}_v$	$\bar{M}_n$	$\bar{M}_w$	$\bar{M}_z$	$\bar{M}_w/\bar{M}_n$	$\bar{M}_z/\bar{M}_w$
14	rac-EBTHIZrCl$_2$	4.9		472 100	216 200	520 200	942 300	2.41	1.81
16	rac-EBDMIZrCl$_2$	3.8	3.7	321 900	143 600	365 200	829 600	2.54	2.27
21	Me$_2$Si(2-MeInd)$_2$ZrCl$_2$	4.0	3.6	308 000	143 600	347 800	769 000	2.42	2.21
22	Me$_2$Si(4,7-Me$_2$Ind)$_2$ZrCl$_2$	6.3	5.4	526 800	279 500	579 000	1 090 000	2.02	1.86
24	Me$_2$Si(Me$_4$Cp)$_2$ZrCl$_2$	3.6	3.1	247 000	134 800	270 500	488 400	2.00	1.80
25		3.5	2.8	214 200	108 300	233 700	408 900	2.16	1.75
26		3.5	3.2	256 300	124 500	282 000	514 200	2.27	1.82
27			2.3	165 700	81 900	182 000	337 800	2.22	1.85
28		3.1	2.5	181 500	84 600	198 900	359 800	2.35	1.81
30		3.6	3.4	280 000	144 200	307 200	557 200	2.13	1.81
31		4.0	3.7	321 300	147 000	358 400	718 100	2.44	2.00
32		4.1	3.7	315 500	136 300	351 400	684 900	2.58	1.95
33		3.6	3.1	249 100	91 900	279 300	560 400	3.04	2.00
34		3.3	2.8	211 400	98 200	231 200	403 500	2.35	1.74

Table 5 Ethylene polymerization with different cocatalytic systems based on AlR_3/H_2O reaction products[a]

Entry	Metallocene	Cocatalyst	Amount mg	Amount μ mol	Amount (mmol)	Al:H₂O molar	Al:Zr molar	Yield (g)	Activity kg/g_cat	Activity kg/(g_zr h)	[η] (dl/g)
35	Cp₂ZrCl₂	AlMe₃/H₂O	0.29	0.99	5	2	5000	19.1	65.9	211.1	2.6
36		MAO	0.29	0.99	5		5000	26.2	90.34	289.6	1.9
37		AlEt₃/H₂O	1	3.42	17.1	2	5000	0			
38		Al("Bu)₃/H₂O	1	3.42	17.1	2	5000	0			
39		Al(ⁱBu)₃/H₂O	0.29	0.99	5	2	5000	0			
40		TIBAO	0.29	0.99	5		5000	0			
41		Al("Oct)₃/H₂O	1	3.42	17.1	2	5000	0.05	0.05	0.2	
42	(MeCp)₂ZrCl₂	AlMe₃/H₂O	0.32	1.00	5	40	5000	1	3.1	11.0	
43		AlMe₃/H₂O	0.32	1.00	5	2	5000	37	115.6	406.2	3.2
44		MAO	0.13	0.41	5		12000	37.5	288.5	1013.4	2.6
45		AlEt₃/H₂O	1	3.12	15.6	2	5000	0			
46		Al("Bu)₃/H₂O	1	3.12	15.6	2	5000	0			
47		Al(ⁱBu)₃/H₂O	0.32	1.00	5	2	5000	9	28.1	98.8	3.8
48		IBAO	0.32	1.00	5		5000	0			
49	(Me₅Cp)₂ZrCl₂	AlMe₃/H₂O	0.43	0.99	5	40	5000	2	4.7	22.1	2.4
50		AlMe₃/H₂O	0.43	0.99	5	2	5000	34	79.1	375.1	2.0
51		MAO	0.43	0.99	5		5000	25	58.1	275.8	3.2
52		AlEt₃/H₂O	1	2.31	11.6	2	5000	44.5	44.5	211.1	3.3
53		Al("Bu)₃/H₂O	1	2.31	11.6	2	5000	4	4.0	19.0	3.4

	Catalyst	Cocatalyst									
54		Al(^{i}Bu)$_3$/H$_2$O	0.43	0.99	5	2	5000	38	88.4	419.2	3.6
55		TIBAO	0.43	0.99	5		5000	15.2	35.4	167.7	5.4
56		IBAO	0.43	0.99	5		5000	0			
57	(4, 7 − Me$_2$ − Ind)$_2$ZrCl$_2$	MAO	0.18	0.40	2		5000	55.5	308	1516.3	5.9
58		Al(^{i}Bu)$_3$/H$_2$O	0.45	1.00	5	2	5000	55.7	123.8	608.7	5.2
59		TIBAO	0.45	1.00	5		5000	41	91.1	448.1	6.5
60	rac − EBIZrCl$_2$	AlMe$_3$/H$_2$O	0.17	0.41	2	2	5000	19	111.8	512.9	1.4
61		MAO	0.17	0.41	2		5000	33	194.1	890.8	1.6
62		Al(^{i}Bu)$_3$/H$_2$O	0.42	1.00	5	2	5000	19	45.2	207.6	1.5
63		TIBAO	0.42	1.00	5		5000	15	35.7	163.9	1.7
64		Al(^{i}Bu)$_2$H/H$_2$O	0.5	1.19	6.0	2	5000	11	22.0	100.9	1.3
65		IBAO	0.42	1.00	5		5000	0			
66		Al(^{n}Oct)$_3$/H$_2$O	1	2.39	12.0	2	5000	10	10.0	45.9	1.0
67	rac − EBDMIZrCl$_2$	AlMe$_3$/H$_2$O	0.095	0.20	2	2	10000	11.8	124.2	646.4	2.5
68	(rac : meso = 94 : 6)	MAO	0.095	0.20	2		10000	28.5	300.0	1561.2	2.7
69		AlEt$_3$/H$_2$O	0.3	0.63	3.2	2	5000	0	0		
70		Al(^{n}Bu)$_3$/H$_2$O	0.3	0.63	3.2	2	5000	22	73.3	381.6	2.4
71		Al(^{i}Bu)$_3$/H$_2$O	0.095	0.20	2	2	10000	24.2	254.7	1325.6	2.9
72		TIBAO	0.095	0.20	2		10000	21.5	226.3	1177.7	2.8
73		Al(^{n}Hex)$_3$/H$_2$O	0.3	0.63	3.2	2	5000	0			
74		Al(^{n}Oct)$_3$/H$_2$O	1	2.11	10.5	2	5000	0			

[a] Polymerization conditions: 2.3 l autoclave, hexane 1 l, 80 °C, ethylene partial pressure 9.6 bar, 1 h; metallocene and Al-alkyl were aged 5 min before injection into the autoclave; H$_2$O when used, was introduced into the autoclave before the solvent.

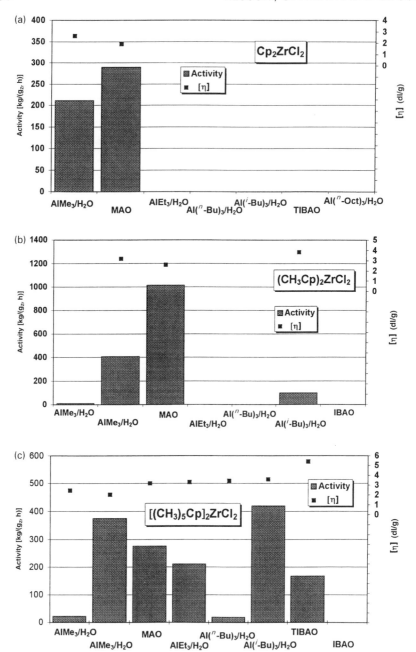

Figure 1 Activity and intrinsic viscosity of HDPE from different zirconocenes and different cocatalytic systems. (a) Cp_2ZrCl_2; (b) $(MeCp)_2ZrCl_2$; (c) $(Me_5Cp)_2ZrCl_2$; (d) rac-EBIZrCl$_2$; (e) rac-EBDMIZrCl$_2$; (f) $(4,7-Me_2Ind)_2ZrCl_2$

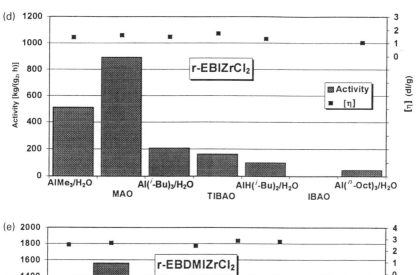

(d) r-EBIZrCl₂

Activity [kg/(g$_{Zr}$ h)] vs. AlMe₃/H₂O (MAO), Al(ⁱ-Bu)₃/H₂O, AlH(ⁱ-Bu)₂/H₂O (TIBAO), Al(ⁿ-Oct)₃/H₂O (IBAO)

Legend: ▦ Activity, ■ [η]

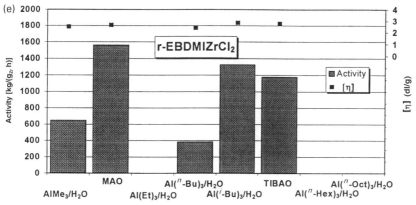

(e) r-EBDMIZrCl₂

Activity [kg/(g$_{Zr}$ h)] vs. AlMe₃/H₂O, MAO, Al(Et)₃/H₂O, Al(ⁿ-Bu)₃/H₂O, Al(ⁱ-Bu)₃/H₂O, TIBAO, Al(ⁿ-Hex)₃/H₂O, Al(ⁿ-Oct)₃/H₂O

Legend: ▦ Activity, ■ [η]

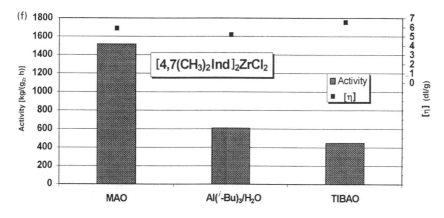

(f) [4,7(CH₃)₂Ind]₂ZrCl₂

Activity [kg/(g$_{Zr}$ h)] vs. MAO, Al(ⁱ-Bu)₃/H₂O, TIBAO

Legend: ▦ Activity, ■ [η]

with $Al(^iBu)_3/H_2O$ cocatalyst [Figure 1(b)]. Interestingly, no activity is observed when the R group of AlR_3 is a linear alkyl group. The inactivity of polyisobutyl-alumoxane (IBAO) confirms the results presented above.

By increasing the number of alkyl substituents around the unbridged cyclopent-adienyl ring, as is in $(Me_5Cp)_2ZrCl_2$, more AlR_3/H_2O cocatalysts become active [Figure 1(c)]. Under our polymerization conditions, the activity with $Al(^iBu)_3/H_2O(Al:H_2O = 2)$ is higher than with MAO. This metallocene shows interesting activity even with the cocatalyst obtained from $AlEt_3$ (about 75 % of MAO activity). In this particular case, the freshly prepared alumoxane from $AlMe_3$ shows higher activity than a commercial MAO product. Also in this case, IBAO is not active. The viscosity average molecular weights of all HDPE produced with this metallocene, independently of the cocatalyst, are similar.

The influence of cocatalyst type was studied also in some indenyl-based metallocenes. The systems studied were rac-EBIZrCl$_2$, rac-EBDMIZrCl$_2$ and BDMIZrCl$_2$. rac-EBIZrCl$_2$ shows low activity with the $Al(^iBu)_3$-based cocatalysts independently of their method of preparation, be it freshly prepared $Al(^iBu)_3/H_2O$, aged TIBAO or the product obtained from $Al(^iBu)_2H$ and water. Similar activities (about 20 % in comparison with MAO) are obtained in all cases. Again, the system based on IBAO is totally inactive [Figure 1(d)]. Much higher polymerization activities are obtained with rac-EBDMIZrCl$_2$ owing to the inductive effect of the electron-releasing methyl groups (see above), with both MAO and $Al(^iBu)_3$-based cocatalysts, with a great improvement for $Al(^iBu)_3/H_2O$ and TIBAO that in these cases reach about 80–90 % of MAO activity [Figure 1(e)]. The aluminum alkyls in which R is a linear chain confirm their very poor activity also with the indenyl-based systems: from low, for R = n-butyl, to none, for R = ethyl, n-hexyl and n-octyl.

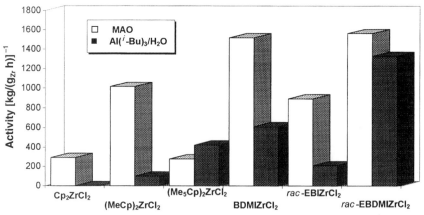

Figure 2 Comparison of the cocatalytic efficiencies of MAO and $Al(^iBu)_3/H_2O$ with different zirconocenes

The positive influence of methylation of the indenyl system is confirmed by the results obtained with the analogous unbridged $BDMIZrCl_2$. Ethylene polymerization activity is very high, as for the bridged system, when MAO is used and lower for the $Al(^iBu)_3$-based cocatalyst (30–40 % of MAO activity) [Figure 1(f)]. In summary, also under these polymerization conditions we confirm: (a) the inactivity of Cp_2ZrCl_2 with TIBAO and AlR_3/H_2O (R $\neq$ Me) cocatalysts; (b) the inactivity of all tested zirconocenes with the IBAO cocatalyst; (c) the dependence of polymerization activity on the bulkyness of the ancillary ligands; and (d) the higher activity of the catalyst systems based on AlR_3/H_2O cocatalysts when R is a branched group, this effect being especially notable with rac-$EBDMIZrCl_2$.

As expected, the molecular weight decreases with increasing T_p. The cocatalytic efficiency of MAO and $Al(^iBu)_3/H_2O$ are compared in Figure 2.

5 ETHYLENE–BUTENE COPOLYMERIZATION

It is well known that, besides polymerization activity, also the type of copolymers produced (degree of comonomer incorporation and its distribution along the polymer chain) depend strongly on the number, type and relative positions of the substituents on the cyclopentadienyl (Cp) ligands, i.e. are functions of both the shape of the zirconocene (Cp—Zr—Cp angle, symmetry, etc.) and the electron density at the Zr atom [39,40].

Adding alkyl groups to the Cp ligands also increases the steric bulk around the metal center; hence an increase in Cp alkylation above a certain level produces an adverse effect on activity, especially for olefins larger than ethylene. In fact, we found that $(Me_5Cp)_2ZrCl_2$ is ill-suited for LLDPE synthesis, as the highest 1-butene incorporation (liquid butene, 50 °C, 4 bar ethylene) is about 2 % molar. As discussed in Section 4, these considerations initially led us to the choice of $Me_2Si(Me_4Cp)_2ZrCl_2$, a zirconocene with the same maximum degree of alkylation as in $(Me_5Cp)_2ZrCl_2$ (and hence a high electron density on the Zr center) but a larger wedge angle [and hence an expected higher comonomer incorporation rate with respect to $(Me_5Cp)_2ZrCl_2$].

These two characteristics were expected to increase polymerization activity and comonomer incorporation and lower the molecular weight of the resulting polymer. In fact, $Me_2Si(Me_4Cp)_2ZrCl_2$ proved far more active, gave lower molecular weights and incorporated comonomers (1-butene, propylene and 4-methyl-1-pentene) more readily than $(Me_5Cp)_2ZrCl_2$, while maintaining its high activity also in aliphatic solvents or liquid monomer. In particular, in ethylene/1-butene copolymerization $Me_2Si(Me_4Cp)_2ZrCl_2$ produced 'super-random' LLDPE [38].

Several ethylene/1-butene copolymers have been synthesized with the $Me_2Si(Me_4Cp)_2ZrCl_2$–TIBAO system (Table 6). The polymerizations were performed with different relative monomer concentrations, by working either in propane (lower 1-butene concentration) or in liquid 1-butene (higher 1-butene concentration), at a 50–80 °C polymerization temperature. As shown in Figure 3,

Table 6 Synthesis of LLDPE (1-butene) by using $Me_2Si(Me_4Cp)_2ZrCl_2$ and TIBAO as catalytic system[a]

Entry	Polymerization solvent	Catalyst (mg)	TIBAO (mmol)	Al:Zr molar ratio	1-Butene (l)	P_p ethylene (bar)	$P_p H_2$ (bar)	T_p (°C)	Time (min)	Yield (g)	Activity [kg/(g_{Zr} h)]	$[\eta]$ (dl/g)	Melt Index E (g/10 min)	Melt Index F (g/10 min)	F/E	1-Butene (wt%)	Density (g/ml)	DSC $T_m(II)$ (°C)	ΔH (J/g)	XSRT (wt%)
75	Propane	1.8	3.9	2000	1.31	13	0.01	70	120	160	224.4	1.98	0.77	14.6	20.0	1.8	0.9283	118	114	0.5
76		1.8	3.9	2000	1.24	14	0.01	80	120	440	617.2	1.83	1.31	22.1	16.9	1.7	0.9273	118	110	0.4
77		1.8	3.9	2000	1.47	9	0.05	50	120	86	120.65	2.24	0.5	8.1	16.1	2.1	0.9261	118	103	0.4
78		1.8	3.9	2000	1.55	7	0.04	50	120	123	172.5	2.48	0.7	11.7	16.7	2.8	0.9230	117	98	0.3
79	1-Butene	1.2	3.9	3000	2.06	10	0.15	50	120	112	235.7	2.64	0.24	3.8	16.0	3.1	0.9215	119.6	125	0.2
80		2.4	3.9	1500	2.06	8	0.15	50	180	266	186.6	2.04	0.8	11.2	14.0	4.9	0.9172	113.7	120	0.3
81		2.4	3.9	1500	2.06	6	0.14	50	180	240	168.3	1.56	3.96	67.0	16.9	8.6	0.9112	107.6	99	1.2
82		2.4	3.9	1500	2.06	5	0.12	50	180	249	174.6	1.53	3.28	52.7	16.1	9.6	0.9067	103	92.5	1.7
83[b]		14.4	23.4	1500	12.2	8	0.15	50	240	2130	186.7	1.95	0.88	13.0	14.8	5	0.9171	113.5	111	0.4

[a] Polymerization conditions: 4.25 l stainless-steel autoclave; the metallocene was dissolved in 5 ml of toluene and aged 5 min in the presence of TIBAO at 25 °C. Propane, 1.6 l.
[b] 2 2 l autoclave.

two different comonomer reactivity behaviors can be observed by working in liquid propane or in liquid 1-butene. In this way, LLDPE containing about 1–5 mol% of 1-butene can be produced.

Polymers with higher comonomer contents were difficult to obtain with significant yields owing to the extreme conditions necessary for their synthesis. In fact, also by working in neat 1-butene as the polymerization solvent and a low ethylene partial pressure (entry 82 in Table 6) with a relative concentration of 1-butene higher than 0.9, only 9.6 wt% of 1-butene was introduced in the copolymer. In order to increase the comonomer concentration, a lower ethylene partial pressure is necessary, but the activity in this case will be dramatically reduced. The molecular weight is well controlled by using hydrogen, while the MWD, typical of metallocene based catalysis, is very narrow, as can be observed in the SEC diagram shown in Figure 4 ($M_w/M_n = 2.3$). Similar results can be deduced also from the rheological behavior of the molten polymers: the MIR value (melt index ratio; see section 9), which is dependent on the molecular weight distribution, is lower than 20 and indicates a very narrow MWD (Table 6).

The chemical comonomer distribution is also narrow. This conclusion is drawn from different analyses:

(a) Polymer fractionation by TREF shows a narrow extraction peak, without any peak over 100 °C as can be usually observed in polymers coming from

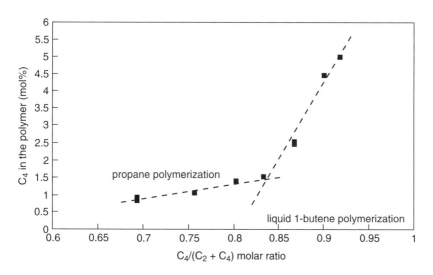

Figure 3 LLDPE (1-butene) with $Me_2Si(Me_4Cp)_2ZrCl_2$/TIBAO: correlation between 1-butene concentration in solution and in the polymer

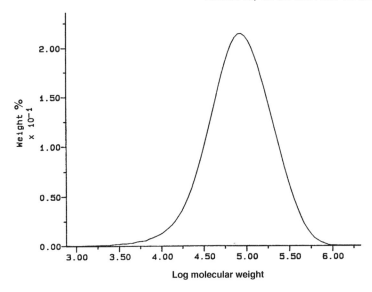

Figure 4 SEC trace for an LLDPE (1-butene) obtained with the $Me_2Si(Me_4Cp)_2ZrCl_2$/TIBAO catalyst (sample 80 in Table 6)

$MgCl_2$/Ti systems of the same comonomer contents, as is clearly apparent in Figure 5.

(b) The content of extractables in xylene at room temperature is very low (XSRT in Table 6).

(c) Both melting temperature and density decrease more rapidly with increasing comonomer content (Figure 6), in comparison with the behavior observed for analogous copolymers from $MgCl_2$/Ti systems.

(d) Only traces of comonomer homosequences were detected in the ^{13}C NMR spectra; for a 1-butene content of 3 mol%, we evaluated a cluster index of 0.6, that is, an average ethylene sequence length of 32 units [39]. On the hypothesis of a first-order Markovian distribution, from the diad distribution calculated from the ^{13}C NMR spectrum (Figure 7), it was possible to estimate (although with high approximation) $r_1 = 150, r_2 = 0.0014$ and $r_1 \times r_2 = 0.21$ [40]. These values explain the good comonomer distribution, the very low value of $r_1 \times r_2$ and also the difficulty of obtaining polymers with a high comonomer content, given the very high value of r_1.

The comonomer (propylene, 1-butene, 4-methyl-1-pentene) incorporation ability of $Me_2Si(Me_4Cp)_2ZrCl_2$, although higher than that found for $(Me_5Cp)_2ZrCl_2$, is still unsatisfactory. Chiral, C_2-symmetric zirconocenes, in comparison, incorporate larger amounts of comonomer, producing copolymers from LLDPE to plastomers (VLDPE) to elastomers (EPR), with both MAO [41] and TIBAO cocatalysts [31].

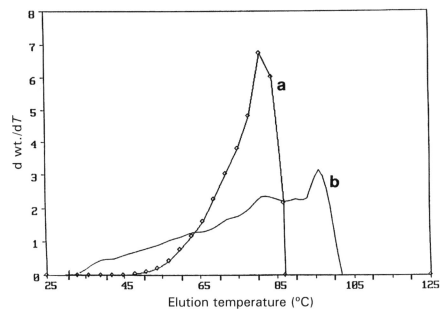

Figure 5 TREF diagram of (a) an LLDPE (1-butene) obtained with the $Me_2Si(Me_4Cp)_2ZrCl_2$/TIBAO catalyst (sample 80 in Table 6) and (b) an LLDPE of similar composition from an $MgCl_2$/Ti catalyst

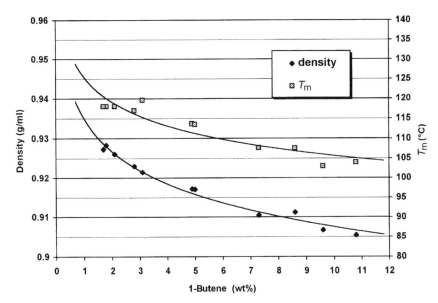

Figure 6 Density and T_m (DSC) in ethylene/1-butene copolymers synthesized from $Me_2Si(Me_4Cp)_2ZrCl_2$/TIBAO catalyst (see Table 6 for polymerization conditions)

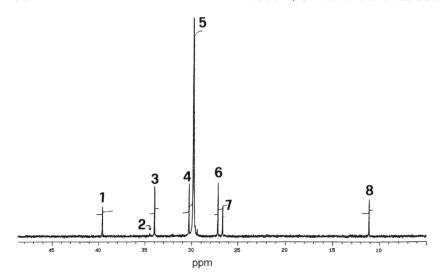

Figure 7 ^{13}C NMR of an LLDPE (1-butene) obtained with the Me$_2$Si(Me$_4$Cp)$_2$ZrCl$_2$/TIBAO catalyst (sample 80 in table 6). The peaks were assigned to the following sequences: 1, T$\delta\delta$ (EBE); 2, S$\alpha\gamma$ (EBEB); 3, S$\alpha\delta$ (EBEE); 4, S$\gamma\delta$ (BEEE); 5, S$\delta\delta$ (EEE); 6, S$\beta\delta$ (EBEE); 7, CH$_2$ (branch) (EBE); 8, P$\delta\delta$ (EBE)

The molecular weight of ethylene copolymers depends mainly on the rate of chain transfer after both an ethylene insertion and after insertion of the α-olefin, which, in the case of α-olefin homopolymerization, is always much higher. However, the molecular weights of HDPE obtained with different zirconocenes provide an indication of the molecular weight in ethylene copolymers. In fact, as already pointed out, rac-EBDMIZrCl$_2$, which gives HDPE with a higher molecular weight, but produces iPP with a much lower molecular weight than rac-EBIZrCl$_2$, produces also a higher molecular weight EPR and LLDPE than rac-EBIZrCl$_2$.

6 GENERAL CONSIDERATIONS

The most surprising evidence that emerges from the results of ethylene homo-polymerization with metallocenes and TIBAO reported in Tables 1–4 is the extreme sensitivity of the activity to the structure of the π-ligands and to the nature of the transition metal (hafnocenes are inactive and titanocenes show much lower activities than zirconocenes), and the impressive variation of catalytic activity observed for Cp$_2$ZrCl$_2$ by replacing MAO with TIBAO as cocatalyst. Indeed, Cp$_2$ZrCl$_2$ is almost inactive with TIBAO whereas it becomes one of the most active zirconocenes in the presence of MAO. By increasing alkyl substitution on the Cp ligands of the zirconocenes, the difference between the activities of the catalyst systems based on the two cocatalysts becomes less evident, and with some zirconocenes TIBAO gives a higher activity than MAO. Therefore, when studying the influence of π-ligands substituents on catalyst performance, in addition to polymerization temperature and Al:Zr ratio, one must also take into account the nature of the cocatalyst.

In order to try and explain the different behaviors of the above two cocatalysts, it seems useful to recall the role of the alumoxanes and their suggested mode of action. Spectroscopic evidence [42–46] suggests that MAO first alkylates the metallocene dihalide to generate the species Cp'_2MMeCl and Cp'_2MMe_2 and then, owing to its Lewis acidity, generally related to the presence of three coordinate Al atoms, abstracts the methide or halide, forming a catalytically active cationic metal center, stabilized by association with the weakly coordinating counteranion in more or less tight ion pairs:

$$[Cp'_2MMe]^+[MAOX]^- (X = Cl \text{ or } Me)$$

The highly electrophilic cationic species can also be stabilized by coordination with the nucleophilic species present in solution, such as Cp_2MMe_2 and $AlMe_3$, always present in MAO solutions, to give homodinuclear complexes such as $[(Cp'_2MMe)_2(\mu\text{-Me})]^+$ or heterodinuclear cations such as $[(Cp'_2M(\mu\text{-Me})_2 AlMe_2]^+$ [47,48]. The coordination of an oxygen atom from MAO to the metal center has been also suggested [8,49–51].

All the above complexes have been proposed [47] as temporarely inactive resting states of the catalyst in equilibrium with the active $[Cp'_2MMe]^+$ species; the formation of such complexes should also prevent bimolecular reduction of the catalyst. The stabilization of the cation through one of the above mechanisms has been considered essential, to obtain an active catalyst, for the unstable, highly electrophilic unshielded $[Cp_2ZrMe]^+$ cation [47], whereas it appears unnecessary for the more stable $[Me_2Si(Ind)_2ZrMe]^+$.

The influence of substituents on the Cp rings on the activity and stereospecificity of the related catalysts and on the molecular weight of the resulting polymers has been the object of many studies [52,53]. Electronic and steric effects have been considered, but the lack of kinetic data, the different reaction conditions used by the authors that hinder an accurate comparison of the results and the difficulty of singling out the steric from the electronic effects, do not allow sound conclusions to be drawn, not even about the influence on activity. Nevertheless, in general, electron-releasing substituents on Cp are reported [10,26–28,52] to enhance catalytic activity, at least until steric hindrance inhibits monomer coordination. The presence of electron-donating alkyl groups on the Cp ligands should reduce the positive charge on the metal and consequently facilitate the formation of the active cationic species (because the R^- or Cl^- donor strength of the metal is increased) and should also promote its stabilization. Furthermore, the increased steric hindrance of the π-ligand should reduce the ion pairing and prevent possible bimolecular reactions that can cause irreversible deactivation. All the above effects are expected to increase the catalytic activity. On the other hand, the reduced electrophilicity of the metal should decrease its coordination power with respect to the monomer and consequently cause a lower propagation rate [54]. From the results reported in Table 1, it is apparent that TIBAO is able to activate ring-substituted metallocene dihalides; therefore, assuming an activation mechanism analogous to that with MAO, we can infer that TIBAO is able to alkylate these zirconocenes and presents a Lewis acidity sufficiently high to abstract the R^- or Cl^- ions, thus generating the active cationic species. In order to explain the inadequacy of TIBAO to activate Cp_2ZrCl_2, we propose the following hypotheses:

(a) TIBAO is not sufficiently Lewis acidic to abstract the X^- ion from Cp_2ZrX_2 ($X = Cl$ or ^iBu). Only when the X^- donor strength of the zirconocene increases, by the effect of electron-releasing groups present on the Cp ring, does this reaction become possible. In accordance with this hypothesis, Barron and co-workers [16] reported that $[(^tBu)_2AlOAl(^tBu)_2]_2$, a dialumoxane with a dimeric structure analogous to that suggested for TIBAO, shows no reaction with $Cp_2ZrX_2(X = Cl, Me)$; steric hindrance to the close approach of the two reagents and to the formation of the alumoxane anion have been excluded. Furthermore, all hafnocenes tested, even those containing alkyl-substituted Cp ligands, are completely inactive in ethylene polymerization in the presence of TIBAO. Marks and co-workers, on the basis of thermodynamic and kinetic data assert that, for the reaction

$$(C_5H_3Me_2)_2MMe_2 + B(C_6F_5)_3 \longrightarrow [(C_5H_3Me_2)_2MMe]^+[MeB(C_6F_5)_3]^-$$

the stability of the ion pairs with respect to the constituent neutrals is much higher for Zr than Hf [55], probably owing to the different strengths of the respective M—Me bonds [56].

(b) The second hypothesis is related to the claimed [47] instability of the Cp_2ZrMe^+ cation if not stabilized by coordination with the alumoxane or $AlMe_3$. TIBAO and $Al(^iBu)_3$ could be too sterically hindered to coordinate with the cation. It is known that $Al(^iBu)_3$, unlike $AlMe_3$ and $AlEt_3$, is mainly monomeric in solution, and therefore it should be sterically hindered to form complexes such as $[Cp_2Zr(\mu-^iBu)_2Al(^iBu)_2]$. This hypothesis, however, cannot explain the inactivity of hafnocenes. We cannot, however, rule out the possibility of a too strong cation–anion pairing as the cause of the inactivity of the Cp_2ZrCl_2–TIBAO catalyst system. Indeed, a dramatic dependence of catalytic activity on ancillary ligand bulk, analogous to that observed when using TIBAO as cocatalyst, has been found in ethylene polymerization with the catalyst system $(L_2ZrMe)^+PBA^-$ [$PBA^- =$ tris(2,2′,2″-nonafluorobiphenyl)fluoroaluminate anion] [57]. This behavior has been correlated to the weakening of cation–anion pairing via the Zr—F—Al bridge by increasing ancillary ligand bulkiness.

We note that iPP samples made with rac-EBIZrCl₂ have the same identical microstructure whether the cocatalyst is MAO or TIBAO [58], although the catalyst activity is lower by two orders of magnitude for the latter cocatalyst; this finding is against the second hypothesis, as differences in ion pairing or cocatalyst coordination ability should in turn modify the microstructural details of iPP [59].

A comparison between the Lewis acidities of MAO and TIBAO and the investigation of the reaction between zirconocenes and TIBAO will hopefully shed light on this problem.

7 CONCLUSIONS

The $Al(^iBu)_3/H_2O$ cocatalysts represent a novel, simple alternative of industrial interest to methylalumoxane (MAO) in metallocene-catalyzed polyolefin production, owing to the much lower price and lower pyrophoricity of $Al(^iBu)_3$ and isobutyl-alumoxanes compared with $AlMe_3$ and MAO. In addition, aromatic solvents can be avoided when using the $Al(^iBu)_3/H_2O$ cocatalyst.

In the present investigation we have better defined the scope of these novel systems: ethylene homopolymerization with several zirconocene-based catalysts has been investigated in hexane in the absence of hydrogen to assess the correlation between catalyst structure and activity/molecular weight. With some of the most interesting zirconocenes, different cocatalysts and experimental conditions have been studied. In the field of HDPE catalysts, the two main results are:

(a) several zirconocene catalysts have been identified which show excellent poly-merization activity upon activation with $Al(^iBu)_3/H_2O$ cocatalysts;
(b) HDPEs with narrow molecular weight distributions (MWD $\approx 2-3$) can be produced with molecular weights ranging from medium-low ($[\eta] \approx 2$) to ultra-high ($[\eta] \approx 21$) by changing the zirconocene.

The $Al(^iBu)_3/H_2O$ cocatalyst represents the simplest solution to the problem of replacing MAO in metallocene-based catalysts. However, its chemical composition(s), controlled synthesis and the reaction products with the zirconocene catalysts have still to be unveiled. Hence a broader study aimed at clarifying these aspects could lead to better catalysts and is currently being pursued. Work in this direction has already produced improved systems and practical applications [60].

8 EXPERIMENTAL

8.1 CATALYSTS

Standard Schlenk-tube techniques for handling air- and moisture-sensitive compounds were employed. The purity of all metallocenes was confirmed by ^{1}H NMR ($CDCl_3$, 200 MHz, room temperature). Bis(1,2,3,4,5-pentamethylcyclo-pentadienyl)zirconium dichloride ($Cp_2^*ZrCl_2$) was purchased from Strem and used as received. Ind_2ZrCl_2 [61], $(MeCp)_2ZrCl_2$ [62], $(H_4Ind)_2ZrCl_2$ [62], rac-EBIZrCl$_2$ [63], rac-EBTHIZrCl$_2$ [64], $Me_2SiCp_2ZrCl_2$ [65], $Me_2Si(Me_4Cp)_2ZrCl_2$ [37a] and rac-$Me_2SiInd_2ZrCl_2$ [66] were prepared according to literature procedures. Cocatalysts: TIBAO was a standard Witco $Al(^iBu)_3/H_2O$ $(Al:H_2O = 2)$, 1.55 M (Al) in cyclohexane; MAO was methylalumoxane from Witco, dried *in vacuo* to a free-flowing powder (residual $AlMe_3 \approx 3-5mol\%$); polyisobutylalumoxane (IBAO) was prepared by hydrolysis of $Al(^iBu)_3$ and used as a solution in hexane, $Al:H_2O = 1$; $Al(^iBu)_3/H_2O$ was isobutylalumoxane prepared *in situ*, H_2O being added to the hexane solvent into the autoclave prior to addition of the catalyst/cocatalyst solution. Ethylene and hexane (residual water content 2 p.p.m.) were from the Montell pilot plants.

8.2 POLYMERIZATION PROCEDURE

8.2.1 Catalyst Preparation

The metallocene was weighed out in a 5 mm NMR tube under nitrogen (usually 3–10 mg), transferred into a 10 ml Schlenk tube and dissolved in 10 ml of toluene. The required amount of metallocene solution was added to the required amount of TIBAO in 10 ml of toluene and the resulting solution was aged for 5 min and then charged into an autoclave containing 0.4 l of purified hexane at 45 °C. In the case of $Al(^iBu)_3/H_2O$ mixtures, H_2O was dispersed into the hexane solvent prior to catalyst/$Al(^iBu)_3$ addition, and $Al(^iBu)_3$ was prereacted with the zirconocene. Alternatively, the required amount of $Al(^iBu)_3$ (*ca* 1.5 M solution in toluene) was added to a rapidly stirred suspension of H_2O in toluene ($Al:H_2O$ ratio from 2 to 10) at room temperature and, after 30 min of reaction, the required amount of metallo-cene solution was added to it and the resulting solution was aged for 5 min and then charged into the autoclave containing hexane/ethylene at 45 °C. When fluorenyl-containing metallocenes were used, they were dissolved directly (as a solid) in the aluminum alkyl solution.

8.2.2 Reactor Set-up and Polymerization (50 °C)

A Büchi 1 l glass reactor was chosen for its easy handling, high versatility and transparency. Prior to the polytest, the reactor was purified by washing with a dilute solution of $Al(^iBu)_3$ in hexane and dried in a nitrogen stream at 60 °C. Hexane (0.4 l) and ethylene (1 bar) were charged into the reactor, then the catalyst solution added to the autoclave through a PTFE tube with nitrogen pressure under ethylene flow, and the autoclave rapidly brought to the polymerization temperature (*ca* 1–5 min). After degassing the monomer, the polymer was isolated by filtration and dried *in vacuo* at 60 °C.

8.2.3 Reactor Set-up and Polymerization (80 °C)

A 2 l stainless-steel autoclave, thermostated with H_2O/steam and purified by purg-ing with ethylene at 80 °C, was used with 0.76 l *n*-hexane (0.75 l toluene) or 1.1 l

n-hexane, 9.6 bar-a ethylene, P_{tot} – 11 bar-a, at 80 °C for 1 h (n-hexane contains 2 p.p.m. water). The catalyst solution was added to the autoclave through a stainless-steel vial with ethylene overpressure, and the autoclave was rapidly brought to the polymerization temperature (ca 1–5 min). After degassing unreacted ethylene, the polymer was isolated by filtration and dried *in vacuo* at 60°C. Given Al:H$_2$O ratios do not include residual water in the solvent, except for TMA at Al/H$_2$O ≈ 40, where no water was added.

8.2.4 Reactor Set-up and 1-Butene Copolymerization (50 °C)

Copolymer tests were performed in a 4.5 l stainless-steel autoclave connected to a thermoregulating system and equipped with a helicoidal magnetic stirrer, monomer feeding lines, valve and a stainless-steel vial for injecting the catalyst/cocatalyst solution. The autoclave was purified with an ethylene stream at 80 °C. The desired amounts of water when the Al(^{i}Bu)$_3$/H$_2$O cocatalyst system was used, propane (when used), 1-butene, ethylene and hydrogen (see Table 6) were introduced into the autoclave at room temperature and then the temperature was raised to 5 °C below the polymerization temperature. Catalyst and aluminum alkyl [Al(^{i}Bu)$_3$ or TIBAO] were mixed in 10 ml of toluene, aged for 5 mins and added to the autoclave through the vial with an ethylene overpressure. The temperature was set to 50 °C and the total polymerization pressure was maintained constant by continuous feeding of an ethylene/1-butene mixture. Finally, after degassing the monomers (and propane), the polymer was isolated and dried under vacuum at 60 °C.

8.3 POLYMER ANALYSIS

The 1-butene content of each polymer sample was calculated from fourier transform IR spectra.

8.3.1 ^{13}C NMR Analysis

The spectrum was recorded at 130 °C on a Varian UNITY-300 spectrometer operating at 75.4 MHz in the Fourier transform mode, 6000 transients were accumulated with a 90° pulse and 12 delay period between pulses. The sample was dissolved in 1,1,2,2-tetrachloro-1,2-dideuteroethane to give an 8% (w/v) concentration. The assignments were made according to Hsieh and Randall [39].

The ethylene average sequence length n_e was calculated according to the equation:

$$n_e = (2[EE]/[EB]) + 1$$

The monomer reactivity ratio product, r_1, r_2 and $r_1 \times r_2$, were determined from the dyad distribution as described by Soga [40].

8.3.2 DSC Analysis

Calorimetric measurements were performed by using a Perkin-Elmer DSC-7 differential scanning calorimeter. The instrument was calibrated with indium and tin standards. Weighed samples (5–10 mg) were sealed into aluminum pans, heated to 180 °C and kept at that temperature for 4 min to allow complete melting of all the crystallites, to remove any influence of the previous thermal history. Successively, after slow cooling at 10 °C/min to 0 °C, the samples were heated to 180 °C at a rate of 10 °C/min. In the second heating run, the peak temperature was assumed as the melting temperature and the area as global melting enthalpy.

8.3.3 Intrinsic Viscosity

Intrinsic viscosity was measured in tetrahydronaphthalene at 135 °C, (ASTM D-1601/78).

8.3.4 Size-exclusion Chromatography (SEC)

The analyses were performed by using a Waters 150-C GPC system equipped with a TSK column set (type GM-HT$_{xl}$), working at 135 °C with 1,2-dichlorobenzene stabilized with 0.1 wt% of BHT. Universal calibration was performed by using the Mark–Houwink constants calculated from those of polyethylene following the method of Scholte [68].

8.3.5 Temperature Rising Elution Fractionation (TREF)

This was performed as reported [67].

8.3.6 Xylene Soluble at Room Temperature (XSRT)

This is the weight percentage of polymer soluble in xylene at 25 °C after complete dissolution at 135 °C.

8.3.7 Melt index (MI)

The melt index was measured at 190 °C following ASTM D-1238 by using a Göttfert grader, Model MP/E, over a load of 2.16 kg (MI E) and 21.6 kg (MI F). The melt index ratio is then defined as MIR = (MI F)/(MI E).

8.3.8 Density

The absolute density was measured using a density gradient column (ASTM D-1505) on a portion of polymer extruded from the melt index apparatus.

9 ACKNOWLEDGMENTS

We thank R. Mazzocchi, A. Coassolo, F. Piemontesi, D. Balboni, R. L. Jones and V. A. Dang for the syntheses of some of the ligands and the zirconocenes, G. Franciscono, M. Colonnesi and F. Guglielmi for the polymerization tests, I. Mingozzi for the solution polymer analysis, R. Zeigler for the ^{13}C NMR analysis and G. Balbontin and R. Agosti for DSC and SEC analyses.

10 REFERENCES

1. Natta, G., Pino, P., Mazzanti, G. and Giannini, U., *J. Am. Chem. Soc.*, **79**, 2975 (1957).
2. Breslow, D. S. and Newburg, N. R., *J. Am. Chem. Soc.*, **79**, 5072 (1957),
3. Reichert, K. H. and Meyer, K. R., *Makromol. Chem.*, **169**, 163 (1973).
4. Long, W. P. and Breslow, D. S., *Liebigs Ann. Chem.*, 463 (1975).
5. Andresen, A., Cordes, H. G., Herwig, J., Kaminsky, W., Merck, A., Mottweiler, R., Pein, J., Sinn, H., and Vollmer, H. J., *Angew. Chem., Int. Ed. Engl.*, **15**, 630 (1976).
6. Sinn, H., Kaminsky, W., and Vollmer, H. J., *Angew. Chem., Int. Ed. Engl.*, **19**, 396 (1980).
7. Ewen, J. A., and Welborn, H., *U. S. Pat.*, 5 324 800, to Exxon (1987); Kaminsky, W., in R. B. Seymour and T. Cheng (Eds) *History of Polyolefins*, 1986, p. 257; Gupta, V. K., Satish, S., and Bhardwaj, I. S., *J. Macromol. Sci. Rev. Macromol. Chem. Phys.*, **C34** 439 (1994), Brintzinger, H,-H., Fischer, D., Mulhaupt, R., Rieger, B., and Waymouth, R. M., *Angew. Chem., Int. Ed. Engl.*, **34**, 1143 (1995).
8. (a) Kaminsky, W., Miri, M., Sinn, H. J. and Woldt, R. *Makromol. Chem., Rapid Commun.*, **4**, 417 (1983); (b) Woldt, R. Dissertation, Hamburg (1982), Table 15, p. 64.
9. Reddy, S. S., *Polym. Bull.*, **36**, 317 (1996).
10. Möhring, P. C. and Coville, N. J., *J. Mol. Catal.*, **77**, 41 (1992).
11. Resconi, L., Giannini, U. and Albizzati, E., *US, Pat.*, 5 126 303, to Himont (1992). At the same time we observed comparable, relatively high activities when using simple AlR$_3$, as cocatalysts, with activities higher for substituted versus unsubstituted zirconocenes, and for Al(^{i}Bu)$_3$ versus AlEt$_3$ and AlMe$_3$. We reported our findings in a patent [12] and in two publications, [13,14]. Attempts to reproduce these experiments at later times gave capricious results. Finally, using the more reliable polymerization reactor systems available at the Centro Ricerche G. Natta, we unambiguously established that the activities observed with simple AlR$_3$. as cocatalysts were due to trace amounts of water contamination. L. R. apologizes to those who wasted their time trying to reproduce those results.
12. Resconi, L., Giannini, U. and Albizzati, E., *US, Pat.*, 5 049 535, to Himont (1992).
13. Resconi, L., Bossi, S. and Abis, L., *Macromolecules*, **23**, 4489 (1990).
14. Resconi, L., Giannini, U., Albizzati, E., Piemontesi, F. and Fiorani, T., *ACS Polym. Prepr.*, **32**, 463 (1991).
15. Lasserre, S. and Derouault, J., *Nouv. J. Chim.*, **7**, 659 (1983).
16. (a) Mason, M. R., Smith, J. M., Bott, S. G. and Barron, A. R. *J. Am. Chem. Soc.*, **115**, 4971 (1993); (b) Harlan, C. J., Mason, M. R. and Barron, A. R., *Organometallics*, **13**, 2957 (1994); (c) Harlan, C. J., Bott, S. G. and Barron, A. R., *J. Am. Chem. Soc.*, **117**, 6465 (1995), (d) Barron, A. R., *Organometallics*, **14**, 3581 (1995).
17. Tritto, I., Sacchi, M. C., Locatelli, P. and Li, S. X., *Macromol. Chem. Phys.*, **197**, 1537 (1996).
18. Uhl, W., Koch, M., Hiller, W. and Heckel, M., *Angew. Chem., Int. Ed. Engl.*, **34**, 989 (1995).

19. Pasynkiewicz. S., *Macromol. Symp.*, 971 (1995).
20. Boleslawski, M. and Servatowski, J., *J. Organomet. Chem.*, **254**, 159 (1983).
21. Heyema, N., Araki, T. and Tani, H., *Inorg. Chem.*, **12**, 2218 (1973).
22. Storr, A., Jones, K. and Laubengayer, A. W., *J. Am. Chem. Soc.*, **90**, 3173 (1968).
23. These activities were lower than that reported by Kaminsky (175 $kg_{PE} \, g_{Zr}^{-1} h^{-1} bar^{-1}$ at 80 °C). Comparison with the original data in Woldt's thesis[8b] showed that polymerization activities were actually in $kg_{PE} mol_{Zr}^{-1} h^{-1} bar^{-1}$ (toluene, 80 °C, 8 bar), hence the figure to be compared with our results is an estimated $15 kg_{PE} g_{Zr}^{-1} h^{-1}$, in the range of values found by us at 70 °C.
24. Aged TIBAO solution in cyclohexane from Witco, Al = 5.4%(w/w), 1.55 $mmolAl \, l^{-1}$, nominal Al-^{i}Bu$_3$: H$_2$O = 2, containing *ca* 13% free Al(^{i}Bu)$_3$ by ^{1}H NMR.
25. Kaminsky, W., Engehausen, R., Zoumis, K., Spaleck, W. and Rohrmann, J., *Makromol. Chem.*, **193**, 1643 (1992).
26. Miya, S., Harada, M., Mise, T. and Yamazaki, H., *Polym. Prepr., Jpn.*, **36**, 189 (1987).
27. Tait, P., Booth, B. and Jejelowo, M., *ACS Symp. Ser.*, **496**, 78 (1992).
28. Lee, I., Gauthier, W., Ball, J., Iyengar, B. and Collins, S., *Organometallics*, **11**, 2115 (1992).
29. Winter, A., Antberg, M., Dolle, V., Rohrmann, J. and Spaleck, W., *US Pat.*, 5 304 614, to Hoechst (1994).
30. Resconi, L., Piemontesi, F., Camurati, I., Balboni, D., Sironi, A., Moret, M., Rychlicki, H., and Zeigler, R., *Organometallics*, **15**, 5046 (1996).
31. Resconi, L., Galimberti, M., Piemontesi, F., Guglielmi, F. and Albizzati, E, *Eur. Pat. Appl.*, 575 875, to Himont (1993).
32. Spaleck, W., Küber, F., Winter, A., Rohrmann, J., Bachmann, B., Antberg, M., Dolle, V. and Paulus, E., *Organometallics*, **13**, 954 (1994).
33. Ewen, J., in T. Keii and K. Soga (Eds), *Catalytic Polymerization of Olefins*, Elsevier, Amsterdam, 1986, pp. 271–292. The systems used were Cp$_2$TiPh$_2$/Cp$_2^*$ZrMe$_2$/MAO (bimodal, MWD 5.5) and Cp$_2$TiPh$_2$/Cp$_2$ZrMe$_2$/MAO (bimodal, MWD 5.4–7.8).
34. Ahlers, A. and Kaminsky, W., *Makromol. Chem., Rapid Commun.*, **9**, 457 (1988). The systems used were Cp$_2$ZrCl$_2$/Cp$_2$HfCl$_2$/MAO, Cp$_2$ZrCl$_2$/*rac*-EBIZrCl$_2$/MAO, Ind$_2$ZrCl$_2$/*rac*-EBIZrCl$_2$/MAO and *rac*-EBIZrCl$_2$/Cp$_2$HfCl$_2$/MAO, with the last producing the best results (bimodal distributions with MWD up to 10).
35. Dall'Occo, T., Resconi, L., Balbontin, G. and Albizzati, E., *World Pat.*, WO 95/35333, to Montell (1994).
36. Piccolrovazzi, N., Pino, P., Consiglio, G., Sironi, A. and Moret, M., *Organometallics*, **9**, 3098 (1990).
37. (a) Jutzi, P. and Dickbreder, R., *Chem. Ber.*, **119**, 1750 (1986); (b) Fendrick, C. M., Mintz, E. A., Schertz, L. D., Marks, T. J. and Day, V. W. *Organometallics*, **3**, 819 (1984).
38. By 'super-random' we mean a copolymer (here, LLDPE) for which the 'Cluster Index' as defined by Hsieh and Randall [39] lower than 1, that is, the comonomer homosequences are present at concentrations lower than those of a statistical, or random, copolymer; Resconi, L., Dall'Occo, T., Piemontesi, F., Guglielmi, F. and Albizzati E., *Eur. Pat. Appl.*, 589 364, to Montell (1994).
39. Hsieh, E. T. and Randall, J. C., *Macromolecules*, **15**, 353 (1982).
40. Uozomi, T. and Soga, K., *Makromol. Chem.*, **193**, 823 (1992).
41. Hoel, E. L., *Eur. Pat. Appl.*, 347, 128, to Exxon (1989); Hoel, E. L. and Floyd, S., *Eur. Pat. Appl.*, 347 129, to Exxon (1989); Zambelli, A., Grassi, A., Galimberti, M., Mazzocchi, R. and Piemontesi, F., *Makromol. Chem., Rapid Commun.*, **123**, 523 (1991).
42. Gassman, P. G. and Callstron, M. R. *J. Am. Chem. Soc.*, **109**, 7875 (1987).
43. Shishta, C., Hathorn, R. N. and Marks, T. J., *J. Am. Chem. Soc.*, **114**, 1112 (1992).

44. Siedle, A. R., Newmark, R. A., Lamanna, W. M. and Schroepfer, J. N. *Polyhedron*, **9**, 301 (1990).
45. Tritto, I., Li, S. X., Sacchi, M. C., Locatelli, P. and Zannoni, G., *Macromolecules*, **28**, 5358 (1995).
46. Cam, D. and Giannini, U., *Makromol. Chem.*, **193**, 1049 (1992).
47. Bochmann, M. and Lancaster, S. J., *Angew. Chem., Int. Ed. Engl.*, **33**, 1634 (1994); Bochmann, M. and Lancaster, S. J., *J. Organomet. Chem.*, **497**, 55 (1995).
48. Haselwander, T., Beck, S. and Brintzinger, H.-H., in Fink, G., Mülhaupt, R. and Brintzinger, H.-H., (Eds), *Ziegler Catalysts*, Springer, Berlin, 1995, p.181.
49. Giannetti, E., Nicoletti, G. and Mazzocchi, R., *J. Polym. Sci. Polym. Chem.*, **23**, 2117 (1985).
50. Chien, J. C. W. and Wang, B.-P. *J. Polym. Sci. A, Polym. Chem.*, **27**, 1539 (1989).
51. Erker, G., Albrecht, M., Werner, S. and Kruger, C., *Z. Naturforsch., Teil B*, **45**, 1205 (1990).
52. Möhring, P. C. and Coville, N. J., *J. Organomet. Chem.*, **479**, 1 (1994).
53. Janiak, C., Versteeg, U., Lange, K. C. H., Weimann, R. and Hahn, E., *J. Organomet. Chem.*, **501**, 219 (1995).
54. Farina, M. and Puppi, C., *J. Mol. Catal*, **82**, 3 (1993).
55. Deck, P. A. and Marks, T. J., *J. Am. Chem. Soc.*, **117**, 6128 (1995).
56. Schock, L. E. and Marks, T. J., *J. Am. Chem. Soc.*, **110**, 7701 (1988).
57. Chen, Y. -X.; Stern, C. L. and Marks, T. J., *J. Am. Chem. Soc.*, **119**, 2582 (1997).
58. Both in terms of molecular weights and microstructure; Resconi, L., unpublished results.
59. Chien, J. C. W., Tsai, W.-M. and Rausch, M. D., *J. Am. Chem. Soc.*, **113**, 8570 (1991); Giardello, M. A., Eisen, M. S., Stern, C. L. and Marks, T. J., *J. Am. Chem. Soc.*, **115**, 3326 (1993); Giardello, M. A., Eisen, M. S., Stern, C. L. and Marks, T. J., *J. Am. Chem. Soc.*, **117**, 12114 (1995); Chen, Y. X., Stern, C. L. and Marks, T. J., *J. Am. Chem. Soc.*, **119**, 2582 (1997).
60. Dall'Occo, T., Galimberti, M., Resconi, L., Albizzati, E. and Pennini, G., *PCT Int. Appl.*, WO 96 02 580, to Montell (1996); Galli, P., Collina, G., Sgarzi, P., Baruzzi, G. and Marchetti, E., *J. Appl. Polym. Sci.*, **66**, 1831 (1997).
61. Samuel, E. and Setton, R. *J. Organomet. Chem.*, **4**, 156 (1965).
62. Samuel, E., *Bull. Soc. Chim. Fr.*, 3548 (1966).
63. Piemontesi, F., Camurati, I., Resconi, L., Balboni, D., Sironi, A., Moret, M., Zeigler, R. and Piccolrovazzi, N., *Organometallics*, **14**, 1256 (1995).
64. Wild, F., Wasiucionek, M., Huttner, G., and Brintzinger, H.-H., *J. Organomet. Chem.*, **288**, 63 (1985).
65. Bajgur, C., Tikkanen, W. and Petersen, J., *Inorg. Chem.*, **24**, 2539 (1985).
66. Hermann, W., Rohrmann, J., Herdtweck, E., Spaleck, W. and Winter, A., *Angew. Chem.*, Int. Ed. Engl., **28**, 1511 (1989).
67. Balbontin, G., Camurati, I., Dall'Occo, T., Finotti, A., Franzese, R. and Vecellio, G., *Angew. Makromol. Chem.*, **219**, 139 (1994).
68. Scholte, T. G., Meijerink, N. L. J., Schoffeleers, H. M. and Brands, A. M. G., *J. Appl. Polym. Sci.* **29**, 3763 (1984).

4

Influence of Metallocene Structures on Molecular and Supermolecular Architectures of Polyolefins

D. FISCHER†, S. JÜNGLING‡, M.J. SCHNEIDER‡, J. SUHM‡
AND R. MÜLHAUPT
Institut für Makromolekulare Chemie, Albert-Ludwigs Universität, Freiburg,
Germany
†Present address: Targor GmbH, Mainz, Germany
‡Present address: BASF AG, Ludwigshafen, Germany

1 INTRODUCTION

During the pioneering days of transition metal-catalyzed 1-olefin and 1,3-diene polymerization, it was well recognized that there exists a very close relationship between the structures of the catalytically active sites, polymer microstructures and polymer properties. Most early homogeneous and heterogeneous catalyst generations were composed of complex mixtures of active, inactive and temporarily inactive transition metal compounds. Therefore, catalysts were developed empirically in order to improve catalyst activities and to achieve high regio- and stereoselectivity. Especially in the case of ethene/1-olefin copolymerization, the presence of multi-site catalysts with various catalytically active sites of greatly different reactivities in copolymerization accounted for the formation of heterogeneous product mixtures, with higher 1-olefins being preferentially incorporated in wax-like low molecular weight fractions. Only industrial catalysts tailored for ethene solution polymerization above 220 °C and special, purely academic vanadium-based catalysts for living propene polymerization at very low temperatures qualified as 'single-site' catalysts

Metallocene-based Polyolefins   Edited by J. Scheirs and W. Kaminsky
© 2000 John Wiley & Sons Ltd

containing exclusively one type of active sites. The discovery of metallocene-based catalysts represents a major breakthrough in single-site catalyst technology.

Modern single-site metallocene catalysts produce very uniform homo- and copolymers with narrow molecular weight distribution (polydispersities M_w/M_n around 2), controlled stereo- and regioregularities and random incorporation of comonomers such as 1-olefins, cycloolefins and even styrene. Progress in polymerization mechanisms and opportunities in polymer synthesis were reviewed by Brintzinger *et al.* [1], Kaminsky [2] and Nakamura and co-workers [3]. Variation of the ligand substitution pattern of metallocene catalysts is playing an important role in the development of industrial catalysts. Correlations between polypropylene stereoregularity, e.g. iso- and syndiotacticity, and metallocene structures are well established. Models for steric control are based either on metallocene symmetry or, according to a recent universal model proposed by Fink, on lowest energy conformers of propene complexes taking into account the preferred position of the polymer chain [4]. Here we give an overview of our research on metallocene-catalyzed olefin polymerization by means of methylaluminoxane (MAO)-activated dimethylsilylene-bridged metallocenes with a bisindenyl ligand framework and 2-methyl substitution and benzannelation. Special emphasis is placed on the correlations between indenyl ligand substitution pattern, polymerization mechanisms and polyolefin molecular and supermolecular architectures.

2 PROPENE HOMOPOLYMERIZATION: THE ROLE OF DORMANT SITES

Most of the first isoselective metallocene catalysts, discovered during the mid-1980s, gave low molecular weights and poor stereoselectivity. The systematic variation of metallocene molecular architectures led the groups of Brintzinger [5] and Spaleck [6] independently to the development of 2-methyl-substituted silylene-bridged bisindenyl zirconocenes, which gave much higher molecular weights and stereoselectivities in propene polymerization. We have compared the influence of 2-methyl substitution and benzannelation of the indenyl ligand framework using MAO-activated Me$_2$Si[indenyl]$_2$ZrCl$_2$(I), Me$_2$Si[benz[e]indenyl]$_2$ZrCl$_2$(BI), Me$_2$Si[2-methylindenyl]$_2$ZrCl$_2$(MI) and Me$_2$Si[2-methylbenz[e]idenyl]$_2$ZrCl$_2$ (MBI). For comparison the non-bridged and non-stereoselective Cp$_2$ZrCl$_2$ (CP) and the non-stereoselective Me$_2$Si(CpMe$_4$)(N^{t}Bu)TiCl$_2$, known as 'constrained geometry' catalysts (CGC), were included for better comparison of the above-mentioned *ansa*-metallocenes with other catalyst families. The metallocene structures are displayed in Figure 1. Catalyst activities and polypropylene microstructure analyses of polypropylenes prepared by polymerizing propene in toluene at 40 °C and 2 bar pressure using homogeneous MAO-activated catalysts are summarized in Table 1. Both CP and CGC, in accord with observations by Waymouth and co-workers [7], produced mainly atactic polypropylene with slight syndiotactic enrichment in the cases of CGC and were not included in this comparison. From investigations reported by Brintzinger, Jüngling and co-workers [5,8–12] and Spaleck *et al* [6], it is obvious that benzannelation promotes high catalyst activities, whereas 2-methyl substitution affords substantially higher molecular weights at the expense of catalyst activity. Moreover, 2-methyl substitution gives higher stereoregularity, which is reflected by a higher content of the (mmmm) pentad in ^{13}C NMR spectroscopic microstructure analysis and a higher polypropylene melting temperature. Comparison of the polymerization rate–time plots, displayed in Figure 2, illustrates that 2-methyl substitution (MBI and MI, Figure 1) gives lower activity but also much less deactivation during polymerization, as evidenced by a constant polymerization rate for prolonged periods of time.

Catalyst activity, referred to moles of Zr, was increased by lowering the total Zr concentration and by increasing the MAO and propene concentrations. Although the catalyst activity was always referred to the propene feed concentration, higher

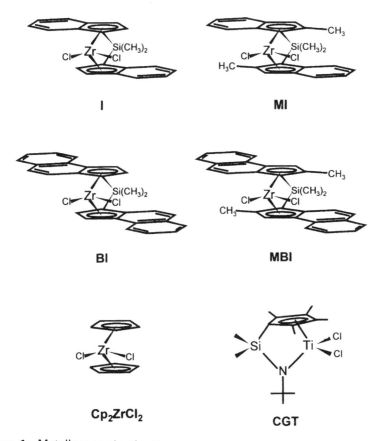

Figure 1   Metallocene structures

Table 1   Propene polymerization using homogeneous MAO-activated metallocenes[a]

Metallocene type	Catalytic[b] activity	$M_n{}^c$ (10^3 g/mol)	$M_w/M_n{}^c$	T_m (DSC) ($^\circ$C)	(mmmm)[d] (%)	2,1-Insertion[d] (%)	Bulk density[e] (g/ml)
I	24	47	1.7	146	92.2	0.5	0.22
BI	62	27	1.7	147	92.6	0.9	0.41
MI	6	122	1.7	154	94.7	0.2	0.08
MBI	31	127	1.7	155.4	96.1	0.4	0.10

[a] Data from Jüngling *et al.* [8]. Polymerization was performed in toluene at 40 $^\circ$C and 2 bar total pressure, [Zr] = 2 μmol/l; [Al] = 20 μmol/l, [propene] = 0.91 mol/l.
[b] Maximum activity was measured as 10^6 g PP/[(mol Zr) (mol/l propene) h].
[c] Number-average molecular weight and polydispersity M_w/M_n were determined by means of size-exclusion chromatography.
[d] (mmmm) pentads and head-to-head units resulting from 2,1-insertion were determined by means of ^{13}C NMR.
[e] Bulk density of poylpropylene powder.

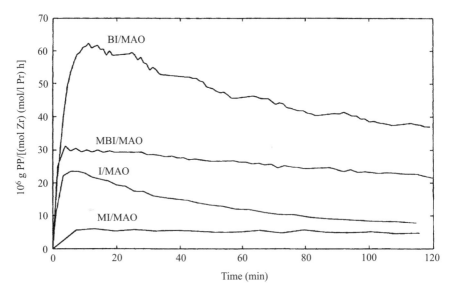

Figure 2   Propene polymerization using metallocenes with different indenyl ligand frameworks Polymerization was performed in toluene at 40 $^\circ$C and 2 bar total pressure using [Zr] = 2 μmol/l, [Al] = 20 mmol/l, [propene] = 0.91 mol/l

propene concentrations gave much larger activities, which decreased when the propene concentration was lowered. Propene concentration also affected molecular weight and stereoselectivity. At low propene pressures, both MB and BI systems (Figure 3, right) gave much lower stereoselectivity with respect to that at high propene concentrations. Most likely at low propene pressures and with a low

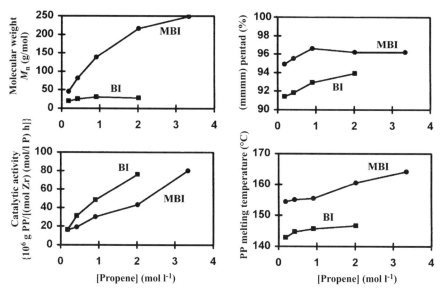

Figure 3 Influence of propene concentration on molecular weight (left) and isoselectivity, as determined by (mmmm) pentad content

polymerization rate, racemization can take place in the absence of monomer. Since propene pressure did not affect the molecular weights of polypropylenes prepared with BI systems, thus indicating that propene monomer was involved in both chain propagation and chain termination reactions, the 2-methyl substitution appeared to prevent chain termination via hydride transfer to the monomer. In liquid propene, very high molecular weights can be observed. Spaleck *et al.* [6] reported *ansa*-metallocene producing ultrahigh molecular weight polypropylene with molecular weight $>10^6$ g/mol when using 2-methyl-substituted metallocenes.

The comparison of various substitution patterns provides strong experimental evidence that metallocene-catalyzed polymerization involves numerous dynamic equilibria between active and dormant sites which are schematically displayed in Figure 4. In order to achieve high catalyst activity, it is important to increase the concentration of active cationic metallocene complexes at the expense of inactive or temporarily inactive ('dormant') sites. The presence of dormant sites was proposed originally by Reichert, Fink and others [1] when they studied elementary reactions in aluminum alkyl-activated polymerization with Cp_2TiCl_2. Most likely neutral and cationic complexes can dimerize to form dormant dimers. Recently, Brintzinger and co-workers [13] identified such binuclear zirconocene cations with μ-CH_3 bridges. Moreover, Fischer and Mülhaupt [14] observed reversible second-order deactivation processes to be important in propene polymerization kinetics. The reaction order of propene polymerization with MBI/MAO and BI/MAO was found to be 1.7, typical for the presence of simultaneous equilibria involving propene. Determination of the

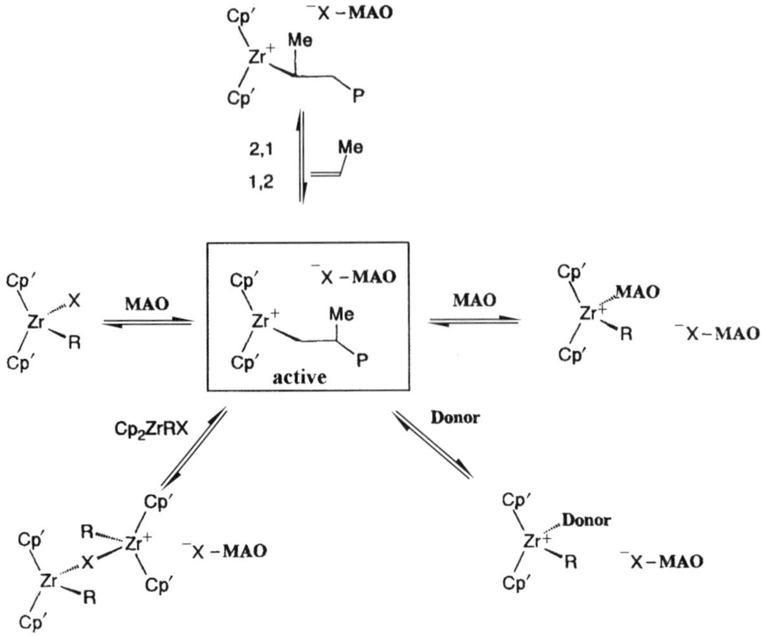

Figure 4 Equilibria between dormant and active sites during metallocene-catalyzed propene polymerization

number of active sites by Woo and co-workers [15], who used the CO inhibition method, indicated that CO was complexed to both active and inactive and dormant sites.

At low MAO concentration, MAO addition increases catalyst activity because MAO promotes the formation of cationic metallocene at the expense of neutral metallocene and also functions as a scavenger to remove catalyst poisons. At high MAO concentrations, MAO can complex at the active site to form dormant sites [10]. Such Lewis acid–Lewis base interactions are strongly dependent upon the Lewis acidity of the transition metal, which is reduced drastically when steric hindrance is introduced, e.g. by benzannelation of the bisindenyl ligand framework.

A dormant site is formed when 2,1-insertion takes place (Figure 5). The sterically hindered transition metal alkyl, resulting from 2,1-insertion, appears to be much less reactive with respect to the sterically much less hindered transition metal alkyls formed by 1,2-insertion. The presence of such dormant secondary alkyl groups can be monitored by means of hydrogen addition and also by ^{1}H-NMR spectroscopic end-group analysis [8,9]. Investigations of propene homopolymerization [8,9] and propene/1-octene copolymerization [16] shows the exclusive formation of vinyl-idene-terminated polymers in the case of the MBI/MAO catalyst, whereas the BI/MAO system gives considerable amounts of the 2-butenyl end group, which results from β-hydride elimination involving dormant sites. When 1,2-insertion was

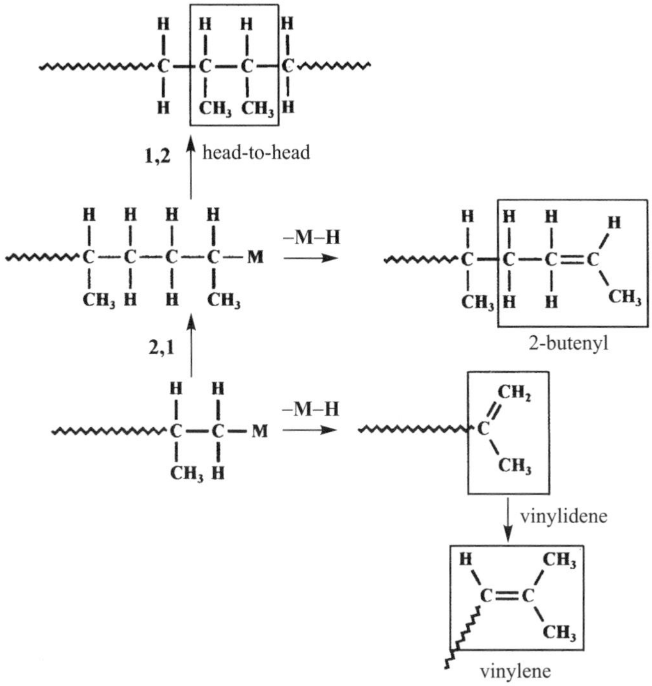

Figure 5 Olefin end groups formed via β-hydride elimination following 1,2- and 2,1-insertion

promoted with respect to 2,1-insertion, the regioselectivity increased and also the catalyst activity and molecular weight were increased. Since dormant sites are able to cause chain termination via β-hydride elimination, the formation of dormant sites is associated with low molecular weight. Most likely, in BI/MAO systems the dormant sites can be converted into active sites when transition metal alkyls are formed via hydride transfer to the propene monomer. With increasing propene pressure the content of 2-butenyl end groups increased [8,9]. The formation of 2-butenyl end groups and dormant sites was also observed by Resconi et al. [17] for ethylene-bridged bisindenyl metallocene catalysts.

3 COMPARISON OF HOMOGENEOUS AND HETEROGENEOUS METALLOCENES

Owing to very low bulk densities, homogeneous catalysts are not suitable for application in industrial slurry and gas-phase processes. Therefore, homogeneous catalysts are heterogenized on an SiO_2 support, which is pretreated with MAO, in order to allow immobilization without sacrificing the single-site nature. In Figure 6

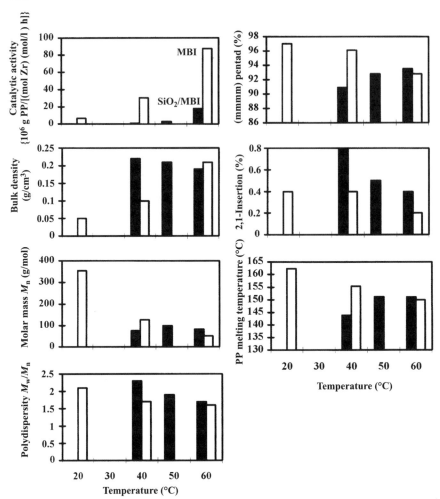

Figure 6 Comparison of homogeneous MBI/MAO/toluene and heterogeneous SiO$_2$/MAO/MBI/MAO/n-hexane catalyst systems in propene polymerization

the influence of polymerization temperature on catalyst activity, stereo- and regioregularity and polymer properties such as bulk density, molecular weight, molecular weight distribution and melting temperature are displayed for propene polymerization using MAO-activated homogeneous MBI in toluene and heterogeneous SiO$_2$/MAO/MBI in n-hexane. Although heterogeneous catalysts are less active than homogeneous catalysts, they afford much higher bulk densities. However, polymer properties at elevated temperature are very similar. At a lower temperature of 20 °C, heterogeneous catalysts give much lower stereo- and regioregularity. This was attributed to severe diffusion limitation accounting for very low local propene concentration in the catalyst particles, thus reducing steroregularity and increasing chain termination with respect to chain propagation in accord with the above-mentioned observations in propene homopolymerization.

At low temperature, the morphology of the polypropylene resembles that of the catalyst particles, indicating the template effect taking place during fragmentation of the catalyst particles. At elevated temperatures the morphology changes and is very similar to that of polypropylene produced with homogeneous catalysts. Most likely leaching of the MAO/MBI catalyst, which is removed from the support, could account for this behavior [11].

The key role of catalyst particle fragmentation and propene diffusion became evident when l-octene was added together with propene during polymerization on heterogeneous catalysts. The influence of l-octene concentration is displayed in Figure 7. Significant increases in both catalyst activity and stereoselectivity were

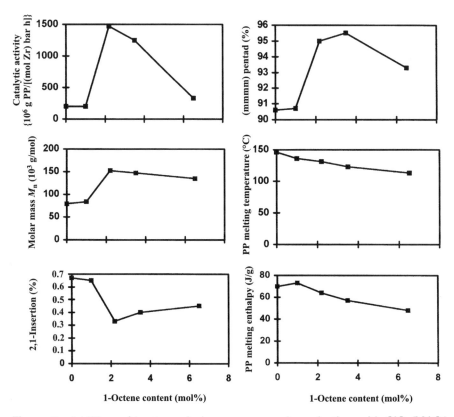

Figure 7 Addition of l-octene during propene polymerization with SiO_2/MAO/MBI/MAO/n-hexane

observed when small amounts of l-octene were added. Most likely, traces of l-octene are sufficient to influence catalyst particle deagglomeration in the initial phase of the polymerization and to prevent the formation of a dense shell which hampers propene diffusion. This observation is in accord with a model proposed by Fink and co-workers [18], who also assumed that encapsulation of the catalyst in a dense shell may account for peculiarities in propene polymerization kinetics observed especially in the early stages of the polymerization.

4 METALLOCENE-CATALYZED OLEFIN COPOLYMERIZATION

The influence of indenyl ligand substitution patterns was investigated in propene/ l-olefin[16] and ethene/l-olefin [19–21] copolymerization. The performance of homogeneous catalysts in ethene/l-octene copolymerization is compared in Figure 8. It is apparent that CGC gives the highest l-octene incorporation. In the family of the 2-methyl-substituted and benzannelated metallocenes, benzannelation promotes l-olefin incorporation, whereas 2-methyl substitution affords copolymers with high molecular weight. Force field calculations were applied successfully by Prosenc and co-workers to model the effect of benzannelation in ethene/l-octene copolymeriza-tion [19]. As a function of l-octene concentration and polymerization temperature, it was possible to vary the copolymer molecular weight and composition independently over a very wide range [21]. In accord with earlier observations by Fink and co-workers [22], long l-olefins can also be incorporated to produce polyethylene with long alkyl branches. Only the CGC catalyst is capable to incorporate large amounts of styrene comonomers to produce tercopolymers with attractive property

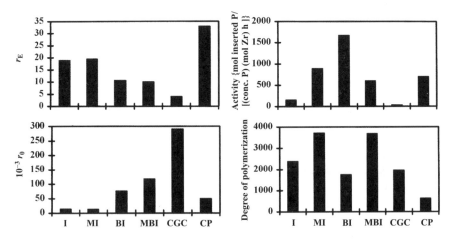

Figure 8 Copolymerization of ethene with l-octene using different MAO-acti-vated metallocene catalysts

ranges [23]. In contrast to many conventional catalysts, highly stereoregular copolymers, e.g. syndiotactic poly(propene-co-1-octene), can be obtained [24].

5 STERIC IRREGULARITIES AND CONTROLLED POLYOLEFIN CRYSTALLIZATION

Stereoselective olefin homo- and copolymerization and ethene copolymerization by means of single-site metallocene catalysts offer attractive opportunities in polymer synthesis. Short- and long-chain branches and also stereo- and regio-irregularities can be placed randomly in polymer backbones. This is of particular interest for controlling polymer crystallization, e.g. of polyethylene and polypropylene. In contrast to conventional Ziegler catalysts, metallocene catalysts produce ethene/ 1-olefin copolymers covering the entire feasible copolymer composition range without sacrificing the narrow molecular weight distribution. With increasing 1-olefin content it is possible to lower the melting temperature and crystallinity. According to Minick et al. [25] and Suhm et al. [26], polyethylene crystallization at comonomer contents below 10 wt% involves chain folding, whereas above 20 wt% fringed micelles are formed which were imaged by means of atomic force microscopy (AFM)[26].

Isotactic polypropylene can crystallize to form different crystal modifications (α, β, γ and smectic). The α-form is considered to be the preferred modification formed during crystallization of isotactic polypropylenes which were prepared with conventional catalyst systems, although calculations indicate that the γ-form is more stable. Brückner and co-workers [27–29] and Lotz and co-workers [31,32] have used oligopropenes to determine the crystal structure of the γ-modification where the orthorhombic unit cell is composed of bilayers of two parallel helices (Figure 9). The direction of the chain axis in adjacent bilayers is tilted at an angle of 80°. This represents a unique packing arrangement for polymers and was previously known only for fatty acids.

With metallocene catalysis it became possible to produce high molecular weight isotactic polypropylene which can crystallize to form the γ-form exclusively. It was Fischer and Mülhaupt [33] who discovered a linear correlation between the content of the γ-form and the average isotactic segment length between two steric irregularities such as 2,1- and 1,3-insertion or isolated stereoirregular insertion, as reflected by (mrrm) pentads in the ^{13}C NMR spectra. As shown in Figure 10, this average segment length n_{iso} also controls the melting temperature of polypropylene, which can be varied between 120 and 165 °C.

Kressler and co-workers [34–36] used metallocene-based polymers to study the crystallization behavior of polypropylene and succeeded in crystallizing 100% of the γ-form. Since the γ-form does not form spherulites, the corresponding polypropylenes exhibit much improved optical clarity with respect to that of conventional isotactic polypropylene.

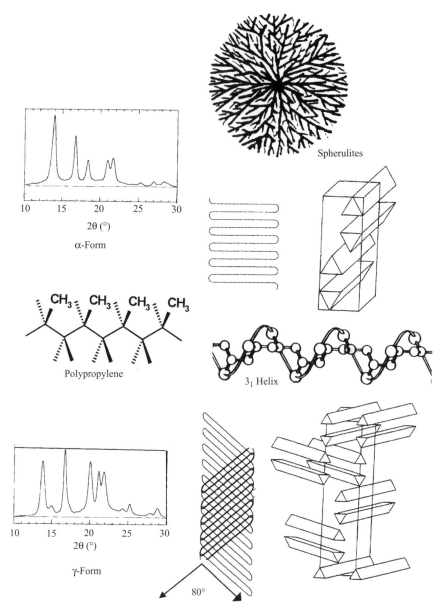

Figure 9 Crystallization of polypropylene in α- and γ-modifications

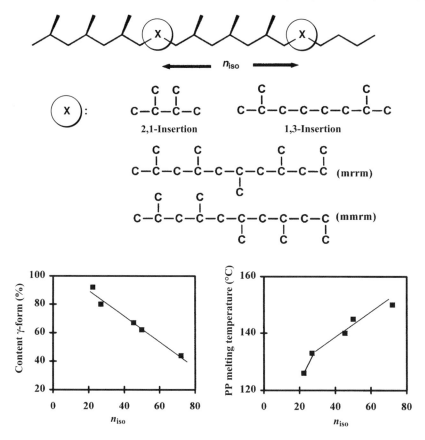

Figure 10 Influence of the isotactic segment length on the formation of the γ-modification (data taken from Fischer and Mülhaupt [33]).

A controlled polypropylene molecular architecture is the key to controlled polymer properties such as stiffness, strength, optical clarity, permeability and processing. With metallocene-based polypropylene a large variety of stereoisomer blends were formed, exploiting the phase separation of such stereoisomers in bulk and melt [37]. Also, the processing properties such as melt flow are affected by the molecular architectures of polypropylene and ethylene copolymers prepared with metallocenes [38,39]. Owing to the very uniform nature of metallocene-based homo- and copolymers it became possible to establish basic correlations between molecular and supermolecular polyolefin architectures, rheology and mechanical, thermal and optical properties. This basic knowledge, combined with a highly effective and environmentally friendly process, is the key to a large variety of novel polyolefin materials which represent a solid modification of oil and are readily recycled.

6 ACKNOWLEDGMENTS

The authors thank the Bundesminister für Bildung und Forschung (BMBF) for supporting their research as part of the BMBF projects Nos 03M40719 and 03N10280. They also thank BASF AG for their support.

7 REFERENCES

1. Brintzinger, H. H., Fischer, D., Mülhaupt, R., Rieger, B. and Waymouth, R. M., *Angew. Chem., Int. Ed Engl.*, **34**, 1143 (1995).
2. Kaminsky, W., *Macromol. Chem. Phys.*, **197**, 3907 (1996).
3. Mashima, K., Nakayama, Y. and Nakamura, A., *Adv. Polym. Sci.*, **133**, 1 (1997).
4. Vanderleck, Y., Angermund, K., Reffke, M. and Kleinschmidt, R., *Chem. Eur. J.*, **34**, 585 (1997).
5. Stehling, U., Diebold, J., Kirsten, R., Röll, W., Brintzinger, H. H., Jüngling, S., Mülhaupt, R. and Langhauser, F., *Organometallics*, **13**, 964 (1994).
6. Spaleck, W., Antberg, M., Aulbach, M., Bachmann, B., Dolle, V., Haftka, S., Küber, F., Rohrmann, J. and Winter, A., In Fink, G., Mülhaupt, R. and Brintzinger, H. H. (Eds), *Ziegler Catalysts*, Springer, Berlin, 1995, pp. 83–97.
7. McKnight, A. L., Masood, M. A. and Waymouth, R. M., *Organometallics*, **16**, 2879 (1997).
8. Jüngling, S., Mülhaupt, R., Stehling, U., Brintzinger, H. H., Fischer, D. and Langhauser, F., *Macromol. Symp.*, **97**, 205 (1995).
9. Jüngling, S., Mülhaupt, R., Stehling, U., Brintzinger, H. H., Fischer, D. and Langhauser, F., *J. Polym. Sci. Part A: Polym. Chem.*, **33**, 1305 (1995).
10. Jüngling, S. and Mülhaupt, R., *J. Organomet. Chem.*, **497**, 27 (1995).
11. Jüngling, S., Koltzenburg, S. and Mülhaupt, R., *J. Polym. Sci. Part A: Polym. Chem.*, **35**, 1 (1997).
12. Mülhaupt, R., Fischer, D. and Jüngling, S., *Macromol. Symp.*, **66**, 191 (1993).
13. Beck, S., Prosenc, M. H., Brintzinger, H. H., Goretzki, R., Herfert, N. and Fink, G., *J. Mol. Cat. A-Chemical*, **111**, 67 (1996).
14. Fischer, D. and Mülhaupt, R., *J. Organomet. Chem.*, **417**, C7 (1991).
15. Han, T. K. H., Ko, Y. S., Park, J. W. and Woo, S. I., *Macromolecules*, **29**, 7305 (1996).
16. Schneider, M. J. and Mülhaupt, R., *Macromol. Chem. Phys.*, **1985**, 1121 (1997).
17. Resconi, L., Fait, A., Piemontesi, F., Colonnesi, O., Rychlicki, H. and Zeigler, R., *Macromolecules*, **28**, 6667 (1995).
18. Steinmetz, B., Tesche, B., Przybyla, C., Zechlin, J. and Fink, G., *Acta Polym.*, **48**, 392 (1997).
19. Schneider, M. J., Suhm, J., Mülhaupt, R., Prosenc, M. H. and Brintzinger, H. H., *Macromolecules*, **30**, 3164 (1997).
20. Suhm, J., Schneider, M. J. and Mülhaupt, R., *J. Mol. Catal.*, in press (1998).
21. Suhm, J., Schneider, M. J. and Mülhaupt, R., *J. Polym. Sci. Part A: Polym. Chem.*, **35**, 735 (1997).
22. Koivumaki, J., Fink, G. and Seppälä, J. V., *Macromolecules*, **27**, 6254 (1994).
23. Sernetz, F. G. and Mülhaupt, R., *J. Polym. Sci. Part A: Polym. Chem.*, **35**, 2549 (1997).
24. Jüngling, S., Mülhaupt, R., Fischer, D. and Langhauser, F., *Angew. Makromol. Chem.*, **229**, 93 (1995).

25. Minick, J., Moet, A., Hiltner, A., Baer, E. and Chum, S. P., *J. Appl. Polym. Sci.*, **58**, 1371 (1995).
26. Suhm, J., Heinemann, J., Thomann, Y., Thomann, R., Maier, R. D., Schleis, T., Okuda, J., Kressler, J. and Mülhaupt, R. *J. Mater. Chem.*, **84**, 553 (1998).
27. Meille, S. V., Brückner, S. and Porzio, W., *Macromolecules*, **23**, 4114 (1990).
28. Brückner, S. and Meille, S. V., *Nature (London)*, **340**, 455 (1989).
29. Brückner, S., Meille, S. V., Sozzani, P. and Torri, G., *Macromol. Chem. Phys.*, **11**, 55 (1990).
30. Marigo, A., Marega, C., Zanetti, R., Paganetto, G., Canossa, E., Coletta, F. and Gottardi, F., *Macromol. Chem. Phys.*, **190**, 2805 (1989).
31. Lotz, B., Wittmann, J. C. and Lovinger, A. J., *Polymer*, **37**, 4979 (1996).
32. Lotz. B., Graff. S., Straupé, O. and Wittmann, J. C., *Polymer*, **32**, 2902 (1991).
33. Fischer, D. and Mülhaupt, R., *Macromol. Chem. Phys.*, **195**, 1433 (1994).
34. Thomann, R., Wang, C., Kressler, J. and Mülhaupt, R., *Macromolecules*, **29**, 8425 (1996).
35. Thomann, R., Kressler, J., Setz, S., Wang, C. and Mülhaupt, R., *Polymer*, **13**, 2627 (1996).
36. Thomann, R, Kressler, J, Rudolf, B. and Mülhaupt, R., *Polymer*, **37**, 2635 (1996).
37. Maier, R. D., Thomann, R, Kressler, J, Mülhaupt, R. and Rudolf, B., *J. Polym. Sci. Part B: Polym. Phys*, **35**, 1135 (1997).
38. Eckstein, A., Friedrich, C., Lobbrecht, A., Spitz, R. and Mülhaupt, R., *Acta Polym.*, **48**, 41 (1997).
39. Eckstein, A., Suhm, J., Freidrich, C., Maier, R. D., Sassmannshausen, J., Bochmann, M. and Mülhaupt, R., *Macromolecules*, **31**, 1335 (1998).

5

Olefin Polymerization by Monocyclopentadienyl Compounds of Titanium, Zirconium and Hafnium

SEAN W. EWART AND MICHAEL C. BAIRD
Queen's University, Kingston, Ontario, Canada

1 INTRODUCTION

Considerable research has gone into the study and development of dicyclopenta-dienyl (metallocene) complexes of the Group 4 metals (Ti, Zr, Hf) and the development of their uses as olefin polymerization initiators (see elsewhere in this volume). On the other hand, only much more recently has the development and utilization of monocyclopentadienyl compounds of these metals been examined. Typically the Group 4 monocyclopentadienyl compounds normally used are of the general formula $CpMR_3$($Cp = \eta^5\text{-}C_5H_5$, R = alkyl group), which assume so-called 'piano stool' structures **1**.

Compounds of this type, and also derivatives in which one or more alkyl groups are replaced by other anionic ligands, contain a formally 12-electron metal center and are therefore electronically less saturated as well as sterically less hindered than are their 16-electron, metallocene counterparts. For these reasons, catalysts based on compounds of type **1** would be expected to result in higher catalytic activities for olefin polymerization. In fact, the actual active species in most olefin polymeriza-tions by this type of compound has been shown to be the formally 10-electron, cationic species $[CpMR_2]^+$ (**2**; R = alkyl, anionic ligands). These are prepared in a variety of ways, as outlined below. Complexes of type **2** have been used to polymerize a wide variety of olefins via both Ziegler–Natta and carbocationic

Metallocene-based Polyolefins   Edited by J. Scheirs and W. Kaminsky
© 2000 John Wiley & Sons Ltd

mechanisms, and thus exhibit capabilities as yet not recognized in metallocene chemistry. In accord with the above-mentioned steric and electronic factors, catalysts of type **2** do indeed exhibit very high catalytic activities.

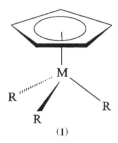

(1)

2 SYNTHESES OF CATALYST PRECURSORS

Syntheses of Group 4 monocyclopentadienyl compounds began in the 1950s with reports of $CpTiCl_3$ and $CpTiCl_2Br$[1]. Since much of the early synthetic work has been reviewed elsewhere [2], however, we shall discuss only the syntheses of compounds which have been successfully used as precursors for the polymerization of olefins. The most commonly used precursors are the compounds $CpTiCl_3$ and $Cp*TiCl_3$($Cp* = \eta^5\text{-}C_5Me_5$), which were originally synthesized via reactions of alkali metal salts of the cyclopentadienyl or pentamethylcyclopentadienyl anions with titanium tetrachloride [3]. However, this method gave low yields and often resulted in the reduction of the Ti(IV) to Ti(III). An improved methodology utilizes trimethylsilyl-substituted cyclopentadienyl compounds, which are less reducing than alkali metal salts [4]:

$$C_5H_6 + KH \longrightarrow KCp + \tfrac{1}{2}H_2 \qquad (1)$$
$$KCp + Me_3SiCl \longrightarrow CpSiMe_3 + KCl \qquad (2)$$
$$CpSiMe_3 + TiCl_4 \longrightarrow CpTiCl_3 + Me_3SiCl \qquad (3)$$

This approach can be used to synthesize a wide range of cyclopentadienyl and substituted cyclopentadienyl titanium trihalides in excellent yields. By using indenyl rather than cyclopentadienyl precursors, this method may also be used to synthesize many indenyl and substituted indenyl [5,6] titanium trichlorides. The trihalides are used as precursors for catalysis or can be further modified by substituting the halide ligands to make a wide variety of different precursors. They may be reacted with appropriate amounts of methyl lithium to make $Cp'TiMe_3$[7] ($Cp' =$ substituted cyclopentadienyl ring) or the mixed halo alkyl compounds $Cp'TiMe_2Cl$[8] and $Cp'TiMeCl_2$[8]. They can also be reacted with the Grignard reagent $ClMgCH_2Ph$ to make the benzyl complexes $Cp'Ti(CH_2Ph)_3$[7].

Syntheses of phenoxy and alkoxy complexes are rather more complicated, but two methods have been developed [9]. One involves the reaction of a trichloride with the appropriate alcohol, with triethylamine added to remove the HCl produced:

$$Cp'TiCl_3 + ROH \longrightarrow Cp'TiCl_2OR \qquad (4)$$

An equally useful procedure involves direct substitution of $Cp'TiCl_3$ by alkoxide ion:

$$NaOR + Cp'TiCl_3 \longrightarrow CpTiCl_2OR \qquad (5)$$

A wide variety of phenoxy and alkoxy compounds has been made utilizing these methods [9], while mixed Cp* titanium perfluorophenyl/methyl and perfluoro-phenoxy/methyl compounds have also been synthesized by reacting the appropriate mixed methyl/chloro compound with a perfluorophenyl or perfluorophenoxy lithium salt [10].

There has been less work on the development of useful zirconium and hafnium compounds. $CpMCl_3$ and $Cp*MCl_3$ (M = Zr, Hf) are usually synthesized by direct reaction of LiCp or LiCp* with MCl_4[11]. Because zirconium (IV) and hafnium (IV) compounds are much less readily reduced than are the corresponding titanium (IV) analogs, there is no need to use the trimethylsilyl-substituted cyclopentadienyls. The compound $CpZrCl_3$ has also been prepared via treatment of the metallocene Cp_2ZrCl_2 with Cl_2 gas [12]. Alkylation of $Cp'MCl_3$ compounds is readily achieved by utilizing Grignard reagents [11].

3 ACTIVATION OF PRECURSORS

The compounds discussed in the previous section do not generally exhibit catalytic activity, and must be 'activated' in order to form an olefin polymerization initiator. This is generally accomplished by reacting the precursor with a Lewis acidic cocatalyst to generate species of the type $[Cp'TiR_2]^+$ (**2**). A commonly used cocatalyst is methyl aluminoxane (MAO), an incompletely characterized oligomeric mixture of the general formula $(-AlMeO-)_n$. MAO reacts with a number of halo- and alkoxytitanium compounds to give active catalytic systems, and is believed in effect to replace all the ligands with the exception of the cyclopentadienyl ligand with methyl groups and then to extract one of the methyls to produce the active species, *i.e.*

$$Cp'TiCl_3 + MAO \longrightarrow [Cp'TiMe_2]^+ \qquad (6)$$

To be effective, the MAO must be used in a large excess and the presumed cationic products have not, in fact, been isolated or characterized. Hence, although much work has been done with MAO, the actual mechanism of activation still lacks solid evidence.

In order to characterize better the presumed cationic active species, there have been developed a number of stoichiometric reactions of precursors of type **1** with Lewis acids such as $B(C_6F_5)_3$ and $[Ph_3C]^+[B(C_6F_5)_4]^-$ and, to a lesser extent, with Brønsted acids such as $[NMe_2PhH]^+[B(C_6F_5)_4]^-$:

$$Cp'TiR_3 + B(C_6F_5)_3 \longrightarrow [Cp'TiR_2]^+ + RB(C_6F_5)_3^- \qquad (7)$$

$$Cp'TiR_3 + Ph_3C^+ \longrightarrow [Cp'TiR_2]^+ + CMePh_3 \qquad (8)$$

$$Cp'TiR_3 + HNMe_2Ph^+ \longrightarrow [Cp'TiR_2]^+ + RH + NMe_2Ph \qquad (9)$$

The most commonly used of these cocatalysts is $B(C_6F_5)_3$, which reacts in analogous fashion with dialkylzirconocene compounds [13]. Cp^*TiMe_3 has been shown to react with $B(C_6F_5)_3$, releasing a methyl carbanion to give the bridged methyl borate species $Cp^*TiMe_2(\mu\text{-}Me)B(C_6F_5)_3$ (**3**) [14]. This methyl-bridged compound is unstable in CD_2Cl_2 at temperatures above about $-10\,^\circ C$, but has been fully characterized by ^1H and ^{13}C NMR spectroscopy at lower temperatures. Double irradiation experiments show that the bridged species **3** is in equilibrium with the solvent-separated ion pairs $[MeB(C_6F_5)_3]^-$ and $[Cp^*TiMe_2]^+$, resulting in the latter 10-electron species existing in solution. However, the equilibrium [equation (10)] lies well to the left [14].

$$Cp^*TiMe_3(\mu\text{-}Me)B(C_6F_5)_3 \rightleftharpoons [Cp^*TiMe_2]^+ + [MeB(C_6F_5)_3]^- \qquad (10)$$

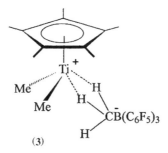

(3)

Similar reactions are observed with the precursors $Cp^*TiMe_2OC_6F_5$, $Cp^*TiMe_2C_6F_5$ and Cp^*TiMe_2Cl[10,15], all three of which react with $B(C_6F_5)_3$ to give the methyl-bridged species $Cp^*TiMeR(\mu\text{-}Me)B(C_6F_5)_3$($R = Cl$, C_6F_5, OC_6F_5). The product with the OC_6F_5 ligand is found to exist in equilibrium with the corresponding solvent-separated ion pairs in a manner similar to $Cp^*TiMe_2(\mu\text{-}Me)B(C_6F_5)_3$. However, double-irradiation experiments with the bridged species formed in the reactions of $Cp^*TiMe_2C_6F_5$ and Cp^*TiMe_2Cl with $B(C_6F_5)_3$ suggest that any exchange between the bridged species and the separated ions is very slow. Hence these species, which would probably have the most electron-deficient titanium centers because of the electron-withdrawing abilities of the C_6F_5 and Cl ligands, apparently form much tighter ion pairs.

The alkyl abstraction reactions are complicated by observations that Cp^*TiMe_3 and $Cp^*TiMe_2OC_6F_5$ react with a deficiency of $B(C_6F_5)_3$ to form methyl-bridged cationic dititanium species $[(Cp^*TiMeR)_2(\mu\text{-}Me)]^+$ [14,15]. These complexes probably arise from the initial formation of the expected species $[Cp^*TiMeR]^+$, followed by reaction with another molecule of neutral species:

$$[Cp^*TiMeR]^+ + Cp^*TiMe_2R \longrightarrow [(Cp^*TiMeR)_2(\mu\text{-}Me)]^+ \qquad (11)$$

Thus, surprisingly, the neutral Cp^*TiMe_2R appear to have a higher affinity for the cationic species $[Cp^*TiMeR]^+$ than does the borate anion $[MeB(C_6F_5)_3]^-$. However, methyl-bridged species do not form with $Cp^*TiMe_2C_6F_5$ or Cp^*TiMe_2Cl when they are reacted with a deficiency of borane; reaction of these with 0.5 equiv. of $B(C_6F_5)_3$ results in the formation of equimolar amounts of $Cp^*TiMeR(\mu\text{-}Me)B(C_6F_5)_3(R = Cl, C_6F_5)$ and unreacted $Cp^*TiMe_2C_6F_5$ or Cp^*TiMe_2Cl. It appears that, in these two cases, the cations $[Cp^*TiMeR]^+$ are not released from the methyl borate counterion and are therefore unable to react with the neutral species in solution.

Also of some interest are the reactions of $Cp^*TiMe_2(\mu\text{-}Me)B(C_6F_5)_3$ with a variety of aromatic molecules such as toluene or mesitylene to form η^6-arene complexes $[Cp^*TiMe_2(\eta^6\text{-}arene)]^+[MeB(C_6F_5)_3]^-$ **(4)** [16,17]. Equilibria are observed to exist between the neutral bridged borate complex and the cationic arene complexes, which can thus behave as precursors for the 10-electron species $[Cp^*TiMe_2]^+$:

$$Cp^*TiMe_2(\mu\text{-}Me)B(C_6F_5)_3 + arene \rightleftharpoons$$
$$[Cp^*TiMe_2(\eta^6\text{-}arene)]^+[MeB(C_6F_5)_3]^- \qquad (12)$$

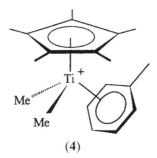

(4)

Similar arene compounds are not formed with $Cp^*TiMe_2C_6F_5$ or $Cp^*TiMe_2OC_6F_5$, presumably because of steric factors. The zirconium and hafnium complexes Cp^*ZrMe_3 and Cp^*HfMe_3, however, do react with $B(C_6F_5)_3$ and a wider variety of arenes to form η^6-arene coordinated complexes [17,18]. In many cases these complexes may be isolated as stable solids and are of some interest as they then represent single component catalysts capable of initiating polymerizations. The complexes $[Cp^*HfMe_2(\eta^6\text{-}toluene)]^+[MeB(C_6F_5)_3]^-$ and $[Cp''HfMe_2(\eta^6\text{-}toluene)]^+$

$[MeB(C_6F_5)_3]^- [Cp'' = C_5H_3(SiMe_3)_2]$ have been isolated and crystal structures confirm the expected η^6-coordination of the arenes, as in **4** [17,18].

Interestingly, reaction of the benzyl complex $CpZr(CH_2Ph)_3$ with $B(C_6F_5)_3$ results in formation of the complex $[CpZr(CH_2Ph)_2]^+[B(\eta^6\text{-}PhCH_2)(C_6F_5)_3]^-$ [19]. Thus a benzyl group is extracted by the borane as are the methyl groups discussed above, but the cation–anion interaction involves η^6-arene coordination of the BCH_2Ph group to the zirconium, much as in **4**. $Cp^*Zr(CH_2Ph)_3$ reacts with borane in a similar manner to produce the complex $[Cp^*Zr(CH_2Ph)_2]^+$ $[BCH_2Ph(C_6F_5)_3]^-$, but in this case the electronic unsaturation of the zirconium ion is relieved by strong π coordination of both benzyl ligands, one in an η^3 manner and one in very unusual η^7 manner [20]. Both π benzyl complexes are fluxional in solution, exchange of coordinated and non-coordinated benzyl groups being observed. Interestingly, the compound $[Cp^*Zr(CH_2Ph)_2]^+[BCH_2Ph(C_6F_5)_3]^-$ has been found to react with propylene under mild conditions to immediately and quantitatively form the single insertion adduct $[Cp^*Zr(CH_2CHMeCH_2Ph)$ $(CH_2Ph)]^+[BCH_2Ph(C_6F_5)_3]^-$ [21]. Although this system does not behave as an olefin polymerization catalyst, the single insertion adduct is important in that it provides evidence for the 1,2-insertion process by which propylene polymerization is initiated by similar compounds (see below).

Another cocatalyst which has often been used in conjunction with the monocyclopentadienyl precursors is trityl borate, $[Ph_3C]^+[B(C_6F_5)_4]^-$ [equation (8)]. This cocatalyst has the advantage over the borane in that the resulting counteranion $[B(C_6F_5)_4]^-$ is less coordinating than the $[RB(C_6F_5)_3]^-$ counteranion provided by the borane [22]. The trityl borate is expected to react with compounds such as $CpTiMe_3$ by abstraction of a methyl group by the trityl cation to give the active species in solution:

$$Cp'MMe_3 + [Ph_3C]^+[B(C_6F_5)_4]^- \longrightarrow$$
$$[Cp'MMe_2]^+ + Ph_3CMe + [B(C_6F_5)_4]^- \qquad (13)$$

The lower proclivity of the anion to coordination should provide a less sterically hindered species for monomer coordination and subsequent polymerization.

It has been found, however, that the expected cationic species is not always the product formed. When the compounds Cp^*TiMe_2R ($R = Cl, C_6F_5, OC_6F_5, Me$) react with an equivalent amount of trityl borate, the products are, in fact, the same type of methyl-bridged dititanium complexes, $[(Cp^*TiMeR)_2(\mu\text{-}Me)]^+$ [23], discussed above as forming when some of these same precursors are reacted with a deficiency of $B(C_6F_5)_3$ [equation (11)]. However, here they are formed even when one full equivalent of the trityl borate is reacted, and it would seem that the unreacted neutral trialkyl compound coordinates preferentially to the cationic center produced in the absence of a strongly coordinating anion.

Double irradiation NMR experiments on these complexes show that, for the complexes with Me and OC_6F_5 ligands, there is an equilibrium between this bridged species and the separate cation/neutral species pair:

$$[(Cp^*TiMeR)_2(\mu\text{-Me})]^+ \rightleftharpoons [Cp^*TiMeR]^+ + Cp^*TiMe_2R \qquad (14)$$

For the compounds with $R = Me, OC_6F_5$, this equilibrium lies well to the left, whereas when $R = Cl, C_6F_5$, the equilibrium is not observed by irradiation experiments, exchange being very slow [23]. Interestingly, reaction of $(C_5Me_4CH_2CH_2Ph)TiMe_3$ with $[Ph_3C]^+[B(C_6F_5)_4]^-$ results in abstraction of the methyl group but, in the absence of a good coordinating anion, the benzyl arm on the cyclopentadienyl ring chelates with the titanium center in an η^6-manner [24].

As suggested by equation (9), precursor activation can also be effected by reaction of neutral alkyl compounds with a Brønsted acid and thus the compound $CpZr(CH_2Ph)_3$ reacts with $[HNPhMe_2][B(C_6F_5)_4]$ to give an active catalyst [equation (15)] [25]. Although no NMR experiments have been reported with this system in order to identify the intermediates completely, it is believed to involve simple protonation of a benzyl ligand.

$$CpZr(CH_2Ph_3 + [HNPhMe_2]^+[B(C_6F_5)_4]^- \longrightarrow$$
$$PhMe + NMe_2Ph + [CpZr(CH_2Ph)_2]^+ + [B(C_6F_5)_4]^- \qquad (15)$$

4 POLYMERIZATION STUDIES

Cationic monocyclopentadienyl compounds of titanium, zirconium and hafnium are possibly the most diverse group of initiators in terms of their abilities to polymerize a wide variety of olefins, both Ziegler–Natta and carbocationic processes having been observed for these systems. We discuss below first general applications of both Ziegler–Natta and carbocationic processes. We then discuss a number of specialized topics, styrene and diene polymerization, styrene–ethylene copolymers and constrained geometry catalysts.

4.1 POLYMERIZATIONS VIA ZIEGLER–NATTA PROCESSES

Ziegler–Natta or coordination polymerization is the most important process by which organometallic compounds catalyze the polymerization of simple olefins; a general mechanism for ethylene polymerization by a catalyst of the type $[CpTiMeR]^+$ is shown in Scheme 1. Thus initiation involves coordination of the olefin *cis* to, say, a methyl group and migratory insertion of the olefin into the Ti—Me bond to give an *n*-propyl group. Propagation involves a sequence of such steps, the coordinated alkyl group growing into a polymer chain.

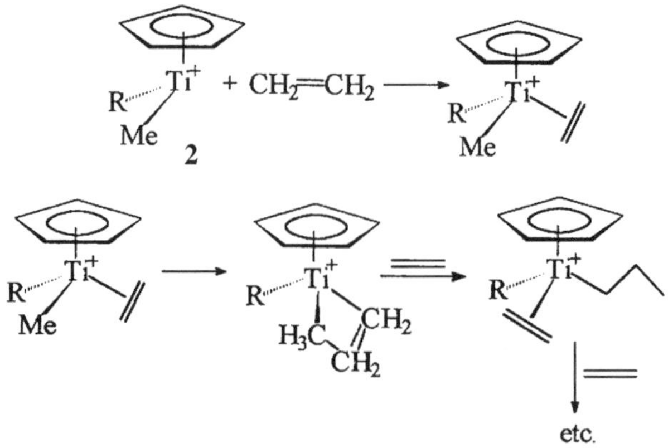

Scheme 1

There have been numerous investigations of the polymerizations of simple olefins such as ethylene and propylene by monocyclopentadienyl systems, as it was hoped that the sterically less crowded and electronically less saturated nature of these compounds, with respect to metallocenes, would lead to higher activities. One of the earliest such investigations involved comparison of the two compounds Cp_2ZrCl_2 and $CpZrCl_3$ as ethylene polymerization catalysts when activated with methyl aluminoxane [26]. It was found that the two catalysts exhibited similar polymerization behaviors, both exhibiting an optimum temperature for polymerization. It was also found, in both cases, that increasing the [Al]/[Zr] ratio resulted in increased activity, and that increasing the polymerization temperature resulted in decreased molecular weights of the resulting polymers but increased polydispersity due to more prevalent chain transfer mechanisms at higher temperatures. At high catalyst concentrations, the two catalysts gave almost identical activities although, at low concentrations, the activity of the monocyclopentadienyl compound was much less, possibly a result in part of lower reactivity between $CpZrCl_3$ and MAO. For $CpZrCl_3$, values of M_w and M_w/M_n for polyethylene were ca 3×10^5 and 3.5, respectively, at ambient temperature.

A substantial amount of work on the polymerizations of olefins by supported monocyclopentadienyl compounds, for instance the polymerization of propylene with silica-supported $CpTiCl_3$ activated with either MAO or $Al(i\text{-}C_4H_9)_3$, has been carried out [27]. Interestingly, the atactic polypropylene produced contained a substantial amount of regioerrors arising from abnormal 2,1-misinsertions, unlike the polypropylene obtained with metallocene systems. The ability of the mono-cyclopentadienyl compound to allow regioerrors probably arises from its lower steric bulk, which would have less of an effect on the geometry of the incoming olefin. Also investigated was the use of Cp^*TiCl_3 supported on SiO_2 to copolymerize

ethylene and 1-octene [28], making the industrially important low-density polyethylene. It was found that the copolymers produced contained randomly distributed
octene units, with a high octene content of up to 40% and narrow molecular weight
distributions and chemical compositions.

The polymerization of ethylene and propylene by a series of monocyclopenta-
dienylbenzyl compounds has been investigated, and it was found that activation of
$CpZr(CH_2Ph)_3$ with $[HNPhMe_2][B(C_6F_5)_4]$ gave a catalyst with activities for
ethylene polymerization which were comparable to activation by MAO even when
using 200-fold excesses of MAO [25]. This study also assessed the compounds
$CpZr(CH_2Ph)_3$, $Zr(CH_2Ph)_4$, $Cp*Zr(CH_2Ph)_3$, $Ti(CH_2Ph)_4$, $Cp*Ti(CH_2Ph)_3$ and
$Cp*TiMe_3$ activated with $B(C_6F_5)_3$ [29]. The catalysts all showed high activities for
ethylene polymerizations, although molecular weight data were not given. The
results suggested that cyclopentadienyl derivatives are more active than tetrabenzyl
compounds, but that changing from cyclopentadienyl to Cp* or changing from
benzyl to methyl ligands on the metal center makes very little difference. The same
systems were also used to polymerize propylene, with the result that the catalysts
based on the tetrabenzyl compounds gave mixtures of isotactic and atactic material
whereas the monocyclopentadienyl complexes gave only atactic products. In addition, the monocyclopentadienyltitanium compound was much more active than its
zirconium counterpart.

A broad investigation of the use of a variety of cyclopentadienyl and Cp*
compounds of titanium, zirconium and hafnium to polymerize ethylene in toluene
found little or no activity for neutral compounds of the type $CpMR_3$ or for cationic
species of the type $[CpMR_2L]^+$ or $[CpMR_2L_2]^+$ (M = Ti, Zr, Hf; R = alkyl; L =
amine, phosphine) [30a]. The neutral species appeared to have little affinity for
olefins, while the 12- and 14-electron species were presumably inactive because the
necessary olefin coordination sites were occupied by better ligands. However, high
activities were found with the complexes $[Cp*MMe_2]^+[MeB(C_6F_5)_3]^-$, obtained by
treating $Cp*MMe_3$ with $B(C_6F_5)_3$ as discussed above. The polyethylene produced
with the titanium compound was linear with molecular weights in excess of 300 000
and melting-points up to 141 °C. The zirconium and hafnium analogs were less
active, possibly because they formed the relatively stable η^6-toluene complexes as
mentioned above. The $Cp*TiMe_3/B(C_6F_5)_3$ system has also been shown to be
highly active for the polymerization of norbornene via both Ziegler and ring-opening
metathesis processes but, surprisingly, not for the polymerization of cyclohexene or
cycloheptene [30b]. The same system is fairly active for the polymerization of 1,5-
hexadiene via both Ziegler and cyclopolymerization processes [30b] and for the
homopolymerization of 1-hexene [31].

The compounds $Cp*TiMe_2R$ (R = Me, Cl, C_6F_5) or OC_6F_5) have all been found
to polymerize ethylene when activated with either $B(C_6F_5)_3$ or $[Ph_3C]^+[B(C_6F_5)_4]^-$
[32,33]. High molecular weight linear polyethylene was formed in all cases, but
higher activities were found when the compounds were activated with the trityl
borate, probably because of the lesser coordinating ability of $[B(C_6F_5)_4]^-$ relative to

$[MeB(C_6F_5)_3]^-$. All four of these systems were found to polymerize propylene when activated under rigorously dry conditions [32] and the nature of the polymer produced appears to be independent of the cocatalyst used [23], although the activities were greater with $[B(C_6F_5)_4]^-$. As expected, the molecular weights decreased and polydispersity increased with increasing temperature, while the more electron-deficient compounds $[Cp*TiMeCl]^+$ and $[Cp*TiMeC_6F_5]^+$ gave products of higher molecular weights, especially at low temperatures. Indeed, at low temperatures molecular weights of greater than 10^6 were obtained, with polydispersities approaching 1. The more electron-deficient compounds also give lower activities for both ethylene and propylene polymerizations, possibly because of stronger coordination with the counter anion.

The polypropylene produced by all of these systems is atactic, with some slight stereoregulation due to chain end control evident for the more sterically demanding species $Cp*TiMe_2OC_6F_5$ and $Cp*TiMe_2C_6F_5$. Interestingly, the polymers produced above $0\,°C$ were extremely regioirregular with a significant number of 2,1-insertions similar to the heterogeneous catalysts described above [27]. However, when the polymerization temperature was reduced to $-78\,°C$, no evidence of regioerrors was observed.

4.2 POLYMERS PRODUCED BY CARBOCATIONIC MECHANISMS

Unlike conventional metallocene initiators, the compounds $Cp*TiMeR(\mu\text{-Me})$ $B(C_6F_5)$ initiate polymerization of some monomers via a carbocationic mechanism. For instance, $Cp*TiMe_2(\mu\text{-Me})B(C_6F_5)$ has been shown to polymerize vinyl ethers [34], N-vinylcarbazole [34], 2,3-dihydrofuran [34] and isobutylene [35], all by carbocationic mechanisms. None of these monomers, which are generally polymerized by Lewis acid initiators, had previously been polymerized by metallocene-like complexes, but the polymerization processes are believed to occur at the same active site as Ziegler–Natta polymerization on the cationic titanium center of the $[Cp*TiMe_2]^+$ cation (Scheme 2). Instead of coordinating in an η^2 fashion, the monomer coordinates in a non-classical η^1 fashion, the metal–olefin interaction being stabilized by a complementary borate–olefin interaction. The cationic charge is then located at the β-position of the coordinated olefin, making this site susceptible to attack by a second monomer molecule (propagation) in the same manner. The methylborate anion probably remains closely associated with the growing polymer chain, stabilizing the positive charge.

A number of vinyl ethers have been polymerized by this system, including methyl (MVE), ethyl (EVE) and isobutyl (IBVE) vinyl ethers [34]. MVE and EVE were both polymerized in toluene at $-78°C$ to give products with molecular weights approaching 50 000 and polydispersities of less than 2, results comparable to those for polymers produced using classical Lewis acid initiators. IBVE gave even higher

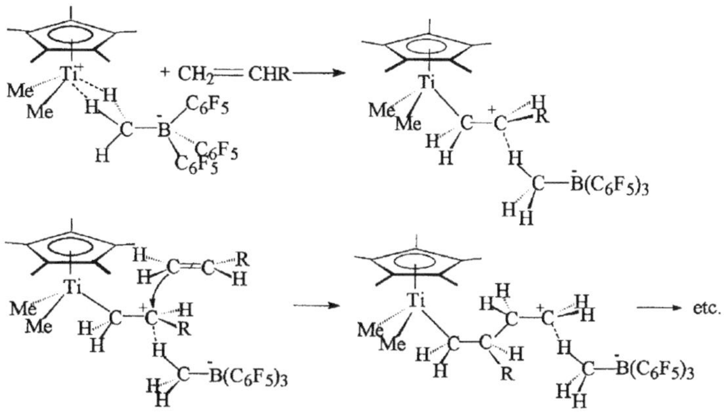

Scheme 2

molecular weights of up to 100 000 and still gave low polydispersities. Interestingly the poly(IBVE) obtained was predominantly syndiotactic in nature. The end groups of the poly(EVE) also gave more evidence for the proposed carbocationic mechanism. Upon quenching the reaction with methanol then water, the presence of aldehydic end groups was detected. These were formed via hydrolysis of the acetal groups formed on nucleophilic attack by the methanol on the carbenium center of the growing polymer.

Polymerization of poly(2,3-dihydrofuran) was also accomplished with this system. It was found that polymerization occurs through head-to-tail addition of the olefinic carbon atoms, keeping the ring intact rather than by a ring-opening mechanism. It is interesting that no activity occurred with 2,5-dihydrofuran, consistent with the requirement that monomers must contain an ether linkage α to the olefinic group, in a position capable of stabilizing a carbenium ion center. This result also helps confirm the carbocationic process believed to be occurring. N-Vinylcarbazole (NVC) was also polymerized by this same system and again results are consistent with those for other conventional carbocationic initiators [34].

The industrially most interesting carbocationic process which the $Cp^*TiMe_3/B(C_6F_5)_3$ system has been used for is in the polymerization of isobutylene and the copolymerizations of isobutylene and isoprene. The polyisobutylene produced was very regioregular, and molecular weights as high as 500 000 and polydispersities as low as 2.1 were obtained. This system has also been shown to produce copolymers of isobutylene/isoprene. By using an isobutylene: isoprene ratio of 99:1, a high molecular weight, low polydispersity product was obtained. This product contains 1 % isoprene inserted by the desired *trans*-1,4 addition. The ^1H and $^{13}C\{^1H\}$ NMR spectra of the homo-and copolymers were virtually indistinguishable from the spectra of a commercial sample of the copolymer [35].

4.3 POLYMERIZATION OF STYRENE AND CONJUGATED DIENES

The ability of monocyclopentadienyl compounds to polymerize styrene specifically to syndiotactic polystyrene (s-PS) has been one of the most interesting and important contributions of this class of compounds. Ishihara et al. [36] first reported s-PS in 1986, using an unnamed titanium catalyst and an organoaluminum cocatalyst. Later the formation of s-PS using tetrabenzyltitanium and -zirconium activated with MAO was also reported [37], but the first examples of monocyclopentadienyl compounds being used for the production of s-PS were $CpTiCl_3$ and Cp^*TiCl_3, which, when activated with MAO, gave highly, syndiotactic polystyrene ($M_w \approx 10^5$, $M_w/M_n \approx$ 2 at 20 °C) with conversions 10 times higher than for metallocenes or compounds without cyclopentadienyl ligands [38].

Activities were found to be generally dependent on time, temperature and Al:Ti ratio, the systems having an optimal polymerization temperature of 50 °C, above and below which the activities decreased. Increasing the Al:Ti ratio by increasing the MAO concentration resulted in an increase in the yields of s-PS but a decrease in the molecular weights, implying that the MAO not only had an important role in producing the active species but also acted as a chain transfer agent. A number of ring-substituted styrenes were also polymerized, including methyl-, butyl-, chloro-, fluoro- and bromostyrenes. It was found that increasing the electron density on the ring resulted in increased activity, suggesting that the mechanism involved attack of an electrophilic center (probably cationic) on the electron-rich olefin [38]. A further study of the polymerization activities of a variety of different titanium and zirconium compounds, all activated with MAO, found that titanium compounds of oxidation states 1–4 all gave s-PS with similar activities, suggesting that reduction or oxidation is taking place with some of these species [39]. In contrast, it was found that Zr(IV) compounds exhibited little or no activity, possibly because of the higher stability of Zr(IV) to reduction. The lower reactivity of zirconium compounds was subsequently confirmed [40].

EPR spectra of $CpZrCl_3$ and $CpTiCl_3$ activated by MAO exhibited doublets attributable to Ti(III) or Zr(III) hydrides; the Zr hydride appeared more slowly because of the greater stability of the Zr(IV) compound towards reduction [41a]. The suggestion that hydride species were formed was strengthened by collapse of the doublets to singlets upon addition of D_2, presumably because of H/D exchange. In contrast, the titanium compound Cp^*TiCl_3 gave a more complicated spectrum. Rather than a doublet, a sextet was observed, consistent with a Ti(III) center interacting with five equivalent protons. The suggested formula for this complex was $C_5Me_3(CH_2)_2TiHCl$, with an η^3-coordinated $C_5Me_3(CH_2)_2$ ring and the hydride proton able to exchange with all four of the methylene protons. This sextet was also observed to collapse to a singlet upon treatment with D_2. In an EPR study of the $CpTi(OBu)_3/MAO$ system [41b], a doublet was observed, consistent with a Ti(III) hydride and similar to the spectrum observed in the $CpTiCl_3/MAO$ reaction (see above). Upon addition of styrene to the mixture, the doublet disappeared and was replaced by a singlet, consistent with styrene insertion into the Ti— H bond. Note that none of these assignments has been confirmed by isolation of the putative species, although the important role of titanium (III) in s-PS formation has been demonstrated by the greater activity of $Cp*Ti(OMe)_2/MAO$ than of $Cp*Ti(OMe)_3/MAO$ [41c].

A mechanistic investigation of the chemistry of $CpTiCl_3$ and $Ti(CH_2Ph)_4$ activated with MAO [42] found comparable activities at low Al:Ti ratios of 50–100, but much higher activity for $CpTiCl_3$ at an Al:Ti ratio of 900:1. In a follow-up on earlier work (see above) [39], the polymerization of p-methyl- and p-chlorostyrene by $CpTiCl_3/MAO$ was also reinvestigated. Activities were found to increase with an electron-donating methyl substituent but to decrease with an electron-withdrawing chloro substituent, suggesting that the catalyst is an electrophile and confirming the earlier results. Solvent effects on the polymerization of styrene were also assessed, with the finding that aromatic solvents with electron-withdrawing groups on the ring resulted in increased activity whereas those with electron-donating substituents exhibited decreased activity. This result implied that the solvent may coordinate to the metal center and hinder reactivity, and it was tentatively suggested that the species responsible for the formation of s-PS is the complex cation $[CpTi(III)(P)(styrene)]^+$ (**5**, where P is the growing polymer chain coordinated in an η^3 fashion), in which a styrene molecule is at least η^4 coordinated. It was suggested that this structure could then give the steric hindrance around the Ti center needed to control the stereochemistry of propagation.

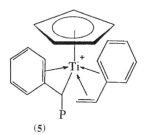

(5)

Further work on the mechanism of the formation of s-PS involved a thorough study of the styrene polymerization by the $CpTi(OBu)_3/MAO$ system [43]. The proportions of active species in solution were measured by terminating the polymerizations with CH_3O^3H in order to label the chain end with tritium. It was found that the Al:Ti ratio had a large effect on the percentage of titanium that was active, since up to 92% of the Ti centers were found to be active towards styrene polymerizations at low [Ti] (83 μM) and high Al:Ti ratios (10^3), whereas only 18% of the Ti centers were active at higher [Ti] (83 mM). It was also shown that 79% of the total active species produced s-PS and that only 18% produced atactic PS. The rate of polymerization to s-PS was also much higher, and it was found that the activity to s-PS reached a maximum at 45 °C but that the molecular weight decreased while the number of stereoerrors increased at higher temperatures. In contrast to previous work [42], changing aromatic solvent polarity had no effect on activities. A complementary EPR study of this system [41b] was discussed above.

An investigation of the effects of changing the nature of the cyclopentadienyl ligands on styrene polymerization involved the compounds $Cp'Ti(O^iPr)_3$ ($Cp' = C_5H_5, C_5Me_4H, C_5Ph_4H$) activated with MAO [44]. It was found that the C_5Me_5 compound gave the highest activity, the C_5Ph_5 the lowest, but the reasons for the differences are not clear. A series of compounds of the type (substituted indenyl)-$TiCl_3$/MAO, with different electron-donating and steric properties, was also investigated [5,6], and it was found that the activities increased as the electron-donating properties of the ligands increased but decreased with increased steric bulk.

Research on the use of Cp^*TiCl_3 supported on alumina surfaces and activated with alkylaluminum compounds showed that yields of s-PS were very low, and it appeared that heterogeneous catalysts will not compete with homogeneous catalysts in this field [45a]. Later work involving $CpTiCl_3$ supported on silica and activated with MAO gave conflicting results with respect to activities, but EPR signals were observed in both cases, similar to solution studies [45b,c].

The first MAO free systems for s-PS formation was reported in 1992 [46a], the compounds Cp^*TiMe_3 and $Cp^*MCH_2Ph)_3$ ($M = Ti, Zr$) being activated with $B(C_6F_5)_3$ and $[HNPhMe_2]^+[B(C_6F_5)_4]^-$. It was found that activation of Cp^*TiMe_3 and $Cp^*Ti(CH_2Ph)_3$ with $B(C_6F_5)_3$ resulted in highly active compounds for polymerization to s-PS, but that the Zr compounds or the compounds activated with the ammonium salt gave only low yields of atactic PS. The likely reason for the low activities when using $[HNPhMe_2]^+[B(C_6F_5)_4]^-$ as cocatalyst is that the amine which is produced coordinates to the vacant site and hinders monomer coordination. A more exhaustive investigation of the $Cp^*TiMe_3/B(C_6F_5)_3$ system showed that styrene is converted predominantly to s-PS above about $-15\,^\circ C$ in toluene or in the absence of solvent, but to atactic PS in methylene chloride or in toluene below about $-15\,^\circ C$ [30a]. The molecular weights and degrees of syndiotacticity of the s-PS formed were both very high, and use of Cp^*TiMe_3 labeled with ^{13}C in the Ti-methyl groups resulted in enrichment of the terminal methyl groups of the polymer. The results were consistent with an initiation step involving a 2,1-migratory insertion of a styrene molecule into a Ti—$^{13}CH_3$ bond [30a,46b]. The atactic PS was formed via carbocationic initiation, as outlined above. In this case, the use of $Cp^*Ti(^{13}CH_3)_3$ resulted in no $^{13}CH_3$ enrichment at any position of the polymer [30a].

Polymerization of styrene by Cp^*TiMe_3 activated with $[Ph_3C]^+[B(C_6F_5)_4]$ has also been investigated [47a]. Similar to other olefins, the use of this cocatalyst increased the polymerization activity by a factor of two over $B(C_6F_5)_3$ and by a factor of 10 over MAO, in part, at least, because of the poorer coordinating ability of the $[B(C_6F_5)_4]^-$ anion. A related, detailed kinetic investigation of s-PS formation induced by systems involving Cp^*TiR_3 ($R = Me, PhCH_2$) activated by MAO, $B(C_6F_5)_3$ and $[Ph_3C]^+[B(C_6F_5)_4]$ confirmed that the catalytic species in these three cases were very similar and probably identical, and that the concentrations of active species were similar to the concentrations of Ti(III) species derived from EPR data [47b]. The latter were believed to be of the type $[Cp*TiMe]^+$.

Related to the above, an extensive magnetic resonance study of the $CpTi(CH_2Ph)_3/B(C_6F_5)_3$ styrene polymerization system involved a combined EPR/NMR study in chlorobenzene and found that the ionic Ti(IV) species $[CpTi(CH_2Ph)_2]^+[BCH_2Ph(C_6F_5)_3]^-$ was formed within seconds when equimolar amounts of these two compounds were mixed at $25°C$; there was neither anion nor solvent coordination to the cation [48]. The initially formed product was unstable, however, and only 66% of the original amount was still present after 7 min. Although the addition of p-chlorostyrene to a freshly prepared solution resulted in no immediate change, polymerization began to occur after 15 min. In contrast, addition of an aluminum alkyl to a freshly prepared solution resulted in immediate disappearance of the Ti(IV) species and the rapid onset of p-chlorostyrene polymerization to syndiotactic product. These findings strongly suggested that $[CpTi(CH_2Ph)_2]^+[BCH_2Ph(C_6F_5)_3]^-$ was not the active species in styrene polymerization, but rather that it decomposed to the active species, a process which was aided by aluminum alkyls.

EPR monitoring of this reaction showed that, within a few minutes after mixing, two singlets, attributable to two Ti(III) species, appeared at $g = 1.967$ and 1.970. Aging resulted in an increase in the total amount of Ti(III) present as the amount of the Ti(IV) species disappeared according to the NMR experiment. Also over time the ESR signal at $g = 1.970$ increased in intensity whereas that at $g = 1.967$ decreased. The system exhibited no activity for styrene polymerization when only the former was left, and it was therefore proposed that the active species in s-PS production is a Ti(III) species resulting from the decomposition of $[CpTi(CH_2Ph)_2]^+$ $[BCH_2Ph(C_6F_5)_3]^-$ and giving rise to an EPR signal at $g = 1.967$. This in turn decomposes to an inactive Ti(III) species with a signal at $g = 1.970$.

The reactions between $Cp*Ti(CH_3)_3$ and $Cp*Ti(^{13}CH_3)_3$ with $B(C_6F_5)_3$ (mentioned above [30a,46b]) and $[Ph_3C]^+[B(C_6F_5)_4]^-$ have also been monitored by EPR spectroscopy [49] and conclusions have been reached that the original Ti(IV) species produced by the reaction between $Cp*TiMe_3$ and borane decomposed to the active Ti(III) species $[Cp*TiMe]^+$. The resonance of this putative species appeared as a singlet when $Cp*TiMe_3$ was used but as a doublet when the corresponding ^{13}C-labeled complex was used because of hyperfine coupling from the Ti to the ^{13}C nucleus. There was also observed a triplet, which was tentatively attributed to the zwitterionic species $CpTi(^{13}CH_3)(\mu\text{-}^{13}CH_3)B(C_6F_5)_3$, in which the two $^{13}CH_3$ groups exchanged and thus exhibited equal coupling to the titanium center. With $[Ph_3C]^+[B(C_6F_5)_4]^-$, the same doublet was observed but not the triplet as a methyl-bridged species now could not form. Addition of styrene to the system resulted in an increase in the rate of reduction and a collapse of the doublets and triplet into singlets as, presumably, a styrene monomer inserted into the Ti— $^{13}CH_3$ bond. There was no indication of the types of putative hydride species mentioned above as arising in the $Cp*TiCl_3/MAO$ system [41a].

Attempts to polymerize a variety of 4-alkylstyrenes with the CpTiCl$_3$/MAO system resulted in the formation of syndiotactic products with styrene and 4-methylstyrene but atactic products when longer alkyl chains were used [50]. It was suggested that steric factors which inhibit simultaneous coordination of the last inserted unit and an incoming monomer are the cause of the stereoirregularity.

Another conjugated monomer which has been successfully polymerized by the CpTiCl$_3$/MAO system is (Z)-1,3-pentadiene [51]. The E-isomer had previously been polymerized by a variety of homogeneous and heterogeneous catalysts [52], but polymerization of the Z-isomer is much rarer. Very interesting temperature effects were found for this system, since the polymer prepared at 20 °C contained cis-1,4-units whereas that formed at −28 °C or below consisted almost exclusively of 1,2-units. Between these temperatures a mixture was observed, a possible indication of different species being active at high and low temperatures. Interestingly, copolymerization of this monomer with styrene resulted in a block copolymer of styrene and pentadiene containing almost exclusively 1,2-inserted pentadiene units [53].

An extensive investigation of copolymerizations by the CpTiCl$_3$/MAO system with styrene and the conjugated dienes isoprene, butadiene and 4-methyl-1,3-pentadiene has also been made [54,55]. Reactivity ratios and reactivities of the different monomers for homopolymerization were assessed and, for these monomers, the rates for homopolymerization were found to increase in the order isoprene ≪ styrene < butadiene ≪ 4-methyl-1,3-pentadiene whereas the reactivities of the different monomers toward any given reactive chain end increased in the order styrene < isoprene < butadiene < 4-methyl-1,3-pentadiene.

4.4 STYRENE–ETHYLENE COPOLYMERS

The interesting thermoelastic properties of ethylene/styrene copolymers have resulted in considerable interest in their synthesis. Early work with monocyclopentadienyl compounds involved use of the CpTiCl$_3$/MAO system, and it was reported that low molecular weight copolymer was formed along with s-PS and polyethylene [56]. Although this report was later disputed by another group [57] a further, more definitive, study showed that this system was indeed able to produce some copolymer [58]. The copolymer could be separated from the two homopolymers by sequential extractions with methyl ethyl ketone and benzene, and it was found that the styrene concentration in the homopolymer was 27–36 %, depending on the feed composition. Most of the styrene had inserted in a 2,1-fashion, similar to the s-PS homopolymer although some primary insertion was also observed.

A perfectly alternating ethylene/styrene copolymer was produced at temperatures above 50 °C using the Cp*Ti(CH$_2$Ph)$_3$/B(C$_6$F$_5$)$_3$ system; up to 63 % of a perfectly alternating copolymer was obtained, along with some polyethylene and s-PS [59]. The styrene was again inserted in a 2,1-manner, and it was suggested that, if the last inserted unit is styrene, it coordinates to the metal center in an η^n manner ($n \geq 2$), similar to that postulated earlier for styrene homopolymerization (see above). If this is the case when the last inserted unit is styrene, the metal center could be sufficiently sterically hindered that it might well coordinate and insert the smaller ethylene monomer preferentially. On the other hand, if the last inserted unit is ethylene, this steric hindrance would not occur and the more electron-rich styrene monomer might preferentially coordinate to the electrophilic metal center. This type of discrimination would lead to the perfectly alternating structure observed.

Styrene/ethylene copolymers were also formed by the $CpTi(OPh)_3$/MAO system [60], copolymers being prepared with as much as 55 % styrene content. It was found that the ability to produce copolymer was dependent on the Al:Ti ratio, copolymer only being formed at ratios below 1000; s-PS was formed at higher Al:Ti ratios. Again, EPR measurements were interpreted in terms of a Ti(III) species being responsible for the copolymerization.

4.5   CONSTRAINED GEOMETRY CATALYSTS

The most recent development in the field of monocyclopentadienyl complexes has been the evolution of so-called 'constrained geometry' catalysts, illustrated in a general way as **6**, containing a pendant arm on the cyclopentadienyl ring which can chelate to the metal center through a lone pair on a nitrogen atom. The donor atom can either occupy a vacant coordination site in the neutral molecule, as shown, or it can be free and hence available to coordinate to the cationic active metal. This class of compounds has generated considerable industrial interest because of the ability both to polymerize ethylene to polyethylene containing long-chain branching and to copolymerize ethylene with α-olefins to give polymers containing high proportions of the α-olefin [61].

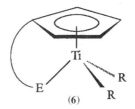

(6)

Although much of the relevant information is to be found only in the patent literature [62], we discuss below only those catalytic systems which have been reported in the open literature.

The compounds $(C_5R_4CH_2CH_2NMe_2)TiCl_3$ (R = H, Me) are examples of constrained geometry catalysts which contain a pendant arm in which the donor atom is a tertiary amine nitrogen atom. When activated with either MAO or a combination of triisobutylaluminum and $[Ph_3C]^+[B(C_6F_5)_4]^-$, these compounds were found to exhibit much lower activities than $CpTiCl_3$/MAO and Cp^*TiCl_3/MAO for the polymerization of styrene to s-PS but much higher activities for the polymerization of ethylene and propylene [63]. It was suggested that the nitrogen atoms donate electrons to the metal centers, making them less susceptible to reduction to Ti(III), which is believed to be responsible for the preparation of s-PS. The pendant arms may also provide steric hindrance to the metal centers, not allowing the type of multi-hapto coordination of styrene (**5**) which is possibly necessary for the formation of s-PS. The higher activities towards ethylene and propylene polymerizations compared to the other monocyclopentadienyl complexes may also be due to the greater stabilities of the Ti(IV) species responsible for the polymerization of these monomers. In related work, methoxy-substituted compounds of the type $(C_5H_4CH_2CH_2OMe)TiCl_3$, activated with MAO, were found to be much less active for the polymerization of ethylene, propylene and styrene [64]. It was suggested that MAO coordinates to the methoxy group through an Al–O interaction, causing virtual deactivation of the catalysts as the large, oligomeric MAO sterically hinders the active site. The polyethylene formed using the amino systems appeared to be branched, while the polypropylene formed was atactic.

Constrained geometry catalysts in which the tether of **6** contains an amide nitrogen, as in **7**, have recently been attracting increasing attention. In contrast to the above-mentioned $(C_5R_4CH_2CH_2NMe_2)TiCl_3$, which would form the 12-electron species $[(C_5R_4CH_2CH_2NMe_2)TiR_2]^+$ on activation, compounds of type **7** would form the chiral, 10-electron species $[(C_5H_4SiR'_2NR'')MR]^+$ (M = Ti, Zr, R, R', R'' = alkyl, aryl). The latter would be sterically less hindered and electronically less saturated than are metallocene catalysts, and thus may be expected to form very active catalyst systems.

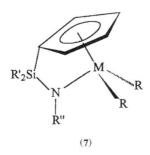

(7)

An early investigation with this class of compounds involved the polymerization of ethylene by the compound $[(C_5Me_4SiMe_2N'Bu)TiCl_2$ (**8**) activated with triiso-butylaluminum and $[Ph_3C]^+[B(C_6F_5)_4]^-$ in toluene [65]. Very high molecular weight polyethylene was obtained, a result confirmed later by work with the similar compounds $[(C_5Me_4SiMe_2N'Bu)M(CH_2Ph)_2$ (M = Ti, Zr) [66]. When activated with $B(C_6F_5)_3$, $B(C_{12}F_9)_3$ and $[Ph_3C]^+[B(C_6F_5)_4]^-$, these compounds formed very active catalysts for ethylene and propylene polymerization, yielding ultrahigh molecular weight linear polyethylene and high molecular weight, atactic polypropylene. In addition, the titanium compound **8**, when activated with MAO, formed a very reactive catalyst for ethylene/1-hexene copolymerization [65]. The copolymer product contained up to about 70 % 1-hexene compared with only 10 mol% for the Cp_2ZrCl_2/MAO system, and it was suggested that this class of catalysts can oligomerize ethylene to long-chain α-olefins, which can then copolymerize with ethylene to produce branched products. The ability to insert higher olefins as efficiently as ethylene is probably due to the relatively low degree of steric hindrance at the metal center of these catalysts as compared to metallocenes.

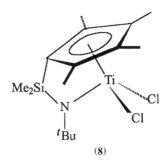

(**8**)

The catalyst system $(\eta^5\text{-}C_5Me_4SiMe_2N^tBu)TiCl_2$/MAO, of type **7**, has been investigated, as have analogous indenyl and fluorenyl compounds of titanium and zirconium, for the polymerization of propylene [67]. In general, high molecular weight atactic polypropylene was obtained with substitution of the group on nitrogen having very little effect on the polymer produced. Interestingly, catalysts containing the more electron-donating C_5Me_4 and fluorenyl ligands produced higher molecular weight polypropylene than did those containing the indenyl ligand, and it was suggested that the better electron-donating ligands might both decrease the rate of β-elimination and also stabilize the catalyst towards reduction. The C_5Me_4 and indenyl compounds were found to give 2–5 % 2,1-propylene misinsertions, higher than most metallocene catalysts but lower than other monocyclopentadienyl compounds (see above). The fluorenyl compound gave no misinsertions.

There have been several studies of copolymer formation utilizing constrained geometry catalysts, in addition to the ethylene/1-hexene investigation mentioned above [65], and it is now clear that monocyclopentadienyl titanium compounds will be found to be very effective catalysts for the formation of copolymers involving a variety of monomers. Thus copolymerization of ethylene and styrene was readily effected by $(\eta^5\text{-}Me_4C_5SiMe_2N^tBu)TiCl_2$ and a number of related indenyl and fluorenyl compounds when activated with MAO [68], up to 34.6 mol% styrene being incorporated. Monomer concentration and substituent effects on polymer compositions, yields and molecular weight distributions were assessed. On the basis of NMR data, it was deduced that no more than two styrene units followed each other and that only head-to-tail styrene coupling occurred. Subsequent work also investigated the terpolymerization of ethylene, styrene and 1-hexene and ethylene, styrene and 1,5-hexadiene by $(\eta^5\text{-Me}_4\text{C}_5\text{SiMe}_2\text{N}^t\text{Bu})\text{TiCl}_2/\text{MAO}$, finding that a range of monomer incorporation was possible [69].

Finally, the polymerization of ethylene by constrained geometry catalysts has been studied by a combination of density functional theory and molecular dynamics calculations on the model tri- and tetravalent systems $(\eta^5\text{-C}_5\text{H}_4\text{SiH}_2\text{NH})\text{MMe}$ and $[(\eta^5\text{-C}_5\text{H}_4\text{SiH}_2\text{NH})\text{MMe}]^+$ (M = Ti, Zr, Hf) [70]. In a study of the polymer initiation stage, modest activation barriers for insertion of ethylene into the Ti(III)—Me and Ti(IV)— Me bonds were found (3.3 and 3.8 kcal/mol respectively), and higher barriers for insertion into the Zr(IV)—Me and Hf(IV)—Me bonds (both 5.1 kcal/mol). The kinetic products for all four complexes were found to be γ-agostic n-propyl complexes, and the exothermicities for conversion of the initial cationic M(IV) complexes plus ethylene to the n-propyl products were comparable (33–34 kcal/mol). Interestingly, however, the γ-agostic n-propyl complex $[(\eta^5\text{-C}_5\text{H}_4\text{SiH}_2\text{NH})\text{Ti}(^n\text{Pr})]^+$ was found to take part in rapid exchange between the initially formed γ-agostic structure and a β-agostic isomer of comparable stability.

Chain propagation was also modeled, the energy profiles of the reactions of ethylene with the n-alkyl complexes $[(\eta^5\text{-C}_5\text{H}_4\text{SiH}_2\text{NH})\text{Ti}(\text{R})]^+$ (R = ^{n}Pr,^{n}Bu) being investigated. The calculations suggested that an intermediate η^2-ethylene complex of the type $[(\eta^5\text{-C}_5\text{H}_4\text{SiH}_2\text{NH})\text{Ti}(\text{R})(\text{C}_2\text{H}_4)]^+$ would be stabilized by a β-agostic alkyl group, and that the transition state is stabilized by an α-agostic interaction. The overall propagation barrier was found to be the insertion process, which was calculated to have a free energy barrier of 24.3 kcal/mol. Chain transfer, on the other hand, appeared to involve direct β-hydrogen transfer to monomer rather than to metal, yielding a long-chain terminal olefin which could insert into a subsequent chain to give the long-chain branching which appears to be the normal type of polyethylene formed with this type of catalyst system [70c].

5 ACKNOWLEDGMENTS

Financial support from the Natural Sciences of Canada, through Research, Strategic and CRD Grant Programs, made possible our research in this area.

6 REFERENCES

1. EI du Pont de Nemours *US Pat.*, 3 038 915 (1953).
2. (a) Bottrill, M., Gavens, P. D., Kelland, J. W. and McMeeking, J., *Comp. Organomet. Chem.*, **3**, 332 (1982); (b) Gomez-Sal, M. P., Mena, M., Royo, P. and Serrano, R., *J. Organomet. Chem.*, **358**, 147 (1988).

3. (a) Sloan, C. L. and Barber, W. A. *J. Am. Chem. Soc.*, **81**, 1364 (1959); (b) Nesmeyanov, A. N., Nekrasov, Y. S., Sizoi, V. F., Nogina, O. V., Dubovitsky, V. A. and Sirotkina, Y. I., *J. Organomet. Chem.*, **61** 225 (1975).
4. Cardoso, A. M., Clark, R. J. H. and Moorhouse, S., *J. Chem. Soc., Dalton Trans.*, 1156 (1980).
5. (a) Ready, T. E., Day, R. O., Chien, J. C. W. and Rausch, M. D., *Macromolecules*, **26**, 5822 (1993); (b) Foster, P.,Chien, J. C. W. and Rausch, M. D., *Organometallics*, **15**, 2404 (1996).
6. Kim, Y., Koo, B. H. and Do, Y., *J. Organomet. Chem.*, **527**, 155 (1997).
7. Mena, M., Royo, P., Serrano, R., Pellinghelli, M. A. and Tiripicchio, A., *Organomettallics*, **8**, 476 (1989).
8. Martin, A., Mena, M., Royo, P., Serrano, R., Pellinghelli, M. A. and Tiripicchio, A, *J. Chem. Soc., Dalton Trans.*, 2117 (1993).
9. Fussing, I. M. M., Pletcher, D. and Whitby, R. J., *J. Organomet. Chem.*, **470**, 109 (1994).
10. Tremblay, T. L., Ewart, S. W., Sarsfield, M. J. and Baird, M. C., *J. Chem. Soc., Chem. Commun.*, 831 (1997).
11. (a) Wolczanski, P. T. and Bercaw, J. E., *Organometallics*, **1**, 793 (1982); (b) Schock, L. E. and Marks, T. J. *J. Am. Chem. Soc.*, **110**, 7710 (1988).
12. Erker, G., Berg, K., Treschanke, L. and Engel, K., *Inorg. Chem.*, **21**, 1277 (1982).
13. Yang, X., Stern, C. L. and Marks, T. J., *Organometallics*, **13**, 10015 (1994).
14. Wang, Q., Quyoum, R., Gillis, D. J., Jeremic, D. and Baird, M. C., *J. Organomet. Chem.*, **527**, 7 (1997).
15. Sarsfield, M. J., Ewart, S. W., Tremblay, T. L., Roszak, A. W. and Baird, M. C., *J. Chem. Soc., Dalton Trans.*, 3097 (1997).
16. Gillis, D. J., Tudoret, M.-J. and Baird, M. C., *J. Am. Chem. Soc.*, **115**, 2543 (1993).
17. Gillis, D. J., Quyoum, R., Tudoret, M.-J., Wang, Q., Jeremic, D., Roszak, A. W. and Baird, M. C., *Organometallics*, **15**, 3600 (1996).
18. Lancaster, S. J., Robinson, O. B., Bochmann, M., Coles, S. J. and Hursthouse, M. B., *Organometallics*, **14**, 2456 (1995).
19. Pellecchia, C., Immirzi, A., Grassi, A. and Zambelli, A., *Organometallics*, **12**, 4473 (1993).
20. Pellecchia, C., Immirzi, A., Pappalardo, D. and Peluso, A., *Organometallics*, **13**, 3773 (1994).
21. Pellecchia, C., Immirzi, A. and Zambelli, A., *J. Organomet. Chem.*, **479**, C9 (1994).
22. Jia, L., Yang, X., Stern, C. L. and Marks, T. J., *Organometallics*, **16**, 842 (1997).
23. Ewart, S. W., Sarsfield, M. J. and Baird, M. C., unpublished results.
24. Flores, J. C., Wood, J. S., Chien, J. C. W. and Rausch, M. D., *Organometallics*, **15**, 4944 (1996).
25. Pellecchia, C., Proto, A., Longo, P. and Zambelli, A., *Makromol. Chem. Rapid Commun.*, **12**, 663 (1991).
26. Chien, J. C. W. and Wang, B. P., *J. Polym. Sci., Polym. Chem.*, **28**, 15 (1990)
27. Park, J. R., Shiono, T. and Soga, K., *Macromolecules*, **25**, 521 (1992).
28. Uozumi, T., Toneri, T., Soga, K. and Shiono, T, *Macromol. Rapid Commun.*, **18**, 9 (1997).
29. Pellecchia, C., Proto, A., Longo, P. and Zambelli, A., *Makromol. Chem. Rapid Commun.*, **13**, 277 (1992).
30. (a) Wang, Q., Quyoum, R., Gillis, D. J., Tudoret, M. J., Jeremic, D., Hunter, B. K. and Baird, M. C., *Organometallics*, **15** 693 (1996); (b) Jeremic, D., Wang, Q., Quyoum, R. and Baird, M. C., *J. Organomet. Chem.*, **497**, 143 (1995).
31. Murray, M. C. and Baird, M. C. *J. Mol. Catal.*, **128**, 1 (1998).
32. Ewart, S. W., Sarsfield, M. J., Tremblay, T. L. and Baird, M. C. *Organometallics*, submitted for publication.

33. Quyoum, R., Wang, Q., Tudoret, M.-J., Baird, M. C. and Gillis, D. J., *J. Am. Chem. Soc.*, **116**, 6435 (1994).
34. Wang, Q. and Baird, M. C., *Macromolecules*, **28**, 8021 (1995).
35. Barsan, F. and Baird, M. C., *J. Chem. Soc., Chem. Commun.*, 1065 (1995).
36. Ishihara, N., Kuramoto, M., Seimiya, T. and Uoi, M. *Macromolecules*, **19**, 2464 (1986).
37. Pellecchia, C., Longo, P., Grassi, A., Ammendola, P. and Zambelli, A., *Makromol. Chem. Rapid Commun.*, **8**, 277 (1987).
38. Ishihara, N., Kuramoto, M. and Uoi, M., *Macromolecules*, **21**, 3356 (1988).
39. Zambelli, A., Oliva, L. and Pellecchia, C. *Macromolecules*, **22**, 2129 (1989).
40. Longo, P., Proto, A, and Oliva, L., *Macromol. Rapid Commun.*, **15**, 151 (1994).
41. (a) Bueschges, U. and Chien, J. C. W., *J. Polym. Sci., Polym Chem.*, **27**, 1525 (1989); (b) Chien, J. C. W., Salajka, Z. and Dong, S., *Macromolecules*, **25**, 3199 (1992); (c) Newman, T. H. and Malanga, M. T., *J. M. S.—Pure Appl. Chem.*, **A34**, 1921 (1997).
42. (a) Zambelli, A., Pellecchia, C., Oliva, L., Longo, P. and Grassi, A, *Makromol. Chem.*, **192**, 223 (1991). (b) Longo, P., Proto, A. and Zambelli, A., *Makromol. Chem. Phys.*, **196**, 3015 (1995).
43. Chien, J. C. W. and Salajka, Z. *J. Polym. Sci., Polym. Chem.*, **29**, 1253 (1991).
44. Kucht, A., Kucht, H., Barry, S., Chien, J. C. W. and Rausch, M. D., *Organometallics*, **12**, 3075 (1993).
45. (a) Soga, K., Koide, R. and Uozumi, T., *Makromol. Chem., Rapid Commun.*, **14**, 511 (1993); (b) Yim, J. H., Chu, K. J., Choi, K. W. and Ihm, S.-K., *Eur. Polym. J.*, **32**, 1381 (1996); (c) Xu, J., Zhao, J., Fan, Z. and Feng, L., *Makromol. Rapid Commun.*, **18**, 875 (1997).
46. (a) Pellecchia, C., Longo, P., Proto, A. and Zambelli, A., *Makromol. Chem. Rapid Commun.*, **13**, 265 (1992); (b) Pellecchia, C., Pappalardo, D., Oliva, L. and Zambelli, A., *J. Am. Chem. Soc.*, **117**, 6593 (1995).
47. (a) Kucht, A., Kucht, H., Chien, J. C. W. and Rausch, M. D., *Appl. Organomet. Chem.*, **8**, 393 (1994). (b) Grassi, A., Lamberti, C., Zambelli, A. and Mingozzi, I., *Macromolecules*, **30**, 1884 (1997).
48. Grassi, A., Pellecchia, C. and Oliva, L., *Macromol. Chem. Phys.*, **196**, 1093 (1995).
49. Grassi, A., Zambelli, A. and Laschi, F., *Organometallics*, **15**, 480 (1996).
50. Nakatini, H., Nitta, K. H., Takata, T. and Soga, K., *Polym. Bull.*, **38**, 43 (1997).
51. Ricci, G., Italia, S. and Porri, L. *Macromolecules*, **27**, 868 (1994).
52. (a) Natta, G., Porri, L., Corradini, P., Zanini, G. and Ciampelli, F., *J. Polym. Sci.*, **51**, 463 (1961); (b) Natta, G., Porri, L., Carbonara, A. and Stoppa, G., *Makromol. Chem.*, **77**, 114 (1964).
53. (a) Longo, P., Proto, A., Oliva, P., Sessa, I. and Zambelli, A., *J. Polym. Sci., Polym. Chem.*, **35**, 2697 (1997); (b) Longo, P., Proto, A., Oliva, P. and Zambelli, A., *Macromolecules*, **29**, 5500 (1996).
54. Pellecchia, C., Proto, A. and Zambelli, A., *Macromolecules*, **25**, 4450 (1992).
55. Zambelli, A., Proto, A., Oliva, P. and Longo, P., *Makromol. Chem. Phys.*, **195**, 2623 (1994).
56. Longo, P., Grassi, A. and Oliva, L., *Makromol. Chem.*, **191**, 2387 (1990).
57. (a) Aaltonen, P. and Seppala, J., *Eur. Polym. J.*, **30**, 683 (1994); (b) Aaltonen, P. and Seppala, J., *Eur. Polym. J.*, **31**, 79 (1995).
58. Oliva, L., Mazza, S. and Longo, P., *Macromol. Chem. Phys.*, **197**, 3115 (1996).
59. Pellecchia, C., Pappalardo, D., D'Arco, M. and Zambelli, A., *Macromolecules*, **29**, 1158 (1996).
60. Xu, G. and Lin, S. *Macromolecules*, **30**, 685 (1997).
61. See, for instance, *Chem. Eng. News*, March, 33, 1997.

62. For recent examples of American patents, see (a) Canich, J. A. M., *US Pat.*, 5 026 798 (1991); (b) Canich, J. A. M., *US Pat.*, 5 055 438 (1991); (c) Canich, J. A. M. and Licciardi, G. F., *US Pat.*, 5 057 475 (1991); (d) Canich, J. A. M., *US Pat.*, 5 096 867 (1992); (e) Canich, J. A. M. and Licciardi, G. F., *US Pat.*, 5 227 440 (1993); (f) Canich, J. A. M., *US Pat.*, 5 547 675 (1996); (g) Canich, J. A. M., *US Pat.*, 5 631 391 (1997); (h) Canich, J. A. M., Turner, H. W. and Hlatky, G. G., *US Pat.*, 5 621 126 (1997); (i) Neithamer, D. R. and Stevens, J. C., *US Pat.*, 5 350 723 (1994); (j) Neithamer, D. R. and Stevens, J. C. *US Pat.*, 5 399 635 (1995); (k) Rosen, R. K. Nickias, P. N., Devore, D. D., Stevens, J. C. and Timmers, F. J., *US Pat.*, 5 532 394 (1996); (I) Devore, D. D., Stevens, J. C., Timmers, F. J. and Rosen, R. K., *US Pat.*, 5 539 068 (1996); (m) Rosen, R. K., Devore, D. D., Nickias, P. N., Stevens, J. C. and Timmers, F. J., *US Pat.*, 5 494 874 (1996); (n) Lai, S., Wilson, J. R., Knight, G. W. and Stevens, J. C., *US Pat.*, 5 665 800 (1997).

63. (a) Flores, J. C., Chien, J. C. W. and Rausch, M. D., *Organometallics*, **13**, 4140 (1994); (b) Flores, J. C., Chien, J. C. W. and Rausch, M. D., *Macromolecules*, **29**, 8030 (1996).

64. Foster, P., Chien, J. C. W. and Rausch, M. D. *J. Organomet. Chem.*, **527**, 71 (1997).

65. Soga, K., Uozumi, T., Nakamura, S., Toneri, T.,Teranishi, T., Sano, T., Arai, T. and Shiono, T., *Macromol. Chem. Phys.*, **197**, 4237 (1996).

66. Chen, Y-X. and Marks, T. J., *Organometallics*, **16**, 3649 (1997).

67. McKnight, A. L., Masood, M. A., Waymouth, R. M. and Straus, D. A., *Organometallics*, **16**, 2879 (1997).

68. (a) Sernetz, F. G., Mulhaupt, R. and Waymouth, R. M., *Macromol. Chem. Phys.*, **197**, 1071 (1996); (b) Sernetz, F. G., Mulhaupt, R., Amor, F., Eberle, T. and Okuda, J., *J. Polym. Sci., Part A*, **35**, 1571 (1997); (c) Shiono, T., Moriki, Y., Ikeda, T. and Soga, K., *Macromol. Chem. Phys.*, **198**, 3229 (1997).

69. (a) Sernetz, F. G. and Mulhaupt, R. *J. Polym. Sci., Polym. Chem.*, **35**, 2549 (1997); (b) Sernetz, F. G., Mulhaupt, R. and Waymouth, R. M., *Polym. Bull.*, **38**, 141 (1997).

70. (a) Fan, L., Harrison, D., Woo, T. K. and Ziegler, T., *Organometallics*, **14**, 2018 (1995); (b) Woo, T. K., Margl, P. M., Lohrenz, J. C. W., Blöchl, P. E. and Ziegler, T., *J. Am. Chem. Soc.*, **118**, 13021 (1996); (c) Woo, T. K., Margl, J. C. W., Ziegler, T and Blöchl, P. E., *Organometallics*, **16**, 3454 (1997).

6

Molecular Architecture: a New Application Design Process Made Possible by Single-site Catalysts†

ANTONIO TORRES, KURT SWOGGER, CHE KAO
AND STEVE CHUM
Dow Europe SA., Horgen, Switzerland; Dow Chemical Co., Freeport, TX, USA

1 INTRODUCTION

One of the key consequences of the discovery and use of the constrained geometry (CGC) single-site catalysts (SSC) is that they permit a new era of polymer design flexibility that allows for the modeling and subsequent control of polymer molecular architecture. The CGC catalyst not only gives rise to new materials science understanding and molecular and process control, but also extends the capability of olefin polymers and processes to new markets and applications. Interestingly, these catalysts can be said to move olefin polymerization from the hands of the chemists to those of the engineers and material scientists. The concept of molecular architecture for olefins using the chemistry introduced in 1993 [1] is now reality.

These catalysts have unique characteristics that lead to a new level of polymerization control. In addition to being single-site, they also have high activity and they readily incorporate new comonomers. This results in kinetic equations that can be derived and solved with relative ease, at least when compared with the conventional

---

†This chapter is dedicated to our good friend Gerry Lancaster. His unencumbered vision was instrumental in the development of these design capabilities and INSITE Technology commercialization, the fruits of his relentless work. For this we thank him and we will all miss his comradery and perseverance. Heaven will never be the same.

Metallocene-based Polyolefins   Edited by J. Scheirs and W. Kaminsky
© 2000 John Wiley & Sons Ltd

144

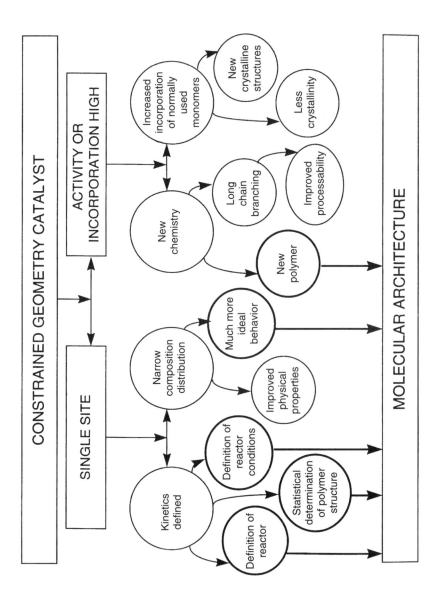

Figure 1 Implications of constrained geometry catalyst technology

Ziegler–Natta (Z–N) catalysts. The previous catalysts based on Z–N chemistries contain a mixture of different active catalyst sites, within a given support, which can vary from batch to batch, making it difficult to model. In addition to the ability to derive kinetic equations, single-site catalysts give very narrow molecular weight and comonomer distributions. These narrow distributions lead to nearly ideal behavior of the polymer, making the modeling task much less cumbersome.

Incorporation of higher α-olefins, including cyclic olefins and even long, vinyl-terminated polymer chains, extends the usefulness of these catalysts. As part of the molecular architecture concept, ready incorporation of different building blocks is essential to extending the use of olefin polymers. Because of the single-site nature of the catalysts, these comonomers are incorporated homogeneously, which again results in systems that can be predicted statistically. A schematic diagram of the many implications of the CGC catalyst is given in Figure 1.

2   REACTOR AND POLYMER DESIGN

Understanding the kinetics and having the necessary models lead to the obvious aspects of reactor and polymer design. Using kinetics in conjunction with mass and energy balances, a reactor model can be constructed that for a given reactor can be used to predict operating conditions for making a polymer. These models using CGC are very accurate, as can be see in Figures 2, 3 and 4, in which predicted versus actual values for melt index (MI, I^2), density and I_{10}/I_2, respectively, are plotted. The ratio I_{10}/I_2 is one way of representing the degree of long-chain branching (LCB) of the polymer and is discussed in later chapters. The data scatter is well within the analytical error of the methods used.

The ideal nature of the CGC polymer, which results in model compounds, simplifies the ability to isolate specific physical features for further analysis and prediction. This new capability was previously difficult with polymers made by Z–N catalyst systems. These Z–N materials resulted in complex fractions. Unfortunately, given the heterogeneity of the polymer fractions, the effects of comonomer concentration, molecular weight (MW), molecular weight distribution (MWD) and comonomer length could not always be separated, through either conventional or advanced fractionation techniques. This typically led to certain assumptions which restricted the validity of the analysis and in many cases resulted in less than accurate predictions of performance. Common examples of this difficulty were seen when analyzing the comonomer content in the low molecular weight fraction, and also the composition or presence of a high-density fraction, in the Z–N-catalyzed polymers. Figures 5 and 6 illustrate typical molecular weight and comonomer distributions for polymers made using CGC single-site catalysts compared with polymers made using Z–N catalysts.

It is important to note that vanadium catalysts can also be utilized to make narrow distribution polymers [2]. However, they are limited in their ability to incorporate

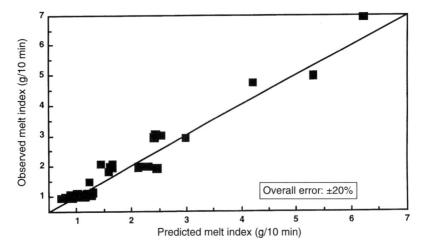

Figure 2   Prediction of melt index in CGC polymers

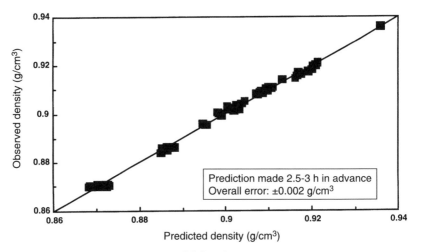

Figure 3   Prediction of density in CGC polymers

high levels of higher α-olefins such as octene, and generally have low catalyst efficiency that results in high catalyst residues in the polymer. Given these limitations, these catalysts are restricted to use in very high value applications, e.g. wire and cable coatings. Nevertheless, vanadium systems can be modeled using the same concepts as CGC catalysts, but their range of capability is severely limited.

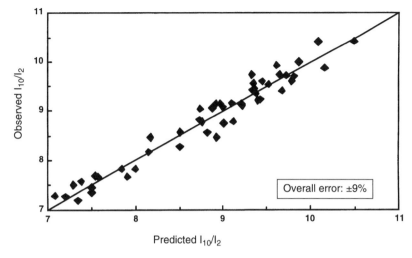

Figure 4 Prediction of I_{10}/I_2 in CGC polymers

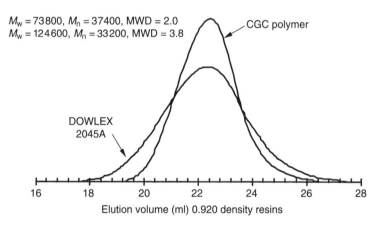

Figure 5 Molecular weight comparisons of CGC polymers and heterogeneous DOWLEX linear low-density polyethylene

3 UTILIZING IDEAL POLYMERS TO MODEL PERFORMANCE

The ability to make model polymers allows the polymer designer to determine the effect of structure on the polymers' physical properties. Performance properties such as modulus, impact, mar resistance, clarity, heat-seal temperatures, tear and many others can be established and then predicted with the given knowledge of polymer structure. Since the inter-relationships between these various properties can be

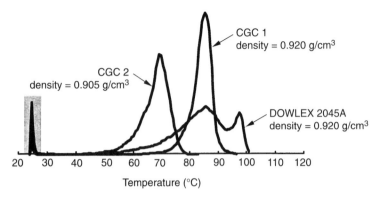

Figure 6 Short-chain branching distribution analysis of CGC polymers and heterogeneous DOWLEX linear low-density polyethylene. Analytical temperature-rising elution fractionation.

modeled, the polymer designer begins to develop numerous design options that were previously impossible or non-intuitive.

A number of important polymer microstructures control the desired performance properties. Many of these can be easily modeled deriving from kinetic models: molecular weight, molecular weight distribution, comonomer type and comonomer concentration and the degree of long-chain branching. All are now controllable parameters that enhance the design and application breadth. Given this new capability, which also enhances manufacturing control and consistency, the impact in a given application insures that success can be accomplished in fewer iterations, resulting in a faster development time and less trial and error.

The possibilities are infinite, particularly when one further expands into polymer systems that are created through multi-modal distributions. Multi-modal distributions are molecular blends of polymers and are powerful tools to achieve a proper balance of properties and performance. Each compositional fraction can have different molecular weights, comonomer concentrations, degree of long-chain branching and molecular weight distributions. Understanding the overall effects of these structures on physical properties is the key to molecular architecture and design.

Comonomer flexibility is very interesting since not only can current α-olefins such as 1-butene, 1-hexene and 1-octene be used, but also new comonomers such as styrene or vinyl-terminated long polymer chains can be incorporated into the polymer by the CGC system. These new types of polymer structures have very unique features giving new properties typically not available in olefin-based materials. Using styrene, for example, new interpolymers of styrene and ethylene and/or propylene can give new properties such as dead-fold, enhanced noise damping and improved stress relaxation, which were previously available only in

Table 1 Comparison of % stress relaxation and T_g for CGC produced ethylene/octene (E/O) and ethylene/styrene (E/S) copolymers vs flexible poly (vinyl chloride) (f-PVC)

Polymer	Stress relaxation[a] (%)	$T_g(^\circ C)$
f-PVC	76	~ 0
E/O copolymer ($D0.87^{b)}$	29	~ -40
E/S copolymer (30% S^c)	30	~ -25
E/S copolymer (40% S)	38	~ -25
E/S copolymer (80% S)	93	$\sim +20$

[a] At 50% strain and 10 min.
[b] D = density in g/cm^3.
[c] S = styrene (wt%).

filled, plasticized or non-polyolefin-based systems. The key material parameters for these are stress relaxation (which also controls dead-fold) and the glass transition temperature (T_g) (which is active in noise abatement). Table 1 lists these key properties in comparison with other materials. The unique ability to vary the T_g and the stress relaxation of these new interpolymers through the use of compositional changes with CGC expands the options for the materials engineer.

High incorporation of vinyl-terminated polymer, called long-chain branching, results in a unique polymer system that is highly shear sensitive, allowing for better extrusion processability and excellent melt strength, which significantly enhance fabrication.

Incorporation of >19% α-olefins in a homogeneous manner results in the development of a fringed micellar morphology. A particularly exciting outcome from the work with CGC systems was the confirmation of this theory once proposed by Flory [3]. Figure 7 demonstrates the fringed micelle morphology. This structure has resulted in new polyolefin-based elastomers that have performance characteristics very similar to those of existing natural and synthetic counterparts. The fringed micelle morphology is analogous to the hard–soft segment morphology seen in more traditional urethane-based elastomer systems, thereby, for the first time, providing high elastomeric performance with an olefin-based material.

All the above examples are just some of the design tools made possible by the new catalyst systems. New structures, and the ability to control them, make possible molecular design and the creation of many new product families.

4 MODELS AND THEIR USE

There are three specific aspects of modeling which are critical to the commercial success of any new technology. The key to success is the ability to (a) design or model products that can meet end-use performance requirements, within the (b)

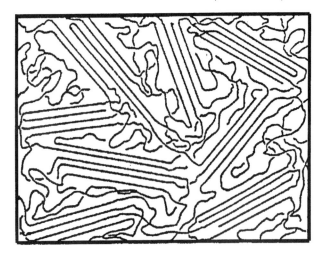

Figure 7 Fringed micelle model of semi-crystalline polymer

manufacturing capability and at (c) economics that allow for expedient market penetration. Modeling, or the automation of the above process, is critical if one is to drive speed-to-market, which today differentiates the winners from the average producer in the ever more capable market.

Meeting end-use performance requirements is critical. The ability to be first-to-market and to reduce the trial-and-error time at a customer is a key differentiation tool and has significant value. The development of performance requirements requires a very close relationship with the customer and end-user. Defining performance requirements, in terms of materials science, is critical to design expediency and will be the only way that one can achieve a real competitive advantage that will accurately capture all of a customer's needs. Historically, it has been shown that customers do not adequately focus on their specific performance requirements when they interact with the polymer suppliers. This has resulted from years of little or no polymer design flexibility, where a customer often adjusted existing products or equipment to meet their application needs and was not accustomed to asking for the key performance requirements they needed. Working with the customer to help determine and measure key performance requirements is critical to molecular architecture and design expediency.

Given the performance requirements, the design or modeling system must now be capable of determining the conditions under which this new material can be manufactured. Given the flexibility provided by the new single-site catalyst systems, the manufacturing process represents millions of 'micro-assemblers' or replicators that assemble the homogeneous material mixtures to provide the needed polymer performance. The inherent accuracy of these new catalyst systems will one day result in the elimination of off-line quality control systems, since the variability of the instruments themselves will be higher than that of the catalyst–polymer assembly process.

Having manufactured the polymer from the proposed polymer design completes two of the three critical steps of the modeling process. Evaluating the economics of the proposed polymer design and the resulting manufacturing set-up is the last but most critical step. If the combination of polymer design and manufacturing conditions does not result in a polymer with the right economics, the designer must re-evaluate the design and develop secondary or tertiary options.

Only when the performance requirements are met through the manufacture of a product at the right cost can the designer be assured of a successful trial. It is through the capabilities developed from the new single-site catalyst systems that this kind of model development will, in the end, differentiate the new olefin products and drive market growth.

5 COMPUTER-BASED MODEL FOR MOLECULAR DESIGN

Single-site and metallocene catalyst technology resulted in the production of polyolefin copolymers with well defined molecular structure as previously discussed. The Dow INSITE technology, based on the constrained geometry catalyst chemistry, coupled with Dow's flexible process technology, will further expand the capability and flexibility of molecular architecture by allowing more flexibility in molecular designs, such as comonomer choices, long-chain branching structure and morphology control in the polymer. Because of the increase in the design flexibility, it became essential for the product designer to take advantage of modern computer technology to increase the efficiency of polymer design. Based on this philosophy, Dow Chemical developed and currently utilizes a computer-based product design software system. The system consists of a performance requirement-driven product design software based on the logic as described in Figure 8. In addition, it utilizes advanced on-line control systems to monitor and control basic polymer structure variables such as rheology, density, molecular weight and molecular weight distribution during the manufacturing process. This has resulted in enhanced polymer property control capability and allows Dow to manufacture INSITE technology polymers consistently and efficiently. For this approach, the most critical element to enhance the success for the product design exercise is to acquire precise performance requirements for the applications defined by the customers. Once the performance requirements have been defined, the computer-based materials science models can be used to relate polymer structure to performance requirements, and the kinetic and process models can be applied to optimize the manufacturing process to produce the defined polymer structure.

Examples of materials science models that related performance attributes to molecular structure are listed in Table 2. Using the computer-driven, model-based

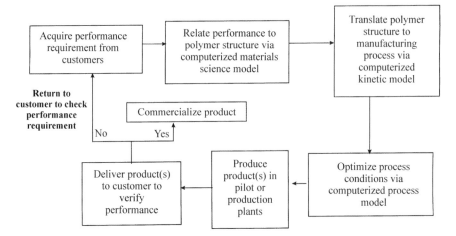

Figure 8   Performance requirement-driven product design logic

Table 2   Performance attributes and related molecular structure for product design

Performance attributes	Related molecular structure
Stiffness/modulus	Polymer density, level of comonomer, crystal morphology
Toughness/environmental stress crack Resistance	Tie molecules, MW, type of α-olefin
Heat seal performance	Polymer density, comonomer distribution
Melt fracture resistance	LCB
Processability: extruderability and melt strength	LCB, MW and MWD
Optical properties	Crystal morphology, density, comonomer distribution

product design approach, Dow has successfully demonstrated significant improvements in speed-to-market. This has resulted in improved value for both Dow and its customers by meeting performance requirements quickly and efficiently.

5   CONCLUSION

The recent developments in single-site and constrained geometry catalyst technology, such as the INSITE technology, have greatly expanded product capability and allowed the polymer manufacturer to deliver products with new performance attributes to the customer. With this new capability, polymer engineers and product designers can use model-based molecular architecture and computers to design products. Understanding the performance attributes from the customers became the most critical issue for delivering the right product on time.

With the INSITE technology, Dow has successfully demonstrated this capability in recent years.

6 REFERENCES

1. Swogger K. W., presented at the SPE ANTEC '93 Special Management Involvement Forum: Extrusion for the 90s, New Orleans, November 1993.
2. Elston, C., *US Pat.*, 3 645 992 (1972).
3. Flory, P. J., *J. Am. Chem. Soc.*, **184**, 2857 (1962).

7

Generation of Cationic Active Species for Olefin Polymerization from *ansa*-Metallocene Amide Complexes

IL KIM
University of Ulsan, Ulsan, Korea

1 INTRODUCTION

Since the discovery in 1957 that bis(cyclopentadienyl)titanium dichloride in the presence of alkylaluminum chloride acts as a homogeneous Ziegler–Natta catalyst [1] for the polymerization of ethylene, research efforts have been devoted to the development of modified metallocene-based catalysts that not only exhibit high activity but also provide stereochemical control over polymer composition and structure. A significant achievement was made in 1980, when Sinn and Kaminsky [2] reported that Group IV metallocene compounds in the presence of an excess of methylalumoxane (MAO) offer highly active catalysts for olefin polymerization. With the introduction of a short interannular bridge, Brintzinger and co-workers [3] were able to reduce the steric congestion and modify the Lewis acidity at the electrophilic metal center associated with these *ansa*-metallocene complexes.

The most promising *ansa*-metallocene catalysts for the isospecific polymerization of α-olefins contain R_2Si bridged bis(cyclopentadienyl) or bis(indenyl) ligands [4, 5]. Chiral C_2-symmetric metallocenes are normally prepared by chloride

Metallocene-based Polyolefins   Edited by J. Scheirs and W. Kaminsky
© 2000 John Wiley & Sons Ltd

displacement reaction of MCl_x compounds and bis(cyclopentadienyl) dianion reagents. However, these procedures are inefficient and hampered by low yields and complicated separation and purification steps. For example, Brintzinger and co-workers [3e] and Collins and co-workers [6] prepared the important *ansa*-metallocene *rac*-(EBI)ZrCl$_2$ [EBI = 1,2-ethylenebis(1-indenyl)] by reaction of $ZrCl_4(THF)_2$ and (EBI)Li$_2$ and reported low, variable yields (30–50 %). Buchwald and co-workers employed (EBI)K$_2$ and obtained (EBI)ZrCl$_2$ in 70 % yield with a *rac:meso* ratio of 2:1 [7]. Reinmuth prepared Me$_2$Si(1-C$_5$H$_2$-2-Me-4-^{t}Bu)$_2$ZrCl$_2$ in 15% yield with a *rac:meso* ratio of 2:1 by the reaction of Me$_2$Si(1-C$_5$H$_2$-2-Me-4-^{t}Bu)$_2$K$_2$ and ZrCl$_4$(THF)$_2$ [8]. Pure *rac*-Me$_2$Si(1-C$_5$H$_2$-2-Me-4-^{t}Bu)$_2$ZrCl$_2$ was obtained only after repeated recrystallization, in 9 % yield. Following a similar procedure, Brintzinger and co-workers prepared Me$_2$Si(1-C$_5$H$_3$-3-^{t}Bu)$_2$ZrCl$_2$ in 19 % yield with a *rac:meso* ratio of 1:1 and obtained pure *rac*-Me$_2$Si(1-C$_5$H$_3$-3-^{t}Bu)$_2$ZrCl$_2$ in 7 % yield after repeated crystallization [9]. By employing Me$_2$Si(1-C$_5$H$_3$-3-^{t}Bu)$_2$Li$_2$, Mise *et al.* obtained Me$_2$Si(1-C$_5$H$_3$-3-^{t}Bu)$_2$ZrCl$_2$ in 33 % yield as a 1:1 mixture of *rac* and *meso* isomers; recrystallization provided a 2.7:1 *rac–meso* mixture [10]. Salt elimination syntheses of other *ansa*-metallocenes with Me$_2$Si-bridged bis(cyclopentadienyl) ligands typically provide the pure *rac* product in only 5–10 % yield, and separation from the undesired *meso* isomer is not always possible.

Group IV organometallic complexes were prepared by amine elimination reactions of metal dialkylamide compounds [11]. Chandra and Lapert reported that the reaction of Zr(NMe$_2$)$_4$ with excess cyclopentadiene affords Cp$_2$Zr(NMe$_2$)$_4$ and 2 equiv. of Me$_2$H; in contrast, the reaction of Zr(NMe$_2$)$_4$ with excess indene affords the mono(indenyl)compound (η^5-C$_9$H$_7$)Zr(NMe$_2$)$_3$ [12]. Teuben and co-workers showed that the reaction of the cyclopentadienylamine species C$_5$H$_5$(CH$_2$)$_3$NHMe with M(NMe$_2$)$_4$ (M = Zr, Hf) gives chelated {η^5 : η^1-C$_5$H$_4$(CH$_2$)$_3$NMe}M(NMe$_2$)$_2$ complexes in high yield [13]. Petersen and co-workers reported that the homologous series of *ansa*-monocyclopentadienyl amido complexes [(C$_5$H$_4$)SiMe$_2$(N-^{t}Bu)]M(NMe$_2$)$_2$ (M = Ti, Zr, Hf) and [C$_5$Me$_4$SiMe$_2$(N-^{t}Bu)]Zr(NMe$_2$)$_2$ are prepared in 70–85 % isolated yields by heating the neat 1:1 reaction mixture of (C$_5$R$_4$H)SiMe$_2$(N(H)-^{t}Bu), where R = H, Me, and M(NMe$_2$)$_4$ at 110–120 °C for 24–48 h under an N$_2$ purge [14].

Jordan and co-workers reported that the amine elimination reaction (EBI)H$_2$ [1,2-bis(3-indenyl)ethane] and M(NMe$_2$)$_4$ (M = Zr, Hf) provides an efficient streoselective route to *rac*-(EBI)M(NMe$_2$)$_2$ [15]. Me$_2$Si-bridged *ansa*-zirconocene complexes were also prepared in good yield by amine elimination [16]. The reaction of Me$_2$Si(C$_5$H$_4$-3-^{t}Bu)$_2$ and Zr(NMe$_2$)$_4$ gives Me$_2$Si(1-C$_5$H$_3$-3-^{t}Bu)$_2$Zr(NMe$_2$)$_2$ in 95 % NMR yield (*rac:meso* ratio = 1 : 2) and pure *meso*-Me$_2$Si(1-C$_5$H$_3$-3-^{t}Bu)$_2$Zr(NMe$_2$)$_2$ in 38 % isolated yield. The introduction of α-Me Cp substituents improves the *rac:meso* product ratio significantly. The reaction of Me$_2$Si(1-C$_5$H$_3$-2-Me-4-^{t}Bu)$_2$ with Zr(NMe$_2$)$_4$ affords Me$_2$Si(1-C$_5$H$_2$-2-Me-4-

^{t}Bu)$_2$Zr(NMe$_2$)$_2$ in 90 % NMR yield (*rac:meso* ratio = 2.5 : 1) and pure *rac*-Me$_2$Si(1-C$_5$H$_2$-2-Me-4-^{t}Bu)$_2$Zr(NMe$_2$)$_2$ in 52 % isolated yield [16].

Steric crowding in Zr(NR$_2$)$_4$ amide compounds inhibits the amine elimination reaction with Me$_2$Si(1-C$_5$H$_3$-2-Me-4-^{t}Bu)$_2$. The qualitative rate trend for metallocene formation in the reaction of Me$_2$Si(1-C$_5$H$_3$-2-Me-4-^{t}Bu)$_2$ with Zr(NR$_2$)$_4$ compounds is Zr(NMe$_2$)$_4$ ≈ Zr(NC$_4$H$_8$)$_4$ > Zr(NC$_5$H$_{10}$)$_4$ > Zr(NEt$_2$)$_4$. The use of bulky ligands increase the *rac:meso* product ratios in amine elimination reactions of Me$_2$Si(1-C$_5$H$_3$-2-Me-4-^{t}Bu)$_2$ and Zr(NR$_2$)$_4$ compounds. The *rac:meso* ratio trend is Me$_2$Si(1-C$_5$H$_2$-2-Me-4-^{t}Bu)$_2$Zr(NC$_5$H$_{10}$)$_2$ (no *meso* isomer detected) > Me$_2$Si (1-C$_5$H$_2$-2-Me-4-^{t}Bu)$_2$Zr(NMe$_2$)$_2$ (*rac* : *meso* = 2.5 : 1) ≈ Me$_2$Si(1-C$_5$H$_2$-2-Me-4-^{t}Bu)$_2$Zr(NC$_4$H$_8$)$_2$ (*rac* : *meso* = 3 : 1) [16]. In conclusion, the amine elimination reactions offer an attractive alternative to the current salt elimination syntheses, as shown in Figure 1.

Recently, Kim *et al.* [17] showed that amine elimination reactions of pyCAr$_2$OH pyridine alcohols and M(NMe$_2$)$_4$ (M = Ti, Zr, Hf) provide an efficient entry to (pyCAr$_2$O)$_2$M(NMe$_2$)$_2$ chelate complexes. These species adopt C$_2$-symmetric structures in the solid state and in solution but undergo facile inversion of configuration at the metal, with racemization barriers in the range 12–14 kcal/mol. These amide complexes are activated for ethylene polymerization by alkylation with AlR$_3$ reagents and subsequent reaction with MAO; however, the molecular weight distributions of polyethylene are broad and show multimodal behavior characteristic of multiple active sites.

Catalysts derived from chiral C$_2$-symmetric *ansa*-metallocenes, such as *rac*-(EBI)ZrCl$_2$ and *rac*-(SBI)ZrCl$_2$ [SBI = Me$_2$Si(indenyl)$_2$] have been studied extensively for the stereospecific polymerization of α-olefins [4]. It is well established that a cationic complex ^{ch}Cp$_2$Zr(R)$^+$ (^{ch}Cp$_2$Zr = chiral *ansa*-metallocene framework) is the active species of Group IV metallocene catalysts and related systems in olefin polymerization. Standard activation procedures include (i) treatment of ^{ch}Cp$_2$ZrX$_2$, (X = Cl, R, OR) with excess MAO, (ii) reaction of ^{ch}Cp$_2$ZrR$_2$ complexes with ammonium salts e.g. [HNMe$_2$Ph][B(C$_6$F$_5$)$_4$]}, oxidizing agents (e.g. Ag$^+$, Cp$_2$Fe$^+$), or alkyl abstraction reagents [e.g. CPh$_3$, B(C$_6$F$_5$)$_3$] or (iii) *in situ* alkylation of ^{ch}Cp$_2$ZrX$_2$ complexes with AlR$_3$ or other reagents followed by ionization as in (ii) [4].

Even though *ansa*-metallocenes can be synthesized by amine elimination reactions, the practical application of *ansa*-metallocene amide catalysts is limited by the fact that the amide complexes should be converted to the corresponding dichloride or dimethyl complexes, which can then be used in polymerization reactions using standard activation procedures. However, the efficiency of the amine elimination route would be more fully exploited if the amide derivatives could be used directly in catalyst formulations. In this paper, the detailed activation procedures of ^{ch}Cp$_2$Zr(NMe$_2$)$_2$ are described by using MAO or non-coordinating anionic compounds as co-activators. The activation procedures of conventional halide complex ^{ch}Cp$_2$ZrCl$_2$ are also investigated for the comparison.

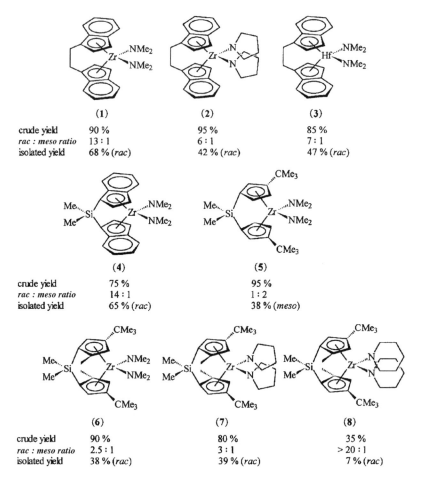

Figure 1 Summary of chiral *ansa*-metallocenes prepared by amine elimination. Crude yields were determined by NMR spectroscopy. The *rac*: *meso* ratios are those for the crude products. Isolated yields are recrystallized yields of the pure isomer indicated [16b]

2 EFFICIENT SYNTHESIS OF METALLOCENES BY AMINE ELIMINATION

2.1 ACHIRAL METALLOCENES

Group IV metal amide complexes are useful precursors to a wide range of derivatives [11,18]. For example, reaction of $M(NR_2)_4$ complexes with other amines (NR'_2H) provide routes to either $M(NR'_2)_4$ or mixed amide $M(NR_2)_{4-x}(NR'_2)_x$ complexes, and amine elimination reactions with alcohols, thiols and acidic hydrocarbons afford routes to alkoxide, sulfide and organometallic derivatives [11,18,19]. Group IV metal amides may be converted to halide derivatives via protonolysis (anhydrous HX or $NR_2H.HX$) [13,20] or amide–halide exchange reaction (e.g. reaction with $M'Cl_4$, $M' = Ti$, Zr, Hf, Si, Ge, Sn) [21,22]. Amine elimination reactions of Group IV metal dialkylamide compounds with protic reagents have been used to prepare a wide variety of organometallic and inorganic complexes [11,18]. With cyclopentadiene (CpH) and indene (IndH) (which are acidic hydrocarbon, forming stable aromatic anions), organometallic derivatives of Sn [equation (1)], Ti [equation (2)], Zr [equation (3) and (4)], Hf [equation (3), M = Zr, Hf], Nb [equation (5)] and U [equation (6)] are obtained [11]. It is interesting to note that with excess CpH, only one NR_2 group is directly replaced in the case of Ti or Nb, whereas two NR_2 groups are replaced for Zr, Hf or U. The differences between Ti on the one hand and Zr or Hf on the other are undoubtedly attributable to steric effects which clearly are more restricting on the smaller (Ti) reacting center: Ti^{4-} 0.74 A, Zr^{4+} 0.84 A [23]; M---NMe_2 bond lengths: Ti 1.91–1.92 A [24] and Zr 2.03–2.11 A) [11, 25].

$$R_3SnNR'_2 + HA \longrightarrow R_3SnA + HNR'_2 \tag{1}$$

$$[Ti(NR_2)_4] + 3CpH \longrightarrow [Ti(\eta^5\text{-}C_5H_5)(NR_2)_3] + HNR_2 \tag{2}$$

$$[M(NR_2)_4] + 3CpH \longrightarrow [M(\eta^5\text{-}C_5H_5)_2(NR_2)_2] + 2HNR_2 \tag{3}$$

$$[Zr(NMe_2)_4] + 2IndH \longrightarrow [Zr(\eta^5\text{-}Ind)(NMe_2)_3] + HNMe_2 \tag{4}$$

$$[Nb(NMe_2)_5] + 4CpH \longrightarrow [Nb(\eta^5\text{-}C_5H_5)(NMe_2)_3] + HNMe_2 \tag{5}$$

$$[U(NEt_2)_4] + 2CpH \longrightarrow [U(\eta^5\text{-}C_5H_5)_2(NEt_2)_2] + 2HNEt_2 \tag{6}$$

Hermann *et al.* reported the synthesis of the achiral Me_2Si-bridged metallocene $Me_2Si(\eta^5\text{-}C_5H_4)_2Zr(NEt_2)_2$ via the reaction of $Me_2Si(C_5H_5)_2$ with $Zr(NEt_2)_4$ [26]. Bridged cyclopentadienylamide derivatives of the type $\{\eta^5, \eta^1\text{-}(CH_2)_3NMe\}$ $M(NMe_2)_2$ (M = Zr, Hf) [13] and $\{\eta^5, \eta^1\text{-}Cp'SiMe_2NR'\}M(NR_2)_2$ (Cp' = C_5H_4, C_9H_6; R' = Ph, 'Bu; M = Ti, Zr, Hf; R = Me, Et) [27] have also been prepared via amine elimination reactions.

2.2 ANSA-METALLOCENES

The most widely used and highly developed chiral *ansa*-metallocenes contain bridged bis-indenyl ligands. Amine elimination affords an efficient approach to the synthesis of *rac*-(EBI)Zr(NMe_2)_2 complexes [15]. The reaction of $Zr(NMe_2)_4$ and (EBI)H_2 in toluene at 100 °C gives (EBI)Zr(NMe_2)_2 [equation (7)].

The *rac:meso* ratio of the product could be controlled by adjusting the reaction conditions; a 1:1 *rac*-**1**:*meso*-**1** ratio is the kinetic product, *rac*-**1** is the thermodynamic product and the *meso*-**1** to *rac*-**1** isomerization is catalyzed by the NMe_2H co-product [15]. The reversibility of the amine elimination is the key to the stereoselectivity of the reaction. Under static N_2 purge conditions, *rac*-(EBI)Zr(NMe_2)_2 was isolated in 68 % yield (see Figure 1). The reaction of (EBI)H_2 and the bulky pyrrolidide complex $Zr(NC_4H_8)_2$ in *m*-xylene at 140 °C

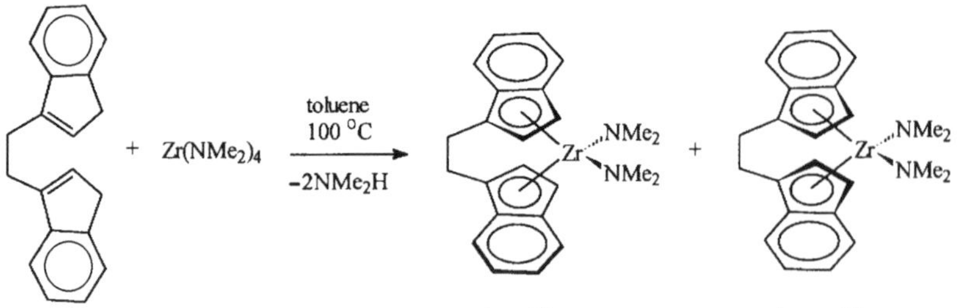

toluene
100 °C
+ Zr(NMe₂)₄ ————→
−2NMe₂H

Zr...NMe₂, NMe₂   + Zr...NMe₂, NMe₂   (7)

(rac-1)   :   (meso-1)

N₂ purge   1   :   1
static N₂ (open to bubbler)   13   :   1

for 7 h, with a slow nitrogen flow through the reaction vessel, produced $(EBI)Zr(NC_4H_8)_2$ in 95 % NMR yield with a *rac:meso* ratio of 6:1 [16a]. Recrystallization from diethyl ether afforded pure *rac*-$(EBI)Zr(NC_4H_8)_2$(*rac*-**2**) in 42 % isolated yield. $(EBI)Zr(NC_4H_8)_2$ was obtained in high yield only when the reaction was performed at higher temperatures or when the nitrogen purge was used to help remove the pyrrolidide from the reaction vessel. The reaction of $(EBI)H_2$ and sterically bulky $Zr(NC_4H_8)_2$ demonstrates the importance of the reversibility of the amine elimination reactions and the volatility of the amine byproduct.

The reaction of $(EBI)H_2$ and $Ti(NR_2)_4$ did not give *ansa*-titanocene products [16a]. It is likely that increased steric crowding around the smaller Ti (versus Zr or Hf) disfavors *ansa*-metallocene formation. However, amine elimination does offer an attractive route to *rac*-$(EBI)Hf(NMe_2)_2$ complexes. The reaction of $Hf(NMe_2)_4$ and $(EBI)H_2$ in *m*-xylene at 140 °C for 21 h gives $(EBI)Hf(NMe_2)_2$ in 85 % NMR yield (6:1 *rac:meso* ratio) and pure *rac*-$(EBI)Hf(NMe_2)_2$ in 47 % isolated yield (see Figure 1) [16a].

Me_2 Si-bridged *ansa*-zirconocene complexes of interest for α-olefin polymerization catalysis can be prepared in high yield by amine elimination. The reaction of $Zr(NMe_2)_4$ and $(SBI)H_2$ in refluxing hexanes under nitrogen for 8 h, in a reaction flask equipped with a water-cooled fractional distillation column packed with glass helices, affords $(SBI)Zr(NMe_2)_2$ (*rac*-**4**) in 75 % NMR yield with a *rac:meso* ratio of 14:1. [16c] Crystallization from hexanes affords pure *rac*-**4** in 65 % isolated yield.

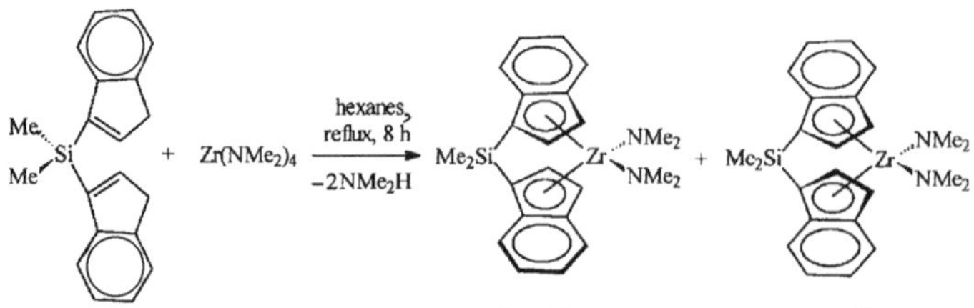

Me...Si, Me   +   Zr(NMe₂)₄   hexanes, reflux, 8 h ————→ −2NMe₂H   Me₂SiZr...NMe₂, NMe₂   +   Me₂SiZr...NMe₂, NMe₂   (8)

(rac-4)   :   (meso-4)

14   :   1

The reaction of $Me_2Si(1-C_5H_3-2-Me-4-{}^tBu)_2$ with $Zr(NMe_2)_4$ in toluene at $100\,°C$ for 5 h (system open to oil bubbler) affords the desired *ansa*-zirconocene $Me_2Si(1-C_5H_2-2-Me-4-{}^tBu)_2Zr(NMe_2)_2$ in 90 % NMR yield with a *rac : meso* ratio of 2.5:1 and pure *rac*-$Me_2Si(1-C_5H_2-2-Me-4-{}^tBu)_2Zr(NMe_2)_2$ in 52 % isolated yield (see also Figure 1) [16]. Steric crowding in $Zr(NR_2)_4$ amide compounds inhibits the amine elimination reaction. Thus, the reaction of $Me_2Si(1-C_5H_3-2-Me-4-{}^tBu)_2$ with the piperidide complex $Zr(NC_5H_{10})_4$ in *m*-xylene at high temperatures ($140\,°C$, 24 h, system open to oil bubbler) afforded *rac*-**8** in 35 % NMR yield, with no *meso*-**8** detected. In contrast, the pyrrolidide complex $Zr(NC_4H_8)_4$, in which the CH_2 groups of the amides are tied back in five-membered ring, reacts with $Me_2Si(1-C_5H_3-2-Me-4-{}^tBu)_2$ under milder conditions. Reaction at $90\,°C$ for 4 h in *m*-xylene solution, with a flow of nitrogen bubbling through the reaction solution system, affords $Me_2Si(1-C_5H_2-2-Me-4-{}^tBu)_2Zr(NMe_2)_2$ in 80 % NMR yield with a *rac:meso* ratio of 3:1 and pure *rac*-$Me_2Si(1-C_5H_2-2-Me-4-{}^tBu)_2Zr(NC_4H_8)_2$ in 39 % yield (recrystallization from hexane). Equation (9) summarizes the reactions of $Me_2Si(1-C_5H_3-2-Me-4-{}^tBu)_2$ with $Zr(NR_2)_4$ in various solvents.

In summary, compared with the current syntheses of ${}^{ch}Cp_2MCl_2$ via salt elimination reactions, the amine elimination route has several advantages: (i) generation of $(EBI)^{2-}$ reagent is not required, (ii) high dilution and simultaneous addition procedures are not required and (iii) no acid wash steps, Soxhlet extractions or recrystallization are needed [15,16].

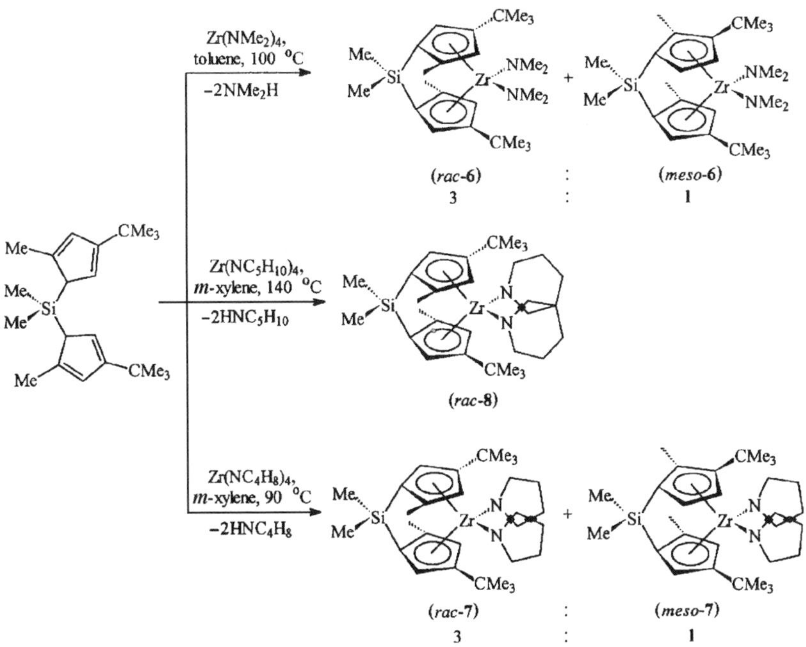

3 IN SITU ACTIVATION OF ANSA-METALLOCENE AMIDE COMPOUNDS

3.1 BY MAO

When a toluene solution of Cp_2ZrCl_2 is treated with MAO, a fast ligand exchange reaction takes place to produce primarily $Cp_2Zr(Me)Cl$ [28]:

$$Cp_2Zr\overset{Cl}{\underset{Cl}{<}} + \left[\underset{\underset{n}{\overset{|}{Al-O}}}{\overset{Me}{}}\right] \longrightarrow Cp_2Zr\overset{Cl}{\underset{Me}{<}} + \left[\underset{\underset{n}{\overset{|}{Al-O}}}{\overset{Cl}{}}\right] \qquad (10)$$

However, the Cp_2ZrCl_2/MAO system only becomes catalytically active when the [Al]:[Zr] ratio is $>200 : 1$, or even higher. The way in which excess MAO induces this activity has been studied in detail, mainly by spectroscopic means [28,29]. It is now recognized that either methyl or chloride abstraction by Al centers in the MAO takes place to produce the metallocene cation, $[Cp_2ZrMe]^+$:

$$Cp_2Zr\overset{Me}{\underset{Cl}{<}} + \left[\underset{\underset{n}{\overset{|}{Al-O}}}{\overset{X}{}}\right] \longrightarrow Cp_2Zr\overset{+}{\underset{\square}{<}}\overset{Me}{} + \left[\underset{\underset{n}{\overset{\diagdown\diagup}{Al-O}}}{\overset{X\ X}{}}\right]^- \qquad (11)$$

If reactions (10) and (11) are true, MAO plays many roles in the metallocene polymerizations: (i) it acts as an alkylating agent for the generation of a transition metal–alkyl molecule, (ii) it acts as a Lewis acid for anion abstraction from the metal–alkyl molecule, generating an electrophilic $[Cp_2ZrMe]^+$ species, and (iii) it acts as a scavenger for the removal of impurities such as water.

In order to establish whether ansa-metallocene diamide compounds are activated for the α-olefin polymerization by MAO, Kim and co-workers [30] carried our sequential reactions of rac-1 with MAO in an NMR tube. With sequential increases in the amount of MAO from a MAO : rac-1 ratio of 10 to 40, the ^1H NMR spectrum of each solution mixture was recorded. On increasing the amount of MAO from 10 to 40 equiv., NMe_2 in rac-1 is first methylated by MAO and/or free $AlMe_3$ contained in MAO, followed by activation to afford cationic active species. By reacting rac-1 with 10 equiv. of MAO (MAO : rac-1 = 10 : 1), unreacted rac-(EBI)Zr(NMe_2)_2, rac-(EBI)Zr(NMe_2)(Me), rac-(EBI)ZrMe_2 and cationic $[(EBI)ZrMe]^+[MAO]$ species are observed with ratios of $1 : 1$, $1 :$, $2 : 1$ and $3 : 1$, respectively. All free $AlMe_3$ contained in MAO is transformed into Al_2Me_4 $(NMe_2)_2$, demonstrating that the methyl ligand in rac-(EBI)Zr(NMe_2)(Me), rac-(EBI)ZrMe_2 and cationic $[(EBI)ZrMe]^+[MAO]^-$ species originates from free $AlMe_3$ contained in MAO [MAO donated by Albemarle as a 10% solution in toluene contains 1.85 wt% $AlMe_3$ and 8.15 wt% MAO (4.49 wt% total Al)]. On adding a further 10 equiv. of MAO (MAO : rac-1 = 20 : 1) to the solution mixture, unreacted *rac*-(EBI)Zr(NMe$_2$)$_2$ and *rac*-(EBI)Zr(NMe$_2$)(Me) are transformed in to *rac*-(EBI)ZrMe$_2$ and cationic [(EBI)ZrMe]$^+$[MAO]$^-$ species. On further increasing the amount of MAO to MAO:*rac*-1 = 40:1, *rac*-(EBI)ZrMe$_2$ is completely transformed into [(EBI)ZrMe]$^+$[MAO]$^-$ species. Hence the procedure for the formation of cationic zirconium species by the reaction of *rac*-1 with an excess amount of MAO can be summarized as shown in Scheme 1.

The structure of the cationic species and the nature of the interaction between MAO and zirconium cations under catalytic conditions, in other words in solution, remain uncertain. In order to confirm the formation of an active complex, a small amount of liquid propylene was introduced into the NMR tube containing the reaction mixture (MAO:*rac*-1 = 40:1) at −78 °C. White *i*PP was precipitated from the solution mixture by slowly increasing the temperature of reaction mixture to room temperature. The *meso* pentad [mmmm] value of *i*PP isolated from the NMR tube was 80.2 %.

Similar results were obtained by the sequential NMR-scale reactions of *rac*-(EBI)Zr(NC$_4$H$_8$)$_2$ (*rac*-2), *rac*-(SBI)Zr(NMe$_2$)$_2$ (*rac*-4), and *rac*-Me$_2$Si(1-C$_5$H$_2$-2-Me-4-^{t}Bu)$_2$Zr(NMe$_2$)$_2$ with MAO: ^{ch}CpZr(NR$_2$)$_2$ compounds are trans-

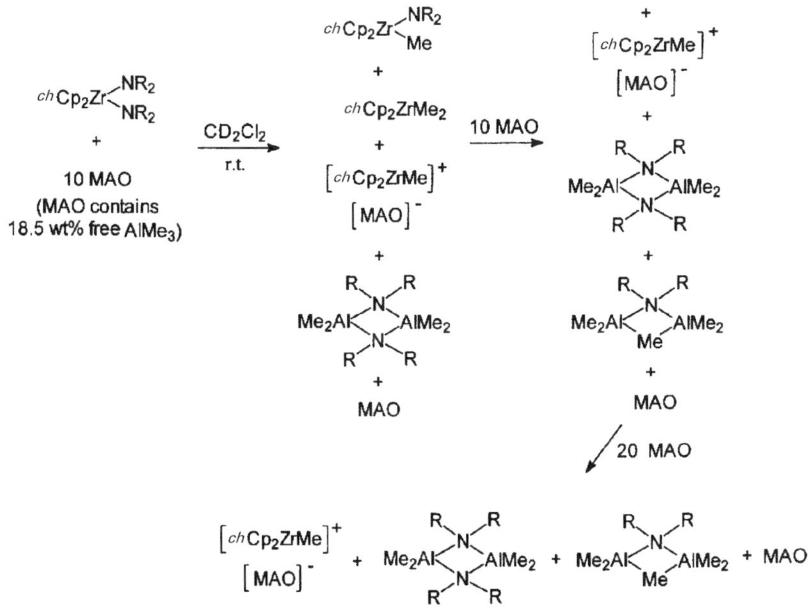

Scheme 1

formed into ^{ch}CpZrMe$_2$ mainly by the reaction with free AlMe$_3$ contained in MAO, and then MAO combines the removal of a methyl ligand to form a coordinatively unsaturated cation, $[^{ch}$CpZrMe]$^+$, with the introduction of the non-coordinating anion [30]. Introduction of liquid propylene into the NMR tube containing *in situ*-generated $[^{ch}$CpZrMe]$^+$ species affords *i*PP of high [mmmm] value.

Interestingly, the analogous compound *rac*-(EBI)ZrCl$_2$ could not be activated under the same reaction condition, i.e. on reacting *rac*-(EBI)ZrCl$_2$ with 40 equiv. of MAO no conspicuous methyl hydrogen peaks representing the formation of *rac*-(EBI)ZrMe$_2$ and cationic [(EBI)ZrMe]$^+$ species were observed [30]. Even after increasing the reaction temperature to 70 °C or increasing the amount of MAO (MAO : *rac*-(EBI)ZrCl$_2$ = 100 : 1), the same results were observed. Only 10 % of *rac*-(EBI)ZrCl$_2$ was transformed into *rac*-(EBI)Zr(Cl)(Me) and 90 % remained unreacted. These results demonstrate that the generation of [(EBI)ZrMe]$^+$ species from the reaction of *rac*-(EBI)ZrCl$_2$ and MAO is more difficult than that from *rac*-**1**.

There are some differences between ^{ch}CpZr(NR$_2$)$_2$ and ^{ch}CpZrCl$_2$ in the formation procedures of cationic $[^{ch}$CpZrMe]$^+$ species by the reaction with MAO. These differences are presumably caused by the alkylation capability of each compound by MAO and/or free AlMe$_3$, i.e. *ansa*-metallocene amides are easily alkylated by MAO and/or free AlMe$_3$, whereas *ansa*-metallocene chlorides are not easily alkylated.

Kim and co-workers [30] demonstrated the alkylation capability of *ansa*-metallocene amides by the reaction of *rac*-**1** and AlMe$_3$. As the amount of AlMe$_3$ increases from Al$_2$Me$_6$: *rac*-**1** = 1:1 to 2:1, *rac*-**1** is partly methylated to form *rac*-(EBI)Zr(NMe$_2$)(Me) and then completely methylated to form *rac*-(EBI)ZrMe$_2$. This methylation reaction occurs stoichiometrically:

$$rac\text{-}(EBI)Zr(NMe_2)_2 + Al_2Me_6 \longrightarrow$$
$$rac\text{-}(EBI)Zr(NMe_2)(Me) + Al_2Me_5(NMe_2) \quad (12)$$

$$rac\text{-}(EBI)Zr(NMe_2)(Me) + Al_2Me_6 + Al_2Me_5(NMe_2) \longrightarrow$$
$$rac\text{-}(EBI)ZrMe_2 + 2Al_2Me_5(NMe_2) \quad (13)$$

Part of the Al$_2$Me$_5$(NMe$_2$) seems to be transformed into Al$_2$Me$_4$(NMe$_2$)$_2$:

$$2Al_2Me_5(NMe_2) \rightleftharpoons Al_2Me_6 + Al_2Me_4(NMe_2)_2 \quad (14)$$

In the reaction mixture of *rac*-**1** with 2 equiv. of Al$_2$Me$_6$, three kinds of dimeric methylaluminum compounds, Al$_2$Me$_5$(NMe$_2$), Al$_2$Me$_6$ and Al$_2$Me$_4$(NMe$_2$)$_2$, were observed with ratios of 5:1, 2:1 and 1:1 respectively. Similar results were obtained by the reactions of *rac*-(EBI)Zr(NC$_4$H$_8$)$_2$ (*rac*-**2**), *rac*-(SBI)Zr(NMe$_2$)$_2$ (*rac*-**4**) and *rac*-Me$_2$Si(1-C$_5$H$_2$-2-Me-4-^{t}Bu)$_2$Zr(NMe$_2$)$_2$ with AlMe$_3$ [30].

In the reactions of *rac*-(EBI)ZrCl$_2$ with 4 equiv. of AlMe$_3$, *rac*-(EBI)ZrCl$_2$ was not alkylated to afford *rac*-(EBI)ZrMe$_2$ or even *rac*-(EBI)Zr(Me)Cl [30]. Even after increasing the amount of AlMe$_3$ to 40 equiv., the alkylation reaction was incomplete. These results provide direct evidence that ^{ch}CpZr(NR$_2$)$_2$ is completely alkylated by common alkyl aluminums, but ^{ch}CpZrCl$_2$ is not. Fink and co-workers [31] suggested a mechanism involving a cationic electronically unsaturated intermediate as the catalytically active species in the reaction of CpTiCl$_2$ and AlR$_3$ [equation (15)]. The existence of such species was supported further by the synthesis of the first isolable cationic titanium alkyl complexes [32].

$$Cp_2M\begin{smallmatrix}Cl\\Cl\end{smallmatrix} + AlR_3 \longrightarrow Cp_2M\begin{smallmatrix}R\\Cl\end{smallmatrix}\text{-}AlR_2Cl \rightleftharpoons \left[Cp_2M\begin{smallmatrix}R\\\square\end{smallmatrix}\right]^+ AlR_2Cl_2^- \quad (15)$$

3.2 BY NON-COORDINATING ANIONS

In contrast with most MAO-containing catalyst systems, which start with metallocene dihalide complexes, non-aluminoxane-containing catalyst systems can start from alkylated metallocene complexes such as Cp$_2$ZrR$_2$ and avoid the alkylation requirement. Non-MAO cocatalysts also combine the formation of the coordinatively unsaturated cation with the introduction of a non-coordinating anion. Three examples of types of reactions that the MAO uses can be given. First, a strong Lewis acid can remove an anionic ligand (such as alkyl group) from the metallocene [33].

$$Cp_2MMe_2 + B(C_6F_5)_3 \longrightarrow [Cp_2M(Me)]^+ + [CH_3B(C_6F_5)_3]^- \quad (16)$$

The Lewis acid may also be the cation part of a salt [34]:

$$Cp_2MMe_2 + [Ph_3C][B(C_6F_5)_4] \longrightarrow [Cp_2M(Me)]^+ + [B(C_6F_5)_4]^- + Ph_3CMe \quad (17)$$

A second approach, to remove the fourth ligand, generate the metallocene cation and introduce a non-coordinating anion, is to use the salt of the non-coordinating anion with a cationic oxidizing agent [35]:

$$2Cp_2MMe_2 + 2Ag^+ \longrightarrow 2[Cp_2M(Me)]^+ + CH_3CH_3 + 2Ag^0 \quad (18)$$

The third method is to use a proton source such as an ammonium salt of the non-coordinating ligand [36]:

$$Cp_2MMe_2 + [R_3N][B(C_6F_5)_4] \longrightarrow$$
$$[Cp_2M(Me)]^+ + [B(C_6F_5)_4]^- + R_3N + MeH \quad (19)$$

As substitutes for MAO, bulky anions are generally used to stabilize cationic metallocene complexes in order to maintain their activity of polymerization. The bulky anion must be non-coordinating and must be chemically very stable so that it does not react with the highly reactive metallocene cation. As the most effective bulky anions for the generation of metallocene cations, non-coordinating anions such as [HNMe$_2$Ph][B(C$_6$F$_5$)$_4$], [HNMePh$_2$][B(C$_6$F$_5$)$_4$] and [Ph$_3$C][B(C$_6$F$_5$)$_4$] have been known to be effective. In order to use the *ansa*-metallocene dichloride

$^{ch}Cp_2ZrCl_2$ for the generation of cationic zirconium species by using bulky anions, it must be previously alkylated in a separate synthetic process to obtain $^{ch}Cp_2ZrR_2$. However, since the *ansa*-metallocene diamide $^{ch}CpZr(NR_2)_2$ is stoichiometrically alkylated by common alkyl aluminums (such as $AlMe_3$), it can be used directly to afford methylzirconium cations without transforming it into the corresponding dialkyl complex by utilizing non-coordinating anions during polymerization. Kim and co-workers [30] carried out sequential reactions of *rac*-1, $AlMe_3$ and various non-coordinating anions in an NMR tube in order to prove *in situ* generation of methylzirconium cations. Addition of 1 equiv, of $[HNMePh_2][B(C_6F_5)_4]$ to the solution ($Al_2Me_6 : rac$-1 = 2:1) containing *rac*-(EBI)ZrMe$_2$, resulted from reaction of *rac*-1 and 2 equiv. of Al_2Me_6 [reactions (12 and (13)], affords the immediate formation of $[rac$-(EBI)Zr(μ-Me)$_2$AlMe$_2]^+$, the adduct of the base-free *rac*-[(EBI)ZrMe]$^+$ cation and $AlMe_3$, which was previously identified by Bochmann and Lancaster as the principal component in mixtures of these species. The resonances of the cationic heterodinuclear complexes and the Al amide species are broadened, presumably owing to the reversible formation of $NMePh_2$ adducts. The cationic complex $[rac$-(EBI)Zr(μ-Me)$_2$AlMe$_2]^+$ may undergo loss or displacement of $AlMe_3$, ultimately leading to *rac*-[(EBI)ZrMe]$^+$ or *rac*-[(EBI)Zr(Me)(propene)]$^+$ species. This was indirectly confirmed by the NMR-scale polymerization of propylene. Liquid propylene (0.5 ml) was added to the mixture containing $[rac$-(EBI)Zr(μ-Me)$_2$AlMe$_2]^+$ at $-78\,^\circ C$ and then the temperature was slowly increased to room temperature. White *i*PP showing a *meso* pentad value of 84.7 % was isolated in this procedure.

In similar sequential reactions of *rac*-1, 2 equiv. of Al_2Me_6 and 1 equiv. of $[Ph_3C][B(C_6F_5)_4]$ (instead of $[HNMePH_2][B(C_6F_5)_4]$), the same cationic species $[rac$-(EBI)Zr(μ-Me)$_2$AlMe$_2]^+$ were obtained in stoichiometric yield [30]. By changing the *ansa*-metallocene amide to *rac*-(EBI)Zr(NC$_4$H$_8$)$_2$, *rac*-(SBI)Zr(NMe$_2$)$_2$ or *rac*-Me$_2$Si(1-C$_5$H$_2$-2-Me-4-^{t}Bu)$_2$Zr(NMe$_2$)$_2$, and by using various kinds of non-coordinating anions such as $[Ph_3C][B(C_6F_5)_4]$, $[HNMePh_2][B(C_6F_5)_4]$, $[HNMe_2Ph]$ $[B(C_6F_5)_4]$ and $[HNEt_2Ph][B(C_6F_5)_4]$, Kim and co-worikers [30] found thaat cationic $[rac$-(EBI)Zu(μ-Me)$_2$AlMe$_2]^+$ species are generated *in situ* Hence the formation of base-free methylzirconium cations by non-coordinating anions can be summarized as shown in Scheme 2.

The solution mixture containing cationic $[rac$-(EBI)Zr(μ-Me)$_2$AlMe$_2]^+$species is very stable, so that the mixture was not decomposed after storage for about 1 month at room temperature and even afer heating to 70°C for 3 h. The structure of $[rac$-(EBI)Zr(μ-Me)$_2$AlMe$_2]^+$ in solution is assumed to be different from that of cationic species $[rac$-(EBI)ZrMe]$^+$ generated by using MAO. The chemical shift of methyl in $[rac(EBI)Zr(\mu$-Me)$_2$AlMe$_2]^+$ is -0.63 ppm, whereas that of $[rac$-(EBI)ZrMe]$^+$ is -0.67 ppm [30].

Similar NMR-scale reactions performed by using coordinating anions such as $[HNMe_2Ph][BPh_4]$ and $[HNBu_3][BPh_4]$ showed that the effectiveness of combining the formation of $[rac$-(EBI)Zr(μ-Me)$_2$AlMe$_2]^+$ cations with the introduction of a

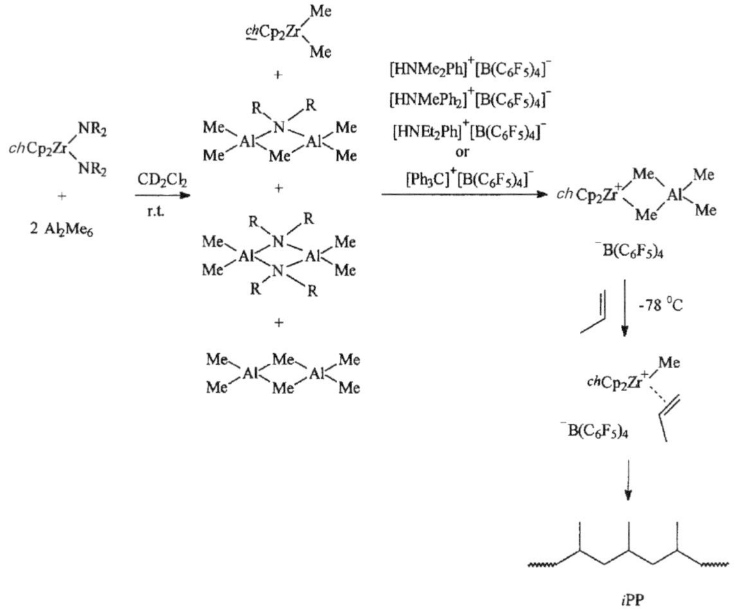

Scheme 2

non-coordinating anion is relatively poor. By using [HNMe$_2$Ph][BPh$_4$] and [HNBu$_3$][BPh$_4$] as co-activators, 2 and 75 % of *in situ*-generated *rac*-(EBI)ZrMe$_2$ (Al$_2$Me$_6$: *rac*-**1** = 2) were activated to fom base-free methylzirconium cations, respectively. These results suggest that a considerable amount of [BPh$_4$] anions is not coordinated to the zirconium, but instead is present as a mixture of solvated and ion-paired species.

Since MAO-containing catalyst systems may also start with the alkylated metallocene complex, a different amount of MAO was introduced into the solution mixture (Al$_2$Me$_6$: *rac*-**1** = 2 : 1) containing *in situ*-generated *rac*-(EBI)ZrMe$_2$. With 10 equiv. of MAO, 65% of *rac*-(EBI)ZrMe$_2$ was converted into [*rac*-(EBI)ZrMe]$^+$, and the remaining *rac*-(EBI)ZrMe$_2$ was completely activated to give [*rac*-(EBI)ZrMe]$^+$ by a further equiv. of MAO. These results demonstrate that a relatively smaller amount of MAO is needed for the activation of *rac*-**1** if it is preliminarily methylated by AlMe$_3$. In this way the amount of MAO required for the activation of *rac*-**1** may be decreased.

Since *in situ* generation of cationic species by using non-coordinating anions from an *ansa*-metallocene diamide ^{ch}CpZr(NR$_2$)$_2$ is made via its prior methylation with Al$_2$Me$_6$, it is expected that changing the order of reaction of the three reactants will influence the formation route of the cationic species. The effect of changing the order of reaction was investigated by first reacting *rac*-**1** with [HNMePh$_2$][B(C$_6$F$_5$)$_4$] before adding Al$_2$Me$_6$ [30]. Cationic Zr species containing an amide as a ligand,

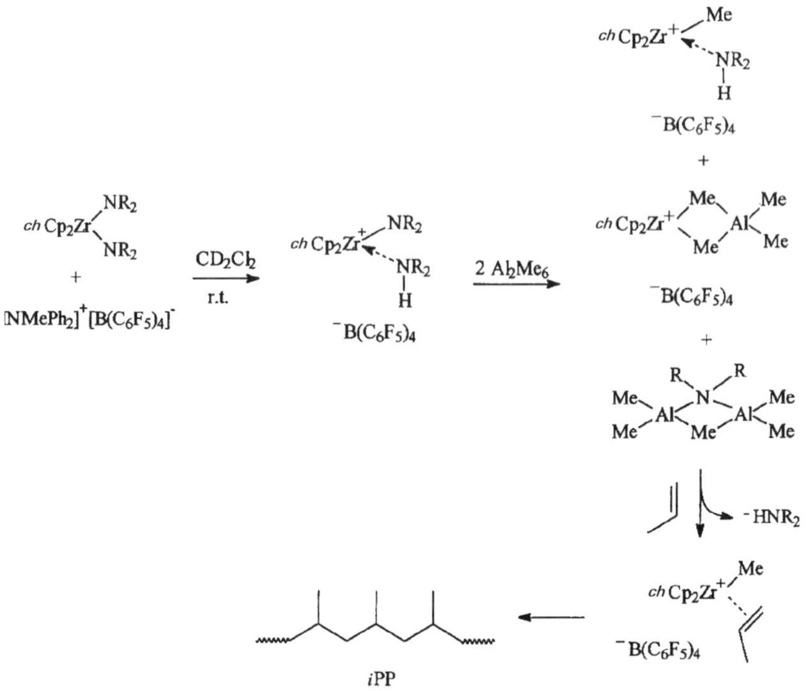

Scheme 3

$[(EBI)Zr(NMe_2)]^+$, were formed. NMR-scale polymerization of propylene has been tried by adding liquid propylene to this solution mixture containing $[(EBI)Zr(NMe_2)]^+$, but resulted in no iPP. This is because the nitrogen atom of the $HNMe_2$ byproduct is strongly coordinated to the Zr center. The cationic species $[(EBI)Zr(NMe_2)]^+$ were subsequently alkylated by adding 2 equiv. Al_2Me_6. However, the NMe_2 ligands are not completely transformed into methyl groups to form cationic $[(EBI)ZrMe]^+$ species, i.e. about 50 % of it remains unreacted, presumably owing to the strong coordination of $HNMe_2$ to the metal center. $HNMe_2$ remains coordinated even to the alkylated cationic $[(EBI)ZrMe]^+$ species and a small amount of rac-$(EBI)Zr(\mu$-$Me)_2AlMe_2]^+$ is also formed. Hence, the generation of cationic active species by changing the order of reaction is inefficient, and can be summarized as shown in Scheme 3.

4 POLYMERIZATION OF PROPYLENE

To determine if $^{ch}CpZr(NR_2)_2$ compounds could be directly utilized for the olefin polymerization in the presence of MAO or non-coordinating anions, Kim and

Run	Catalyst[a]	μmol of catalyst	AlR_3	mmol of AlR_3	Cocatalyst[b,c]	μmol of cocatalyst	Time (min)	$\bar{R}_p \times 10^{-6}$ [g PP/(mol Zr h)]	% mmmm	T_m (°C)	$\bar{M}_w \times 10^{-3}$	$\bar{M}_w/\bar{M}_n$
1	rac-1	5.7			MAO	5600	60	0.27	74.9	135.1	29.4	1.9
2	rac-1	5.7			MAO	5800	60	0.26				
3	rac-4	5.4			MAO	5200	60	0.97	79.0			
4	rac-(EBI)ZrCl$_2$	5.9			MAO	5800	96	6.7	73.4			
5	rac-(SBI)ZrCl$_2$	5.5			MAO	4300	106	6.6	81.5			
6	rac-1	5.7	$AlMe_3$	520	MAO	4300	90	5.9	79.8			
7	rac-1	5.7	$AlMe_3$	2080	MAO	5000	60	6.1	76.9	128.8	11.9	1.9
8	rac-1	5.7	$Al(^iBu)_3$	200	MAO	4300	90	6.6				
9	rac-1	5.7	$Al(^iBu)_3$	2000	MAO	5000	60	8.1	82.9	134.8	25.4	2.1
10	rac-4	5.4	$AlMe_3$	2080	MAO	4300	90	8.7				
11	rac-4	5.4	$Al(^iBu)_3$	790	MAO	4200	60	6.3	82.1	135.9	33.9	2.2
12	rac-1	5.7	$AlMe_3$	520	CPh_3^+	5.7	126	2.6	69.2			
13	rac-1	5.7	$Al(^iBu)_2H$	280	CPh_3^+	5.4	85	6.3	72.2			
14	rac-1	5.7	Al^iPr_3	260	CPh_3^+	5.4	100	5.4	80.0			
15	rac-1	5.7	$AlMe_3$	520	$HNMePh_2^+$	5.7	96	2.1	76.2			
16	rac-1	5.7	$AlMe_3$	520	$HNMePh_2^+$	5.7	66	4.8	78.8			
17	rac-1	5.7	$Al(^iBu)_2H$	560	$HNMe_2Ph^+$	5.7	150	2.1	86.3			
18	rac-4	5.6	Al^iPr_3	2080	CPh_3^+	6.3	66	9.9	79.6			
19	rac-4	5.4	Al^iPr_3	260	CPh_3^+	5.4	60	5.9	81.5			
20	rac-4	5.4	$Al(^iBu)_3$	1000	CPh_3^+	6.0	90	6.1				
21	rac-4	5.4	$AlMe_3$	2080	$HNMe_2Ph^+$	7.5	77	7.5				
22	rac-4	5.3	$AlMe_3$	1000	$HNMePh_2^+$	5.2	60	8.5	85.2	141.3	29.4	1.9
23[d]	rac-2	3.4			MAO	12000	60	4.2	82.4	137.8	18.4	1.8
24[d]	rac-2	3.4	$Al(^iBu)_3$	340	CPh_3^+	3.4	60	8.9	84.9	135.8	36.2	2.1
25	rac-3	2.5	$Al(^iBu)_3$	25	CPh_3^+	1.4	60	4.0		129.0	170.0[e]	
26	rac-6	3.1			MAO	15000	60	2.5	87.0	142.2	3.1	1.6
27	rac-6	3.1	$Al(^iBu)_3$	1800	CPh_3^+	3.1	60	5.9	87.1	141.3	3.6	1.6

[a] rac-(EBI)Zr(NMe$_2$)$_2$ (rac-1); rac-(EBI)Zr(NC$_4$H$_8$)$_2$ (rac-2); rac-(EBI)Hf(NMe$_2$)$_2$ (rac-3); rac-(SBI)Zr(NMe$_2$)$_2$ (rac-4); rac-Me$_2$Si(1-C$_5$H$_2$-2-Me-4-^{t}Bu)$_2$Zr(NMe$_2$)$_2$ (rac-6).
[b] mmol of MAO = mmol of Al, of which 82 wt% = 'AlMeO' and 18 wt% = AlMe$_3$.
[c] Anion = B(C$_6$F$_5$)$_4^-$ for CPh$_3^+$ and HNR$_3^-$ cocatalysts.
[d] Polymerization temperature = 30 °C.
[e] Viscosity-average molecular weight.

co-workers compared the activities of propylene polymerization by rac-(EBI)Zr(NMe$_2$)$_2$ (rac-1), rac-(EBI)Zr(NC$_4$H$_8$)$_2$ (rac-2), rac-(EBI)Hf(NMe$_2$)$_2$ (rac-3), rac-(SBI)Zr(NMe$_2$)$_2$ (rac-4) and rac-Me$_2$Si(1-C$_5$H$_2$-2-Me-4-^{t}Bu)$_2$Zr-(NMe$_2$)$_2$ (rac-6) with those by rac-(EBI)ZrCl$_2$/MAO and rac-(SBI)ZrCl$_2$/MAO [30]. Table 1 shows the results of polymerizations. The ^{ch}CpZr(NR$_2$)$_2$ compounds are significantly less active than conventional chloride-derived catalysts under the same conditions (48 °C, 1 atm of propylene, Al : Zr = 1000 : 1), although the stereoselectivity is comparable [31]. GPC results for the polymers produced by ^{ch}CpZr(NR$_2$)$_2$/ MAO and ^{ch}CpZr(NR$_2$)$_2$/AlR$_3$/non-coordinating anions are consistent with single-site behavior.

One possible reason for the lower activities observed for the catalysts based on rac-(EBI)Zr(NMe$_2$)$_2$ and rac-(SBI)Zr(NMe$_2$)$_2$ is that these amide complexes are not efficiently alkylated by MAO (or the AlMe$_3$ contained therein). In an effort to circumvent this potential problem, rac-1 and rac-4 were pretreated with excess AlR$_3$ prior to activation with MAO, such that the total Al:Zr ratio was maintained in the range 800:1 to 1300:1. The results of these polymerizations using AlMe$_3$ or Al(^{i}Bu)$_3$ are summarized in runs 6–11. This procedure produces catalysts with activities and stereoselectivities which are comparable to those of rac-(EBI)ZrCl$_2$/MAO or rac-(SBI)ZrCl$_2$/MAO catalysts.

The amide compounds ^{ch}CpZr(NR$_2$)$_2$ could also be used with non-coordinating anions (CPh$_3$$^+$, HNR$_3$$^+$, etc.) after prealkylating with AlR$_3$, as illustrated by runs 12–27. The amide compounds are alkylated with small amount of AlR$_3$ and subsequent addition of anionic compounds leads to high-activity catalytic systems. GPC results for representative examples are characteristic of single-site behavior.

5 REFERENCES

1. (a) Breslow, D. S. and Newburg, N. R. *J. Am. Chem. Soc.*, **79**, 5072 (1957); (b) Natta, G., Pino, P., Mazzanti, G. and Giannini, U., *J. Am. Chem. Soc.*, **79**, 2976 (1957).
2. Sinn, H. and Kaminsky, W. *Adv. Organomet. Chem.*, **18**, 99 (1980).
3. (a) Smith, J., von Seyerl, J., Huttner, G. and Brintzinger, H. H., *J. Organomet. Chem.*, **173**, 175 (1979); (b) Schnutenhaus, H. and Brintzinger, H. H., *Angew. Chem., Int. Ed. Engl.*, **18**, 777 (1979); (c) Smith, J. A. and Brintzinger, H. H., *J. Organomet. Chem.*, **218**, 159 (1981); (d) Wild, F. R. W. P., Zsolnai, L., Hutter, G. and Brintzinger, H. H., *J. Organomet. Chem.*, **232**, 233 (1982); (e) Wild, F. R. W. P., Wasiucionek, M., Hutter, G. and Brintzinger, H. H., *J. Organomet. Chem.*, **288**, 63 (1985); (f) Wochner, F., Zsolani, L., Hutter, G. and Brintzinger, H. H., *J. Organomet. Chem.*, **288**, 69 (1985); (g) Burger, P., Hortmann, K., Diebold, J. and Brintzinger, H. H., *J. Organomet. Chem.*, **417**, 9 (1991).
4. Reviews: (a) Thayer, A. M., *Chem. Eng. News*, **73** (37), 15 (1995); (b) Brintzinger, H. H., Fischer, D., Mulhaupt, R., Rieger, B. and Waymouth, R. M., *Angew. Chem., Int. Ed. Engl.*, **34**, 1143 (1995); (c) Sinclair, K. B. and Wilson. R. B., *Chem. Ind.* (London) 857 (1994); (d) Mohring, P. C. and Coville, N. J., *J. Organomet. Chem.*, **479**, 1 (1994); (e) Horton, A. D., *Trends Polym. Sci.*, **2**, 158 (1994); (f) Spaleck, W., Antberg, M., Dolle, V., Klein, R., Rohrmann, J. and Winter, A., *New. J. Chem.*, **14**, 499 (1990).

5. (a) Stehling, U., Diebold, J., Kirsten, R., Roll, W., Brintzinger, H. H., Jungling, S., Mulhaupt, R. and Langhauser, F., *Organometallics*, **13**, 964 (1994); (b) Spaleck, W., Kuber, F., Winter, A., Rohrmann, J., Bachmann, B., Antberg, M., Dolle, V. and Paulus, E. F., *Organometallics*, **13**, 964 (1994); (c) Spaleck, W., Antberg, M., Rohrmann, J., Winter, A., Bachmann, B., Kiprof, P., Behm, J. and Hermann, W. A., *Angew. Chem. Int. Ed. Engl.*, **31**, 1347 (1992); (d) Chacon, S. T., Coughlin, E. B., Henling, L. M. and Bercaw, J. E., *J. Organomet. Chem.*, **479**, 171 (1995).
6. (a) Collins, S., Kuntz, B. A., Taylor, N. J. and Ward, D. G., *J. Organomet. Chem.*, **342**, 21 (1988); (b) Collins, S., Kuntz, B. A. and Hong, Y., *J. Org. Chem.*, **54**, 4154 (1989).
7. Grossman, R. B., Doyle, R. A. and Buchwald, S. L., *Organometallics*, **10**, 1501 (1991).
8. Reinmuth, A., PhD Dissertation, University of Konstanz (1992).
9. Wisenfeldt, H., ReinMuth, A., Barsties, E., Evertz, K. and Brintzinger, H. H. *J. Organomet. Chem.*, **369**, 359 (1989).
10. Mise, T., Miya, S. and Yamazaki, H., *Chem. Lett.*, 1853 (1989).
11. (a) Lappert, M. F., Power, P. P., Sanger, A. R. and Srivastava, R. C., *Metal and Metalloid Amides*, Ellis Horwood, Chichester, 1980; (b) Bradely, D. C., *Adv. Inorg. Chem. Radiochem.*, **15**, 259 (1972); (c) Jenkins, A. D., Lappert, M. F. and Srivastava, A. C., *J. Organomet. Chem.*, **23**, 165 (1970).
12. Chandra, G. and Lapert, M. F., *J. Chem. Soc. A*, 1940 (1986).
13. Hughes, A. K., Meetsma, A. and Teuben, J. H., *Organometallics*, **12**, 1936 (1993).
14. Carpenetti, D. W., Kloppenburg, L., Kupec, J. T. and Petersen, J. L., *Organometallics*, **15**, 1572 (1996).
15. (a) Jordan, R. F. and Diamond, G. M., *Int. Pat. Appl.*, WO 9 532 979 (1995); (b) Diamond, G. M., Rodewald, S. and Jordan, R. F., *Organometallics*, **14**, 5 (1995); (c) Diamond, G. M., Jordan, R. F. and Petersen, J. L., *J. Am. Chem. Soc.*, **118**, 8024 (1996).
16. (a) Diamond, G. M., Jordan, R. F. and Petersen, J. L., *Organometallics*, **15**, 4030 (1996); (b) Diamond, G. M., Jordan, R. F. and Petersen, J. L., *Organometallics*, **15**, 4045 (1996); (c) Christopher, J. N., Diamond, G. M., Jordan, R. F. and Petersen, J. L., *Organometallics*, **15**, 4038 (1996).
17. Kim, I., Nishihara, Y., Jordan, R. F., Rogers, R. D., Rheingold, A. L. and Yap, G. P. A., *Organometallics*, **16**, 3314 (1997).
18. (a) Bowen, D. E., Jordan, R. F. and Rogers, R. D., *Organometallics*, **14**, 3630 (1995); (b) Black, D. G., Swenson, D. C., Jordan, R. F. and Rogers, R. D., *Organometallics*, **13**, 3539 (1995); (c) Duan, Z. and Verkade, J. G., *Inorg. Chem.*, **34**, 4311 (1995); (d) Galakhov, M., Martin, A., Mena, M. and Yelamos, C., *J. Organomet. Chem.*, **496**, 217 (1995); (e) Lemke, F. R., Szalda, D. J. and Bullock, R. M., *J. Am. Chem. Soc.*, **113**, 8466 (1991).
19. (a) Bradley, D. C. and Thomas, I. M., *Proc. Chem. Soc.*, 225 (1959); (b) Bradley, D. C. and Thomas, I. M., *J. Chem. Soc.*, 3857 (1960).
20. Gauthier, W. J., Corrigan, J. F., Taylor, N. J. and Collins, S., *Macromolecules*, **28**, 3771 (1995).
21. Benzing, E. and Kornicker, W., *Chem. Ber.*, **94**, 2263 (1961).
22. Wades, S. R. and Wiley, G. R., *J. Chem. Soc. Dalton Trans.*, 1264 (1981).
23. Shannon, R. D., *Acta Crystallogr. Sect. A*, **32**, 751 (1976).
24. (a) Haaland, A., Rypdal, K., Volden, H. V. and Andersen, R. A., *J. Chem. Soc., Dalton Trans.*, 891 (1991); (b) Martin, A., Mena, M., Yelamos, C., Serrano, R. and Raithby, P. R., *J. Organomet. Chem.*, **469**, 79 (1994).
25. (a) Pupi, R. M., Coalter, J. N. and Petersen, J. L., *J. Organomet. Chem.*, **497**, 17 (1995); (b) Bai, Y., Roesky, H. W., Noltemeyer, M. and Witt, M., *Chem. Ber.*, **125**, 825 (1992); (c) Sartin, W. J., Huffman, J. C., Lundquist, E. G., Streib, W. G. and Caulton, K. G., *J. Mol. Catal.*, **56**, 20 (1989); (d) Chisholm, M. H., Hammond, C. E. and Huffman, J. C., *Polyhedron*, **7**, 2515 (1988).

26. Hermann, W. A., Morawietz, M. J. A. and Priermeier, T., *Angew. Chem. Int. Ed. Engl.*, **33**, 1946 (1994).
27. Hermann, W. A. and Morawietz, M. J. A., *J. Organomet. Chem.*, **482**, 169 (1994).
28. Kaminsky, W., Bark, A. and Steiger, R., *J. Mol. Catal.*, **74**, 109 (1992).
29. Siedle, A. R., Lamanna, W. M., Olofson, J. M., Nerad, B. D. and Newmark, R. A., *ACS Symp. Ser.*, **517**, 156 (1993).
30. (a) Kim, I. and Jordan, R. F., *Macromolecules*, **29**, 491 (1996); (b) Jordan, R. F., Diamond, G. M., Christopher, J. N., and Kim, I., *Polym. Prepr.*, **37**, 256 (1996); (c) Kim, I. and Jordan, R. F., in *Proceedings of the 4th International Congress on Metallocene Polymers (Metallocenes Asia '97)*, Schotland Business Research, Skillman, 1997, p. 253; (d) Kim, I. and Jordan, R. F., *Polym. Bull.*, **39**, 325 (1997); (e) I. Kim, *J. Macromol. Sci. Pure Appl. Chem.*, **A35**, 293 (1998); (f) I. Kim, *J. Appl. Polym. Sci.*, **71**, 875 (1999).
31. (a) Fink, G., Fenzl, W. and Mynott, R., *Z. Naturforsch., Teil B*, **40**, 158 (1985); (b) Fink, G., Mynott, R., and Fenzl, W., *Angew. Makromol. Chem.*, **154**, 1 (1987).
32. Bochmann, M. and Wilson, L. M., *J. Chem. Soc., Chem. Commun.*, 1610 (1986).
33. Sishta, C., Hathorn, R. and Marks, T., *J. Am. Chem. Soc.*, **114**, 1112 (1992).
34. Chien, J. C. W., Tsui, W.-M. and Rausch, M. D., *J. Am. Chem. Soc.*, **113**, 3623 (1991).
35. Jordan, R., Dasher, W. and Echols, S., *J. Am. Chem. Soc.*, **108**, 1718 (1986).
36. Yang, X., Stern, C. L. and Marks, T. J., *J. Am. Chem. Soc.*, **113**, 3623 (1991).
37. (a) Bochmann, M. and Lancaster, S. J., *Organometallics*, **12**, 663 (1993); (b) Bochmann, M. and Lancaster, S. J., *Angew. Chem. Int. Ed. Engl.*, **33**, 1637 (1993).
38. (a) Tsai, W. -M. and Chien, J. C. W., *J. Polym, Sci. A: Polym. Chem.*, **32**, 194 (1994); (b) Busico, V. and Cioullo, R., *J. Am. Chem. Soc.*, **116**, 9329 (1994); (c) Fischer, D. and Mulhaupt, R., *Macromol. Chem. Phys.*, **159**, 1433 (1994); (d) Herfert, N. and Fink, G., *Makromol. Chem., Rapid Commun.*, **14**, 91 (1993); (e) Tsai, W.-M., Rausch, M. D. and Chien, *J. Appl. Organomet. Chem.*, **7**, 71 (1993); (f) Kaminsky, W., Engehausen, R., Zoumis, K., Spaleck, W. and Rohrmann, J., *Makromol. Chem.*, **193**, 1643 (1992); (g) Huang, J. and Rempel, G. L., *Prog. Catal.*, 169 (1962); (h) Collins, S., Gauthier, W. J., Holden, D. A., Kuntz, B. A., Taylor, N. J. and Ward, D. G., *Organometallics*, **10**, 2061 (1991); (i) Rieger, B., Mu, X., Mallin, D. T., Rausch, M. D. and Chien, J. C. W., *Macromolecules*, **21**, 617 (1988); (j) Mise, T., Miya, S. and Yamazaki, H., *Chem. Lett.*, 1853 (1989); (k) Grassi, A., Zambelli, A., Resconi, L., Albizzati, E. and Mazzocchi, R., *Macromolecules*, **21**, 617 (1988); (l) Tsutsui, T., Ishimaru, N., Mizuno, A., Toyota, A., and Kashiwa, N., *Polymer*, **30**, 1350 (1989).

8

Supported Metallocene Polymerization Catalysis

JAMES C. W. CHIEN
Amherst Polymer Technology, Inc., Amherst, MA, USA

1 INTRODUCTION

Supported catalysts are used in industry to manufacture isotactic polypropylene by $TiCl_3$ supported on $MgCl_2$, high-density polyethylene by CrO_3/silica, etc. Although metallocene possesses unprecedented olefin polymerization activity (A) and selectivity, it must be heterogenized for drop-in replacement in slurry or gas-phase plant.

Chien and Hsieh (1976) prepared catalyst systems of R_4Ti, R_3TiCl (R = benzyl, p-methylbenzyl), $Cp_2M_{tr}(CH_3)_2$ (M_{tr} = Ti, Zr, Hf) supported on alumina, zeolite, silica, alumina, $Mg(OH)Cl$, $Ca(OH)_2$ and poly(vinyl alcohol) and used it to polymerize both ethylene and propylene, the latter to isotactic polypropylene. Subsequent supported catalysts reported are Cp_2TiCl_2/SiO_2 by Slotfeldt-Ellingsen et al. (1980) and chemically anchored analogs (Jackson et al., 1977; Reissová et al., 1986). $AlEt_nCl_{3-n}$ was the cocatalyst employed in these studies.

Supported metallocene/MAO catalysts was first suggested by Sinn et al. (1983) and Kaminsky et al. (1985). Many types of supported metallocene catalyst have since been reported; they have been surveyed recently by Ribeiro et al. (1997).

The advantages and disadvantages of supported metallocene polymerization catalyst vis-à-vis the homogeneous analog are common knowledge. The improvements a supported catalyst need to achieve in order for it to be usable in existing manufacturing facilities are the following.

This contribution has been reprinted from the journal Topics in Catalysis (ISSN 1022-5528, www. baltzer.nl) Volume 7 (1999) entitled Advances in Polymerization Catalysis. Catalysts and Processes, Editors: Tobin J. Marks and James C. Stevens. The editors gratefully acknowledge the permission given by J. C. Baltzer Science Publishers to reproduce this chapter.

Metallocene-based Polyolefins   Edited by J. Scheirs and W. Kaminsky
© 2000 John Wiley & Sons Ltd

(i) Homogeneous metallocenes can have maximum activity for ethylene polymerization in excess of 10^{10} g PE (mol Zr [C_2H_4] h) and exceeding 10^9 g PP/(mol Zr [C_3H_6] h) for propylene polymerization (Tsai *et al.*, 1993; Tsai and Chien, 1994; Chen *et al.*, 1995a,b). The supported version should have activity approaching the homogeneous value.

(ii) The activity of some homogeneous metallocene catalysts suffer significant loss during the course of a polymerization. The supported system needs to be more stable.

(iii) A very large quantity of methylaluminoxane (MAO) cocatalyst is required to achieve high activity. This amount has to be greatly reduced for economical viability.

(iv) Most metallocenes in solution produce polyolefins of very low molecular weight (MW), especially at high polymerization temperature (T_p). The supported catalyst should do better.

(v) The polyolefins obtained with homogeneous processes have a very fine particle size with a broad size distribution and low bulk density. Supported catalyst must be able to control the polymer morphology.

(vi) Metallocenes in solution give polymers too narrow in MW distribution (MWD) and consequently poor processability–physicomechanical property balance. The support system should have the capability of giving polymers having desired broad or bimodal MWD for good rheological and physical properties.

(vii) Some supported metallocene catalysts have been reported to cause fouling of gas-phase reactors. Corrective measures must be found.

(viii) The steric control of metallocenes is determined by their molecular structure. It would be advantageous if this can be controlled or even altered by the support.

In this chapter we review the present state of supported metallocene catalysis with respect to the goals set forth above, speculate about the reasons for their observed polymerization activity and specificity and discuss the prospects of future improvements. The patent literature is not discussed to avoid issues about validity, priority, etc. The notation used for a supported catalyst, D/C/B/A–E, is an abbreviation for support A, treated in turn with B, C and D, while E is the activator/cocatalyst employed for polymerization.

2 INORGANIC SUPPORT MATERIALS

The most commonly employed inorganic supports for Ziegler–Natta catalysts are silica (SiO_2), alumina (Al_2O_3) and magnesium chloride ($MgCl_2$). Some other materials have been investigated such as zeolite, MgO, MgF_2, CaF_2 and AlF_3, but they are not yet of commercial consequence.

Typical reactive surface functionalities of a support are Brønsted acidic OH groups, Lewis basic oxide groups and Lewis acidic metal centers. Reaction of zirconocene precursor (L_2ZrX_2, $X = CH_3$, Cl) with surface hydroxyl ($HO—M—$, $M = Si$, Al) fixes the former via a μ-oxo bond to the surface:

$$L_2ZrX_2 + HO—M—\ \longrightarrow\ L_2Zr(X)—O—M— + HX \qquad (1)$$

A similar μ-oxo complex was proposed by Marks (1992) to be formed upon the reaction of zirconocene with a Lewis basic metal oxide group:

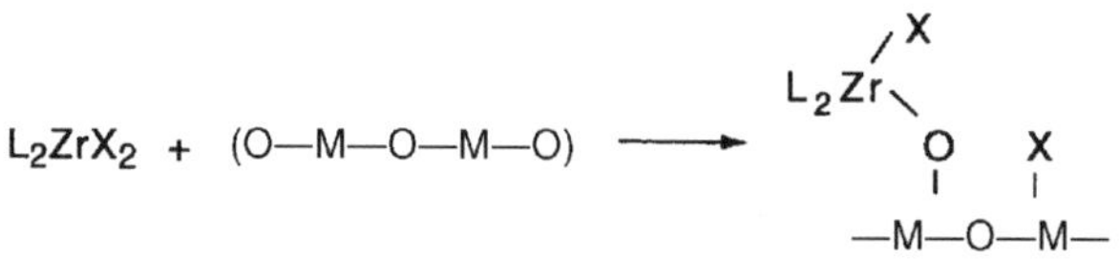

$$
L_2ZrX_2 + (O—M—O—M—O) \longrightarrow
\begin{array}{c}
\quad\; _{/}X \\
L_2Zr_{\diagdown} \\
O\quad X \\
|\quad\;\; | \\
—M—O—M—
\end{array}
\qquad (2)
$$

The extraction of X^- by a cocatalyst affords $L_2Zr^+ —O—M$; this species cannot catalyze olefin insertion because it lacks a Zr–alkyl bond.

The chemistry of the Lewis acidic metal center in Al_2O_3 and $MgCl_2$ has been investigated. The ^{13}C CP/MAS NMR spectra of $Cp_2Th(^{13}CH_3)_2$ (Toscano and Marks, 1986; Hedden and Marks, 1988; Finch et al., 1990) and of $Cp_2Zr(^{13}CH_3)_2$ (Dahmen et al., 1988) exhibit resonance characteristics of 'cation-like' species. The spectra led to a remarkable conclusion: a methide anion has been transferred from the Th or Zr center to a Lewis acid site on the surface. The former becomes a 'cation-like' (zirconium cation, zirconocenium) species:

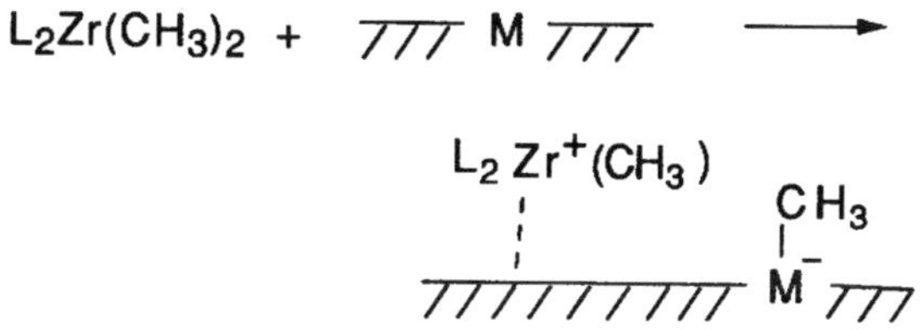

$$
L_2Zr(CH_3)_2\ +\ /\!/\!/\ M\ /\!/\!/\ \longrightarrow
$$

$$
\begin{array}{c}
L_2\,Zr^+(CH_3) \\
| \\
| \qquad\qquad\quad CH_3 \\
\qquad\qquad\qquad\;\; | \\
\overline{/\!/\!/\!/\!/\!/\!/\!/}\ M\ \overline{/\!/\!/}
\end{array}
\qquad (3)
$$

where M is Mg or Al but not Si, and it applies also to Th complexes. These metallocenium species with σ-bonded alkyls are active for catalysis of olefin polymerization.

2.1　SILICA

Silicas with a wide range of surface area, porosity and pore volume are articles of commerce. The nature of the silanol groups has been investigated with many techniques: paramagnetic probe (Chien, 1971a,b), infrared spectroscopy (Nowlin et al., 1991; Yermakov et al., 1981; Ihm et al., 1994), titrations (Chien, 1971a; Collins et al., 1992), and others.

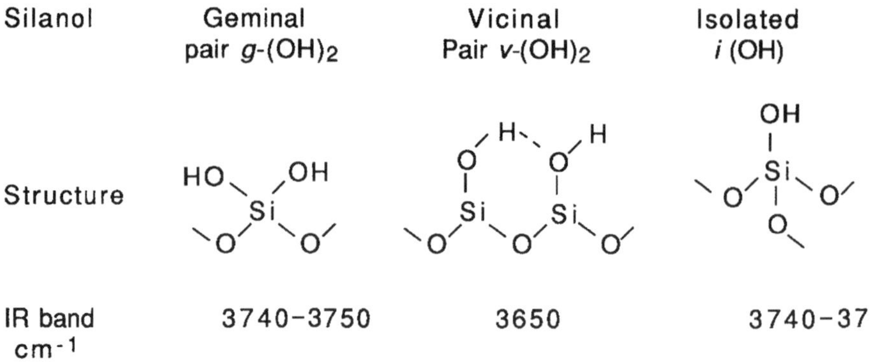

Silanol	Geminal pair g-(OH)$_2$	Vicinal Pair v-(OH)$_2$	Isolated i (OH)
Structure	HO、 ,OH \Si, ᐟO Oᐟ	,H·、 ,H O O \Si, ,Si, ᐟO Oᐟ Oᐟ	OH | \O,ᐟSi,Oᐟ O、
IR band cm^{-1}	3740-3750	3650	3740-37

Figure 1 Various types of silanol groups

Silica gel (HS) has a maximum number of eight Brønsted acidic OH groups per nm^2 (4 mmol OH/g for silica with a surface area) of approximately 300 m^2/g they are present as geminal pairs (Figure 1).

Heat treatment of silica gel at 200 °C (T_d) gave a partially dehydroxylated silica (PDS) having 2.3 mmol OH/g, half of which as geminal hydroxyl pairs and the other half as vicinal pairs. The number of hydroxyl groups decreases steadily upon heating to progressively higher T_d. Above 600 °C, a material referred to as dehydroxylated silica (DS) is obtained, which contains only 0.7 mmol OH/g. Electron paramagnetic resonance studies (Chien, 1971a,b) showed the OHs to be both geminal pairs and isolated groups.

The number of Lewis basic oxide groups is 4, 3.5 and 3.7/nm^2 for HS, PDS and DS silica, respectively. HS can also absorb large amounts of water, which are seen by IR spectroscopy at 3265–3645 cm^{-1}.

2.2 ALUMINA

The surface chemistry of γ-alumina has been studied (Benesi and Winquist, 1978; Knözinger and Ratnasamy, 1978; Kijenski and Baiker, 1989). Dehydroxylated alumina (DA) has ca 0.12 Brønsted acidic OH groups, ca 5.5 Lewis acidic Al^{3+} centers and ca 5.5 Lewis basic oxide groups per nm^2 (Marks, 1992). A partially dehydroxylated alumina would have more Brønsted acid OH groups, but the number of the other two types of surface groups is not appreciably affected by the degree of dehydroxylation.

2.3 MAGNESIUM CHLORIDE

MgCl$_2$ is a common support for third- and fourth-generation Ziegler–Natta catalysts. It has a CdI$_2$ crystal structure with a typical surface area of ca 250 m^2/g. About 24 %

of Ti attached to $MgCl_2$ surface is isospecific and 76 % is aspecific according to isotopic labeling determination of active sites (Chien and Hu, 1988, 1989; Hu and Chien 1988). The former is assigned to five-coordinate Mg sites, while the latter is thought to be highly Lewis acidic four-coordinate Mg sites. In the ^{13}C CP/MAS NMR spectra of $Cp_2'Th(^{13}CH_3)_2/MgCl_2$, about 40 ± 10 % of the surface Mg sites have extracted methide anion from Th resulting in active centers for propylene hydrogenation.

3   MAO-FREE SUPPORTED CATALYSTS

3.1   WITH TRIMETHYLALUMINUM (TMA) COCATALYST

Studies of MAO-free supported catalyst can shed light on the chemical interactions between a metallocene and the support material. Soga and co-workers (Kaminaka and Soga, 1991, 1992; Soga and Kaminaka, 1993) developed a procedure to prepare MAO-free supported catalysts. A 0.74 g amount of support (alumina, $MgCl_2$ or silica, calcined for 6 h at 400 °C and ball-milled for 48 h under N_2) was reacted with 3 mmol of TMA, followed by reaction with 0.084 mmol of zirconocene {Cp_2ZrCl_2 (**1**), $Et[THI]_2ZrCl_2$ (**3**) and $^iPr[CpFl]ZrCl_2$ (**4**)} at room temperature for 10 min. The three catalysts were analyzed and found to contain 6–12 μmol Zr/g Al_2O_3, 16–28 μmol Zr/g $MgCl_2$ and 20–29 μmol Zr/g SiO_2. They were activated with several trialkylaluminum compounds, the best one being TMA; the polymerization results are summarized in Table 1.

The data for the two *ansa*-zirconocenes supported on Al_2O_3 and $MgCl_2$ are found in experiment Nos 2-2, 2-3 and 3-2, 3-3, respectively. Their propylene polymerization activities are 12–22 % of the zirconocenes in solution (Nos 1-2, 1-3). However, supported Cp_2ZrCl_2 catalysts exhibit productivities closer to the polymerization in solution.

The catalysts of *ansa*-zirconocenes **3** and **4** supported on either Al_2O_3 or $MgCl_2$ are both more stereospecific than **3** or **4** in solution. The C_2 and C_s symmetric zirconocenes on the support produce *i*-PP and *s*-PP having higher T_m by 20 and 10 °C, respectively, and greater homosteric pentad population by 20 and 5 %, respectively. Apparently the support enhances the profacial selection by the Zr^+ center and at the same time renders the β-agostic interaction more favorable, as evidenced by the significantly lower MW polymer obtained with the heterogenized zirconocenes.

The MWDs of polypropylene (PP) obtained with Al_2O_3-supported catalysts are $\leqslant 2.0$, but the PP made with $MgCl_2$-supported catalysts are much broader in MWD (3.3–4.2). The latter is probably related to the presence of a multitude of surface sites differing in Lewis acidity as proposed by Chien (1987, 1992) and Pino *et al.* (1981).

A glaring omission in Table 1 is silica-supported systems because they are all inactive when TMA is the cocatalyst. This is consistent with the absence of strong

Table 1 Propylene polymerization^a by MAO-free supported catalysts

Support	None, homogeneous catalyst			Al$_2$O$_3$			MgCl$_2$		
Exp. No.	1-1	1-2	1-3	2-1	2-2	2-3	3-1	3-2	3-3
Zirconocene	Cp$_2$ZrCl$_2$	Et(THI)$_2$ ZrCl$_2$	Pr(CpFl) ZrCl$_2$	Cp$_2$ZrCl$_2$	Et(THI)$_2$ ZrCl$_2$	Pr(CpFl) ZrCl$_2$	Cp$_2$ZrCl$_2$	Et(THI)$_2$ ZrCl$_2$	Pr(CpFl) ZrCl$_2$
No.	1	3	4	1	3	4	1	3	4
Zr (μmol/g catalyst)	1000^c	3000^c	3250^c	10	6	12	27	16	28
TMA:Zr^b				135^b	220^b	220^b	50^b	2200^b	2200^b
Productivity (kg PP/mol Zr)	132	2,100	758	47	259	167	172	354	152
Relative to homogeneous catalysis	–	–	–	0.35	0.12	0.22	1.3	0.17	0.20
T_m (°C)	None	111	123	None	131	138	None	139	134
M_n	330	3300	39 000	390	1730	4600	370	1590	10 700
M_w/M_n	n. r.^{d}	1.9	1.8	n. r.^{d}	1.7	2.0	n. r.^{d}	4.2	3.3
[mmmm] or [rrrr] (%)^e	None	70	(77)^e	None	90	(86)^e	None	91	(81)^e

^{a} Polymerization conditions: 0.74 g catalyst, 40 °C, 18 h.
^{b} TMA is the cocatalyst.
^{c} MAO is the cocatalyst for homogeneous polymerization, TMA is inactive.
^{d} Not reported.
^{e} [rrrr] population.

Lewis acidic center in SiO_2, so no 'cation-like' intermediate is formed upon the absorption of zirconocene by SiO_2. Several investigators (Benesi and Winquist, 1978; Knözinger and Ratnasamy, 1978; Kijenski and Baiker, 1989; Gillespie *et al.*, 1990; Marks, 1992) have given the following order of support Lewis acidity:

$$DA > SiO_2 - Al_2O_3 > MgCl_2 \gg PDA \gg SiO_2 \qquad (4)$$

Sacchi *et al.* (1995) studied the reactions between zirconocene complexes and SiO_2. The results suggest that the attachment of zirconocene to silica occurs mainly through reactions with surface hydroxyl group [equation (1)]. Therefore, Cp_2ZrCl_2 (0.6 wt% Zr) and $Et[Ind]_2ZrCl_2$ (1.35 wt% Zr) on silica were analyzed and showed Cl:Zr molar ratios of 1.1 and 0.9, respectively. This corresponds to the reaction of one metallocene with one silanol group. In the case of $[Ind]_2ZrCl_2$ on SiO_2, analysis gave a Cl:Zr molar ratio of <0.1. This indicates that $[Ind]_2ZrCl_2$ may have reacted with a pair of hydroxyls such as the vicinal $(OH)_2$ pair. Equation (2) does not appear to be significant under the prevailing mild conditions, otherwise the product would have a 2:1 Cl:Zr molar ratio.

There is another reaction which leads to inactive surface-bound zirconium species according to Collins *et al.* (1992). They noted that the absorption of $Et[Ind]_2ZrCl_2$ (**2**) on either PDS or DS was accompanied by significant decomposition to bis(indenyl)ethane, as revealed by ^1H NMR spectroscopy [equation (5)]. Less decomposition was evident for the alumina supports than silica. The absorption of either **2** or **3** on PDS or PDA was not accompanied by detectable decomposition.

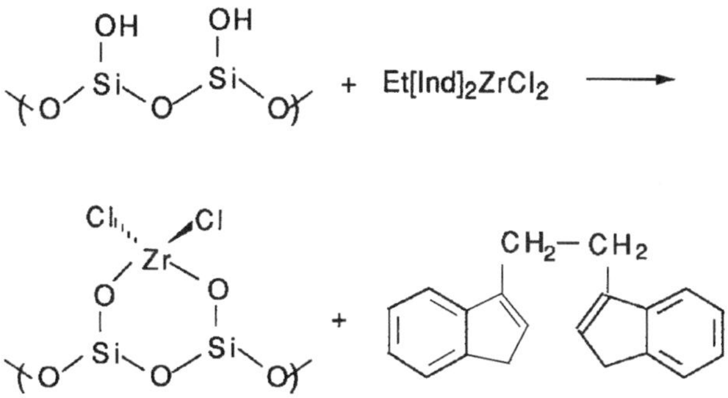

Soga and Kaminaka (1993) investigated other support materials. Only the fluorides afforded active supported *ansa*-zirconocenium catalysts. The observed activities decrease in the order $MgF_2 > CaF_2 > AlF_3$. However, even MgF_2 is an inferior support in comparison with $MgCl_2$. MgO is inactive, like silica.

3.2 COCATALYSTS CONTAINING MAO

The reason why zirconocene supported on silica is inactive is attributed to the formation of μ-oxo bonded species but not a 'cation-like' structure. However, it may be feasible to convert the μ-oxo bonded zirconium complex into catalytically active 'cation-like' species.

Sacchi *et al.* (1995) compared ethylene polymerization catalyzed by Cp_2ZrCl_2 (**1**) homogeneously and on DS at 50 °C (1 atm, MAO cocatalyst, Al : Zr = 300, 500). The supported catalyst has about one tenth the activity of the former, which is 2.1×10^6 g PE/(mol Zr h). The PEs have MW in the range 2.6×10^5–44×10^5 and a broad MWD ($M_w/M_n = 2.9$ to 7.9) for both systems.

Earlier, Kaminsky and Renner (1993) supported **2** on DS. Polymerization of propylene [50 °C, 2 bar and Al(MAO) : Zr = 150] has an activity of 1×10^4 g PP/(mol Zr h). The PP has very high M_w (*ca* 6×10^5), T_m (ca 160 °C) and *mm* ~ 95 %. The activity of **2** in solution [50 °C, Al(MAO) : Zr = 2500, 2 bar] is much greater: 4.2×10^6 g PP/(mol Zr [C_3H_6] h). However, the PP properties are much inferior by comparison: $M_w = 4.3 \times 10^4$, $M_w/M_n = 2.7$, $T_m = 128$ °C and *mm* = 91 % (Rieger *et al.*, 1990).

Sacchi *et al.* (1995) followed the procedure of Kaminsky and Renner (1993) precisely and performed polymerization over a range of conditions, but they were unable to obtain PP with an [mm] triad content higher than 72%. Their catalyst had lower activity [6×10^3 g PP/(mol Zr h)], $M_w(1.7 \times 10^5)$, stereoregularity (*mm* ~ 71 %) and a very broad MWD (14.8). Cp_2ZrCl_2 supported on silica with MAO as cocatalyst does not polymerize propylene.

Collins *et al.* (1992) compared alumina and silica, both partially and fully dehydroxylated, as supports for **2**. They conducted propylene polymerizations under a range of conditions, T_p 25–37 °C, $P_{C_3H_6}$ 3–5 atm and Al : Zr ≈ 1000. The supported catalysts have 1/100 to 1/10 of the homogeneous activity. It is interesting that the DS-supported catalyst is more active than PDS whereas the converse is true for alumina-supported systems.

Chen *et al.* (1995c) attempted the immobilization of *rac*-[MeO]$_2$Si[Ind]$_2$ZrCl$_2$ {**6**, di[(1'S, 2'R, 5'S)-menthoxy]silylenebis[η^5-1-(R, R)-(+)-indenyl]zirconium dichloride on PDS directly. The products had no catalytic activity. This may be due to the instability of this bulky *ansa*-zirconocene.

In the case of alumina, the activity of the PDA support is nine times larger than that of the DA support. This is contrary to expectation because the latter should be more Lewis acidic and more active in promoting the formation of Cp_2Zr^+ species. The apparent contradiction can be obviated if MAO can activate μ-oxo bonded zirconocenes. Reaction (6) is analogous to the condensation dissociation equilibria between TMA and MAO.

3.3 POLYMERIZATION KINETICS

Knowledge of the number of active sites [C*] is one of the keys to the quantitative understanding of Ziegler–Natta catalysis. Radioactive tagging with isotopic I_2 (Chien, 1959), CO (Yermakav and Zakharov, 1975; Mejzlik and Lesna, 1977) and CH_3OH (Burford and Tait, 1972; Schen et al., 1983; Tait, 1983; Chien and Kuo, 1985) had been applied to count C* in various types of heterogeneous Ziegler–Natta catalysts. In the case of metallocene Ziegler catalysts, labeling with CH_3O^3H was found by Chien and Wang, (1989) and Chien et al. (references cited in Chien and Wang, 1989) to be the most reliable and accurate method. When combined with fractionation of polymers (Chien and Hu, 1987, 1988; Chien and Wang, 1990; Chien and Sugimoto, 1991) and oxidation state analysis (Chien et al., 1989), one can also determine the number of different kinds of active sites in a given catalytic system.

The amount of C* in a metallocene catalyst is influenced by the experimental conditions. Under conditions favoring maximum polymerization activity, R_m (low [Zr], high [Al]: [Zr], high T_p), nearly every zirconocene complex participates in the catalysis, as shown by the [C*] : [Zr] molar ratio approaching 1. A smaller number of C* was found under less optimum conditions. Tait et al. (1996) compared C* in homogeneous Cp_2ZrCl_2–MAO with its silica-supported conterpart. The former has C* = 0.91 mol/mol Zr at 4 min after the initiation of ethylene polymerization and C* : Zr decreased to 0.68 after 30 min of polymerization. This is consistent with the observed activity decay. The supported catalyst has only $C_0^* = 0.091$ mol/mol Zr if it is mixed with MAO to initiate ethylene polymerization; this fraction is raised to 0.68 if the catalyst is aged with MAO in the absence of monomer and allowed to stand for 100 min. These results support the conclusion that the lower activities shown by silica-supported catalyst systems arise mainly from the low C*.

A supported metallocene catalyst is much less active for propylene polymerization than the homogeneous analog as compared with ethylene polymerization (Kaminsky and Renner, 1993; Sacchi et al., 1995). Tait and Ediati (1997) supported $Et[Ind]_2ZrCl_2$ on silica (460 °C). This $Et[Ind]_2ZrCl_2/SiO_2$ catalyst exhibited, without any build-up period, an initial rate of propylene polymerization R_m of 10^5 g PP/(mol Zr h) ([Zr] = 0.54 mM, Al : Zr = 110, 6 bar, 60 °C). However, the rate decayed rapidly with a first-order rate constant $k_d = 3.3 \times 10^{-3}$ s. When this catalyst was precontacted with MAO and then used for polymerization with an additional amount of MAO, R_m increased twofold. However, the same rapid decay ($k_d = 5.8 \times 10^{-3}$ s) to zero activity ensued.

The exhaustive and elegant investigations by Sinn and co-workers (Sinn and Patat, 1963; Sinn and Kolk, 1966; Kaminsky et al., 1974; Sinn and Kaminsky, 1980) enabled them to conclude that the formation of alkylene zirconium dimer is the main deactivation process in single-site metallocene catalysis:

$$2L_2(X)ZrC_2H_5 \longrightarrow L_2(X)ZrCH_2CH_2Zr(X)L_2 + C_2H_6 \qquad (7)$$

Therefore, at very low [Zr] and very high [Al] : [Zr], the zirconocene catalyst does not lose much activity with t_p (Sinn *et al.*, 1980; Chien and Razavi, 1988; Chien and Wang, 1989). If this is true, then the rapid decay by first-order kinetics must mean that equation (7) takes place mainly between a pair of zirconocene molecules affixed in close proximity by reactions with a pair of hydroxyls of either the geminal or vicinal type. This process continues until all remaining zirconocenes do not have a neighboring one close enough for equation (7) to occur.

The concentration of zirconocene within an average pore in a typical silica is exceedingly high. Let us assume a typical pore diameter of 100 Å, a pore volume of 1 ml/g and a Zr content of 1 wt%. The volume of each pore (if spherical in shape) is 5.2×10^5 Å^3, which translates to 1.9×10^{18} pores/g and 35 molecules of zirconocene per pore. The concentration of Cp_2ZrCl_2 without a pore is *ca* 0.1 M. The Al content in a typical supported catalyst is *ca* 5 wt%. If the formula weight of MAO is 1114 (Sinn *et al.*, 1988), then there are *ca* 14 molecules of MAO per pore or an MAO concentration within a pore of about 0.05 M. In other words, the numbers of zirconocene and MAO molecules in an average pore are about equal. If the zirconocenium in the pore is deactivated according to equation (7), then the rate of decay will be 10^{13} times faster than the homogeneous polymerization in which the zirconocene concentration is at the μM level. The first-order kinetic decay supports the nearest neighbor deactivation hypothesis. In the case of the Cp_2ZrCl_2/SiO_2 catalyst, there is also a rapid first-order decay of activity with $k_d = 3.4 \times 10^{-3}$ s, but the activity levels off at about a third of R_m.

The polymerization rates of both homogeneous and heterogeneous catalysts first increase sharply with increase in the [Al] : [Zr] ratio, and then reach a limiting value of R_m. This occurs at an [Al] : [Zr] ratio of about 5×10^4 in solution, whereas the corresponding R_m for the supported catalyst is reached at an [Al]:[Zr] ratio of about 10^3. However, since [Zr] is much lower in solution, the actual requirement of MAO concentration is only about three times larger for the supported catalyst.

The effect of the temperature of dehydration of silica was examined by Tait *et al.* (1990). The catalyst of Cp_2ZrCl_2 supported on silica dehydrated at 460 °C has double the activity of the corresponding catalyst using silica heated at a lower temperature of 260 °C, although the shapes of the rate–time profiles of the two catalysts are very similar. This result seems to be at odds with the observation made by Collins *et al.* (1992) that the metallocene catalyst activity diminishes in the order DS > PDS > HS $\approx$ no support, which is also the order of decreasing calcination temperature.

3.4 CATALYST STEREOREGULATION

The finding by Kaminsky and Renner (1993) of significantly enhanced isospecificity of Et[Ind]$_2$ZrCl$_2$ (**2**) when it is supported on silica was discussed in Section 3.2. Sacchi *et al.* (1995) discovered that even a normally aspecific zirconocene can become isospecific in the heterogenized form. Thus $[Ind]_2ZrCl_2$ (**5**) in solution with MAO $(Al:Zr = 300)$ polymerized propylene ($50\,°C$, 1 atm) to an atactic waxy product at an activity of 1.5×10^5 g PP/(mol Zr h), with $[mm] = 0.18$, $[mr] = 0.46$ and $[rr] = 0.36$. However, when **5** was supported on DS, it produced high-M_w (2.15×10^5) i-PP $([mm] = 0.71$, $[mr] = 0.15$, $[rr] = 0.14)$, but with a very poor activity of 4×10^3 g PP/(mol Zr h). In solution the various conformational states of **5** are in rapid equilibrium. If the rate of conformational transition is faster than the formation of the PP molecule, then only atactic product is formed. However, when **5** is immobilized on a support, it assumes fixed conformations on this time-scale. The result shows that a small fraction of **5** on a support can promote stereoregular propagation.

Even the ethylene-bridged **2** can be found in two conformers, Π (indenyl foward) and Y (indenyl backward), which has been proposed to have different stereoregulating abilities and k_p in propylene polymerization (Rieger *et al.*, 1990; Chien and Sugimoto, 1991; Rieger, 1992). These conformations undergo rapid interconversion in solution, as shown by proton NMR (Piemontesi *et al.*, 1995). It is therefore interesting that the supported catalyst of $2/SiO_2$ produces PP having the same steric microstructure as the PP obtained with $5/SiO_2$ (see above). This indicates that the conformations of **5** on SiO_2 may be similar to those of **2** on SiO_2.

A remarkable effect of the support was described by Kaminsky (1995) for C_s symmetric i-Pr(Cp)(Flu)ZrCl$_2$ (**4**) supported on silica. The catalysts produced i-PP $(50\,°C,[Al]:[Zr] = 180)$ with high T_m ($158\,°C$), $[mmmm]$ (90%) and MW (3.5×10^5). Polymerization in solution with $[Al]:[Zr] = 5900$ affords s-PP with low T_m ($131\,°C$), $[mmmm]$ (6%) and MW (4.7×10^4). In this case the support has changed the molecular symmetry of **4** from C_s to C_2. Therefore, heterogenization offers a way to affect stereochemical control in addition to that through the molecular structure of the metallocene. This possibility is hindered at present by the very low activities and polymer yields of the three catalysts described above. In other words, these sites having altered stereoselectivity are atypical.

4 MODIFIED SILICA SUPPORTS

4.1 MODIFICATION WITH (CH₃)₃Al (TMA)

Because of the unfavorable surface chemistry of silica, it is necessary to passivate it. Collins *et al.* (1992) investigated the use of PDA and PDS that had been pretreated with TMA to 0.8 mmol Al/g support. It was then contacted with Et[Ind]$_2$ZrCl$_2$ (**2**). The resulting 2/TMA/PDA has 2.3 times the activity of the unmodified 2/PDA. The benefit of TMA modification is much greater in the case of the PDS support; the increase in activity is 40-fold. This is because the unmodified 2/PDS catalyst is only one-tenth as active as 2/PDA; modification with TMA raised both catalysts to comparable activity of *ca* 3.5×10^5 g PP/(mol Zr h atm).

4.2 MODIFICATION WITH MAO

Modification of supports with MAO is widely practised. It involves simply the treatment of the support with a toluene solution of MAO, followed by washing, drying and impregnation of the appropriate zirconocene (Chien and He 1991). Presumably the absorbed MAO transforms the zirconocene into a 'cation-like' species. The activity of this material is often inadequate but it can be augmented with an activator such as AlR_3 or MAO, which may scavenge impurities, alkylate inactive zirconocene complexes, separate ion pairs, etc.

4.2.1 Activation with R_3Al

Soga and Kaminaka (1993 and references cited therein) prepared a supported catalyst of Cp_2ZrCl_2 (1), $Et[THI]_2ZrCl_2$ (3) and i-Pr(Cp)(Fl)ZrCl$_2$ (4) on MAO/SiO$_2$. Their propylene polymerizations are compared with the homogeneous systems in Table 2. The supported catalyst generally has lower activity than the homogeneous catalyst. In the cases of supported catalysts 3 and 4, they produce polymers with higher T_m and stereoregularity. Each supported catalyst displays the best performance with a particular AIR_3 with an optimal [Al] : [Zr] ratio.

4.2.2 Activation with MAO

A supported catalyst prepared with MAO-modified silica (MAO/SiO$_2$) and activated with more MAO behaves like the corresponding homogeneous systems apart from a modest difference in polymerization activity. Sacchi et $al.$ (1995) and Janiak and Rieger (1994) studied the effect of heterogenization of Cp_2ZrCl_2 (1) in ethylene polymerization. In the former study the supported system of $1/MAO/SiO_2$ activated with 200 : 1, 300 : 1 and 500 : 1 Al(MAO) : Zr ratios had 0.16, 0.32, and 0.26 fractions of the homogeneous activity. The polyethylenes obtained with the heterogeneous and homogeneous catalysis had the same MW and T_m. The latter study

Table 2 Propylene polymerization by zirconocenes 1, 3 and 4 supported on MAO/SiO$_2$

Zirconocene	1		3		4	
Zr (μmol g^{-1} catalyst)	Homogeneous	82	Homogeneous	82	Homogeneous	74
Al(MAO)(mmol/g catalyst)		3.0		2.5		2.4
Activator	MAO	TIBA	MAO	TIBA	MAO	TIBA
Activity [kg PP/(mol Zr h)]	132	99	2070	>1500	758	141
T_m(°C)			111	128	123	133
M_n	330	1800	3300	4700	39 000	45 000
M_w/M_n				3.1		1.9
[mmmm] (%)				89		
[rrrr] (%)						83

Table 3 Propylene polymerizations by zirconocene 7

No.	Support	Cocatalyst	Al:Zr	T_p (°C)	Productivity [10^6 g PP/ (mol Zr · 2 h)]	Relative Productivity[b]	[mmmm] (%)	T_m (°C)	M_w ×10^{-5}	M_w/M_n	Bulk density
1	None[a]	MAO	3000	30	15		95.2	156.0	0.94	1.9	0.08
2	MAO/SiO$_2$	TIBA	312	30	1.2	0.08	95.5	156.1	1.90	2.8	0.34
3	None[a]	MAO	3000	60	7.1		95.8	157.8	0.27	2.8	0.1
4	MAO/SiO$_2$	TIBA	312	60	12.5	0.18	95.6	158.8	1.15	2.3	0.28

[a] Homogeneous polymerization.
[b] Productivity ratio of homogeneous to heterogeneous catalysis.

observed some increase in polyethylene MW. They also compared 2/MAO/SiO$_2$ with Et[Ind]$_2$ZrCl$_2$ (**2**) in solution and found the former to be 0.59 times as active but produced *i*-PP with the same triad distributions, MW and MWD. The same holds true in the comparison of [Ind]$_2$ZrCl$_2$ supported on MAO/SiO$_2$ and in solution both catalyst systems give waxy PP.

Very stereospecific zirconocene maintains its superior characteristics upon immobilization on a support. An example is *rac*-dimethylsilylene(2,4-dimethyl-1-η^5-cyclopentadienyl)(3′,5′-dimethyl-1-η^5-cyclopentadienyl)zirconium dichloride (**7**), developed by Chisso (Miya *et al.*, 1990). Soga and Kaminaka (1994b) prepared a 7/MAO/SiO$_2$ catalyst and employed it to polymerize propylene in comparison with solution polymerization with 7–MAO (Table 3). The heterogeneous polymerization activity at 30 °C is only 8 % of the homogeneous case, but the difference is changed to 18 % at 60 °C. The supported catalyst produces *i*-PP of much higher MW and bulk density. Otherwise the products have the same T_m and homosteric pentad distributions. A distinct advantage of the supported catalyst is that it uses a small amount of TIBA as cocatalyst ([Al]:[Zr] $\approx$ 300), whereas the homogeneous polymerization requires MAO as cocatalyst at 10 times the amount. This supported catalyst should be very useful to the industry for many types of processes.

4.3 CROSS-LINKED SILICA SUPPORTS

ansa-Annelated zirconocenes were supported on cross-linked SiO$_2$ (*c*-SiO$_2$) by Ernst *et al.* (1997). The support was prepared with Grace Davison Sylopol S5N silica, dehydroxylated at 200 °C for 4 h and suspended in toluene. A mixture of MAO and bisphenol A was fed simultaneously at 50 °C; the unreacted and soluble aluminum compounds were removed by toluene extraction. The product comprises a cross-linked network with an average thickness of 20–30 μm around the starting SiO$_2$ particle. Among the several annelated zirconocenes studied, the best one is pseudo racemic (methyl)(trimethylsilyl)silylenebis[1-(2-methyl-4-phenyl-η^5-indenyl)]zirconium dichloride (**8**). The bulk polymerization results with **8** in solution

Table 4 Comparison of propylene polymerizations[a] by *ansa*-annelated zirconocene **8** in solution and on cross-linked SiO_2 support

	Solution[b]	c-SiO_2[d]		Solution[b]	c-SiO_2[d]
Pseudo *rac* (%)	96	96	M_w (kg/mol)	> 1,500	370 $(604)^f$
Al(MAO):SiOH (mol/mol)	n.a.[c]	10	M_w/M_n	n.r.[e]	2.8 $(2.9)^f$
Al(MAO) (mmol/g)	n.a.[c]	40	$T_m(°C)$	n.r.[e]	152
Al(MAO) : bisphenol A (mol/mol)	n.a.[c]	20	[mm] (%)	n.r.[e]	1.0
Zirconocene **8** (wt %)	n.a.[c]	0.75	l. l (%)	97.6	97.5
Zr (mmol/g catalyst)	n.a.[c]	0.01	Xylene cold soluble (%)	0.5	0.7
Cocatalyst MAO (Al:Zr)	10 000	4000	Fine (wt %)	100	1
$AlEt_3$ (Al:Zr)	n.a.[c]	100			
Activity [kg PP/ (mmol Zr h)]	250	200			
Activity [kg PP/g cat·2 h)]	n.a.[c]	$4(8)^c$			

[a] Polymerization conditions: $T_p = 70\,°C$, liquid C_3H_6, no H_2.
[b] Homogeneous catalysis.
[c] Not applicable.
[d] Zirconocene **8** supported on cross-linked SiO_2.
[e] Not reported.
[f] $0.13\,mol\%H_2$.

and in supported form are compared in Table 4. They are very similar or identical for most entries except for Fine (wt%), which is only 1 % using the supported catalyst. The physical properties of the polypropylenes formed with supported **8** are up to 10 % lower than those of the homogeneously polymerized product with respect to flexural modulus, tensile modulus, flexural stress, yield stress, charpy ($+23\,°C$ or $-20\,°C$) Vicat A and ball hardness, but about 10 % larger for MFR and elongation at yield. This shows that a truly stereorigid metallocene, supported on cross-linked MAO, behaves similarly to the homogeneous system.

4.4 EFFECT OF ORDER OF TREATMENT OF SILICA BY METALLOCENE AND MAO

PDS dehydroxylated at $260\,°C$ was treated with a solution of Cp_2ZrCl_2 (**1**) at $70\,°C$ for 18 h, filtered, washed and then treated with MAO solution at $50\,°C$ for 2 h to obtain a catalyst comprising MAO (8.6 % Al)/**1** (1.9 % Zr)/SiO_2. Using the reverse order, another catalyst was obtained comprising **1** (2.3 % Zr)/MAO (7.8 % Al)/SiO_2 (Tait *et al.*, 1996). Comparative rate–time data for ethylene polymerization

($[Zr] = 0.25\,\mu M$, $[Al]:[Zr] = 1570$, $70\,^\circ C$, 1 atm) showed the latter to be about five times more active than the former throughout 1 h runs.

This effect was also investigated (Tait and Ediati, 1997) for propylene polymerization catalysts based on $Et[Ind]_2ZrCl_2$ (**2**). The two catalysts were MAO (7.7% Al)/**2** (0.96% Zr)/SiO_2 and **2** (0.44% Zr)/MAO (7.4% Al)/SiO_2. As was mentioned above, the former catalyst has no build-up period, the initial R_m is ca $2 \times 10^5\,g\,PP/(mol\,Zr\,h)$ ($[Zr] = 0.17\,mM$, $[Al]:[Zr] = 120$, $60\,^\circ C$, 6 bar), which decayed precipitously to zero in a few minutes. In contrast, the **2**/MAO/SiO_2 catalyst required 20 min to reach $R_m \approx 7 \times 10^6\,g\,PP/(mol\,Zr\,h)$ followed by a gradual decay to 2×10^6 activity in 2 h. Therefore, the order of treatment of silica with metallocene and MAO is more crucial for polymerization catalysts for propylene than ethylene.

The surface of a support pretreated with either MAO or TMA may be capable of some but not all of the functions of MAO. It is therefore desirable to react zirconocene and MAO first before it is infused into pores of the support. Tait and Ediati (1997) prepared such a catalyst, $\{Et[Ind]_2ZrCl_2/MAO\}/SiO_2$. Another catalyst was obtained using an MAO-modified support: $\{Et[Ind]_2ZrCl_2/MAO\}/MAO/SiO_2$. The two catalysts have very similar metal contents of $Zr \approx 0.31\%$ and $Al \approx 13\%$. Propylene polymerizations were compared at 6 bar, $60\,^\circ C$ and $Al(MAO):Zr = 4000$. The latter catalyst has a very high activity of $7.5 \times 10^6\,g\,PP/(mol\,Zr\,h)$, which is three times greater than that of the former catalyst and twenty times greater than that of $MAO/Et(Ind)_2ZrCl_2/MAO/SiO_2$.

4.5 POLYMERIZATION KINETICS

It is a well known phenomenon of heterogeneous Ziegler–Natta catalysis that the nascent morphology of the polymer replicates that of the catalyst. A multigrain model was developed by Nagel et al. (1980) to describe the material and thermal diffusion processes for $MgCl_2$-supported $TiCl_3$ catalysts. SiO_2-supported catalysts behave similarly (Münoz-Escalone et al., 1989).

Bonini et al. (1995) studied the kinetics of a rac-$Me_2Si(Ind)_2ZrCl_2/MAO/SiO_2$ catalyst system. They prepared three catalysts with different amounts of Zr on the same MAO/SiO_2, and performed propylene polymerizations ($40\,^\circ C$, 2 bar) without and with addition of TIBA. The resulting variations of the rate–time curves were analyzed by a modified particle growth model in which the spherical catalyst particle fragmented layer by layer as each outer layer was filled with polymers in turn. The experimental curves were faithfully simulated with 17 kinetic parameters.

On the other hand, Tait and Ediati (1997) described a comparison of rate–time profiles of heterogeneous propylene polymerization by $Et[Ind]_2ZrCl_2/MAO/SiO_2$ versus the homogeneous analog. The two curves were superimporable except for a $35\times$ scale factor difference for the rate. This makes one wonder about the necessity and validity of diffusion-limited kinetic modeling in these instances.

4.6 CATALYST LEACHING

The zirconocene in a supported catalyst should be truly immobilized; it must not be leached in a slurry polymerization or otherwise migrate from the interior of a catalyst particle to the surface in a gas-phase polymerization. These occurrences could result in poor polymer morphology, MWD, microstructural dispersion, etc.

Collins *et al.* (1992) examined the possible desorption of zirconocene by precontacting a Et[Ind]$_2$ZrCl$_2$/TMA/PDS with a toluene solution of MAO followed by filtration and washing. The filtrate exhibits no activity but becomes highly active upon the addition of Et[Ind]$_2$ZrCl$_2$. Thus the filtrate contains alkylaluminum compounds but not the zirconocene.

Sacchi *et al.* (1995) prepared Cp$_2$ZrCl$_2$/MAO/SiO$_2$ catalyst then contacted it with MAO at 50 °C in the absence of ethylene. The toluene-soluble filtrate polymerizes ethylene with 18 % of the activity of the catalyst itself. This indicates leaching of Cp$_2$ZrCl$_2$ in these experiments from the MAO-modified support. In contrast, the catalyst of Cp$_2$ZrCl$_2$ supported on unmodified SiO$_2$ resulted in no polymerization by the filtrate as expected [see equations (1) and (2) and the discussions about them].

Tait *et al.* (1996) found that even Cp$_2$ZrCl$_2$ supported directly on unmodified SiO$_2$ can be extracted by MAO. They used SiO$_2$ which had been dehydroxylated at 450 °C and treated with Cp$_2$ZrCl$_2$ to obtain a catalyst with 2.5 wt% Zr. It was contacted with MAO for 100 min, at 22 °C and filtered. The filtrate exhibits 5–7-fold more activity than the solid. At $t_p = 30$ min in these polymerizations, a sample of the solid fraction was added to the former and a sample of the liquid fraction was added to the latter; the final polymerization rates of both systems became the same. It was concluded that significant leaching of the zirconocene took place in the presence of MAO for this supported catalyst system in slurry polymerization.

It is not obvious why the two research groups observed different leaching behaviors. One possible explanation is that Sacchi *et al.* (1995) used thoroughly dehydroxylated silica (650 °C), whereas the silica of Tait *et al.* (1996) was partially dehydroxylated at 450 °C.

4.7 EFFECT OF DEHYDROXYLATION TEMPERATURE (T_d)

The effect of the dehydroxylation temperature (T_d) of the silica was investigated by Tait and Ediati (1997). They used silica which has been dehydroxylated at (260 °C) to prepate Et[Ind]$_2$ZrCl$_2$/MAO/SiO$_2$ 260 °C containing 0.4 % Zr and 4.0 % Al. Another catalyst was prepared the same way using SiO$_2$ heated at $T_d = 460$ °C; it contained 0.39 % Zr and 8.3 % Al. Their propylene polymerization rate–time curves showed much higher rates of polymerization for the former catalyst.

In another investigation, Tait *et al.* (1996) used the above two SiO$_2$ materials of different T_d to support **1**. Ethylene polymerizations ([Zr] = 0.23 μM, [Al] : [Zr] = 1700, 70 °C, 1 atm) showed that the catalyst made with SiO$_2$ ($T_d = 260$ °C) has about twice the activity of that prepared with SiO$_2$ ($T_d = 460$ °C). It seems that the support which has a large number of vicinal hydroxyl pairs is better than the support with fewer. This influence is more pronounced for catalysts of MAO-modified SiO_2 and used to promote propylene polymerization.

Collins et al. (1992) reported the R values for HS-, PDS- and DS-supported catalysts of $Et[Ind]_2ZrCl_2$ to be ca 0, 1×10^4 and 3.6×10^4 g PP/(mol Zr h atm), respectively. Therefore, in this work the support treated at high T_d is more active, contrary to the results of Tait et al. (1996, 1997). However, in the case of alumina, the same zirconocene supported on PDA has a 10-fold higher activity $[1.35 \times 10^5$ g PP/(mol Zr h atm)] than the DA-supported catalyst.

4.8 MODIFICATION WITH SILICON COMPOUNDS

Soga et al. (1993) treated silica with $Cl_2Si(CH_3)_2$ followed by equimolar. $NaHCO_3$ and subsequently contacted it with MAO. This support was mixed with Cp_2ZrCl_2 and various AlR_3 in ethylene polymerizations. The best result was obtained with a support contacted with $\geqslant 5$ mmol of MAO and activated with TMA: Zr = 2000: 1. Other aluminum compounds are much poorer activators: TMA $\gg$ TIBA $>$ TEA $\gg Et_2AlCl \approx 0$. The benefit of the $Cl_2Si(CH_3)_2$ treatment appeared to be an activity enhancement of 2.5-fold.

5 OTHER SUPPORTED CATALYSTS

5.1 ZEOLITE-SUPPORTED CATALYSTS

Crystalline zeolites have a large surface area, a well defined pore structure and a very narrow pore-size distribution (Breck, 1974). Woo et al. (1996) investigated the suitability of this group of materials as supports. In one catalyst, NaY (pore diameter 5 Å) was first modified with MAO, filtered, dried and then impregnated with Cp_2ZrCl_2 using an additional amount of MAO for ethylene polymerization. Without the additional MAO injection there was virtually no polymerization. With the optimum amount of MAO it shows about 1.6 % of the activity of Cp_2ZrCl_2–MAO in solution (Table 5, Nos 1 and 2). The low activity may be due to a small number of cages capable of accommodating both catalyst components, or to a smaller rate constant of propagation or limitation of diffusion monomer to active sites. There is virtually no activity decay at 50 °C, indicating no migration of active species from one cage into another. Polymerization at 70 °C, has a four times greater initial activity but suffers a factor of two loss of activity within 30 min, which was attributed to interchange diffusion and deactivation at this temperature. NaY pretreated with TMA instead of MAO is not an active support material.

Two zeolite-like supports with larger size pores were employed to accommodate rac-$Et(Ind)_2ZrCl_2$ and MAO for stereospecific polymerization of propylene. They are MCM-41 and VPI-5 with pore sizes of 40 and 13 Å, respectively. The results in Table 5 show that these supported catalysts polymerize propylene very much like

Table 5 Polymerization behavior of zeolite-supported metallocene catalysts

No.	Zeolite support	Precursor	Catalyst	Al:Zr	Monomer	Activity[a] [kg polymer (mol Zr atm h)]	M_n $(\times 10^{-4})$	MWD	T_m (°C)	[mmmm] (%)
1	NaY	Cp$_2$ZrCl$_2$	MAO	840	C$_2$	200	13.8	2.9	136.9	n. a.[c]
2	None	Cp$_2$ZrCl$_2$	MAO	17 500	C$_2$	12 500	6.0	2.1	133.4	n. a.[c]
3	MCM-41	Et[Ind]$_2$ZrCl$_2$	MAO	800	C$_3$	1 400	0.73	2.8	131.5	83.5
4	MCM-41	Et[Ind]$_2$ZrCl$_2$	TMA	500	C$_3$	14	0.33	4.4	n. r.[b]	81
5	VPI-5	Et[Ind]$_2$ZrCl$_2$	MAO	2000	C$_3$	1 490	1.0	2.9	135.4	84
6	None	Et[Ind]$_2$ZrCl$_2$	MAO	15 000	C$_3$	3000	0.98[d]		125.1	75

[a] Monomer pressure 8 atm; $T_p = 50$ °C.
[b] Not reported.
[c] Not applicable.
[d] M_v.

they do in solution, with similar activity, molecular weight, T_m, and homosteric m pentad population.

5.2 CYCLODEXTRIN-SUPPORTED CATALYSTS

Cyclodextrins are a family of oligosaccharides derived from starch. Lee and Yoon (1994) treated cyclodextrins with either MAO or TMA. The former contains 9.1 % Al and the latter 2.6 % Al. They were then impregated with Cp$_2$ZrCl$_2$. Activation with AlR$_3$ resulted in modest ethylene polymerization catalysis with activities of only $5 \times 10^5 - 9 \times 10^5$ g PE/(mol Zr h bar).

5.3 ATTACHMENT OF METALLOCENE TO SUPPORT VIA CONNECTOR

Soga et al. (1994) prepared supported catalysts by the following procedures:

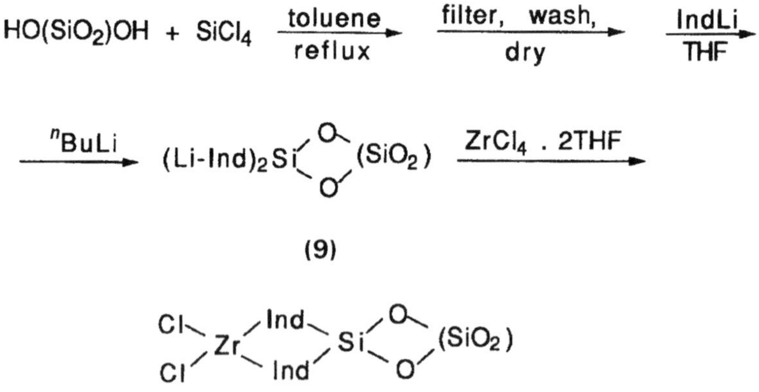

$$HO(SiO_2)OH + SiCl_4 \xrightarrow[\text{reflux}]{\text{toluene}} \xrightarrow[\text{dry}]{\text{filter, wash,}} \xrightarrow[\text{THF}]{\text{IndLi}}$$

$$\xrightarrow{^nBuLi} (Li\text{-}Ind)_2 Si\!\!\begin{array}{c}O\\\diagup\\\diagdown\\O\end{array}\!\!(SiO_2) \xrightarrow{ZrCl_4 \cdot 2THF}$$

(9)   (8)

(10)

Catalyst **10** ($0.59 \, \text{mmol} \, \text{Zr} \, \text{g}^{-1}$) with TIBA as cocatalyst ([Al] : [Zr] = 51) polymerizes propylene with a very low activity of only $10 \, \text{g} \, \text{PP}/(\text{mol} \, \text{Zr} \, [\text{C}_3\text{H}_6] \, \text{h})$ whereas none is expected. The reaction of TIBA with **10** cannot afford the 3d^0 14e Zr^+ catalytic center. The interesting finding is that the PP obtained contains 80 % of heptane-insoluble isotactic products which has T_m, 158.6 °C, [mmmm] 98 % and M_w 720 000.

It is surprising that **10** is not any better a catalyst with MAO as the cocatalyst. Propylene polymerization by **10**–MAO at [Al] : [Zr] = 170, 40 °C and $[\text{C}_3\text{H}_6] = 13 \, \text{M}$ results in an activity of only ca $20 \, \text{g} \, \text{PP}/(\text{mol} \, \text{Zr} \, [\text{C}_3] \, \text{h})$. The product contains 68 % heptane-insoluble material which has T_m, 159.2 °C [mmmm], 94.3 % and M_w 340 000. The very low activity of **10**–MAO may have several causes. Some possibilities are that the dominant zirconocene species may be **10** in the *meso* form, $(\text{Cl}_3\text{Zr})_2 \diagup\!\!\!\!\!\overset{\text{O}}{\underset{\text{O}}{\diagdown}}\!\!\!\!\!\diagup (\text{SiO}_2)$ and other low-activity, aspecific species.

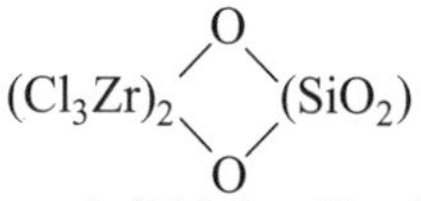

Jin *et al.* (1995) utilized intermediate **9** to prepare silica-supported neodymocene catalysts by the following postulated reactions and proposed the product structures shown:

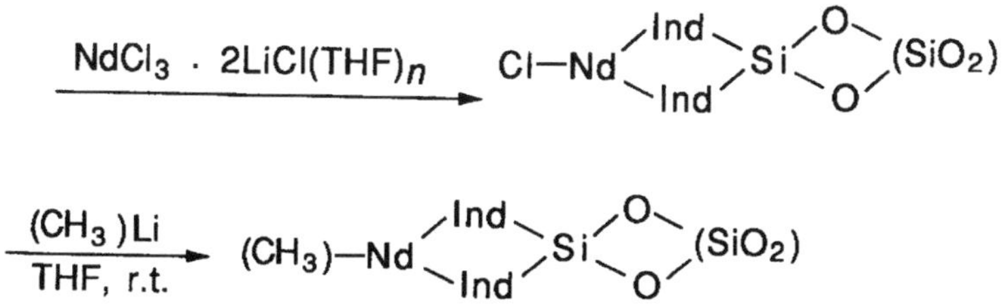

(9)

The ethylene polymerization activity of catalyst **11** is very dependent on the cocatalyst; it increases in the order TMA < MAO < TIBA < TEA < BuLi < BuMgEt. The MW of the PE produced increases in the order BuMgEt < BuLi < TEA < MAO < TMA < TIBA. The polydispersity ranges from 3 to 11, so catalyst **11** cannot be regarded as a 'single-site' catalyst. One remarkable attribute of this system is that the activity, which is $3.1 \times 10^5 \, \text{g} \, \text{PE}/(\text{mol} \, \text{Nd} \, \text{h})$ at 80 °C, remained high at 2.2×10^5 after 30 min of polymerization at 150 °C. However, both the activity and MW are much lower than those obtainable with zirconocene catalysts.

6 POLYMERIC SUPPORTS

Cross-linked styrene–divinylbenzene (20 %) has been used to attach metallocene compounds for hydrogenation of olefins and acetylenes, reduction of dinitrogen, isomerization of allylbenzene and cyclodienes, hydrozirconation of olefins, and epoxidation of olefins Grubbs *et al.*, 1973; (Bonds *et al.*, 1975; Chandrasekaran *et al.*, 1976). This and other polymers have been derivatized with metallocenes for olefin polymerization.

6.1 POLYSILOXANE SUPPORT

A polysiloxane-supported metallocene catalyst was prepared by the following procedure (Soga *et al.*, 1995):

$$LH \xrightarrow[THF]{^nBuLi} L-Li \xrightarrow[THF, 0\,°C]{to\ CH_3SiCl_3} LSi(CH_3)Cl_2 \xrightarrow{H_2O}$$

$$[L(CH_3)SiO]_n \xrightarrow[ZrCl_4.2THF]{^nBuLi} [L(CH_3)SiO]_n ZrCl_2 \quad (10)$$

where L = Ind (**12**) or Flu (**13**). Also synthesized was $[Ind_2SiO]_nZrCl_2$ (**14**).

The polysiloxane-supported catalysts display ethylene polymerization activities of $2.2 \times 10^6 - 4.5 \times 10^6$ g PE/(mol Zr h) (Table 6). Homogeneous catalysis by $(CH_3)_2Si(Ind)_2ZrCl_2$ (**15**) was performed as a reference. A closer analog might be $(CH_3O)_2Si(Ind)_2ZrCl_2$, since $O[(CH_3)_2SiInd]_2ZrCl_2$ (and the Ti compound) were found by Song *et al.* (1995) to have low polymerization activity like catalysts **12**, **13** and **14** (Table 6, Nos 1, 2 and 3). It was suggested that Lewis acid–base interaction between the siloxane group and the Zr^+ center may be partly the cause. A bimolecular deactivation reaction (Kaminsky *et al.*, 1974; Sinn and Kaminsky, 1980) between neighboring zirconocene species on the polysiloxane backbone could be another factor.

The same supported catalysts were examined in propylene polymerizations and found to have only slight activities. All three produced between 40 and 60 % of PP,

Table 6 Polymerization characteristics[a] of polysiloxane-supported zirconocene catalysts

No.	Catalyst	Olefin	Al:Zr	Activity $\times 10^{-6}$ [g polyolefin/ (mol Zr h)]	M_w $\times 10^{-3}$	M_w/M_n	% soln in Et$_2$O	% insol in Et$_2$O	T_m^d (°C)	[mmmm]^d (%)
1	**14**	C$_2$	10^4	2.2	68	2.7	n.a.[c]	n.a.	138.4	n.a.
2	**12**	C$_2$	10^4	4.5	430	6.8	n.a.	n.a.	138.4	n.a.
3	**13**	C$_2$	10^4	3.3	174	14.1	n.a.	n.a.	138.8	n.a.
4	Homo.[b]	C$_2$	4×10^3	19.3	201	2.8	n.a.	n.a.	n.r.[e]	n.a.
5	**14**	C$_3$	10^4	0.1	38	1.9	42	58	146	94.4
6	**12**	C$_3$	10^4	0.003	105	13	61	39	154	86.7
7	**13**	C$_3$	10^4	0.19	38	2.1	49	51	146	87

[a] Polymerization conditions: $T_p = 40\,°C$, solvent = toluene, olefin = 0.31 mol, Zr = 0.2 μmol, [Zr] = 8 μmol.
[b] Homogeneous analog $(CH_3)_2Si(Ind)_2ZrCl_2$.
[c] Not applicable.
[d] For the insoluble fraction.
[e] Not reported.

the product being insoluble in diethyl ether with [mmmm] ca 90% and T_m 142–154 °C. Soga et $al.$ (1995) suggested a certain juxtaposition of the bridging backbone atoms to arrange the zirconocene in some rac-like and '$meso$-like' states which are thought to produce isotactic and atactic polypropylenes, respectively. However, non-bridged zirconocenes, such as bis(indenyl)zirconium dichloride (Chien and Tsai, 1993), produces at 55 % i-PP with T_m 145 °C, M_v5.8 × 10^4 and an activity of 1.4 × 10^6 g PP/(mol Zr [C$_3$] h) at 24 °C. Many asymmetric zircono-cenes are now known to be isospecific catalysts (Chien et $al.$, 1992; Llinas et $al.$, 1992; Chen et $al.$, 1995; Razavi et $al.$, 1995). It is reasonable to expect that any zirconocene-type structure can be biased by the surface environment of a solid support to display some degree of profacial selectivity for propylene monomer, although in neither a quantitative nor predictive manner.

It is worth mentioning that the activity of catalyst **13** is much lower than those of **12** and **14**. One reason may be the instability of $ansa$-bis(fluorenyl)zirconium complexes (Chen et $al.$, 1995; Resconi et $al.$, 1995). The latter C_{2v} symmetric complex produces only atactic PP in solution while catalyst **13** produced about 39 % i-PP.

6.2 POLYSTYRENE SUPPORT

Zirconocene attached to polystyrene was prepared by Nishida et $al.$ (1995). Styrene–divinylbenzene (2 %) copolymer beads were lithiated with n-BuLi catalyzed by TMEDA (Pepper et $al.$, 1953). Catalyst **16** was obtained by reacting it with Cl$_2$Si(Ind)$_2$ followed by standard metallation procedures. The product contained 5.5 µmol Zr/g catalyst and was proposed to have attachment of $\rangle$Si(Ind)$_2$ZrCl$_2$ to two polystyrene phenyl groups. Catalyst **17** was prepared by a similar reaction with Cl(CH$_3$)Si(Ind)$_2$ and metallation; the resulting material is thought to comprise Cl(CH$_3$)Si(Ind)$_2$ZrCl$_2$ attached to a single phenyl ring of the polystyrene matrix at 5.5 µmol g^{-1} catalyst. Propylene polymerizations by catalyst **16** (MAO, Al:Zr = 5500, 40 °C, propylene 3 mol) has an activity of 5 × 10^5 g PP/(mol Zr h), which increases slightly to 6.6 × 10^5 at 70 °C. The product contains 62 % i-PP with T_m 140 °C, [mmmm] 90 %, M_n 2.5 × 10^4 and M_w/M_n 3.1. At 40 °C catalyst **17** is only one third as active as **16**, but its activity reached 3.7 × 10^6 g PP/(mol Zr h) to produce 40% isotactic polypropylene with T_m 124 °C, [mmmm] 83 %, M_w 4000 and M_w/M_n 2.1. The investigators reasoned that the influence of T_p on **17** that was observed but not for **16** can be understood in terms of the mobility of the ligands, i.e. the structure of the catalyst based on **16** is more rigid than that based on **17**. It seems questionable that the styrene–divinylbenzene (2 %) matrix swollen with toluene has significant physical rigidity at the molecular level. The relative rigidities of the Si atoms in the two catalysts should be compared by measuring their ^{29}Si$T_{1\rho}$ relaxation times by NMR spectroscopy.

7 COPOLYMERIZATION REACTIONS

The most remarkable and technically important property of metallocene catalysts is their ability to copolymerize various kinds of olefins and cycloolefins, not only in exceedingly high yields but also giving very random copolymers with narrow MWD and comonomer compositional distribution for ethylene–propylene copolymers (Zambelli *et al.* 1986, 1991; Chien and He, 1991a,b) and also for ethylene–higher α-olefin copolymers (Herfert and Fink, 1992; Chien and Nozaki, 1993; Koiumäki *et al.*, 1994; Arnold *et al.*, 1996; Rossi *et al.*, 1996). The coordination-gap aperture determines the relative magnitude of the copolymerization reactivity ratios. Metallocene catalysts also produce ethylene–propylene–diene terpolymers with very uniform microstructures (Kaminsky and Miri, 1985; Kaminsky and Drögemüller, 1990; Chien and He, 1991d; Chien and Xu, 1993; Herfert *et al.*, 1993; Marques *et al.*, 1995; Yu *et al.*, 1995).

Chien and He (1991c) prepared a supported catalyst of Et[Ind]$_2$ZrCl$_2$ on MAO-mediated silica. It was found to have the same activity and copolymerization reactivity ratio as the homogeneous system.

Soga and Kaminaka (1994a,b) compared the copolymerization of C$_2$–C$_3$, C$_2$–C$_6$ and C$_3$–C$_6$ olefins for Et[THI]$_2$ZrCl$_2$ supported on MAO/SiO$_2$, Al$_2$O$_3$ and MgCl$_2$. Their activities for C$_2$–C$_3$ and C$_2$–C$_6$ copolymerization decrease in the order shown (Table 7). The first two systems have reactivity ratios comparable to those for homogeneous copolymerizations, but the MgCl$_2$-supported catalyst exhibits comparatively higher r_1 values. In general, the Al$_2$O$_3$-supported catalyst produces copolymers having a very narrow MWD, whereas the MgCl$_2$ system gave the most heterogeneous C$_2$–C$_6$ copolymers. These results are useful to guide the further development of supported catalysts.

Table 7   Supported zirconocene catalysis of olefin copolymerization[a]

No.	Support	[3] (μmol)	Activator	Al (mmol)	Monomer 1	Monomer 2	Yield[c] (g)	r_1	r_2	r_1r_2	$M_n \times 10^{-3}$	M_w/M_n
1	MAO/SiO$_2$	2	TIBA	1.5	C$_2$	C$_3$	4.8	7.0	0.15	1.0	19.4	3.9
2	MAO/SiO$_2$	1	TIBA	1.5	C$_2$	C$_6$	1.98	19.6	0.019	0.4	26.3	2.19
3	MAO/SiO$_2$	2	TIBA	1.5	C$_3$	C$_6$	1.2	2.81	0.47	1.3	8.3	2.74
4	MgCl$_2$[b]	24	TMA	10	C$_2$	C$_3$	1.53	10.5	0.074	0.8	16.1	3.2
5	MgCl$_2$[b]	24	TMA	1.5	C$_2$	C$_6$	0.54	45.3	0.023	1.0	18.1	6.1
6	Al$_2$O$_3$[b]	24	TIBA	5	C$_2$	C$_3$	2.76	8.3	0.13	1.1	33.7	1.8
7	Al$_2$O$_3$[b]	24	TIBA	5	C$_2$	C$_6$	1.52	22.7	0.028	0.6	37.8	1.8

[a] Polymerization at 40 °C, 0.5–1 h, 150 ml toluene, 1 atm.
[b] Treated with AlR$_3$.
[c] Run for comonomer feed with the highest rate of copolymerization.
[d] Average value of 3–4 runs at different feeds, calculated from ^{13}C NMR spectra.

8 POLYMERIC ALUMINOXANE AS SUPPORT

MAO materials often form gels on storage, and the molecular weight of the MAO catalyst component seems to play a significant role in determining the zirconocene activity. Pino (1987) suggested the possibility of the formation of microaggregates (micelles) containing the catalytically active component. Therefore, it is logical to employ MAO as a cocatalytic support. Janiak *et al.* (1993) prepared this type of material by the reaction of an aliphatic α,ω-diol dissolved in THF with a toluene solution of MAO. At an Al:OH (decanol) ratio of 5:1 (the approximate MAO:diol molar ratio is 1:1), a macroporous product was obtained with a BET surface area of *ca* $20 \, \text{m}^2/\text{g}$, a pore surface area of $18 \, \text{m}^2/\text{g}$, a pore volume of $0.1–0.2 \, \text{cm}^3 \, \text{g}^{-1}$ and a pore diameter of 300–500 Å. The materials were characterized by ^{13}C and ^{27}Al CP/MAS NMR spectra.

Another type of solid PMAO support was described by Jin *et al.* (1996). It is a coupling product of MAO and *p*-hydroquinone with an Al:OH ratio of 10:1, which corresponds to an MAO:hydroquinone molar ratio of 1:1. This was used as a cocatalyst for Cp_2ZrCl_2, in an *in situ* heterogeneous system which requires TIBA [Al(TIBA):Zr = 300:1] for activation. The best ethylene polymerization activity is $4.36 \times 10^6 \, \text{g PE}/(\text{mol Zr h})$ with M_w 1.48×10^5; its propylene polymerization activity is $1.6 \times 10^5 \, \text{g PP}/(\text{mol Zr h})$ with M_w 2.26×10^4 and T_m $138.6 \, ^\circ\text{C}$. Both products have $M_\text{w}/M_\text{n} < 2.0$. This catalyst system has two critical parameters. One is the stoichiometry of MAO and *p*-hydroquinone. If the ratio is more than 10:1, the yield of solid PMAO is lowered; a smaller ratio adversely affects the polymerization activity. In the latter case more TIBA seems to increase the catalytic activity. Apparently there is cross-linking of MAO when the Al:OH ratio is less than 10:1, thus reducing the number of active oligomeric units (Bliemeister *et al.*, 1995). TIBA, on the other hand, can promote exchange processes with PMAO to cleave the Al—O bonds in the —Al—O—⟨⟩—O—Al— group, which could result in the formation of an active —Al(CH$_3$)^{i}Bu terminus.

9 CONCLUSION

In the Introduction we listed eight objectives for characteristics which were deemed necessary for a viable supported zirconocene catalyst to possess. The first seven goals are all attainable, as shown by the work of Soga and Kaminaka (1994a,b) and Ernst *et al.* (1997). This is to support:

(i) a zirconocene, such as **7** and **8**, designed for homogeneous single-site catalysis which is stable toward bimolecular deactivation [equation (7)], low rate of chain transfer vs β-hydride elimination and high stereo- and regioselectivity in the case of α-olefin polymerizations on

(ii) a substratum, such as MAO, which serves to convert the zirconocene to its cation, to trap the counterion whether it is Cl^- or CH_3^- and to prevent any undesirable reactions between the zirconocene and the

(iii) support foundation, which provides the template for replication during polymerization.

The Zr^+ center are surrounded by MAO and the counterion is trapped by MAO in crown ether-type chelation. The ions are kept apart from deactivation by coulombic forces and ion–dipole interactions. Transport of Zr^+ on the substratum is through thermally activated diffusion with correlated motion of anions. Such a system retains all the desired polymerization characteristics of the homogeneous system, yet with added control of polymer morphology, especially when the substructure is reinforced by cross-links, and replaces MAO by AlR_3 in the polymerization. The best way to prepare such a catalyst beside passivation of the surface of the substructure is to prepare a zirconocene–cocatalyst mixture and infuse this solution into its pores. This will avoid the uncertain process of supplying the cocatalyst separate from the deposition of zirconocene on the support.

The only remaining objective is to enhance or alter the course of polymerization through control by the substructure or the substratum. So far some promising leads have been reported by Kaminsky (1995), Kaminsky and Renner (1993) and Sacchi *et al.* (1995). However, the activities of the described catalysts are exceedingly low, indicating non-typical catalytic sites, which are very few in number, and nothing is known about their structures. The development of a polymeric support, or chemical attachment of the metallocene ligand to the support followed by metallation, etc., are worthwhile, but progress is hampered by the difficulty of having exceedingly clean reactions without side products because there is no way to purify the product after each step of preparation and to characterize what is formed.

In a radiotagging kinetic study on $Et[THT]_2ZrCl_2/MAO$-catalyzed propylene polymerization, Chien and Sugimoto (1991) determined separately the $[C^*]$ and k_p values for C_7 and C_6 soluble stereo-regular fractions (iso specific, *i*) and for C_5 and ether-soluble stereo-irregular fractions (aspecific a). At $T_p = 30\,^\circ C$ and both $[Al]:[Zr] = 3500$ and 350, $k_{p,i} = (15 \pm 1)k_{p,a}$. In other words if one prepared a supported catalyst and it gives, say, 50% *i*-PP, the catalyst probably comprises 94% aspecific sites and only 6% isospecific Zr^+ centers. Chien and co-workers also observed and reported for several $MgCl_2$-supported $TiCl_3$ catalysts that the isotactic sites have k_p values more than 10 times larger than the atactic sites.

10 REFERENCES

Arnold, M., Henschke, O. and Knorr, J. (1996) *Macromol. Chem. Phys.*, **197**, 363.

Benesi, H. A. and Winquist, B. H. C. (1978) *Adv. Catal.*, **27**, 97.

Bliemeister, J., Hagendorf, W., Harder, A., Heitmann, B., Schimmel, I., Schmedt, E., Schnuchel, W., Sinn, H., Tikwe, L., von Thienen, N., Urlass, K., Winter, H. and

Zarncke, O. (1995) in Fink, G., Mülhaupt, R. and Brintzinger, H. H. (Eds), *Ziegler Catalysts*, Springer, Berlin, pp. 57–84.
Bonds, W. D., Jr, Brubaker, C. H., Jr. Chandraskaran, E. S., Gibbons, C., Grubbs, R. H. and Kroll, L. C. (1975) *J. Am. Chem. Soc.*, **97**, 2128.
Bonini, F., Fraaije, V. and Fink, G. (1995) *J. Polym. Sci. Part A: Polym. Chem.*, **33**, 2393.
Breck D. W. (1974) *Zeolite Molecular Sieves*, Wiley, New York.
Burford, D. R. and Tait, P. J. T. (1972) *Polymer*, **13**, 315.
Chandrasekaran, E. S., Grubbs, R. H. and Brubaker, C. H., Jr (1976) *J. Organomet. Chem.*, **120**, 49.
Chang, B.-H. Grubbs, R. H. and Brubaker, C. H., Jr (1985) *J. Organomet. Chem.*, **280**, 365.
Chen, Y. X., Rausch, M. D. and Chien, J. C. W. (1995a) *Macromolecules*, **28**, 5399.
Chen, Y.-X., Rausch, M. D. and Chien, J. C. W. (1995b) *J. Organomet. Chem.*, **497**, 1.
Chen, Y.-X., Rausch, M. D. and Chien, J. C. W. (1995c) *J. Polym. Sci. Part A: Polym. Chem.*, **33**, 2093.
Chien, J. C. W. (1959) *J. Am. Chem. Soc.*, **81**, 86.
Chien, J. C. W. (1971a) *J. Am. Chem. Soc.* **93**, 4675.
Chien, J. C. W. (1971b) *J. Catal.*, **73**, 71.
Chien, J. C. W. (1987) in Seymour, R. B. and Cheng, T. (Eds), *Advances in Polyolefins*, Plenum Press, New York.
Chien, J. C. W. (1992) in Vanderberg, E. J.F and Salamone, J. C. (Eds), *Catalysis in Polymer Synthesis*, ACS Symposium Series, vol. 496, American Chemical Society, Washington, DC, p. 26–55.
Chien, J. C. W. and He, D. (1991a) *J. Polym. Sci. Part A: Polym. Chem.*, **29**, 1585.
Chien, J. C. W. and He, D. (1991b) *J. Polym. Sci. Part A: Polym. Chem.*, **29**, 1595.
Chien, J. C. W. and He, D. (1991c) *J. Polym. Sci. Part A: Polym. Chem.*, **29**, 1603.
Chien, J. C. W. and He, D. (1991d) *J. Polym. Sci. Part A: Polym. Chem.*, **29**, 1613.
Chien, J. C. W. and Hsieh, J. T. T. (1976) *J. Polym. Sci., Polym. Chem. Ed.*, **14**, 1915.
Chien, J. C. W. and Hu, Y. (1987) *J. Polym. Sci., Polym. Chem. Ed.*, **25**, 2847, 2881.
Chien, J. C. W. and Hu, Y. (1988) *J. Polym. Sci. Part A: Polym. Chem.*, **26**, 2973.
Chien, J. C. W. and Hu, Y. (1989) *J. Polym. Sci., Part A: Polym. Chem.*, **27**, 897.
Chien, J. C. W. and Kuo, C. I. (1985) *J. Polym. Sci., Polym. Chem. Ed.* **23**, 731.
Chien, J. C. W. and Nozaki, T. (1993) *J. Polym. Sci. Part A: Polym. Chem.*, **31**, 227.
Chien, J. C. W. and Razavi, A. (1988) *J. Polym. Sci. Part A: Polym. Chem.*, **26**, 2369.
Chien, J. C. W. and Sugimoto, R. (1991) *J. Polym. Sci. Part A: Polym. Chem.*, **21**, 459.
Chien, J. C. W. and Tsai, W.-M. (1993) *Makromol. Chem., Macromol. Symp.*, **66**, 141.
Chien, J. C. W. and Wang, B.-P. (1988) *J. Polym. Sci. Part A: Polym. Chem.*, **26**, 3089.
Chien, J. C. W. and Wang, B.-P. (1989) *J. Polym. Sci. Part A: Polym. Chem.*, **27**, 1539.
Chien, J. C. W. and Wang, B.-P. (1990) *J. Polym. Sci. Part A: Polym. Chem.*, **28**, 15.
Chien, J. C. W. and Xu, B.-P. (1993) *Makromol. Chem., Rapid Commun.*, **14**, 109.
Chien, J. C. W., Weber, S. and Hu, Y. (1989) *J. Polym. Sci. Part A: Polym. Chem.*, **27**, 1514.
Collins, S., Kelly, W. M. and Holden, D. A. (1992) *Macromolecules*, **25**, 1780.
Dahmen, K. H., Hedden, D., Burwell, R. L., Jr, and Marks, T. J. (1988) *Langmuir*, **4**, 121.
Finch, W. C., Gillespie, R. D., Hedden, D. and Marks, T. J. (1990) *J. Am. Chem. Soc.*, **112**, 6221.
Grubbs, R. H., Gibbons, C., Kroll, L. C., Bonds, W. D., Jr, and Brubaker, C. H., Jr. (1973) *J. Am. Chem. Soc.*, **95**, 2373.
Hedden, D. and Marks, T. J. (1988) *J. Am. Chem. Soc.*, **110**, 1647.
Herfert, N. and Fink, G. (1992) *Prep. Mater Sci. Eng.*, **67**, 31.
Herfert, N., Montag, P. and Fink, G. (1993) *Makromol. Chem.*, **194**, 3167.
Hu, Y. and Chien, J. C. W. (1988) *J. Polym. Sci. Part A: Polym. Chem.*, **26**, 2003.

Ihm, S.-K., Chu, K.-J. and Yin, J.-H. (1994) in Soga, K. and Terano, M. (Eds), *Catalyst Design for Tailor-made Polyolefins*, Kodansha, Tokyo, p. 299.

Jackson, R., Ruddlesden, J., Thompson, D. J. and Whelan, R. (1977) *J. Organomet. Chem.*, **125**, 57.

Janiak, C. and Rieger, B. (1994) *Angew Makromol. Chem.*, **215**, 47.

Janiak, C. and Rieger, B., Voelkel, R. and Braun, H.-G. (1993) *J. Polym. Sci. Part A: Polym. Chem.*, **31**, 2959.

Jin, J., Uozumi, T. and Soga, K. (1995) *Macromol. Rapid Commun.*, **16**, 317.

Jin, J., Uozumi, T. and Soga, K. (1996) *Macromol. Chem. Phys.*, **197**, 849.

Kaminaka, M. and Soga, K. (1991) *Makromol. Chem., Rapid Commun.* **12**, 367.

Kaminaka, M. and Soga, K. (1992) *Polymer*, 1105.

Kaminsky, W. (1995) *Macromol. Symp.*, **89**, 203.

Kaminsky, W. and Drögemüller, H. (1990) *Makromol. Chem., Rapid Commun.*, **11**, 89.

Kaminsky, W. and Miri, M. (1985) *J. Polym. Sci., Polym. Chem. Ed.*, **23**, 2151.

Kaminsky, W. and Renner, F. (1993) *Makromol. Chem., Rapid Commun.*, **14**, 239.

Kaminsky, W., Vollmer, H.-J., Heins, E. and Sinn, H. (1974) *Makromol. Chem.*, **175**, 443.

Kaminsky, W., Hähnsen, H., Külper, K. and Woldt, R. (1985) *US Pat.*, 4 542 199, to Hoechst.

Kijenski, J., and Baiker, A. (1989) *Catal. Today*, **5**, 1.

Knözinger, H., and Ratnasamy, P. (1978) *Catal. Rev. Sci. Eng.*, **17**, 31.

Koivumäki, J., Fink, G. and Seppälä, J. V. (1994) *Macromolecules*, **27**, 6254.

Lee, D.-H. and Yoon, K.-B. (1994) *Macromol. Rapid Commun.*, **15**, 841.

Llinas, G. H., Dong, S.-H., Mallin, D. T., Rausch, M. D., Lin, Y.-G., Winter, H. H. and Chien, J. C. W. (1992) *Macromolecules*, **25**, 1242.

Marks, T. J. (1992) *Acc. Chem. Res.*, **25**, 57.

Marques, M., Yu, Z.-T., Rausch, M. D. and Chien, J. C. W. (1995) *J. Polym. Sci. Part A: Polym. Chem.*, **33**, 2707.

Mejzik, J. and Lesna, M. (1977) *Makromol. Chem.*, **128**, 261.

Miya, S., Mise, T., and Yamazaki, H. (1990) in Keii, T. and Soga, K. (Eds), *Catalytic Olefin Polymerization*, Kadansha, Tokyo, p. 531.

Münoz-Escalone, A., Fuentes, A., Liscano, J. and Albornoz, A. (1989) *Stud. Surf. Sci. Catal.*, **56**, 377.

Nagel, E. J., Kirilov, V. A. and Ray, W. H. (1980) *Ind. Eng. Chem. Prod. Res. Dev.*, **19**, 372.

Nishida, H., Uozumi, T., Arai, T., and Soga, K. (1995) *Macromol. Rapid Commun.*, **16**, 821.

Nowlin, T. E., Mink, R. I., Lo, F. Y. and Kumar, T. (1991) *J. Polym. Sci. Part A: Polym. Chem.*, **29**, 1167.

Pepper, K. W., Paisley, H. M., and Young, M. A. (1953) *J. Chem. Soc.*, 4097.

Piemontesi, F., Camurati, I., Resconi, L., Balboni, D., Sironi, A., Moret, M., Zeiger, R. and Piccolrovazzi, N. (1998) *Organometallics*, **14**, 1256.

Pino, P. (1987) in Lemstra, P. J. and Kleintjens, L. A. (Eds), *Integration of Fundamental Polymer Science and Technology*, Elsevier Applied Science, New York p. 3.

Pino, P., Gnastalla, B., Rotzinger, B. and Mülhaupt, R. (1981) *Transition Metal Catalyzed Polymerizations; Unsolved Problems*, MMI International Symposium, Midland, MI.

Razavi, A., Vereecke, D., Peters, L., Dauw, K. D., Nafpliotis, L. and Atwood, J. L. (1995) in Fink, G., Mülhaupt, R. and Brintzinger, H. H. (Eds), *Ziegler Catalysts*, Springe, Heidelberg, p. 111.

Reissov́, A., Bastl, Z. and Capka, M. (1986) *Collect. Czech. Chem. Commun.*, **51**, 1430.

Resconi, L., Fait, A., Piemmtesi, F., Colonnesi, M., Rychlicki, H. and Ziegler, R. (1995) *Macromolecules*, **28**, 6667.

Ribeiro, M. R., Deffieux, A. and Portela, M. F. (1997) *Ind. Eng. Chem. Res.*, **36**, 1224.

Rieger, B. (1992) *J. Organomet. Chem.*, **428**, C33.
Rieger, B., Mu, X., Mallin, D. T., Rausch, M. D. and Chien, J. C. W. (1990) *Macromolecules*, **23**, 3559.
Rieger, B., Jany, G., Fauzi, R. and Steimann, M. (1994) *Organometallics*, **13**, 647.
Rossi, A., Zhang, J. and Odian, G. (1996) *Macromolecules*, **29**, 2331.
Sacchi, M. C., Zucchi, D., Tritto, I., Locatelli, P. and Dall'Occo, T. (1995) *Macromol. Rapid Commun.*, **16**, 581.
Schen, M. A., Karasz, F. E. and Chien, J. C. W. (1983) *J. Polym. Sci., Polym. Chem. Ed.*, **21**, 2787.
Sinn, H. and Kaminsky, W. (1980) *Adv. Organomet. Chem.* **18**, 99.
Sinn, H. and Kolk, E. (1966) *Organomet. Chem.*, **6**, 373.
Sinn, H. and Patat, F. (1963) *Angew. Chem.*, **75**, 805.
Sinn, H. Kaminsky, W., Vollmer, H. and Woldt, R. (1980) *Angew. Chem., Int. Ed. Engl.*, **19**, 390.
Sinn, H., Kaminsky, W. O., Vollmer, H. J. and Woldt, R. (1983) *US Pat.*, 403 344 to BASF.
Sinn, H., Bliemeister, J., Clausnitzer, D., Tikwe, L., Winter, H. and Zarncke, O. (1988) in Kaminsky, W. and Sinn, H. (Eds), *Transition Metals and Organometallics as Catalysts for Olefin Polymerization*, Springer, Berlin, p. 257.
Slotfeldt-Ellingsen, D., Dahl, I. M. and Ellested, O. H. (1980) *J. Mol. Catal.*, **9**, 423.
Soga, K. and Kaminaka, M. (1993) *Makromol. Chem.*, **194**, 1745.
Soga, K. and Kaminaka, M. (1994a) *Macromol. Chem. Phys.*, **195**, 1369.
Soga, K. and Kaminaka, M. (1994b) *Macromol. Rapid Commun.*, **15**, 593.
Soga, K., Shino, T. and Kim, H. J. (1993) *Makromol. Chem.*, **194**, 3499.
Soga, K., Arai, T., Hoang, B. T. and Uozumi, T. (1995) *Macromol. Rapid Commun.*, **16**, 905.
Song, W., Shackett, K., Chien, J. C. W. and Rausch, M. D. (1995) *J. Organomet. Chem.*, **501**, 375.
Tait, P. J. T. (1983) in Quirk, R. P. (Ed.), *Transition Metal Catalyzed Polymerizations*, Harwood Academic Publishers, New York, p. 115.
Tait, P. J. T. and Ediati, R. (1997) presented at Met. Con. '97, Houston, TX.
Tait, P. J. T., Abozeid, A. I. and Paghaleh, A. S. (1995) presented at Metallocene '95, Houston, TX.
Tait, P. J. T., Monteiro, M. G. K., Yang, M. and Richardson, J. L. (1996) presented at *Met Con. '96*, Houston, TX.
Tanabe, K. (1970) *Solid Acids and Bases*, Kodansha, Tokyo and Academic Press, New York.
Toscano, P. J. and Marks, T. J. (1985) *J. Am. Chem. Soc.*, **107**, 653.
TOSCAN. J. and Marks, T. J. (1986) *Langmuir*, **2**, 820.
Tsai, W.-M. and Chien, J. C. W. (1994) *J. Polym. Sci. Part A: Polym. Chem.*, **32**, 149.
Tsai, W.-M., Rausch, M. D. and Chien, J. C. W. (1993) *Appl. Organomet. Chem.*, **7**, 71.
Woo, S. I., Ko, Y. S., Han, T. K., Park, J. W. and Huh, W. S. (1996) in *Proceedings of Metallocenes '96*, Houston, TX, p. 273.
Yermakov, Yu I. and Zakharov, V. A. (1975) in Chien, J. C. W. (ed.), *Coordination Polymerization*, Academic Press, New York, p. 91.
Yermakov, Yu. I., Kuznetsov, B. N. and Zakharov, V. A. (1981) *Catalysis by Supported Complexes*, Elsevier, Amsterdam, p. 59.
Yu, Z.-P., Marques, M., Rausch, M. D. and Chien, J. C. W. (1995) *J. Polym. Sci. Part A: Polym. Chem.*, **33**, 979, 2795.
Zambelli, A., Longo, P., Ammendola, P. and Grassi, A. (1986) *Gazz. Chim. Ital.*, **116**, 731.
Zambelli, A., Grassi, A., Galimberti, M., Mazzocchi, R. and Piemontesi, F. (1991) *Makromol. Chem., Rapid Commun.*, **12**, 523.

9

Supported Metallocene Catalysts for Olefin Polymerization

GREGORY G. HLATKY
Equistar Chemical Company, Cincinnati, OH, USA

1 INTRODUCTION

The commercial success of metallocene catalysts for olefin polymerization will be judged in part by their compatibility with current polymerization processes. The ability to 'drop in' a new catalyst into an existing process and produce polymer efficiently without process upsets and without recourse to expensive equipment modifications can only improve the attractiveness of a new catalyst to polyolefin manufacturers. Metallocene catalysts have proven remarkably versatile, producing a wide range of polymer products in virtually every process in which they have been employed.

Solution processes use homogeneous catalysts to produce polymers that can dissolve in the reaction medium. These plants may operate at low or high pressures and temperatures. For the most part, they are best suited to producing polymers and copolymers of lower density and crystallinity such as VLDPE, plastomers or elastomers, as these will be soluble in the hydrocarbon solvent used or which would melt at the process temperature used. The solvent is then flashed off or steam-stripped to recover the polymer.

Homopolymers and copolymers of higher density and crystallinity, such as LLDPE, HDPE and polypropylene, are usually prepared in continuous slurry, fluidized-bed gas-phase or bulk monomer processes. These resins are insoluble in hydrocarbon solvents at the process temperatures commonly employed. Smooth, continuous operation of these plants requires catalyst particles which can be fed into

Metallocene-based Polyolefins Edited by J. Scheirs and W. Kaminsky
© 2000 John Wiley & Sons Ltd

the polymerization environment without clogging or clumping, do not foul the reactor by clinging to reactor walls, agitators or distributor plates, do not form either large chunks which disturb fluidization nor fines which can be carried into recycle lines, and which produce polymer product with high bulk density which can be easily removed from the reactor periodically. Proper catalyst preparation and process control can produce polymer particles which can be directly used in rotational molding without post-reactor grinding or pelletizing [1]. These requirements imply immobilizing the metallocene catalyst on a morphologically uniform carrier material. A successful supported metallocene catalyst should also retain its 'single-site' nature, affording polymers and copolymers of narrow molecular weight and comonomer distribution like those of the homogeneous counterpart.

2 SUPPORTED METALLOCENE/AIR$_n$X$_{3-n}$ CATALYSTS

Much of the attractiveness of metallocene catalysts, at least while they were still of largely academic interest, was their solubility, which facilitated the investigation of active site characterization and kinetics. While relatively little was done in the period 1956–80 to heterogenize Cp_2TiCl_2/AlR_2Cl ($Cp = \eta^5$-C_5H_5) catalysts, materials such as carbon black [2] and cross-linked rubber [3] were used to produce heterogeneous metallocene polymerization catalysts. An investigation of a silica-supported Cp_2TiCl_2/Et_nAl_{3-n} catalyst showed the activity of the supported catalyst to be similar to that of the homogeneous system; reduction to inactive Ti(III) species took place gradually [4].

Research on supported cyclopentadienyl metal olefin polymerization catalysts activated by simple alkylaluminum compounds continues to the present because of the low cost and wide availability of the cocatalyst. Precipitation of Cp_2TiCl_2 with $MgCl_2$ and activation of the supported metallocene with Al^iBu_3 or $AlEt_2Cl$ have been shown to afford a catalyst with much greater stability than that of the homogeneous system [5]. Supporting a $CpTiCl_3/MgCl_2/THF$ mixture on $AlEt_3$-treated silica leads to an active polymerization system when activated by $AlEt_2Cl$ and $AlEt_3$ [6]. Supporting $Me_2C(C_5H_4)(C_{13}H_8)ZrCl_2$ or Cp_2ZrCl_2 on Al_2O_3, $MgCl_2$ and SiO_2 and activating with $AlMe_3$ give an active propylene polymerization catalyst only for the alumina and magnesium chloride systems [7]; silica affords an active catalyst only when it is pretreated with methylalumoxane [8].

3 SUPPORTED METALLOCENE/ALUMOXANE CATALYSTS

The recognition in the early 1980s of the extremely high catalytic activity of metallocene complexes activated by methylalumoxane (MAO), and the discovery that activities, molecular weight, comonomer incorporation [9] and stereoregularity in α olefin polymerization [10] could be regulated by appropriate substitution on the cyclopentadienyl ring, greatly increased the commercial interest in these catalysts and stimulated extensive research on supported catalysts for use in pilot plants and commercial processes.

Methylalumoxane is an ill-defined mixture of oligomers produced from the reaction of $AlMe_3$ with water under controlled conditions. It is commercially available as a solution in toluene or aliphatic hydrocarbons. The solvent can be evaporated to produce a white pyrophoric solid that can be used as both a cocatalyst and a support. Finely divided MAO solids can be precipitated from a toluene solution by addition of n-decane followed by evaporation. These can be reacted with a toluene solution of $Cp'_2ZrCl_2(Cp' = $ substituted or unsubstituted Cp ligand) to afford a solid catalyst [11]. Polymerization with this catalyst is claimed to give a good particle size distribution with no fouling. Addition of agents such as boroxines [12] and organic peroxides and carbonates [13] also serves to precipitate MAO from toluene.

It is not even necessary, in fact, to isolate solid MAO to achieve particle size control. Atomization of a precontacted metallocene–MAO solution in a gas-phase process [14] or low-temperature prepolymerization of the catalyst solution in liquid propylene followed by bulk polymerization [15] have both been claimed to give uniform particles with good bulk density.

For the most part, though, metallocene polymerization catalysts are heterogenized by combining the catalyst and cocatalyst on a third material that is inert toward the active catalyst species generated. A wide variety of support materials have been used, including starches [16], cyclodextrins [17], $MgCl_2$ [18], clays [19], zeolites [20] and metals and ceramics [21]. The predominant supports used comprise inorganic oxides such as silica and alumina because of their large surface area and pore volume, chemical inertness and low cost.

A remarkable number of methods have been developed to prepare supported metallocene catalysts. In general, these fall in three classes:

- supporting the activator followed by reaction with the metallocene;
- supporting the metallocene, then reacting with the cocatalyst;
- reacting a metallocene–cocatalyst mixture with the support.

The earliest methods of supporting metallocene/alumoxane catalysts involved first evaporating a toluene solution of methylalumoxane on a support, usually silica [22]. Subsequent heating of the supported alumoxane at 150–200 °C is claimed to fix the alumoxane on to the support, avoiding leaching into solution [23]. Adding modifiers such as $Al(OR)_3$ [24], bisphenols [25] and boric acid [26] to the supported alumoxane has been claimed to improve catalyst performance. The solution of the metallocene is then added to the supported activator and the solvent evaporated to form the solid catalyst.

This method has also been used not only for bis(cyclopentadienyl) metal complexes, but also for 'constrained-geometry' compounds such as $Me_2Si(C_5Me_4)(N^t\text{-Bu})TiCl_2$ [27] and monocyclopentadienyl metal complexes such as CpZrCl$_3$ [28]. The metallocene can be dry-blended with the supported MAO [29] or vaporized on to the support [30]. Spray-drying the supported MAO with the catalyst solution is claimed to avoid the formation of larger particles, thus improving the particle size distribution [31].

In another method for preparing supported alumoxanes AlMe$_3$ is reacted with silica containing 5–40% water [32]. Addition of metallocene dichloride and evaporation afford a solid polymerization catalyst, although an alternative method uses separately supported metallocene and MAO to form the active catalyst [33]. Alternatively, a metallocene–AlMe$_3$ mixture can be added to undehydrated silica [34]. AlEt$_3$ can be used as a substitute, with a metallocene–AlMe$_3$ mixture added to the surface-bound ethylalumoxane [35]. Mixtures of AlMe$_3$ with AlEt$_3$ or Al^{i}Bu$_3$ mixtures are also effective for generating the alumoxane [36]. The amount of alumoxane bound to the silica can be adjusted by altering the amount of water on the silica or the ratio of water to trialkylaluminum [37] or by washing the supported metallocene catalyst after preparation [38]. Aging the alumoxane support for 2–4 weeks has been found to improve catalyst activity [39]. An alternative method of generating alumoxanes *in situ* involves hydrolysis of alkylaluminums with water in the presence of the support, either in a hydrocarbon suspension [40] or with the reactive components in the gas phase [41]. MAO may also be post-hydrolyzed by water in the presence of a support material and an emulsifier [42].

With some exceptions [43], most examples of supporting the metallocene component then activating with alumoxane come from cases in which the metallocene is attached to a support through a functional group. A perception exists that the metallocene catalyst can be leached from the support during polymerization [44] and that this 'soluble' catalyst causes an irregular particle size distribution and reactor fouling. In order to overcome this possible cause of poor catalyst performance, researchers have turned to tethering the catalyst chemically to the support.

Direct attachment of cyclopentadienide derivatives can be carried out in a number of ways. Chloromethylated cross-linked polystyrene is reacted with a cyclopentadienyl anion to form a polymer-attached cyclopentadiene. Further synthetic elaboration produces a polystyrene support with pendant Me$_2$Si(C$_5$H$_3$CH$_2$—)(C$_{13}$H$_8$)ZrCl$_2$ groups (Scheme 1) [45].

The hydroxyl groups of silica react with (MeO)$_3$(CH$_2$)$_3$Cp to tether the cyclopentadienyl group. This is then deprotonated and reacted with CpZrCl$_3$ to form a surface-bound metallocene procatalyst [46]. Silica has been chlorinated and reacted with NaCp. Subsequent reaction of the ≡Si—C$_5$H$_5$ functionality with Zr(NMe$_2$)$_4$ produces ≡Si—(C$_5$H$_4$)Zr(NMe$_2$)$_3$, which polymerizes ethylene in the presence of MAO [47]. In a similar preparation, NaCp is reacted with ethylene–methylacrylate copolymer to form what is presumed to be a keto-Cp intermediate. This is then reacted with ZrCl$_4$(THF)$_2$ or CpZrCl$_3$ to form a polymer-bound metallocene (Scheme 2) [48].

Ring substituents also provide a means of chemically incorporating the metallocene in the support matrix. Prepolymerization of bis(3-propenylcyclopentadienyl)-

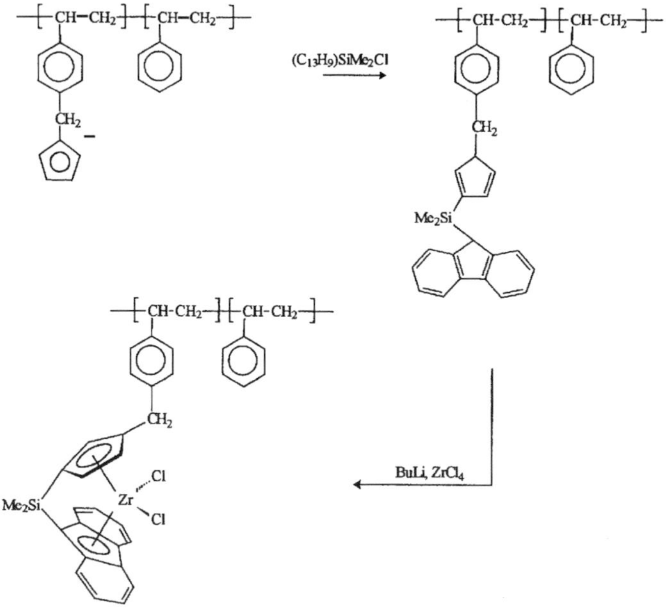

Scheme 1

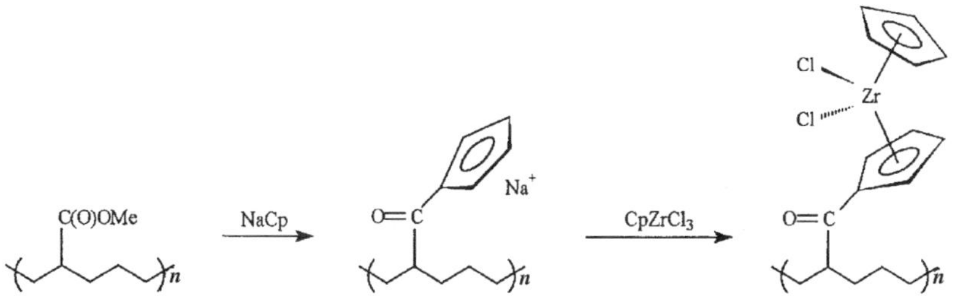

Scheme 2

or bis(3-butenylcyclopentadienyl)zirconium dichloride [49] or $(CH_2{=}CH(CH_2)_2)$ $MeC(C_5H_4)(C_{13}H_8)ZrCl_2$ [50] with ethylene affords a copolymer with the metallocene incorporated in the polymer [Scheme 3(a)]. Similarly, bis(2-vinylcyclopentadienyl)zirconium dichloride has also been copolymerized with styrene in the presence of azobis(isobutyronitrile) (AIBN) to give a metallocene–styrene copolymer [51]. Polysiloxane copolymers of metallocenes are prepared by reacting bis(allylcyclopentadienyl)zirconium dichloride with poly(methylhydrogensiloxane) in the presence of a catalyst [Scheme 3(b)] [52].

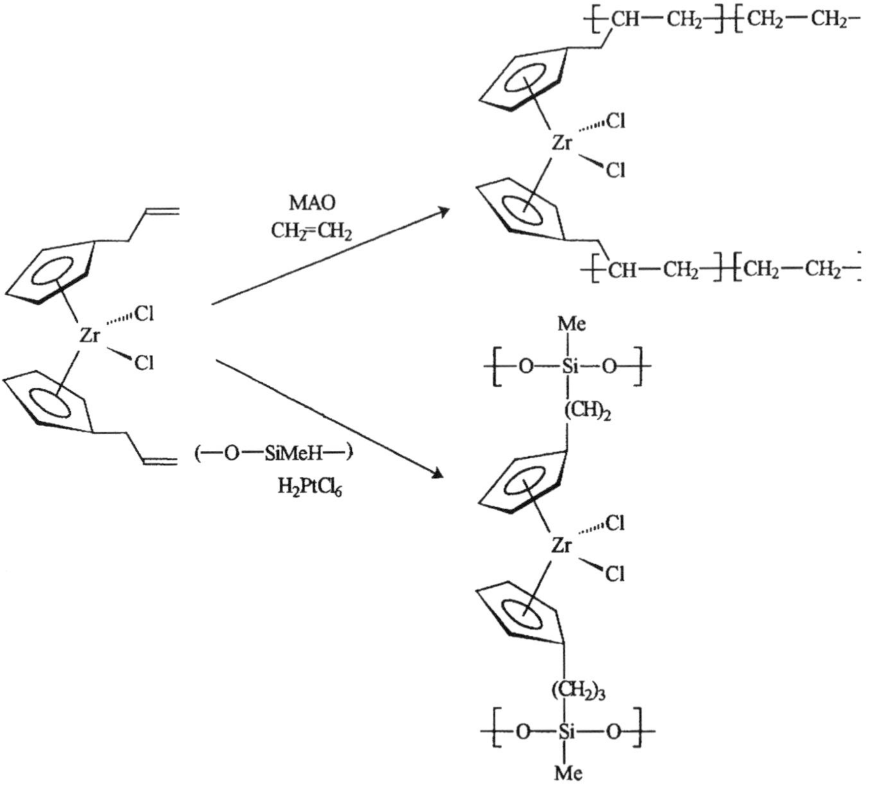

Scheme 3

Reaction of a functionalized silica bearing chloromethyl groups with $Me_2Si(2-$ dimethylaminoindenyl)zirconium dichloride quaternizes the amino group, thereby fixing the metallocene to the support (Scheme 4) [53]. Silica substituted with 3-aminopropyltrimethoxysilane also reacts with $CpTiCl_3$ [54] or $(ClMe_2SiC_5H_4)$ $(Cp)ZrCl_2$ [55] to give a chemically bonded metallocene.

Metallocene complexes containing a bridging group can be immobilized on a support through the bridging group. Soga and co-workers have constructed silyl-bridged bis(indenyl) complexes for stereospecific α-olefin polymerization by reacting partially dehydroxylated silica with $SiCl_4$, adding lithium indenide, aromatizing the supported ligand set and metalating with $ZrCl_4(THF)_2$ (Scheme 5) [56]. $Cl_2Si(C_9H_7)_2$, formed from $SiCl_4$ and lithium indenide, can be reacted with lithiated styrene–divinylbenezene copolymer [57] or hydrolyzed to form a polysiloxane polymer [58]. Deprotonation of the polymer-bound ligand and reaction with $ZrCl_4$ also has been claimed to give a supported metallocene catalyst for isospecific α-olefin polymerization.

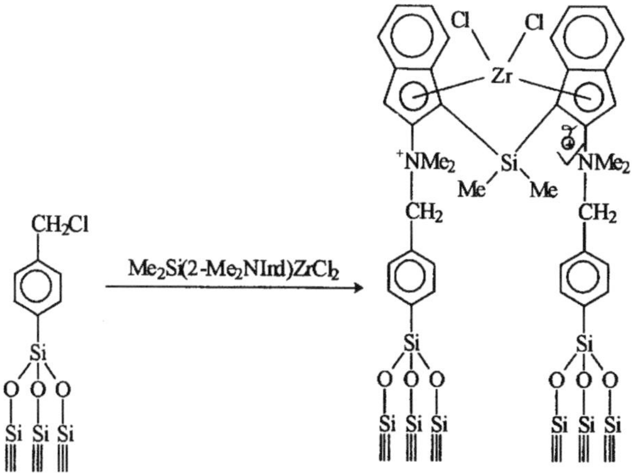

Scheme 4

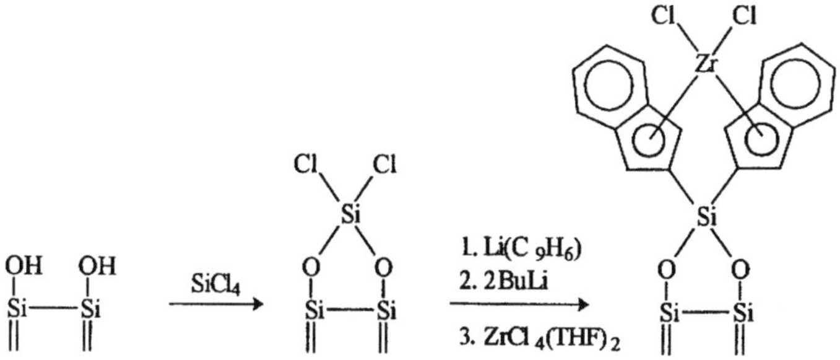

Scheme 5

Syndiospecific propylene polymerization catalysts have been incorporated into polystyrene by copolymerizing styrene with 4-$(C_5H_5)(C_{13}H_9)(Me)CC_6H_4CH=CH_2$ (Scheme 6) [59] or into polysiloxanes by hydrolysis of $Cl_2Si(C_5H_5)(C_{13}H_9)$ [60], aromatizing with BuLi and reacting with $ZrCl_4$.

Investigators at Phillips have reacted silica with $MeClSi(C_5H_5)(C_{13}H_9)$ in the presence of pyridine base to form $\equiv Si-O-SiMe(C_5H_5)(C_{13}H_9)$. The supported ligand was then deprotonated with butyllithium and metalated with $ZrCl_4$ to form a surface-bound metallocene procatalyst [61]. It has also been suggested, although not exemplified, that $(RO)MeSi(C_5Me_4)(N^tBu)TiCl_2$ complexes can be bound to silica through the silyl bridge [62].

The third general method of supporting metallocene catalysts is to add a solution of metallocene and MAO to the support [63]. A very useful refinement of this technique is the incipent impregnation of the support with the catalyst solution [64].

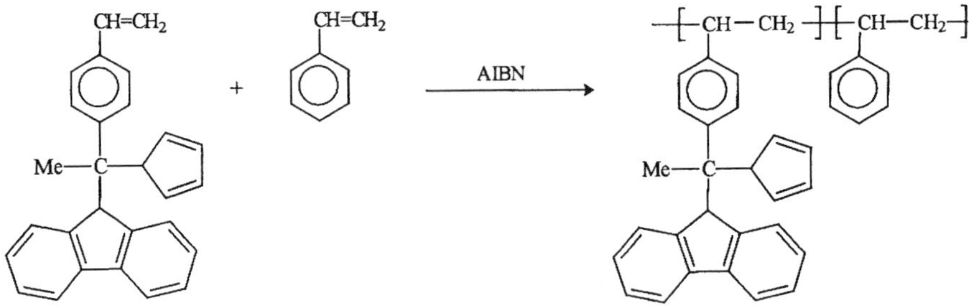

Scheme 6

The pore volume of the support material is measured and a volume of catalyst solution is added which just fills the amount of support used. Volumes greater than the total pore volume of the support can be used if the MAO : metallocene ratio is kept low (below 300) [65] and the total amount of liquid should not form a slurry when contacted with the solids.

There are two perceived advantages of this method: the catalyst occupies the pores of the support and less of the surface—leading to improved practical morphology—and a much smaller amount of liquid solvent needs be used, thus shortening the preparation time and lowering effluents and disposal costs. Equipment suitable for preparing 50–100 lb of supported catalyst by incipient impregnation been described [66].

Supported catalysts have been prepared *in situ* by assembling the metallocene components in the presence of the support. Silica can be impregnated with $TiCl_4$ or $ZrCl_4$ and then reacted with Cp'_2Mg [67], Me_3SiCp' [68] or $LiCp'$ [69]. The metallocene complexes formed on the support are then activated to form the catalyst. Supported metallocenes have also been assembled by mixing the product of $Zr(OR)_4$, $H-Cp'$, and and alkyllithium, -magnesium or -zinc complex with supported MAO [70].

Fouling of the reactor walls and agitators occurs when the catalyst polymerizes too rapidly for proper particle morphology to be maintained. Prepolymerization of the supported metallocene catalyst is a method by which this problem can be minimized. The supported metallocene catalyst is allowed to polymerize monomer under low pressures and temperatures in order to coat the catalyst particle with 0.5–30 g of polymer per gram of support [71]. The prepolymer slows diffusion of monomer to the catalyst sites, resulting in improved particle morphology. Addition of hydrogen in the prepolymerization process has been found to avoid agglomeration of the catalyst particles, especially when the molecular weight of the prepolymer is high [72]. Other methods of reducing reactor fouling include adding polysiloxanes to the reactor system [73] or adding anti-static agents [74] or low molecular weight toluene-soluble polymers to the supported catalyst [75].

One school of thinking holds that the residual surface hydroxyl groups in an inorganic support are deleterious to catalyst performance; these are removed by agents such as Me_3SiCl [76]. To avoid the presence of hydroxyl groups, researchers have used organic polymers and copolymers as supports for metallocene catalysts. Porous polyethylene and polypropylene have been employed [77], but polystyrene [78], especially cross-linked by divinylbenzene (DVB), has attracted the most attention because of the potential to control the porosity of the copolymer by adjusting the level of DVB in the copolymerization.

On the other hand, other investigators assert that some surface functionality is beneficial. By dehydrating silica at a lower temperature (200–400 vs 800 °C), adsorbed water is driven off, but surface hydroxyl groups (1–5 %) remain. These are believed to fix the cocatalyst to the surface, leading to improved performance without reactor fouling [79]. Furthermore, research has indicated that a certain amount of functionality in the polystyrene support is desirable, possibly to fix the alumoxane on to the support. Styrene (55 %), DVB (30 %) and p-acetoxystyrene (15 %) were terpolymerized to form a support material for a Cp_2ZrCl_2/MAO catalyst; the activities were little different from that of silica [80]. Hydrolysis of the terpolymer forms a p-hydroxystyrene derivative which is alo a useful carrier. Investigators at Montell have prepared numerous derivatives of styrene–DVB copolymers, including those with —C(=O) Me,—CH(OH)Me,—CH$_2$Cl,—CO$_2$H and —CH$_2$OH groups, and employed them as supports for 1,2-$C_2H_4(C_9H_6)_2$ $ZrCl_2$/MAO catalysts [81]. Acrylonitrile–divinylbenzene copolymers can also be functionalized and used as support materials for metallocene catalysts [82].

4 ALUMOXANE-FREE SUPPORTED METALLOCENE CATALYSTS

An alternative to MAO as a cocatalyst was developed in the mid-1980s. These boron-based activators, e.g. $[R_3NH][B(C_6F_5)_4]$ [83] and $B(C_6F_5)_3$ [84] (Scheme 7),

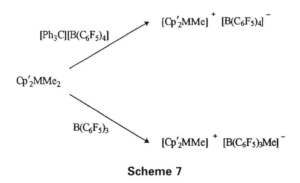

Scheme 7

afford cationic metal catalysts with activities matching those of the MAO-based catalysts.

Silica-supported catalysts based on the boron activators are more problematic than alkylaluminum cocatalysts: in this case, residual surface hydroxyl groups, even on silica dried at $800\,^\circ$C, actually do deactivate the catalyst. Treatment of the silica support with trialkylaluminum serves to passivate the surface by reaction with the surface hydroxyl groups and render it inert to the $Cp'_2MMe_2/[Ct][B(C_6F_5)_4]$ catalyst [85]; the treated silica itself is not an effective cocatalyst. The supported ionic catalysts have been successfully used in slurry, gas-phase and bulk-monomer processes. Similarly treated alumina, $MgCl_2$ and cross-linked polystyrene [86] are also suitable support materials for ionic catalysts.

$Me_2Si(C_5Me_4)(N^tBu)MMe_2$-based ionic catalysts (M$=$Ti, Zr) [87] and $CpZrMe_3$ [88] have also been supported using this technique. Other methods of passivating of supports for ionic metallocene catalysts include capping partially or fully hydroxylated supports, including poly(vinyl alcohol), with butyllithium or butylethylmagnesium and reacting the surface anions with a capping agent such as BrC_6F_5 [89] or by reacting silica with ammonium fluoride [90]. Addition of alkylaluminum-treated silica to the supported ionic catalyst has been claimed to lead to improved catalyst activity, probably by scavenging of adventitious impurities [91].

Ion-pair catalysts can be generated by $in\ situ$ alkylation of metallocene dichlorides followed by reaction with the ionic activator. Supported metallocene catalysts have been prepared using these techniques. For example, solutions of mono- or bis(cyclopentadienyl)metal dichlorides, reacted with alkylaluminums, are evaporated on to Al^iBu_3-passivated silica and are activated by $B(C_6F_5)_3$ [92] or $[Ph_3C][B(C_6F_5)_4]$ [93]. Similarly, MAO has been used as the alkylating agent; the catalyst is prepolymerized before reaction with $[HNMe_2Ph][B(C_6F_5)_4]$ [94]. $CpTiCl_3$ supported on silica is inert when $AlMe_3$ is used as a cocatalyst, but addition of $[Ph_3C][B(C_6F_5)_4]$ or $B(C_6F_5)_3$ activates the metallocene to form a catalyst for the atactic polymerization of propylene [95].

Instead of rendering them inert, surface hydroxyl groups can be exploited in preparing activators for supported ionic metallocene catalysts. Reaction of $B(C_6F_5)_3$ with partially dehydroxylated silica produces $=$Si$-$O$-$(H)B(C_6F_5)_3$ groups which, when reacted with amine bases, form $[R_3NH][=Si-O-B(C_6F_5)_3]$. Reaction with a metallocene forms an active surface-supported ionic catalyst [96]. Triphenylcarbenium analogues of these anions are formed by reacting partially dehydroxylated silica with BuLi, treatment with $B(C_6F_5)_3$ to form $Li[=Si-O-B(C_6F_5)_3]$, and metathesis with Ph_3CCl [97] (Scheme 8).

A supported ionic activator with the cationic portion bound to silica was prepared by reacting p-$(MeO)_3SiC_6H_4NMe_2$ with partially dehydroxylated silica forming the surface-bound amine $\{=Si-O\}_3SiC_6H_4NMe_2$; protonation with HCl and reaction with $[Li][B(C_6F_5)_4]$ affords the supported ionic activator $[\{=Si-O\}_3SiC_6H_4NMe_2H][B(C_6F_5)_4]$ [98].

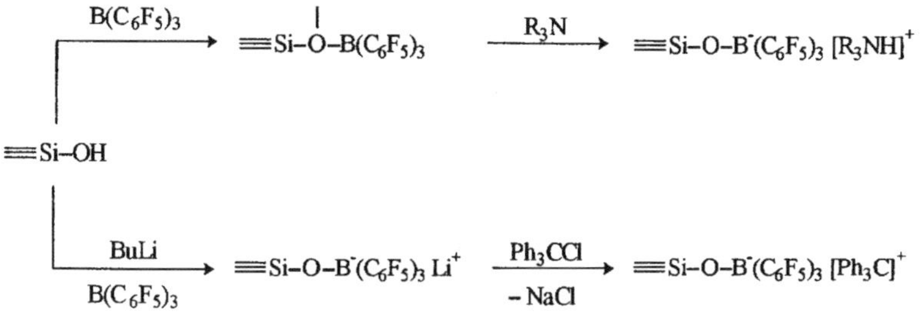

$$\equiv Si-OH \xrightarrow[]{B(C_6F_5)_3} \equiv Si-O-B(C_6F_5)_3 \xrightarrow{R_3N} \equiv Si-O-B^-(C_6F_5)_3 \; [R_3NH]^+$$

$$\xrightarrow[B(C_6F_5)_3]{BuLi} \equiv Si-O-B^-(C_6F_5)_3 \; Li^+ \xrightarrow[- NaCl]{Ph_3CCl} \equiv Si-O-B^-(C_6F_5)_3 \; [Ph_3C]^+$$

Scheme 8

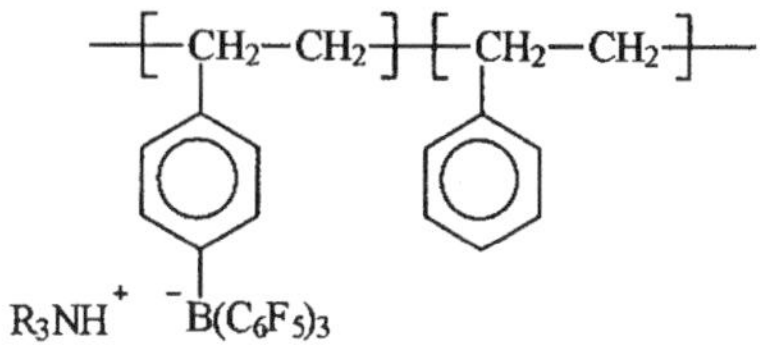

$$-\left[CH_2-CH_2 \right]-\left[CH_2-CH_2 \right]-$$

$$R_3NH^+ \quad {}^- B(C_6F_5)_3$$

Scheme 9

Supports with the anionic portion of the activator chemically bound are also known. Trimethylammonium 4-vinylphenyltris(pentaflurophenyl)borate has been copolymerized with styrene to form a polyanionic copolymer. This can be reacted with $Cp'_2ZrCl_2 - Al^iBu_3$ mixture to form an active catalyst (Scheme 9) [99]. Ion-exchange resins and their triyl salts have also been reacted with zirconocene dichlorides and triisobutylaluminum to form catalysts for olefin polymerization and copolymerization [100].

An unusual MAO-free catalyst is formed by intercalating oxidants such as $[Cp_2Fe][HSO_4]$ and $[HNMe_2Ph][Cl]$ in clays such as montmorillonite. This support serves to activate $Cp'_2ZrCl_2 - Al^iBu_3$ mixtures [101].

5   SUPPORTED METALLOCENE CATALYSTS WITHOUT COCATALYSTS

Of course, the most notable supported metallocene catalyst requiring no cocatalyst is the Cp_2Cr/silica catalyst developed in the early 1970s [102]. This should be classified as a truly heterogeneous catalyst and not as a supported single-site catalyst. The nature of the active site remains unclear even after 25 years of research, and the polymer produced has a broad molecular weight distribution indicative of multiple active polymerizing sites.

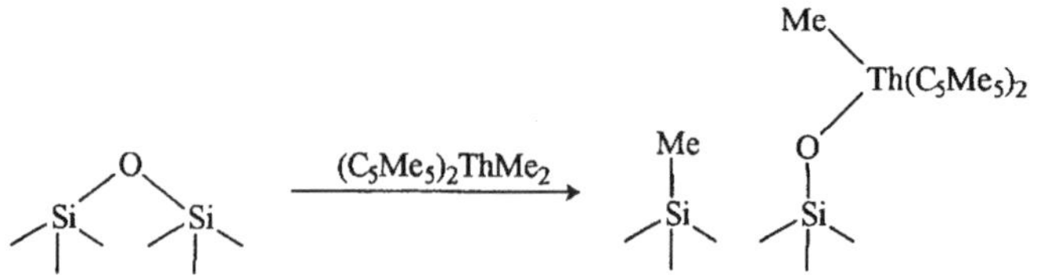

Scheme 10

Supported single-site metallocene catalysts requiring no external activator are known, but not numerous. A sufficiently Lewis acidic surface may itself abstract a σ-bonded substituent from a metallocene and generate an active catalyst. Marks and co-workers found that supporting $(C_5Me_5)MMe_2(M = Th, U)$ on fully dehydroxylated alumina affords a catalyst capable of polymerizing ethylene [103]. Analysis by ^{13}C cross-polarized magic angle spinning (CP/MAS) NMR of the thorocene complex supportod on dehydroxylated silica indicates that surface Si—Me groups are present (Scheme 10) [104]. Even in the absence of any external alkylating agent, $(^nPrCp)_2ZrCl_2$ has been found to have some polymerization activity when supported on silica dried at $850\,^\circ C$ [105].

Magnesium chloride can also function as both support and activator. Evaporating $1,2\text{-}C_2H_4(C_9H_6)_2ZrMe_2$ with $MgCl_2$ affords an active propylene polymerization catalyst in the absence of any further activator [84b].

6 COMPARISONS BETWEEN HOMOGENEOUS AND SUPPORTED METALLOCENE CATALYSTS

From a commercial standpoint, it is most desirable that the polymers produced from supported metallocene catalysts resemble as closely as possible the characteristics of their solution-soluble counterparts, sharing narrow molecular weight distribution, random distribution of comonomers and, in the case of stereospecific polymerization, high stereoregularity and melting-point.

For the most part, there are no differences in the nature of the polymer produced. Isospecific α-olefin polymerization catalysts such as $1,2\text{-}C_2H_4(C_9H_6)_2ZrCl_2$ produce polypropylene with the same isotacticity whether supported or in solution [106], but the effect is dependent on the method of catalyst preparation (Table 1). Supporting the metallocene in a first step without treatment with MAO or $AlMe_3$ can lead to the formation of secondary species which may broaden the molecular weight distribution or even lead to completely different products. For example, the syndiospecific catalyst $Me_2C(C_5H_4)(C_{13}H_8)ZrCl_2$ becomes isospecific when supported on dehydrated but untreated silica [107]. In contrast, Kaminaka and Soga found that the same metallocene supported on $AlMe_3$- or $AlEt_3$-treated alumina or $MgCl_2$ in the

Table 1 Propylene polymerization with homogeneous and supported $1,2$-C_2H_4 $(C_9H_6)_2ZrCl_2[=Et(Ind)_2ZrCl_2]$ catalysts [106e]

Catalyst	Cocatalyst	Activity [kg PP/(mol Zr h)]	M_w	M_w/M_n	[mm] (%)
$Et(Ind)_2ZrCl_2$	MAO	221	9 100	2.1	85
$Et(Ind)_2ZrCl_2 - SiO_2/MAO$	MAO	131	11 300	2.3	85
$Et(Ind)_2ZrCl_2 - SiO_2/MAO$	Al^iBu_3	33	16 500	2.5	85
$Et(Ind)_2ZrCl_2 - SiO_2$	MAO	6	166 000	14.8	71

presence of trialkylaluminums, or silica treated with MAO, afforded syndiotactic polypropylene [108].

The activities of supported metallocene catalysts are usually 10–60 % lower than those of their homogeneous counterparts. Although studies are few, this may be due to deactivation of catalyst sites when contacted with the support or caused by inefficient generation of the active cationic species. It is not a general phenomenon, however, and there are cases in which supporting a metallocene has a positive effect on its activity. For example, in solution the complex $[(C_5Me_5)CrMe(THF)_2][BPh_4]$ is only feebly active in ethylene polymerization, even in the presence of isobutyl-alumoxane. Supporting this complex on silica leads to an almost 100-fold increase in activity [109].

An advantage of supported metallocene/alumoxane catalysts is that the Al : M ratio, which must be 1000 or higher for effective performance in homogeneous polymerization, can be reduced to less than 100 without serious effect. It has been hypothesized that large excesses of alumoxane are needed to suppress bimolecular deactivation of the catalyst; immobilizing the catalyst on a support largely removes this pathway.

Because they are less susceptible to deactivation by adventitious impurities, supported metallocene catalysts show much greater long-term stability, an important consideration when catalyst batches might be stored for several months in a plant. For example, a soluble $Me_2Si(C_9H_{10})_2ZrMe_2/[HNMe_2Ph][B(C_6F_5)_4]$ catalyst in virtually inactive after 24 h whereas the heterogenized counterpart retains a considerable portion of its original activity even after several weeks [85a].

7   PROCESS CONSIDERATIONS

Supported metallocene catalysts have been run successfully in many large-scale plants, such as fluidized-bed gas-phase processes, including those with 'condensed mode' operations [110]. Because they perform differently to conventional Ziegler–Natta catalysts, care must be taken to avoid reactor fouling and process upsets. For example, a scavenger such as $AlEt_3$ is injected into the reactor to remove impurities in the feed stream. With supported metallocene catalysts, the presence of the scavenger can cause fouling. Either operating without scavenger or stopping addition soon after start-up results in improved operability and the preparation of copolymers with densities below $0.90 \, g/cm^3$ [111].

In transitioning between conventional Ziegler–Natta catalysts and metallocene catalysts, the difference in reactivity to hydrogen and comonomers makes the two systems incompatible. Polymerization by one must be stopped and the process conditions modified before introducing the second catalyst. Small amounts of a catalyst killer such as water (injected with the gas stream or added as wet silica) or methanol can be introduced to deactivate irreversibly the catalyst in the reactor. The second catalyst can be introduced after adjustment of feed streams [112].

8 CONCLUSIONS

Much has been done to support metallocene catalysts to form highly active catalysts which operate efficiently in large-scale industrial processes. The number of announcements of the commercial production of polyolefins produced by metallocene catalysts has increased rapidly in the last few years.

Yet with all this research which has been conducted, little has been published about the understanding we have reached on the fundamentals of catalyst–support interactions and the growth of polymer on the supported catalyst. A look at the references to this chapter shows one possible reason: most of the available literature on supported metallocene catalysts is in the form of patents, which need only inform the reader of the best way of using the invention; no explanation of the theories behind the invention or elaboration of the insights achieved during the course of the investigation is required.

An enormous amount of effort has been expended by many companies on developing increasingly elaborate metallocene complexes and catalysts which operate at higher activities and which produce polyolefins with an ever wider range of properties. The homogeneous nature of the active sites has led to an unprecedented level of understanding of the fundamental steps of the polymerization process and of the catalyst–cocatalyst interaction. Yet amorphous materials such as silica continue to be the support of choice for supported catalysts; whatever has been learned is again obscured. Future research should turn to supports which are as 'single-sited' as the catalysts themselves, ones which, in fact, improve the performance of the catalyst. The importance of metallocene catalysts lies in the rational design of catalysts to prepare polymers of targeted structure and properties. It remains to extend this degree of comprehension to supported systems.

9 REFERENCES

1. Kallio K., Palmqvist, U., Knuuttilla, H. and Fatnes, A. M., *PCT Int. Appl.*, 96/34898 (1996).
2. Donnet, J. B., Wetzel, J. P. and Riess, G., *J. Polym. Sci.*, **6**, 2359 (1968).
3. Bocharov, J. N., Kabanov, V. A., Martynova, M. A., Popov, V. G., Smetanyuk, V. I. and Fedorov, V. V., *US Pat.*, 4 161 462 (1979).
4. Slotfeldt-Ellingsen, D., Dahl, I. M. and Ellestad, O. H., *J. Mol. Catal.*, **9**, 423 (1980).
5. Satyanarayana, G. and Sivaram, S., *Macromolecules*, **26**, 4712 (1993).
6. Ali-Huikku, S., Palmqvist, U., Lommi, M. and Iiskola, E., *US Pat.*, 5 324 698 (1994).
7. Kaminaka, M. and Soga, K., *Polymer*, **32**, 310 (1991).
8. (a) Soga K. and Kaminaka, M., *Makromol. Chem.*, **194**, 1745 (1993); (b) Soga K. and Kaminaka, M., *Makromol. Chem.*, **195**, 1369 (1994).
9. Welborn, H. C. and Ewen, J. A., *US Pat.*, 5 324 800 (1994).
10. (a) Ewen, J. A., *J. Am. Chem. Soc.*, **106**, 6355 (1984); (b) Kaminsky, W., Külper, K., Brintzinger H. H. and Wild, F. R. W. P., *Angew. Chem., Int. Ed. Engl.*, **24**, 507 (1985); (c) Brintzinger, H. H., Fischer, D., Mülhaupt, R., Rieger, B. and Waymouth, R. M., *Angew. Chem., Int. Ed. Engl.*, **34**, 1143 (1995).
11. (a) Tsutsui, T. and Kashiwa, N., *Polymer*, **32**, 2671 (1991); (b) Kioka M. and Kashiwa, N., *US Pat.*, 4 923 833 and 4 952 540 (1990).
12. Geerts, R. L., Kufeld, S. E. and Hill, T. G., *US Pat.*, 5 411 925 (1995).
13. Geerts, R. L., *US Pat.*, 5 480 948 (1996).
14. (a) Brady, R. C., III, Karol, F. J., Lynn, T. R., Jorgensen, R. J., Kao, S.-C. and Wasserman, E. P., *US., Pat.*, 5 317 036 (1994); (b) Keller, G. E., Carmichael, K. E., Cropley, J. B., Larsen, E. R., Ramamurthy, A. V., Smale, M. W., Wenzel, T. T. and Williams, C. C., *Eur. Pat Appl.*, 764 665 (1997).
15. (a) Ewen, J. A., *Eur. Pat. Appl.*, 354 893 (1990); (b) Fujita, T., Sugano, T. and Uchino, H., *Eur. Pat. Appl.*, 566 349 (1993); (c) Haspeslagh L. and Maziers, E., *US Pat.*, 5 283 300 (1994).
16. Kaminsky, W., *US Pat.*, 4 431 788 (1984).
17. Lee D. and Yoon, K., *Macromol. Rapid Commun.*, **15**, 841 (1994).
18. (a) Sarma, S. S., Satyanarayana, G. and Sivaram, S., *Polym. Sci.*, **1**, 315 (1994); (b) Bailly, J.-C., Bres, P., Chabrand, C. J. and Daire, E., *US Pat.*, 5 106 804 (1992); (c) Bailly J.-C. and Chabrand, C. J., *Eur. Pat. Appl.*, 435 514 (1991); (c) Milani, F., Labianco, A. and Pivotto, B., *Eur. Pat. Appl.*, 785 220 (1997); (d) Sacchetti, M., Pasquale, S. and Govoni, G., *PCT Int. Appl.*, 95/32995 (1995).
19. (a) Suga, Y., Maruyama, Y., Isobe, E., Suzuki, T. and Shimizu, F., *US Pat.*, 5 308 811 (1994); (b) Suga, Y., Maruyama, Y., Isobe, E., Uehara, Y., Ishihama, Y. and Sagae, T., *Eur. Pat. Appl.*, 683 180 (1995).
20. (a) Woo, S. I., Ko, Y. S. and Han, T. K., *Macromol. Rapid Commun.*, **16**, 489 (1995); (b) Ko, Y. S., Han, T. K. and Woo, S. I., *Macromol. Rapid Commun.*, **17**, 749 (1996).
21. Hayashi, H., Matono, K., Asahi, S. and Uoi, M., *US Pat.*, 4 564 647 (1986).
22. (a) Welborn, H. C., Jr, *US Pat.*, 4 808 561 (1989); (b) Takahashi, T., *US Pat.*, 5 026 797 (1991).
23. Jacobsen, G. B., Spencer, L. and Wauteraerts, P. W., *PCT Int. Appl.*, 96/16092 (1996).
24. Tajima, Y., Nomiyama, K., Kataoka, N. and Matsuura, K., *Eur. Pat. Appl.*, 474 391 (1992).
25. (a) Ernst, E., Reussner, J. and Neissl, W., *Eur. Pat. Appl.*, 685 494 (1995); (b) Ernst E. and Reussner, J. *Eur. Pat. Appl.*, 787 746 (1997).
26. Sugano T. and Yamamoto, K., *Eur. Pat. Appl.*, 728 773 (1996).
27. Canich J. M. and Licciardi, G. F., *US Pat.*, 5 057 475 (1991).

28. (a) Kuramoto, M., Maezawa, H. and Hayashi, H., *US Pat.*, 5 212 232 (1993); (b) Harrison, D. G. and Chisholm, P. S., *US Pat.*, 5 661 098 (1997).
29. Knuuttila, H., Hokkanen, H. and Salo, E., *PCT Int. Appl.*, 96/32423 (1996).
30. Hokkanen, H., Knuuttila, H., Lakomaa, E.-L. and Sormunen, P. *PCT Int. Appl.*, 95/15216 (1995).
31. Wasserman, E. P. Smale, M. W. Lynn, T. R. Brady, R. C. III and Karol, F. J. *US Pat.*, 5 648 310 (1997).
32. (a) Chang, M. *US Pat.*, 4 912 075, 4 914 253 and 4 935 397 (1990); (b) Chang, M. *US Pat.*, 5 008 228 (1991).
33. Jejelowo, M. O. *US Pat.*, 5 466 649 (1995).
34. Chang, M. *US Pat.*, 5 238 892 (1993).
35. Chang, M. *US Pat.*, 4 925 821 (1990).
36. (a) Chang, M. *US Pat.*, 4 937 217 (1990); (b) Chang, M. *US Pat.*, 5 006 500 (1991).
37. Chang, M. *US Pat.*, 5 629 253 (1997).
38. Chang, M. *PCT Int. Appl.*, 96/04318 (1996).
39. Jejelowo, M. O. *US Pat.*, 5 468 702 (1995).
40. (a) Tsutsui T. and Ueda, T. *US Pat.*, 5 234 878 (1993); (b) Gürtzgen, S. *US Pat.*, 5 446 001 (1995); (c) Herrmann, H. F. Bachmann, B. and Spaleck, W. *Eur. Pat. Appl.*, 578 838 (1994).
41. Becker, R.-J. Gürtzgen, S. and Kutschera, D. *US Pat.*, 5 534 474 (1996).
42. Speca, A. N. *US Pat.*, 5 552 358 (1996).
43. Tsutsui, T. Toyota, A. and Kashiwa, N. *Eur. Pat. Appl.*, 250 600 (1987).
44. Semikolenova N. V. and Zakharov, V. A. *Macromol. Chem. Phys.*, **198**, 2889 (1997).
45. Peifer, B. Alt, H. G. Welch, M. B. and Palackal, S. J. *US Pat.*, 5 473 020 (1995).
46. Iiskola, E. Timonen, S. and Pakkanen, T. *Eur. Pat. Appl.*, 799 838 (1997).
47. Gila, L. Proto, A. Ballato, E. Vigliardo, D. and Lugli, G. *Eur. Pat. Appl.*, 725 086 (1996).
48. DiMaio, A. J. *US Pat.*, 5 587 439 (1996).
49. (a) Chabrand, C. J. McNally, J. P. and Little, I. R. *Eur. Pat. Appl.*, 586 167 (1994); (b) Little I. R. and McNally, J. P. *Eur. Pat. Appl.* 586 168 (1994); (c) C. J. Chabrand, Little, I. R. and McNally, J. P. *Eur. Pat. Appl.*, 641 809 (1995).
50. Welch, M. B. Alt, H. G. Peifer, B. Palackal, S. J. Glass, G. L. Pettijohn, T. M. Hawley, G. R. and Fahey, D. R. *US Pat.*, 5 498 581 (1996).
51. Antberg, M. Herrmann, H.-F. and Rohrmann, J. *US Pat.*, 5 134 212 (1992).
52. Antberg, M. Böhm, L. and Rohrmann, J. *US Pat.*, 5 071 808 (1991).
53. Langhauser, F. Fischer, D. Kerth, J. Schweier, G. Barsties, E. Brintzinger, H. H. Schaible, S.and Roell, W. *US Pat.*, 5 627 246 (1997).
54. Uozumi, T. Toneri, T. and Soga, K. *Macromol. Rapid Commun.*, **18**, 9 (1997).
55. Vega, W. M. LaFuente, P. Munoz-Escalona, A. Hidalgo Llinas, G. Sancho Royo, J. and Mendez Llatas, L. *Eur. Pat. Appl.*, 757 992 (1997).
56. Soga, K. *Macromol. Symp.*, **89**, 249 (1995).
57. Soga, K. Kim, H. Lee, S. Jung, M. Thosiya, W. and Hiroro, N. *US Pat.*, 5 610 115 (1997).
58. (a) Soga, K. Arai, T. Ban, H. T. and Uozumi, T.*Macromol. Rapid Commun.*, **16**, 905 (1995); (b) Arai, T. Ban, H. T. Uozumi, T. and Soga, K. *Macromol. Chem. Phys.*, **198**, 229 (1997); (c) Soga, K. Uozumi, T. and Arai, T. *US Pat.*, 5 677 255 (1997).
59. Kitagawa, T. Uozumi, T. Soga, K. and Takata, T. *Polymer*, **38**, 615 (1997).
60. Soga, K. Ban, H. T. Arai, T. and Uozumi, T. *Macromol. Chem. Phys.*, **198**, 2779 (1997).
61. Patsidis, K. Peifer, B. Palackal, S. J. Alt, H. G. Welch, M. B. Geerts, R. L. Fahey, D. R. and Deck, H. R. *US Pat.*, 5 466 766 (1995).
62. Spencer L. and Nickias, P. N. *US Pat.*, 5 688 880 (1997).

63. (a) Burkhardt, T. J. Murata, M. and Brandley, W. B. *US Pat.*, 5 240 894 (1993); (b) Chang, M. *PCT Int. Appl.*, 96/18661 (1996); (c) Razavi A. and Debras, G. *PCT Int. Appl.*, 96/35729 (1996).
64. (a) Nowlin, T. E. Lo, F. Y. Shinomoto, R. S. and Shirodkar, P. P. *US Pat.*, 5 332 706 (1994); (b) Lo F. Y. and Pruden, A. L. *PCT Int. Appl.*, 95/11263 (1995); (c) Kallio, K. Andell, O. Knuuttila, H. and Palmqvist, U. *PCT Int. Appl.*, 95/12622 (1995); (d) Lo, F. Y. Nowlin, T. E. Shinomoto, R. S. and Shirodkar, P. P. *PCT Int. Appl.*, 96/14155 (1996).
65. (a) Speca, A. N. Brinen, J. L. Vaughn, G. A. Brant, P. and Burkhardt, T. J. *PCT Int. Appl.*, 96/00243 (1996); (b) Vaughn, G. A. Speca, A. N. Brant, P. and Canich, J. M. *PCT Int. Appl.*, 96/00245 (1996).
66. Brinen, J. L. Speca, A. N. Tormaschy, K. and Russell, K. A. *US Pat.*, 5 665 665 (1997).
67. Ward, D. G. *PCT Int. Appl.*, 96/13531 (1996).
68. Ward D. G. and Brems, P. *PCT Int. Appl.*, 96/13532 (1996).
69. Spitz, R. Pasquet, V. Dupuy, J. and Malinge, J. *Eur. Pat. Appl.*, 708 116 (1996).
70. Kataoka, N. Numao, Y. Seki, T. Tajima, Y. and Matsuura, K. *US Pat.*, 5 331 071 (1994).
71. (a) Tsutsui, T. Yoshitsugu, K. Toyota, A. and Kashiwa, N. *US Pat.*, 5 126 301 (1992); (b) Kioka, M. Toyota, A. Kashiwa, N. and Tsutsui, T. *US Pat.*, 5 654 248 (1997); (c) Ohno, R. and Tsutsui, T. *Eur. Pat. Appl.*, 582 480 (1994); (d) Burkhardt, T. J. Brinen, J. L. Hlatky, G. G. Spaleck, W. and Winter, A. *PCT Int. Appl.*, 94/28034 (1994).
72. Brinen, J. L. *PCT Int. Appl.*, 96/28479 (1996).
73. Takeuchi, M. Obara, Y. Katoh, M. and Ohwaki, M. *US Pat.*, 5 270 407 (1993).
74. (a) Agapiou, A. K. Kuo, C.-I. Muhle, M. E. and Speca, A. N. *PCT Int. Appl.*, 96/11960 (1996); (b) Speca A. N. and Brinen, J. L. *PCT Int. Appl.*, 96/11961 (1996).
75. Brant, P. *PCT Int. Appl.*, 96/34020 (1996).
76. Kioka M. and Kashiwa, N. *US Pat.*, 4 874 734 (1989).
77. Sugano, T. Fujita, T. and Kuwaba, K. *US Pat.*, 5 346 925 (1994).
78. Kioka M. and Kashiwa, N. *US Pat.*, 4 921 825 (1990).
79. (a) Ueda T. and Okawa, K. *US Pat.*, 5 252 529 (1993); (b) Tsutsui, T. Yoshitsugu, K. and Yamamoto, K. *US Pat.*, 5 308 816 (1994); (c) Cheruvu, S. Lo, F. Y. and Ong, S. C. *US Pat.*, 5 420 220 (1995); (d) Lo F. Y. and Pruden, A. L. *PCT Int. Appl.*, 94/21691 (1994); (e) Matsushita, F. Nozaki, T. Kaji, S. and Yamaguchi, F. *Eur. Pat. Appl.*, 733 652 (1996).
80. (a) Furtek A. B. and Shinomoto, R. S. *US Pat.*, 5 362 824 (1994); (b) A. B. Furtek and Shinomoto, R. S. *US Pat.*, 5 461 017 (1995).
81. Albizzati, E. Resconi, L. Dall'Occo, T. and Piemontesi, F. *Eur. Pat. Appl.*, 633 272 (1995).
82. Kumamoto, S. Shirashi, H. and Imai, A. *Eur. Pat. Appl.*, 767 184 (1997).
83. Turner, H. W. Hlatky, G. G. and Eckman, R. R. *US Pat.*, 5 198 401 (1993).
84. (a) Yang, X. Stern, C. L. and Marks, T. J. *J. Am. Chem. Soc.*, **115**, 10015 (1994); (b) Ewen J. A. and Elder, M. J. *US Pat.*, 5 561 092 (1996).
85. (a) Hlatky G. G. and Upton, D. J. *Macromolecules*, **29**, 8019 (1996); (b) Hlatky, G. G. Upton, D. J. and Turner, H. W. *PCT Int. Appl.*, 91/09882 (1991).
86. Krause, M. J. Lo, F. Y. and Chranowski, S. M. *US Pat.*, 5 498 582 (1996).
87. (a) Upton, D. J. Canich, J. M. Hlatky, G. G. and Turner, H. W. *PCT Int. Appl.*, 94/03506 (1994); (b) Kolthammer, B.Tracy, J. C. Cardwell, R. S. and Rosen, R. K. *PCT Int. Appl.*, 94/07928 (1994).
88. Matsumoto, J. *US Pat.*, 5 444 134 (1995).
89. Ward, D. G. *PCT Int. Appl.*, 96/40796 (1996).
90. Inatoni, K. Inahara, K. Yano, A. and Sato, M. *Eur. Pat. Appl.*, 628 574 (1994).
91. Jejelowo M. O. and Hlatky, G. G. *PCT Int. Appl.*, 95/14044 (1995).

92. (a) Lynch, J. Fischer, D. Langhauser, F. Görtz, H. H. Kerth, J. and Schweier, G. *Eur. Pat. Appl.*, 700 934 (1996); (b) Fischer, D. Langhauser, F. Kerth, J. Schweier, G. Lynch, J. and Görtz, H. H. *Eur. Pat. Appl.*, 700 935 (1996).
93. Zandona, N. *US Pat.*, 5 612 271 (1997).
94. Takahashi F. and Yano, A. *Eur. Pat. Appl.*, 619 326 (1994).
95. Soga K. and Lee, D. H. *Makromol. Chem.*, **193**, 1687 (1992).
96. Walzer, J. F., *US Pat.*, 5 643 847 (1997).
97. Ward, D. G. and Carnahan, E. M., *PCT Int. Appl.*, 96/23005 (1996).
98. Kaneko, T. and Sato, M., *Eur. Pat. Appl.*, 727 433 (1996).
99. Ono, M., Hinokuma, S., Miyake, S. and Inzawa, S., *Eur. Pat. Appl.*, 710 663 (1996).
100. (a) Furtek, A. B. and Krause, M. J., *US Pat.*, 5 455 214 (1995); (b) Tahiro, T. and Ueki, T., *Eur. Pat. Appl.*, 598 609 (1994).
101. Yano, A. and Sato, M., *Eur. Pat. Appl.*, 658 576 (1995).
102. Karol, F. J., Karapinka, G. L., Wu, C.-S., Dow, A. W., Johnson, R. N. and Carrick, W. L., *J. Polym. Sci., Part A-1*, **10**, 2621 (1972).
103. He, M.-Y., Xiong, G., Toscano, P. J., Burwell, R. L., Jr and Marks, T. J., *J. Am. Chem. Soc.*, **107**, 641 (1985).
104. Toscano, P. J. and Marks, T. J., *Langmuir*, **9**, 820 (1986).
105. Jejelowo, M. O., *US Pat.*, 5 639 835 (1997).
106. (a) Chien, J. C. W. and He, D., *J. Polym. Sci., Polym. Chem.*, **29**, 1603 (1991); (b) Collins, S., Kelly, W. M. and Holden, D. A., *Macromolecules*, **25**, 1780 (1992); (c) Kaminsky, W. and Renner, F., *Makromol. Chem., Rapid Commun.*, **14**, 239 (1993); (d) Soga, K., Kim, H. J. and Shiono, T., *Macromol. Chem. Phys.*, **195**, 3347 (1994); (e) Sacchi, M. C., Zucchi, D., Tritto, I., Locatelli, P. and Dall'Occo, T., *Macromol. Rapid Commun.*, **16**, 581 (1995).
107. Kaminsky, W., Renner, F. and Winkelbach, H., in *Proceedings of MetCon '94, May 25–27, 1994, Houston, TX* (available from Catalyst Consultants, P.O. Box 637, Spring House, PA 19477, USA).
108. Kaminaka, M. and Soga, K., *Polymer*, **33**, 1105 (1992).
109. Carney, M. J. and Beach, D. L., *US Pat.*, 5 418 200 (1995).
110. (a) DeChellis, M. L., Griffin, J. R. and Muhle, M. E., *US Pat.*, 5 405 922 (1995); (b) Griffin, J. R., DeChellis, M. L. and Muhle, M. E., *US Pat.*, 5 462 999 (1995).
111. (a) Brant, P., Griffin, J. R., Muhle, M. E., Litteer, D. L., Agapiou, A. K. and Renola, G. T., *US Pat.*, 5 712 352 (1998); (b) Poirot, E. E., Sagar, V. R. and Jackson, S. K., *US Pat.* 5 712 353 (1998).
112. (a) Agapiou, A. K., Muhle, M. E. and Renola, G. T., *US Pat.*, 5 442 019 (1995); (b) Muhle, M. E., Agapiou, A. K. and Renola, G. T., *PCT Int. Appl.*, 96/39450 (1996).

10

Homogeneous Phosphametallocene-based Catalysts for Olefin Polymerization

MAKOTO SONE
Tosoh Corporation, Mie-ken, Japan

1 INTRODUCTION

Metallocene catalysts are well known as traditional catalysts and were investigated during the late 1950s as model systems of an ethylene polymerization reaction. At the end of the 1970s, Sinn, Kaminsky and co-workers reported polymethylalumoxane (MAO) as a new class of aluminum alkyl cocatalysts and they proposed that zirconocene–MAO catalyst systems are endowed with high activities during ethylene polymerization [1]. Today, metallocene catalysts are of great importance in many fields of science and industry [2, 3]. One of the most significant consequences of the discovers of metallocene catalyst was in the field of organometallic chemistry. Among the many studies, Mathey and co-workers reported transition metal compounds that have pholpholyl ligands which they called phosphametallocene compounds [4]. The pholpholyl ligand is isoelectronic with cyclopentadienyl ligands, so that phosphametallocene compounds can be expected to form easily a reactive site for olefin polymerizations.

We have been engaged in investigations of soluble transition metal catalysts for olefin polymerization, and in 1992 we reported that phosphametallocene compounds have good capability as olefin polymerization catalyst precursors [5]. Following our report, many related patents and studies have been published [6–9].

Metallocene-based Polyolefins   Edited by J. Scheirs and W. Kaminsky
© 2000 John Wiley & Sons Ltd

We report here the synthesis of Group 4 transition metal phosphametallocenes and the results of ethylene, propylene and styrene polymerization using several types of catalyst.

2 SYNTHESIS

2.1 SYNTHESIS OF PHOSPHOLE AND PHOSPHOLYL LIGAND

Phospholes are five-membered ring compounds similar to typical aromatic hetero-cycles such as pyrroles, furans and thiophenes. The first phosphole, 1-phenyl-2,3,4,5-tetraphenylphosphole (**1**), was discovered as late as 1959 [10, 11]. Next, the first unsubstituted phosphole, 1-methylphosphole (**2**), was prepared in 1967 [12] and the parent molecule, 1-H-phosphole (**3**), was characterized for the first time at low temperature in 1983 [13].

However, the chemistry of phospholes is still relatively unknown, in contrast to other five-membered heterocycles which are currently well investigated. The structures, properties and synthesis routes of phospholes have been reported in review form [14], and can be summarized [15] as follows.

(i) The dehydrohalogenation of 1-halophospholenium halides by tertiary amines [15a] [equation (1)] the starting halophospholenium halides can be produced through three different routes, which are also summarized in equation (1).

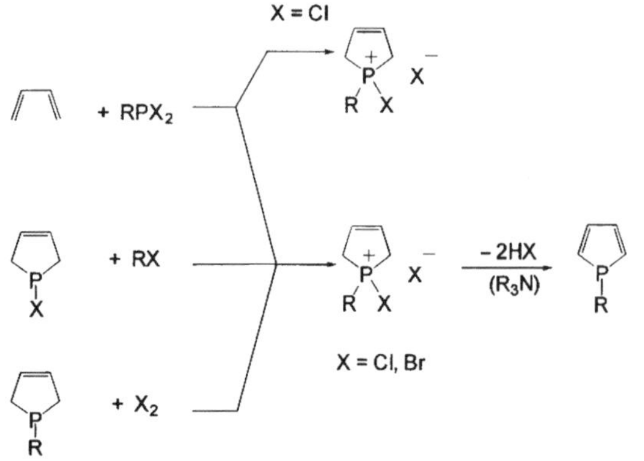

$$X = Cl$$

(1)

(ii) Reversible hydrogen [1, 5] sigmatropic shift [equation (2)].

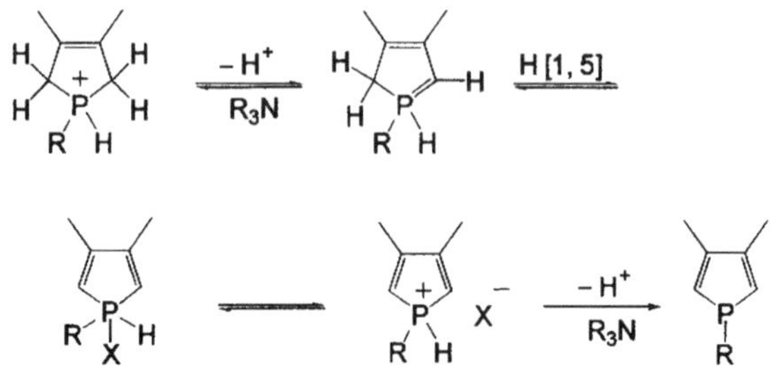

(2)

(iii) Straightforward reaction of $RPCl_2$ with butadienes [15b, c] [equation (3)].

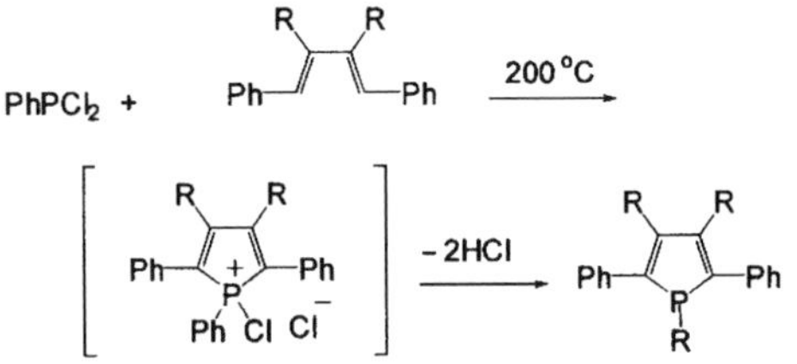

$$PhPCl_2 + \quad [\text{butadiene}] \quad \xrightarrow{200\,^\circ C}$$

$$\left[\text{phospholium chloride} \right] \xrightarrow{-2HCl} \text{phosphole} \tag{3}$$

(iv) Straightforward reaction of RPH_2 with 1,3-diynes [15d, e] [equation (4)].

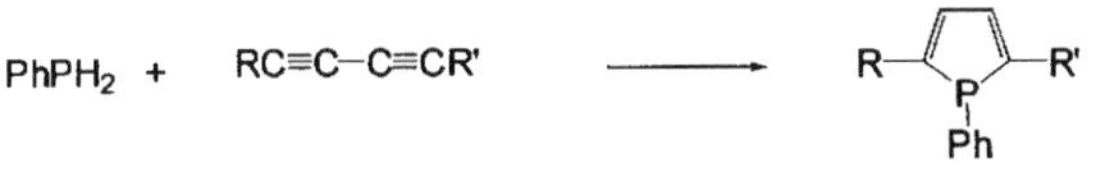

$$PhPH_2 + \quad RC{\equiv}C{-}C{\equiv}CR' \quad \longrightarrow \quad \text{R–phosphole–R'} \tag{4}$$

(v) Condensation of a 1,4-dilithiobutadiene with a dihalophosphine [15f, g] or, conversely, of a 1,4-dihalobutadiene with a dilithio- or disodiophosphine [15h] [equations (5) and (6)].

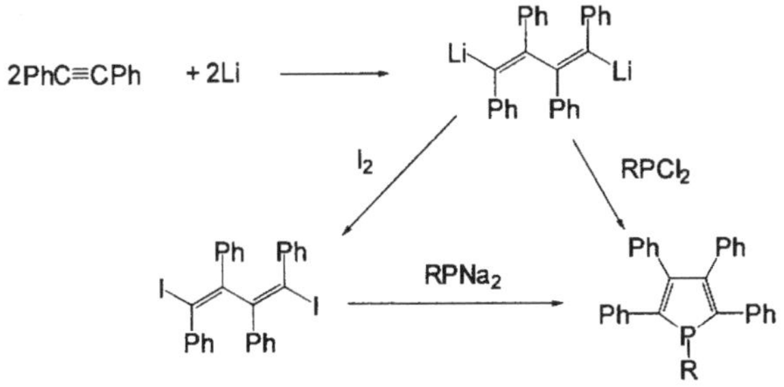

$$2PhC{\equiv}CPh + 2Li \longrightarrow \text{dilithiobutadiene}$$

$$\xrightarrow{I_2} \quad \text{diiodobutadiene} \quad \xrightarrow{RPNa_2} \quad \text{phosphole} \tag{5}$$

$$\xrightarrow{RPCl_2} \quad \text{phosphole}$$

$$\text{1,4-dichlorobutadiene} \xrightarrow{PhPLi_2} \text{phosphole} \tag{6}$$

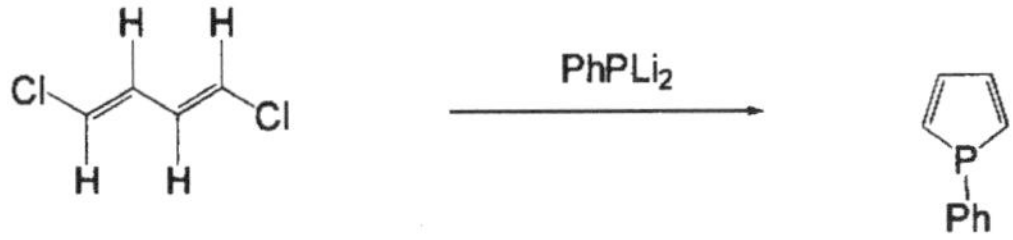

(vi) Reaction of a phenylphosphonous dichloride with a cyclobutadiene aluminate complex [4b] [equation (7)].

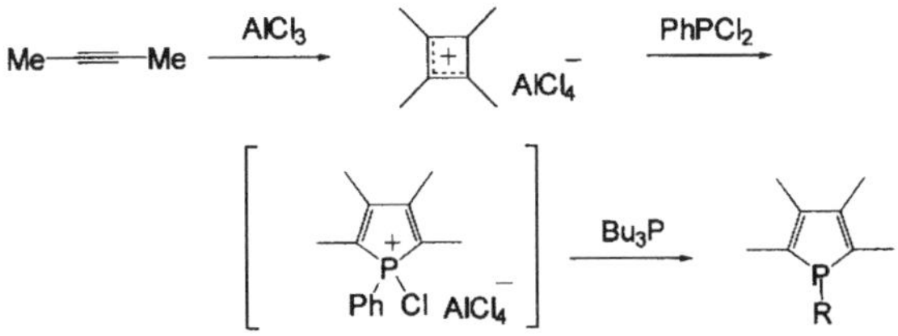

$$Me{-}{\equiv}{-}Me \xrightarrow{AlCl_3} \text{[cyclobutadiene]} AlCl_4^- \xrightarrow{PhPCl_2}$$

$$\left[\text{phospholium} \; Ph \; Cl \; AlCl_4^- \right] \xrightarrow{Bu_3P} \text{phosphole} \tag{7}$$

To synthesize the phosphametallocenes, we used 1-phenyl derivatives for the preparation of phospholide anions. With regard to the 1-phenyl derivatives, not only are the yields often higher than those with 1-alkyl derivatives, but also they are generally stable and resistant to oxidation.

The phospholide anions were discovered by Braye *et al.* in 1971 [16]. The X-ray crystal structure analysis of the phospholide anion Li(TMEDA)PC$_4$Me$_4$ (**4**) was reported in 1989 [17]. The carbon–carbon bond distances within the heterocycle are very similar, from 1.396 to 1.424 Å, indicating a high degree of delocalization, and the phosphorus–carbon bonds (1.75 Å) are significantly shorter than P—C single bonds (*ca* 1.83 Å), consistent with partial double bond character.

They show high aromaticity and are more able to form π-complexes with transition metals than other heterocycles, such as pyrrolyl ligands [18]. For this reason, phospholide anions are successfully used as ligands of transition metals and many complexes, e.g. phosphaferrocenes [19] and phosphacymantrenes (phospha-cyclopentadienylmanganese tricarbonyl) [20]. The main synthesis routes of the phospholide anions are as follows [14, 21, 22].

(i) Cleavage of the phosphorus-phenyl bond of 1-phenylphospholes by alkali metals [equation (8)]: this synthesis route was the first to be reported [16] but is still the only practical route to phospholide anions. The mechanism involves a monoelectronic reduction which yields a phosphole radical anion, and this radical decomposes and finally gives the phospholide anion and phenyl anion [23]. In that case, the phenyl anion is formed as a side product and we should neutralize that anion by using AlCl$_3$ or *tert*-butyl chloride.

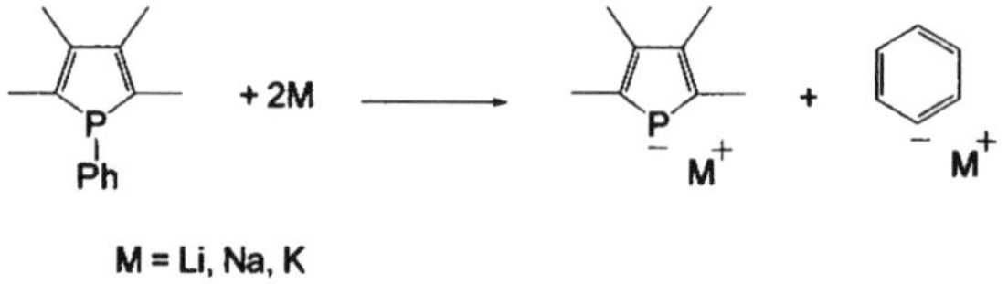

$$\text{(8)}$$

M = Li, Na, K

(ii) Cleavage by alkali metals of the P—P bond or P—CH$_2$—CH$_2$—P bonds [equations (9) and (10)]. No side product is formed in this instance.

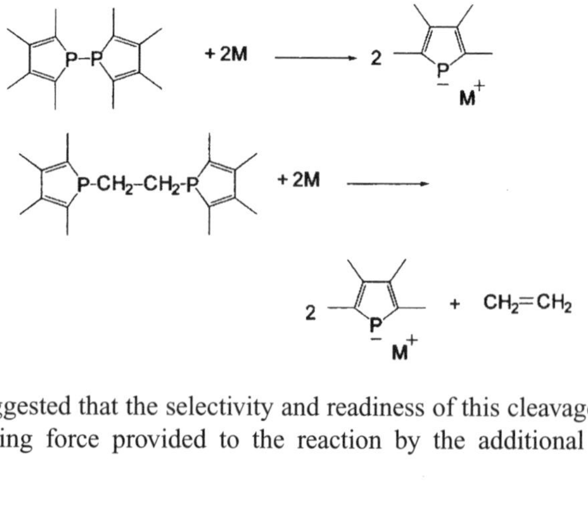

$$\text{(9)}$$

$$\text{(10)}$$

Mathey [11] suggested that the selectivity and readiness of this cleavage were due to the strong driving force provided to the reaction by the additional electronic delocalization occurring within the phosphole ring when the weakly aromatic phosphole was converted into the highly aromatic pholide anion.

In addition to phospholyl anions, 1-stannylphospholes and 1-silylphospholes could be used as phospholyl ligand precursors, in the same way as silyl- and stannylcyclopentadienes have been used as cyclopentadienylation reagents for titanium and zirconium [24, 25]. The stable 1-trimethylstannylphospholes and 1-trimethylsilylphospholes have been synthesized with pholide anions and Me_3SnCl or Me_3SiCl in good yield [5, 26].

2.2 SYNTHESIS OF PHOSPHAMETALLOCENES OF GROUP 4 TRANSITION METALS

The first phosphametallocene of the Group 4 transition metals, bis(3,4-dimethylpho-spholyl)zirconium dichloride (**5**), was reported by Meunier and Gautheron in 1980 [21]; however, it was only partially characterized by ^1H NMR and mass spectro-metry. Later, in 1988, Mathey and co-workers described the synthesis of bis(2,3,4,5-tetramethylphospholyl)zirconium dichloride (**6**) together with the first X-ray crystal structure [equation (11)] [4b].

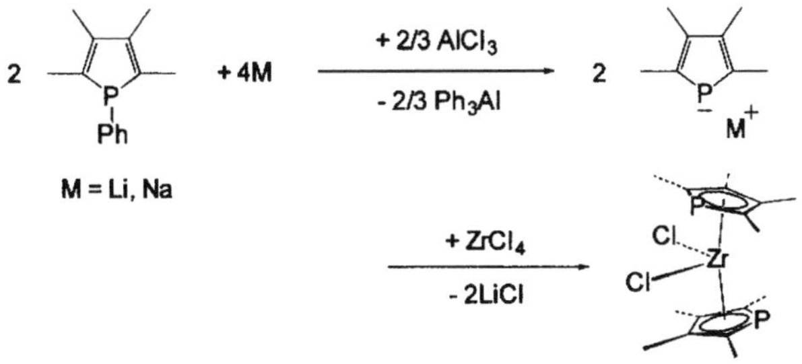

$$\text{(11)}$$

$$M = Li, Na$$

The structures are closely related to that of the corresponding zirconocene complex. In the case of complex **6**, the two phospholyl rings are staggered, with a P—P distance of 3.23 Å, indicating no significant interaction between the two phosphorus atoms. The angle between the ring planes is 48.4°, the centroid phospholyl–Zr–centroid phospholyl angle is 131.6° and the Zr—P bond lengths are 2.72 and 2.73 Å. It was reported that these complexes could be prepared by reaction of the phospholide anion with $ZrCl_4$, but the reaction failed in the case of titanium because monoelectronic reduction of $TiCl_4$ by the phospholide anion and subsequent coupling of the phospholyl radical occurred, thus forming 1,1'-bispho-spholyls [4c]. The same group also succeeded in producing the titanium complex (3,4-dimethylphospholyl)titanium trichloride (**7**) using 1-trimethylstannylphosphole, and similarly we prepared (2,3,4,5-tetramethylphospholyl)titanium trichloride (**8**) using 1-trimethylsilyl-2,3,4,5-tetramethylphosphole [equation (12)] [5]. Since the, many kinds of phosphametallocenes have been synthesized and reported [5–8].

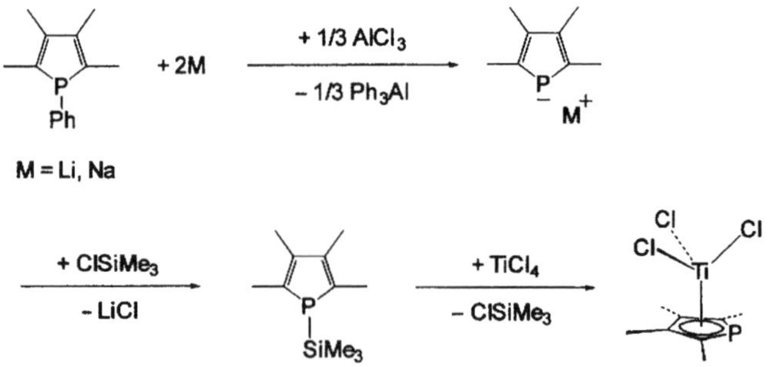

$$(12)$$

3 POLYMERIZATION REACTIVITY

3.1 ETHYLENE POLYMERIZATION

Table 1 shows the activities and analytical data at different Al:Zr ratios. By varying the MAO concentration with a constant concentration of the diphosphametallocene $(Me_4C_4P)_2ZrCl_2$ (6), it was found that the ethylene polymerization rate steadily increased with increasing MAO concentration (Figure 1). In this context, it is interesting to see how the Al:Zr ratio or the MAO concentration acts during the polymerization reactions with the phosphametallocene precursors. We found that the molecular weight distribution of the polyethylenes obtained using the 1/MAO catalyst system becomes broad, and that the molecular weight decreases with increasing MAO concentration (Figure 2).

In Table 2, we list the analytical data for the polymers produced with different Al:Zr ratios. In this case, we examined the copolymerization of ethylene and 1-hexene and the homopolymerization of ethylene. With increasing Al:Zr ratio, the molecular weight of the ethylene–1-hexene copolymer decreases, as in the case of polyethylene. However, the molecular weight distribution of the ethylene–1-hexene

Table 1 Homopolymerization of ethylene: activity and analytical data for polyethylenes obtained with $(Me_4C_4P)_2ZrCl_2$ (6)/MAO at different MAO concentrations[a]

Run	[MAO] (mmol)	Al : Zr (mol/mol^{-1})	Yield (g)	Activity [kg polymer/(mmol Zr h)]	M_w (g mol^{-1})	M_w/M_n
1	0.25	100	27	10.8	—[b]	—
2	1.25	500	73	29.2	490 000	3.3
3	2.50	1000	62	24.8	290 000	4.9
4	5.00	2000	90	36.0	175 000	4.0
5	25.0	10 000	74	59.2	49 000	4.1

[a] Polymerization conditions: [Zr] = 0.0025 mmol; ethylene pressure = 8 bar; solvent = toluene (500 ml); polymerization temperature = 60 °C; polymerization time = 60 min (runs 1–4) or 30 min (run 5).
[b] $[\eta] = 8.8$ dl/g; $M_v = 662\,000$ g/mol.

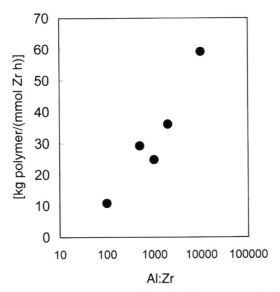

Figure 1 Dependence of ethylene polymerization rate on the Al:Zr ratio

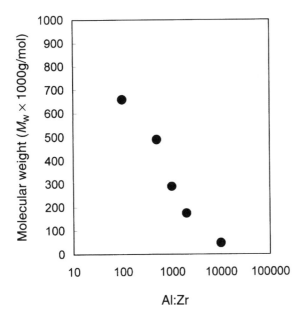

Figure 2 Dependence of ethylene polymerization (molecular weight) on the Al:Zr ratio

Table 2 Analytical data for polyethylenes and ethylene–1-hexene copolymers generated with $(Me_4C_4P)_2ZrCl_2$ (**6**)/MAO at different MAO concentrations[a]

Run	$C^{6=}$ (ml)	Al:Zr (mol/mol)	M.p. (°C)	GPC		FT-IR (per 1000 carbon atoms)			
				M_n	M_w/M_n	CH_3	Vinyl	*trans*-Vinylene	Vinylidene
6	0	1 200	135.2	27 000	3.35	1.16	0.05	0.06	Trace
7	0	12 000	133.9	8 300	3.25	2.31	0.06	0.07	Trace
8	100	1 200	127.1	27 000	2.06	2.96	Trace	Trace	—
9	100	12 000	127.1	8 400	2.21	4.79	Trace	Trace	—

[a] Polymerization conditions: [Zr] = 0.001 mmol; ethylene pressure = 6 bar; solvent = toluene (1000 ml); polymerization temperature = 80 °C; polymerization time = 60 min.

copolymer remains narrow (M_w/M_n = 2.06 and 2.21), as seen in the cases of common metallocenes of the Group 4 transition metals. Further, we hardly observed any vinyl end groups in the polymers with the **6**/MAO catalyst system. With regard to the termination reaction, two major modes of chain transfer prevail in the polymerization reaction: β-H elimination and chain transfer to aluminum. Under typical reaction conditions, the processes compete and result in vinyl end groups and aluminum–polymer groups, respectively. Based on the absence of vinyl end groups, we suggest that chain transfer to aluminum becomes the dominant chain termination reaction.

However, these phenomena have not been understood very well until now. Several causes are possible. First, we suggest the coordination of Lewis acid aluminum atoms and the Lewis base phosphorus atom of the phosphametallocenes. The activated cationic phosphametallocene could result in an interaction between the aluminum components and the lone pairs of the phosphorus atom in the ligand (model 1, 2 or 3), as shown in Figure 3. As a result, the coordinated aluminum group is like a very large functional group in the phosphametallocene component. The more intense interaction between the phosphametallocene and aluminum component with increasing MAO concentration increases the number of active centers; however,

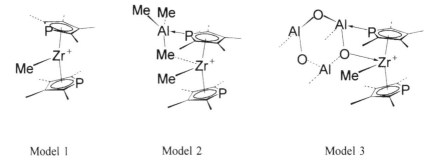

Model 1   Model 2   Model 3

Figure 3 Proposed active sites

the interaction decreases the propagation reactions because of steric hindrance and accelerates the chain termination reactions. In fact, we could not find any evidence of P–Al adduct formation in our NMR study. However, Janiak and co-workers noted the reaction of $(Me_4C_4P)_2ZrCl_2$ (**6**) and MAO/trimethyl aluminum using ^{31}P NMR spectroscopy, and proposed adducts formed from the equilibrium reaction at high Al:Zr ratios [9]. Second, we could postulate that diffusion control gives rise to these phenomena. Under our polymerization conditions, the polymerization temperature was 60 or 80 °C, and the homopolymer of ethylene is insoluble in toluene. As a consequence, polyethylene is produced, and the resulting polymer disturbs the monomer insertion reactions and produces many kinds of heterogeneous active sites in the polymer matrix. On that basis, we might understand the narrow polydispersity of ethylene–1-hexene copolymerization (Table 2). Lastly, we suggest a decomposition reaction or dimerization reaction of the phosphametallocene complex using its phosphorus lone pairs. As a result, these reactions take place between the polymerization reactions, and many types of catalyst species are formed in the polymerization reactions.

In Table 3, details of ethylene polymerization by $(Me_4C_4P)_2ZrCl_2$ (**6**) with a borate catalyst system are presented. In these experiments, we used triisobutyl aluminum as an initiator and scavenger. $[Ph_3C][B(C_6F_5)_4]$ and $[Me_2PhNH]$ $[B(C_6F_5)_4]$ acted as cocatalysts to produce polyethylene with 13 and 9 kg (mmol $Zr)^{-1} h^{-1}$, respectively, comparable to MAO. The molecular weights of the polyethylene with borate systems were too high to measure using gel permeation chromatography (GPC), so we give the viscosity-averaged molecular weight (M_v) in Table 3. With the same Al:Zr ratio (100), these molecular weights were slightly lower than that with MAO.

In the case of the metal-bridged diphosphametallocene $[(Me_4C_4P)_2ZrCl_2]$ $[Fe(CO)_3]$ (**10**), the polymerization activity with MAO is lower than that of the non-bridged diphosphametallocene **6**, as shown in Table 4. At 60 °C, the activity increases with increasing MAO concentration. In addition, the molecular weight of the polymer obtained decreases. These tendencies for **10** / MAO are in line with

Table 3 Homopolymerization of ethylene: activity and analytical data for polyethylenes obtained with $(Me_4C_4P)_2ZrCl_2$ (**6**)$/B(C_6F_5)_4{}^{-a}$

Run	Co-catalyst	Amount (mmol)	C:Zr (mol/mol)	$^iBu_3Al$ (mmol)	Al:Zr (mol/mol)	Yield (g)	Activity [kg polymer/ (mmol Zr h)]	M_v
10	TPC-B	0.0125	5	0.25	100	33	13	569 000
11	DMA-B	0.0125	5	0.25	100	23	9	524 000
1	MAO	0.25	100	0	100	27	11	662 000

^{a} Cocatalyst TPC-B = $Ph_3C[B(C_6F_5)_4]$; cocatalyst DMA-B = $[Me_2PhNH][B(C_6F_5)_4]$. Polymerization conditions: [Zr] = 0.0025 mmol; ethylene pressure = 8 bar; solvent = toluene (500 ml); polymerization temperature = 60 °C; polymerization time = 60 min.

Table 4 Homopolymerization of ethylene: activity and analytical data for polyethylenes obtained with (Me$_4$C$_4$P)$_2$ZrCl$_2$ (**6**)/MAO and (CO)$_3$Fe(Me$_4$C$_4$P)$_2$ZrCl$_2$ (**10**)/MAO[a]

Run	Catalyst	Al:Zr (mol/mol)	Temperature (°C)	Yield (g)	Activity [kg polymer/ (mmol Zr h)]	M_w	M_w/M_n
12	**10**	500	60	14	6	220 000	2.5
13	**10**	10 000	60	69	28	60 000	4.2
2	**6**	500	60	73	29	490 000	3.3
5	**6**	10 000	60	74	59	49 000	4.1
14	**10**	500	90	8	3	230 000	11.8
15	**6**	500	90	15	15	240 000	5.6

[a] Polymerization conditions: [Zr] = 0.025 mmol; ethylene pressure = 8 bar; solvent = toluene (500 ml); polymerization temperature = 60 °C; polymerization time = 60 min.

Table 5 Homopolymerization of ethylene: activity and analytical data for polyethylenes obtained with Cp(Me$_4$C$_4$P)$_2$ZrCl$_2$ (**11**)[a]

Run	Cocatalyst	Amount (mmol)	B:Zr (mol/mol)	^{i}Bu$_3$Al (mmol)	Al:Zr (mol/mol)	Yield (g)	Activity [kg polymer/ (mmol Zr h)]
16	MAO	0.5	—	0	1000	21	42
17	MAO	5.0	—	0	10 000	26	52
18	DMA-B	0.0025	5	0.25	500	22	44

[a] Cocatalyst DMA-B = [Me$_2$PhNH][B(C$_6$F$_5$)$_4$]. Polymerization conditions: [Zr] = 0.0005 mmol; ethylene pressure = 8 bar; solvent = toluene (500 ml); polymerization temperature = 60 °C; polymerization time = 60 min.

those for the non-bridged **6**/MAO as already mentioned. At a higher temperature (90 °C), the polydispersity for **10** becomes broader than that at lower temperature (60 °C), compared with that for **6**. In this case, the weight-average molecular weight (M_w) of the polymer with **10** was retained, i.e. the broader dispersion was due to the bimodal molar mass distribution. Obviously, the higher and lower molecular weight regions in the case of **10** are increased in comparison with those using the non-bridged catalyst system **6**. Therefore, at least two types of reactive centers must be present in this metal-bridged diphosphametallocene catalyst system.

In Table 5, ethylene polymerizations with the Cp(Me$_4$C$_4$P)ZrCl$_2$ (**11**)/MAO system are reported. This catalyst shows good activity for ethylene polymerization, similar to that for the **6**/MAO system.

3.2 PROPYLENE POLYMERIZATIONS

The results of propylene polymerization with phosphametallocenes **6** and **11** appear in Table 6. Cp$_2$ZrCl$_2$ was included as a point of reference. The results indicate that

Table 6 Homopolymerization of propylene: activity and analytical data for polypropylenes obtained with $(Me_4C_4P)_2ZrCl_2$ **(6)** and $Cp(Me_4C_4P)_2ZrCl_2$ **(11)**[a]

Run	Catalyst	[MAO] (mmol)	Al:Zr (mol/mol)	Yield (g)	Activity kg polymer/ (mmol Zr h)]	M_n (g/mol)
19	1	40	4000	3.1	10.8	830
20	3	40	4000	1.6	29.2	470
21	Cp_2ZrCl_2	40	4000	5.1	24.8	n.d.

[a] Polymerization conditions: [Zr] = 0.01 mmol; propylene = 500 ml; solvent = toluene (500 ml); polymerization temperature = 60 °C; polymerization time = 60 min.

these catalysts show low activity and produce low molecular weight oligomers. De Boer also examined propylene polymerization with a series of phosphametallocenes and obtained atactic propylene polymers [8] (Table 7). On increasing the substituted groups, a decrease in the activity was observed: the bis(2,5-substituted-phosphole) derivative $(2,5\text{-}Ph_2C_4H_2P)_2ZrCl_2$ **(13)** showed the highest activity and the bis(per-substituted-phosphole) derivative $(2,3,4,5\text{-}Ph_4C_4P)_2ZrCl_2$ **(16)** did not provide any polymer. Furthermore, the 2,5-substituted-derivative **13**/MAO catalyst system could be used to prepare relatively high molecular weight atactic polypropylene ($M_n > 50\,000$), whereas other phosphametallocenes only produced atactic propylene olygomers (Table 7, experiments 16, 27, 28, and 29). We suggest that the sterically hindered phospholyl ligand prevented the monomer insertion reaction and the growth of the polymer, and De Boer of Shell gave the interpretation that the polymerization catalyst systems prefer one or both of the 2- and 5-positions on the phospholyl ligand being substituted by a bulky substituent and the 3-and 4-positions on the ligand are not substituted by a bulky substituent [8].

Table 7 Homopolymerization of propylene: activity and analytical data for polypropylenes obtained with $(Me_4C_4P)_2ZrMe_2$ **(12)**, $(2,5\text{-}Ph_2C_4H_2P)_2ZrCl_2$ **(13)**, $Cp(2,5\text{-}Ph_2C_4H_2P)ZrCl_2$ **(14)** $(2,5\text{-}Ph_2\text{-}3\text{-}MeC_4HP)_2ZrCl_2$ **(15)** $(Ph_4C_4P)_2ZrCl_2$ **(16)** and $(2,5\text{-}Ph_2\text{-}3,4\text{-}Me_2C_4P)_2ZrCl_2$ **(17)**[a]

Exp.	Catalyst	Polymerization time (min)	Yield	Turnover [mol/(mol h)]	M_n (g/mol)
3	5	52	8 ml	1300	690
4	5	40	5 ml	1200	460
16	6	14	38 g	360 000	>50 000
17	7	30	31 g	140 000	1700
27	8	30	7 g	33 000	450
28	9	30	Trace	—	—
29	10	10	12.5 g	18 000	20 000

[a] Polymerization conditions: see Ref. 8; propylene = 600 kPa, solvent = toluene; polymerization temperature = 45 °C.

Table 8 Homopolymerization of styrene: activity and analytical data for polystyrenes obtained with $(Me_4C_4P)TiCl_3$ $(8)^a$

Run	Catalyst	Yield (mg)	Activity (g polymer/mmol Ti)	T_m (°C)	T_g (°C)
22	**4**	52	5.2	272	93
23	CpTiCl₃	670	67.0	261	92

^{a} Polymerization conditions: [Ti] = 0.01 mmol; [MAO] = 1.0 mol; Al:Ti = 100 mol/mol; styrene = 10 ml; solvent = toluene (10 ml); polymerization temperature = 40 °C; polymerization time = 180 min.

3.3 STYRENE POLYMERIZATIONS

We carried out a brief investigation of styrene polymerization with $(Me_4C_4P)TiCl_3$ **(8)** activated with MAO, i.e. it was active during the polymerization of styrene in the same way as halftitanocene trichloride (Table 8). Although the yield of syndiotactic polystyrene was lower than with the non-substituted halftitanocene trichloride (CpTiCl₃, its syndiotacticity was higher than that of the reference on the basis of the melting-point (261 vs 272 °C).

In addition, De Boer also examined styrene polymerization with (phospholyl)TiCl₃/MAO systems at 60 °C [8] (Table 9). The products were characterized as syndiotactic polystyrenes on the basis of their NMR spectra and melting-points. In this study, (2,5-substituted-phosphole)TiCl₃ was more active than the 3,4-substituted or persubstituted derivatives.

Table 9 Homopolymerization of styrene: activity and analytical data for polystyrenes obtained with $(Me_4C_4P)TiCl_3$ **(8)**, $(3,4-Me_2C_4H_2P)TiCl_3$ **(18)**, $(2,5-Ph_2C_4H_2P)TiCl_3$ **(19)**, $(Ph_4C_4P)TiCl_3$ **(20)**, $(2,5-Ph_2-3,4-Me_2C_4P)TiCl_3$ **(21)**, and $(2,5-^tBu_2C_4H_2P)TiCl_3$ **(22)**^{a}

Exp.	Catalyst	Reaction time (min)	Yield (g)	TOF [mol syndiotactic polystyrene (mol catalyst h)]	M.p. (DSC) (°C)
51	**18**	30	0.01	20	
		90	0.05	30	260, 266
52	**8**	30	0.01	20	
		90	0.06	40	259, 264
53	**19**	20	0.09	260	
		30	0.12	230	
		60	0.16	310^b	258, 266
54	**20**	60	0.04	40^c	
55	**21**	60	0.03	30	256, 265
56	**22**	120	1.50	720^d	253, 264

^{a}Polymerization conditions: see Ref. 8; [Ti] = 0.01 mmol; [MAO] = 5.0 mmol; styrene = 5 ml; solvent = toluene (5 ml); polymerization temperature = 60 °C.
^{b}[Ti] = 0.005 mmol; [MAO] = 20 mmol; solvent = toluene (10 ml).
^{c}[Ti] = 0.0065 mmol; [MAO] = 3.25 mmol.
^{d}[Ti] = 0.02 mmol; [MAO] = 10.0 mmol; styrene = 5 ml; solvent = toluene(10 ml); polymerization temperature = 25 °C.

4 CONCLUSIONS

The phosphametallocenes of the Group 4 transition metals are easily formed and are structurally similar to the metallocenes of the Group 4 transition metals, so that the phosphametallocene/MAO or $B(C_6F_5)_4{}^-$ catalyst systems show good activity during ethylene polymerization. In addition, we found that they have activities in propylene and styrene polymerizations. It has also been found that the molecular weight of the polyethylene given by **6** with MAO is dependent on the concentration of MAO.

5 REFERENCES

1. (a) Sinn, H. and Kaminsky, W., *Adv. Organomet. Chem.*, **18**, 99 (1980); (b) Kaminsky, W., Miri, M., Sinn, H. and Woldt, R., *Makromol. Chem., Rapid Commun.*, **4**, 417 (1983); (c) J. Herwig and W. Kaminsky, *Polym. Bull.*, **9**, 464 (1983); (d) W. Kaminsky and H. Lüker, *Makromol. Chem., Rapid Commun.*, **5**, 225 (1984); (e) W. Kaminsky, K. Külper, H. H. Brintzinger and F. R. W. P. Wild, *Angew. Chem.*, **97**, 507 (1985).
2. H. H. Brintzinger, D. Fischer, R. Mülhaupt, B. Rieger and R. M. Waymouth, *Angew. Chem., Int. Ed. Engl.*, **34**, 1143 (1995).
3. J. C. Stevens, in *MetCon'93 Worldwide Metallocene Conference, May 27–28, 1993*, p. 157.
4. (a) F. Nief, F. Mathey and L. Ricard, *J. Organomet. Chem.*, **384**, 271 (1990); (b) F. Nief, F. Mathey and L. Ricard, *Organometallics*, **7**, 921 (1988); (c) F. Nief and F. Mathey, *J. Chem. Soc., Chem. Commun.*, 770 (1988).
5. Tosoh *US Pat.* 5 434 116 (1900)
6. Mitsubishi Chemical *Jpn. Pat.*, JP7- 188335, (1900)
7. Idemitsu Kosan *Jpn. Pat.*, JP6-340704, JP7-258320 (1900)
8. Shell Internationale Research, WO95-04087 (1900)
9. (a) C. Janiak, U. Versteeg, K. C. H. Lange, R. Weimann and E. Hahn, *J. Organomet. Chem.*, **501**, 219 (1995); (b) C. Janiak, K. C. H. Lange, U. Versteeg, D. Lentz and P.H.M. Budzelaar, *Chem. Ber.*, **129**, 1517 (1996).
10. E. H. Braye and W. Hübel, *Chem. Ind. (London)*, 1250 (1959).
11. F. C. Leavitt, T. A. Manuel and F. Johnson, *J. Am. Chem. Soc.*, **81**, 3163 (1959).
12. L. D. Quin and J. G. Bryson, *J. Am. Chem. Soc.*, **89**, 5984 (1967).
13. C. Charrier, H. Bonnard, G. de Lauzon and F. Mathey, *J. Am. Chem. Soc.*, **105**, 6871 (1983).
14. (a) F. Mathey, *Coord. Chem. Rev.*, **137**, 1 (1994); (b) F. Mathey, *Chem. Rev.*, **88**, 429 (1988); (c) F. Mathey, J. Fischer and J. H. Nelson, *Struct. Bonding (Berlin)*, **55**, 153 (1983).
15. (a) F. Mathey, *C. R. Acad. Sci., Ser., C*, **269**, 1066 (1969); (b) I. G. M. Campbell, R. C. Cookson and M. B. Hocking, *Chem. Ind. (London)*, 359 (1962); (c) I. G. M. Campbell, R. C. Cookson, M. B. Hocking and A. N. Hughes, *J. Chem. Soc.*, 2184 (1965). (d) G. Märkl, and R. Potthast, *Angew. Chem., Int. Ed. Engl.*, **6**, 86 (1967); (e) W. Eagan, R. Tang, G. Zon and K. Mislow, *J. Am. Chem. Soc.*, **93**, 6205 (1971); (f) F. C. Leavitt, T. A. Manuel, F. Johnson, L. U. Matternas and D. S. Lehmann, *J. Am. Chem. Soc.*, **82**, 5099 (1960); (g) E. H. Braye, W. Hübel and I. J. Caplier, *J. Am. Chem. Soc.*, **83**, 4406 (1961); (h) A. J. Ashe, III, S. Mahmoud, C. Elschenbroich and M. Wünsch, *Angew. Chem., Int. Ed. Engl.*, **26**, 229 (1987).
16. E. H. Braye, I Caplier and R. Saussez, *Tetrahedron*, **27**, 5523 (1971).
17. T. Douglas and K. H. Theopold, *Angew. Chem., Int. Ed. Engl.*, **28**, 1367 (1989).
18. A. P. Sadimenko, A. D. Garnovskii and N. Retta, *Coord, Chem. Rev.*, **126**, 237 (1993).

19. F. Mathey and G. deLauzon, *Organomet. Synth.*, **3**, 259 (1986).
20. F. Mathey, A. Mitschler and R. Weiss, *J. Am. Chem. Soc.*, **100**, 5748 (1978).
21. P. Meunier and B. Gautheron, *J. Organomet. Chem.*, **193**, C13 (1980).
22. F. Mathey, F. Mercire, F. Nief, J. Fischer and A. Mitschier, *J. Am. Chem. Soc.*, **104**, 2077 (1982).
23. C. Thomson and D. Kilcast, *Angew. Chem., Int. Ed. Engl.*, **9**, 310 (1970).
24. P. Jutzi and M. Kuhn, *J. Organomet. Chem.*, **173**, 221 (1979).
25. A. M. Cardoso, R. J. H. Clark and S. Moorhouse, *J. Chem. Soc., Dalton Trans.*, 1552 (1979).
26. S. Holand, F. Mathey and J. Fischer, *Polyhedron*, **5**, 1413 (1986).

11

Nickel- and Palladium-based Catalysts for the Insertion Polymerization of Olefins

LYNDA K. JOHNSON
Dupont Central Research and Development, Wilmington, DE, USA

CHRISTOPHER M. KILLIAN
Eastman Chemical Company, Kingsport, TN, USA

1 BACKGROUND

The catalytic conversion of inexpensive olefin feedstocks into high molecular weight polymers is a multi-billion dollar a year business with worldwide production in excess of 80 billion pounds and with polyethylene accounting for more than 50 % of the total polyolefin output [1, 2]. The three major classes of ethylene-based polymers produced and distributed commercially include (i) high-density polyethylene (HDPE), which is a linear semicrystalline ethylene homopolymer ($T_m \approx 135\,^\circ\text{C}$) prepared by Ziegler–Natta-based coordination polymerization technology, (ii) linear low-density polyethylene (LLDPE), which is a random copolymer of ethylene and α-olefins (e.g. 1-butene, 1-hexene or 1-octene) produced commercially using both Ziegler–Natta and metallocene catalysts, and (iii) low-density polyethylene (LDPE), which is a highly branched ethylene homopolymer prepared in a high-temperature and high-pressure free-radical process. Copolymers of ethylene with functionalized olefins such as methyl (meth)acrylate, (meth)acrylic acid and vinyl acetate are also important commercial polymers [3]. Owing to their high oxophilicity, early transition metal catalysts based on titanium, zirconium and chromium are typically poisoned by these functionalized olefins and therefore copolymers of

Metallocene-based Polyolefins   Edited by J. Scheirs and W. Kaminsky
© 2000 John Wiley & Sons Ltd

functionalized olefins with ethylene are still produced commercially by free radical polymerizations [4].

Owing to the reduced oxophilicity and hence greater functional group tolerance of late metals relative to early metals, the development of late metal polymerization catalysts has long been of interest. However, in contrast to the large volume of work describing olefin polymerization catalysts based on early transition metals, there have been relatively few reports of late transition metal catalysts for the insertion polymerization of ethylene and, particularly, α-olefins. Instead, late metal catalysts normally exhibit reduced activities for olefin insertion relative to early metal catalysts, and β-hydride elimination typically competes with chain growth, resulting in the formation of dimers or oligomers. So far, late metal catalysts for the insertion polymerization of olefins based upon iron [5], cobalt [5, 6], rhodium [7, 8], nickel, palladium and platinum [8] have been reported. The following discussion reviews developments in nickel and palladium polymerization catalysis.

2 NICKEL AND PALLADIUM CATALYSTS

2.1 ETHYLENE

A commercially important process that illustrates the reduced olefin insertion activities and competitive β-hydride elimination processes of late metals is the Shell Higher Olefin Process (SHOP) for synthesizing linear α-olefins (C_6-C_{20}) from ethylene [9]. The first step in the process involves the oligomerization of ethylene by a homogeneous nickel catalyst discovered by Keim. The active species is believed to be a neutral nickel olefin hydride complex (**1**), which can be generated *in situ* by combining bis(1,5-cyclooctadiene)nickel with the acid of the phosphine–carboxylate ligand. At 75 °C and 150–1200 psi of ethylene, the nickel catalyst oligomerizes ethylene at a turnover frequency of *ca* 3000 TO h^{-1} [turnover (TO) = moles of olefin consumed, as determined by the weight of the isolated polymer or oligomer, divided by moles of catalyst] to form linear α-olefins (>98 % purity) with a Flory–Schulz chain length distribution [10, 11]. The mechanism of oligomer formation is believed to involve ethylene insertion into the nickel hydride bond of **A** to generate the ethyl complex **B** (Scheme 1). Additional ethylene insertions followed by β-hydride elimination result in the formation of an α-olefin hydride complex **D**. Chain transfer in these systems probably occurs by associative olefin exchange between free ethylene and nickel-ligated α-olefin, regenerating the nickel hydride species **A**. Owing to the selectivity of these systems for ethylene insertion, the formation of branched species that would result from the reinsertion of the α-olefin product is not observed.

In addition to nickel-catalyzed oligomerizations, the production of higher molecular weight oligomers and polymers from ethylene was demonstrated by Keim and co-workers with the modified SHOP catalysts **2** and the 1,3-diphosphaallyl complex **3** [12, 13]. Complexes **2** were synthesized by the oxidative addition of phosphorus

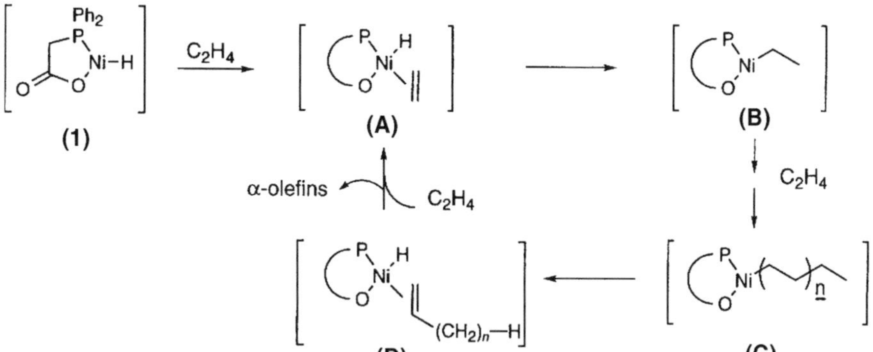

Scheme 1 Proposed mechanism for the preparation of linear α-olefins

ylides to zerovalent nickel compounds in the presence of triphenylphosphine [12]. Linear α-olefins were produced from ethylene by **2** (R = Ph) in toluene solution. However, when the reaction of **2** (R = Ph) with ethylene was carried out as a suspension in hexane, linear high molecular weight polyethylene was obtained [12a]. The molecular weight and polydispersity of the linear α-olefins were increased by increasing the electron density and the steric bulk of the substituent R on phosphorus (Table 1) [12b–d]. A sterically bulky substituent [Ar = 2,4-6-$C_6H_2(^tBu)_3$] was also incorporated in the diphosphallyl compound **3**, which produced linear high molecular weight polyethylene at 70–100 °C and 440 psi [13].

Klabunde and Ittel demonstrated that the coordination of a strongly donating ligand (e.g. L = PR_3 in **4**) to the modified SHOP catalysts was the primary factor in limiting the molecular weight of the products [14]. By using phosphine 'sponges' to scavenge the phosphine ligands and by binding weaker bases such as pyridine to

Table 1 Substituent effects on oligomer product distributions for **2**^{a}

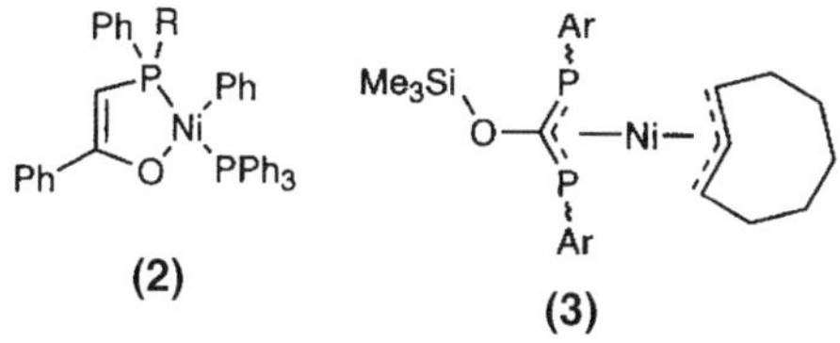

R	Solids(%)	$\bar{M}_n$	$\bar{M}_w$
Phenyl	—	Low molecular weight α-olefins	
Methyl	47	520	1330
Ethyl	74	654	1830
Cyclohexyl	81	484	1550
Neomenthyl	93	862	5470

Conditions: 100 °C, 600 psi.

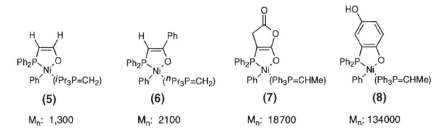

nickel, polyethylene was obtained with weight-average molecular weights ranging from 8000 to 350 000 [equation (1), 25–100 °C, 60–1000 psi; **4**: $R_1 = $ Ph, SO_3Na or H; $R_2 = $ OMe or Ph; L $= PEt_3$, PPh_3 or pyridine). The functional group tolerance of these nickel catalysts was demonstrated by homopolymerizing ethylene in the presence of polar additives and by copolymerizing ethylene with α-olefins containing a functional group in a remote position to the olefin. Low levels of functionalized olefin were incorporated and attempts to copolymerize ethylene with methyl acrylate and vinyl acetate were unsuccessful. The synthesis of block copolymers of ethylene and ethylene/CO was also reported.

In the mid-1980-s, Starzewski and co-workers reported the use of a wide variety of modified SHOP catalysts stabilized by ylides (e.g. **5–8** in Figure 1) for the polymerization of ethylene [15]. By varying the structure of the chelating ligand, a range of oligomers and polymers were synthesized with structures varying from linear α-olefins to HDPE ($M_n > 1.0 \times 10^6$) and with turnover frequencies as high as 50 000 TO/(mol Ni h). Copolymers of ethylene and propylene were also made with these nickel catalysts with up to 6 mol% propylene incorporation. When the nickel catalysts were combined with supported chromium catalysts, branched polyethylene (5.0 methyl-ended branches per 1000 carbon atoms) was produced by chromium-catalyzed copolymerization of ethylene with α-olefins, with the α-olefins being produced *in situ* by the nickel catalyst.

Other interesting examples of nickel-catalyzed ethylene polymerizations that illustrate the effects of ligand modification on catalyst performance include reports by Cavell and co-workers [16] and Wilke and co-workers [17]. Cavell and co-workers explored electronic effects on nickel-catalyzed oligomerizations/polymerizations through the use of a series of substituted pyridine carboxylate complexes

Figure 1 Effect of ligand tuning on the molecular weights of polyethylene produced by ylide-stabilized nickel complexes (1500 psi ethylene, 100–130 °C, cyclohexane)

(9)

(10)

X = C(OMe)(9a) CH (9b) ; N (9c); C(NO₂)(9d)

Oligomers $\xrightarrow[\text{Electron Density}]{\text{Decreasing}}$ Polyethylene

$PR_3 = P(^iPr)_2(^tBu)(10a); P(^tBu)_3 (10b)$

Dimers $\xrightarrow[\text{Steric Bulk}]{\text{Increasing}}$ Polyethylene

Figure 2. Examples of electronic and steric control of molecular weights in nickel-catalyzed ethylene oligomerizations/polymerizations

(9a–d) (Figure 2) [16]. The electron-rich methoxy-substituted catalyst **9a** reacted with ethylene at 600 psi and 80 °C to give largely linear olefins (56% α) with only 10–20 % of the product consisting of higher molecular weight polyethylene. In contrast, the electron-deficient nitro-substituted catalyst **9d** produced between 80 and 100 % high molecular weight HDPE under similar conditions and exhibited the highest activities of the four derivatives [225 gPE/(g Ni h)]. The electron density and reactivity of the pyrazine carboxylate complex **9c** fell between those of **9a** and **9d**, with 55 % of the product consisting of polyethylene. Based on phosphine exchange studies and also on the detection of free phosphine in the presence of ethylene for the complex **9d**, but not for the more electron-rich complex **9b**, two reaction pathways were tentatively proposed. For the electron-deficient pyridine carboxylate complexes, phosphine dissociation was proposed to give a four-coordinate active species [(N^O)Ni(R)(H₂C=CH₂)] in the presence of ethylene, whereas five-coordinate phosphine complexes [N^O)Ni(R)(PPh₃)(H₂C=CH₂)] were proposed as the active species for the more electron-rich complexes such as **9b**. In some instances, it was speculated that both pathways may operate concurrently. Electronic effects were also observed in ethylene/CO copolymerizations with the percentage of polyketone formed varying from 100 % with **9d** to 70 % with **9c** to 10–20 % with **9a**. The remaining portion of the isolated product consisted of ethylene oligomers.

Wilke and co-workers employed nickel allyl phosphine complexes activated by aluminum compounds (**10**, Figure 2) to illustrate the effect of steric and electronic tuning on nickel-catalyzed olefin insertions [17]. A crystal structure of the tricyclo-hexylphosphine choride derivative of **10** where $AlX_3 = AlMeCl_2$ has been reported. The addition of the Lewis acid increases the electrophilicity of the metal center, allowing the extremely rapid dimerization of propylene (up to 1170 TO/s at −45 °C). In the case of ethylene, dimerization was observed for complexes **10** containing ligands as large as diisopropyl-*tert*-butylphosphine (**10a**), while the sterically larger tri-*tert*-butylphosphine-based complex **10b** polymerized ethylene to high molecular weight linear polyethylene. In addition to monodentate phosphine ligands, sterically bulky bidentate phosphine ligands [e.g. $R_2P(CH_2)_nPR_2; n = 1–3; R = {}^iPr, {}^tBu$]

have been reported to support nickel- and palladium-catalyzed ethylene polymerizations and copolymerizations [18].

2.2 α-OLEFINS

Although capable of polymerizing ethylene, most of the above nickel catalysts will only dimerize α-olefins. A notable exception is the oligomerization of α-olefins by the aminobis(imino)phosphorane nickel catalyst **11** developed by Goddard and co-workers for the polymerization of ethylene [19]. This catalyst yields short-chain branched polyethylene at $70\,^{\circ}C$ and 740 psi. Fink and co-workers used this nickel catalyst to polymerize linear and singly branched α-olefins to give polymers with degrees of polymerization between 10 and 20, with the degree of polymerization falling off with an increase in α-olefin chain length ($M_n \approx 1000$) [20].

A remarkable feature of these α-olefin oligomerizations is the exclusive 2,ω-coupling of the olefin (Scheme 2). For example, linear α-olefins form polymers containing only methyl branches [20]. In the case of [1-^{2}H]-1-hexene, the deuterium label is found in the methyl branch. Fink and co-workers proposed a mechanism that involves the exclusive 1,2-insertion of the α-olefin to give the sterically crowded primary nickel alkyl **A**, which does not insert α-olefin [20]. Instead, the metal migrates along the polymer chain through a series of reversible β-hydride eliminations and readditions, forming a series of intermediate tertiary (e.g. **B**) and secondary nickel alkyls (e.g. **C**) until reaching C_ω of the α-olefin. This gives a less crowded primary metal alkyl **D**, which can then insert another equivalent of α-olefin and thus continue the chain propagation. Chain transfer to monomer can occur at any point during the metal's migrations throughout the polymer chain, and the reversible nature of this migration accounts for the drop in degree of polymerization as the length of the α-olefin is increased.

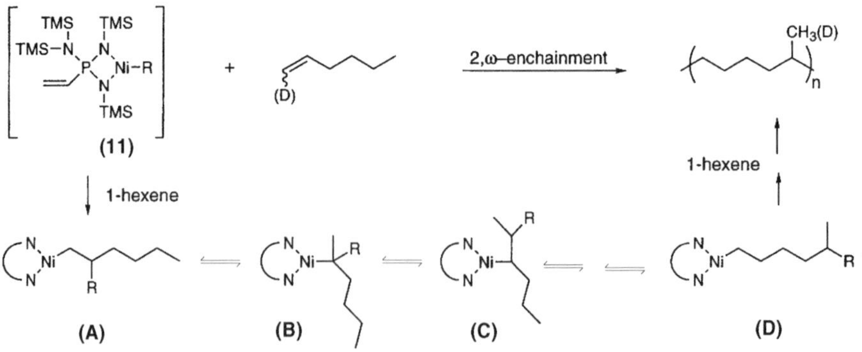

Scheme 2   2,ω-Enchainment of α-olefins by nickel catalyst **11**

2.3 OTHER OLEFINS

In addition to ethylene and α-olefins, other olefins can be polymerized by late metal catalysts. For example, in recent years a number of reports on the use of cationic palladium and nickel catalysts for the insertion polymerization of styrene, substituted cyclopropenes and particularly norbornene derivatives have appeared [21–28]. These reports are reviewed below. Other examples of late metal-catalyzed olefin polymerization reactions include the alternating copolymerization of olefins and carbon monoxide by palladium(II) and nickel(II) complexes [29], the ring-opening metathesis polymerization of cyclic olefins by ruthenium alkylidene complexes [30] and the polymerization of dienes by nickel allyl initiators [31]. The mechanism, scope and functional group tolerance of these reactions have recently been reviewed [29–31].

Styrene can be polymerized by a variety of nickel catalysts including $[(\text{allyl})\text{Ni-}\mu\text{-}(O_2\text{CCF}_3)]_2$ (12), $[(\text{COD})\text{Ni(allyl)}]\text{PF}_6$ (13) and $\text{Ni(acac)}_2/\text{MAO}$ (14) (COD = 1,5-cyclooctadiene; acac = acetylacetonate; MAO = methylaluminoxane) [21–23]. Recent developments include reports by Deming and Novak on the activation of 12 by hexafluoroacetone and hexachloroacetone to give highly active catalysts for the polymerization of styrene and other monomers [21]. Porri and co-workers [23a] reported that the AlMe_3 in MAO decreased the yield and stereospecificity of polystyrene produced by 14. Removal of AlMe_3 from the MAO and the addition of triethylamine to the catalyst system increased the percentage of isotactic units found in the polymer. However, addition of triethylamine also decreased the polymer yield.

Changes in polystyrene microstructure and molecular weight have been observed by Dias and co-workers upon addition of phosphines to 13 [equation (2)] [22].

L = PPh$_3$, P(o-Tol)$_3$, PMe$_3$, P(^{t}Bu)$_3$, PCy$_3$, P(^{t}Bu)$_3$ or P(OPh)$_3$. (P) = polymer chain

Phosphine addition [with the exception of PCy$_3$ and P(^{t}Bu)$_3$] increases the catalytic activity of 13. Within a given class of phosphines, e.g. P(aryl)$_3$ and P(alkyl)$_3$, catalyst activities decrease and molecular weights increase ($M_n \approx 700$–2000) with increasing phosphine cone angle. For these nickel systems, regioregular 2,1-insertion of styrene gives stable π-benzyl intermediates 15, with η^3- to η^1-isomerization of the benzyl group and subsequent β-hydride elimination yielding structure I. With bulky phosphines such as PCy$_3$, 1,2-insertion of styrene may occur and is immediately followed by β-hydride elimination to give structure II. The phosphine-ligated nickel systems in equation (2) produced polystyrenes with varying degrees of isotactic enrichment, with catalysts incorporating sterically large phosphines, e.g. PCy_3 and $P(o\text{-Tol})_3$, yielding highly isotactic polymer.

Various groups have reported the use of several well defined single-component Ni(II) and Pd(II) cationic catalysts for the addition polymerization of norbornene derivatives [24–28, 32]. The catalysts are sterically unhindered with labile ancillary ligands {e.g.[Pd(NCMe)$_4$](BF$_4$)$_2$ and [(COD)Ni(allyl)]PF$_6$ (**16**), equation (3)} and are highly active, yielding soluble norbornene homopolymers with molecular weights in excess of 10^6. The production of high molecular weight polynorbornene results from the inability of the propagating polynobornene chain to chain transfer via β-hydride elimination [the β-hydrogens are inaccessible to the metal center since one β-hydrogen is located at the bridgehead carbon and the other lies *anti* to the metal center; see equation (3)].

(16) PNB = Polynorbornene

The high T_g (380–390°C) of the high molecular weight polynobornene produced by nickel catalyst **16** limits the melt processability of the polymer. The T_g can be lowered and the processability improved by copolymerizing nobornene with 5-alkylnobornenes or by utilizing α-olefins as chain transfer agents [24]. Insertion of a single α-olefin results in rapid β-hydride elimination and liberation of a vinyl-terminated polymer [equation (3).] Some of the engineering thermoplastics accessible by this technology, including polynobornene derivatives containing functional groups, are currently being developed by B. F. Goodrich for use in dielectric, optical and photolithographic applications [24].

Risse and co-workers reported the polymerization of 3,3-disubstituted cyclopropenes to give highly strained polymers with 1,2-*cis*-linked triangular repeating units [equation (4)] [33]. Similarly to polynorbornene, the β-hydrogens of poly(3,3-dialkylcyclopropene) are not accessible to the metal center, limiting chain transfer by this mechanism. However, it was observed that highly electrophilic catalysts such as [Pd(NCMe)$_4$]BF$_4$, which are used for norbornene polymerizations, produced

(17) (18)

(mm)

cyclopropene ploymers with both cyclic and ring-opened repeat units. Ring opening was largely prevented by reducing the electrophilicity of the metal center through the incorporation of bidentate nitrogen-based ligands such as bipyridine, sparteine (**17**) and bisoxazolines (**18**). Polymer produced by the bypyridine-ligated palladium catalyst was atactic, whereas catalysts **17** and **18** gave polymer with a moderate enrichment of isotactic (mm) units.

3 α-DIIMINE-BASED PALLADIUM AND NICKEL CATALYSTS

3.1 CATALYST DEVELOPMENT AND SYNTHESIS

Brookhart and co-workers poineered the use of discrete cationic, electrophilic late metal complexs for catalysis, including the development of cyclopentadienyl-based cobalt and rhodium catalysts for the living polymerization of ethylene and the rapid tail-to-tail dimerization of acrylates, respectively [6, 34]. Recently, they reported the discovery of versatile cationic Pd(II) and Ni(II) α-diimine cataysts (Versipol catalysts) for the polymerization of ethylene, α-olefins, and cyclic olefins and the copolymerization of non-polar olefin with a variety of functionalized olefins [5, 35–44]. These catalysts are now the focus of a joint development effort between the University of North Carolina at Chapel Hill and DuPont.

Three key features of the α-diimine polymerization catalysts are the electrophilic nature of the cationic nickel and palladium complexes, the incorporation of sterically bulky α-diimine ligands and the use of non-coordinating counterions [35]. The electrophilicity of the late metal center in these cationic complexes results in rapid rates of olefin insertion. This was also noted in Brookhart and co-workers' earlier work [6] on cationic cobalt ethylene polymerization catalysts and in the research by Wilke and co-workers [17] and Cavell and co-workers [16] on nickel polymerization catalysts that is mentioned above.

The incorporation of sterically bulky α-diimine ligands in these nickel and palladium systems favors the formation of high molecular weight polymer. Studies by Brookhart and co-workers involving the reactivity of sterically unhindered bipyridine nickel alkyl complexes {e.g. $[(bipy)Ni(Me)(OEt_2)]^+BAr_4'^-$} with ethylene demonstrated rapid rates of olefin insertion; however, only oligomers and low molecular weight polyethylene were produced (e.g. $M_w = 6200, M_n = 1900$ at $0\,°C$). Catalyst tuning was possible by the replacement of bipyridine with the closely related α-diimine ligands. Incorporation of sterically bulky α-diimine ligands in particular resulted in the formation of very efficient catalysts for the production of high molecular weight polyolefins (e.g. see Scheme 3) [35]. This observation finds some precedent in the work cited above by the groups of Keim [12], Wilke [17], Tkatchenko [22] and others [18] on the use of bulky phosphine ligands to increase the molecular weight of nickel-catalyzed ethylene and styrene polymerizations.

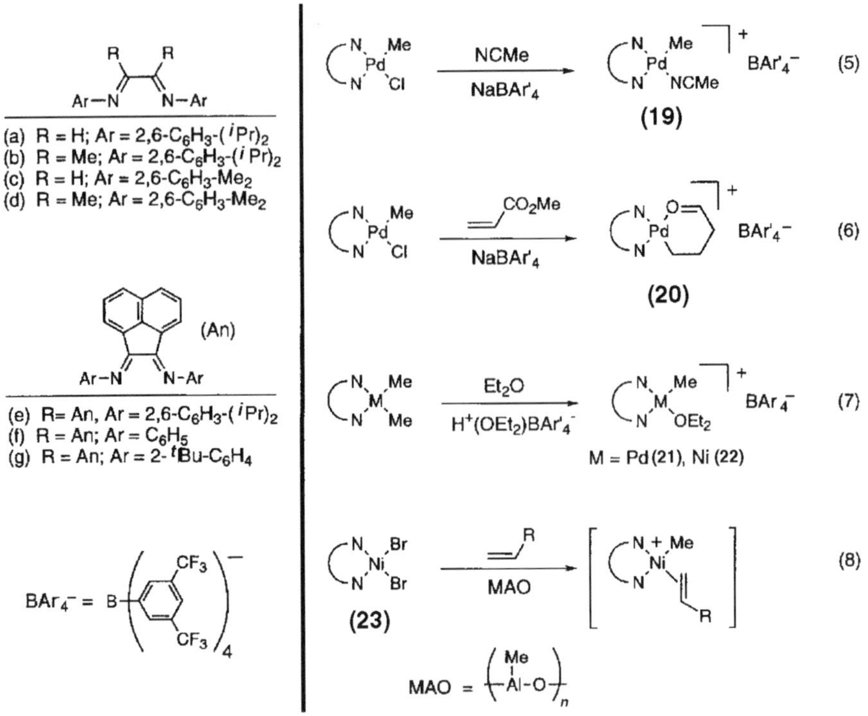

R   R
Ar—N   N—Ar

(a) R = H; Ar = 2,6-C_6H_3-(^{i}Pr)$_2$
(b) R = Me; Ar = 2,6-C_6H_3-(^{i}Pr)$_2$
(c) R = H; Ar = 2,6-C_6H_3-Me$_2$
(d) R = Me; Ar = 2,6-C_6H_3-Me$_2$

(An)

Ar—N   N—Ar

(e) R = An, Ar = 2,6-C_6H_3-(^{i}Pr)$_2$
(f) R = An; Ar = C_6H_5
(g) R = An; Ar = 2-^{t}Bu-C_6H_4

$BAr_4^- = B\left(\begin{array}{c}CF_3\\ \\CF_3\end{array}\right)_4$

$$\begin{array}{ccc}\text{N}\quad\text{Me}\\ \text{Pd}\\ \text{N}\quad\text{Cl}\end{array} \xrightarrow[\text{NaBAr}_4']{\text{NCMe}} \left[\begin{array}{c}\text{N}\quad\text{Me}\\ \text{Pd}\\ \text{N}\quad\text{NCMe}\end{array}\right]^+ \text{BAr}_4'^- \quad (5)$$
(19)

$$\begin{array}{ccc}\text{N}\quad\text{Me}\\ \text{Pd}\\ \text{N}\quad\text{Cl}\end{array} \xrightarrow[\text{NaBAr}_4']{\text{CO}_2\text{Me}} \left[\begin{array}{c}\text{N}\quad\text{O}\\ \text{Pd}\\ \text{N}\end{array}\right]^+ \text{BAr}_4'^- \quad (6)$$
(20)

$$\begin{array}{ccc}\text{N}\quad\text{Me}\\ \text{M}\\ \text{N}\quad\text{Me}\end{array} \xrightarrow[\text{H}^+(\text{OEt}_2)\text{BAr}_4']{\text{Et}_2\text{O}} \left[\begin{array}{c}\text{N}\quad\text{Me}\\ \text{M}\\ \text{N}\quad\text{OEt}_2\end{array}\right]^+ \text{BAr}_4^- \quad (7)$$
M = Pd(21), Ni (22)

$$\begin{array}{ccc}\text{N}\quad\text{Br}\\ \text{Ni}\\ \text{N}\quad\text{Br}\end{array} \xrightarrow[\text{MAO}]{R} \left[\begin{array}{c}\text{N}\quad\text{Me}\\ \text{Ni}\\ \text{N}\end{array}\right] \quad (8)$$
(23)

$$\text{MAO} = \left(\begin{array}{c}\text{Me}\\ \text{Al-O}\end{array}\right)_n$$

Scheme 3 Synthesis of α-diimine nickel and palladium initiators for olefin polymerization

The easily varied steric and electronic properties of the α-diimine ligands are an important feature of the Versipol systems. These ligands are readily tuned owing to their facile synthesis involving the condensation of a diketone with two equivalents of an alkyl or aryl amine. The N-aryl α-diimine ligands that are typically incorporated in the Versipol catalysts were developed by the groups of tom Dieck [45] and Vrieze [46] and the importance of α-diimine ligands in stabilizing organotransition metal complexes has been reviewed [47].

Finally the incorporation of non-coordinating counterions such as B[3,5-$C_6H_3(CF_3)_2]_4^-$ and MAO permits the production of polymers with higher molecular weights than those of polymers prepared with complexes incorporating more coordinating counterions such as BF_4^- and triflate. This trend finds much precedence in early transition metal polymerization chemistry, and reviews on non-coordinating counterions have been published [48, 49]. In addition, Brookhart et al. demonstrated that in late metal systems, the high solubility of cationic complexes incorporating the perfluorinated counterion B[3,5-$C_6H_3(CF_3)_2]_4^-$ is an especially important feature in allowing the low-temperature NMR study and elucidation of the mechanism of olefin insertions and other processes [50].

Polymerization initiators, which include Ni(II) and Pd(II) cationic methyl complexes stabilized by sterically bulky α-diimine ligands, can be prepared by a number of synthetic routes. For preparative-scale polymerizations with palladium, the acetonitrile complexes **19** or chelate complexes **20** are preferred owing to their stability, ease of handling and facile synthesis by chloride abstraction with $NaBAr'_4$ [equations (5) and (6) [42, 43]. Owing to the weak basicity of diethyl ether, attempts to make the palladium ether adducts **21** by chloride abstraction with $NaBAr_4$ resulted instead in the formation of chloride-bridged dimers ($N^\wedge N)PdMe-\mu-Cl]_2BAr'_4$ [5]. Therefore, the cationic palladium and nickel ether adducts **21** and **22** [equation (7)] were synthesized by protonation of the nickel and palladium dimethyl precursors with $H^+(OEt_2)_2BAr'^-_4$ [$Ar' = (3,5-C_6H_3(CF_3)_2]$ in the presence of diethyl ether [35]. The ether ligand is extremely labile, making these precursors ideal for low-temperature NMR mechanistic studies. For preparative-scale polymerizations, the nickel methyl cations are conveniently generated *in situ* by reacting the α-diimine nickel dibromide complex **23** with methylaluminoxane (MAO) [equation (8)] [35, 37, 51].

3.2 ETHYLENE

A typical nickel-catalyzed ethylene polymerization involves the addition of MAO (100–1000 equiv.) to a rapidly stirred suspension of the nickel dibromide complex **23** (ca 1×10^{-6} mol) in toluene under an atmosphere of ethylene [35, 37]. Brookhart and co-workers reported catalyst activities as high as 2×10^6 TO/(mol Ni h), corresponding to productivities of 50 000 kg PE (mol Ni)$^{-1}$ h^{-1}. Polyethylenes with high molecular weights ($M_w \approx 30 000$–$900 000$) and with microstructures ranging from crystalline and strictly linear to amorphous and highly branched can be synthesized with these nickel initiators by varying the catalyst structure and reaction conditions [35, 37]. Exposure of the palladium initiators **19**, **20** and **21** to ethylene also yields high molecular weight polyethylene. However, in contrast to the nickel catalysts, only amorphous, highly branched polyethylenes with densities as low as 0.85 are produced by the palladium catalysts [35, 43]. The ability to form such highly branched polymer from ethylene alone appears to be unprecedented [52].

3.3 MECHANISTIC STUDIES

Chain growth of ethylene and α-olefins as catalyzed by the palladium initiators **21** was monitored by low-temperature NMR spectroscopy [35]. These studies established that alkyl olefin complexes (e.g. **24** in Scheme 4) were the catalyst resting states in these cationic α-diimine systems and that chain growth was zero order in olefin. In addition, it was observed that the exchange of ethylene with free ethylene

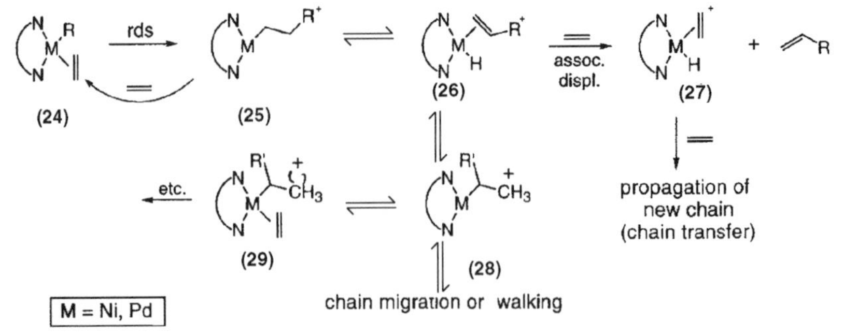

rds

(24)   (25)   (26)   (27)

assoc.
displ.

+

etc.

propagation of
new chain
(chain transfer)

(29)   (28)

M = Ni, Pd

chain migration or walking

Scheme 4 Proposed mechanism for ethylene polymerization and polymer branch formation

was dependent on ethylene concentration and that these associative exchange rates decreased as the steric bulk of the ligand increased.

Based on these low-temperature NMR studies and on the analysis of the microstructures of the branched polymers, the polymerization mechanism represented in Scheme 4 for ethylene was proposed by Brookhart and co-workers [35]. For these cationic α-diimine nickel and palladium systems, migratory insertion of ethylene in the catalyst resting state **24** is the rate-determining step and results in the formation of the primary alkyl intermediate **25**. Complex **25** can then either be trapped by ethylene to regenerate the alkyl olefin complex **24** or, alternatively, **25** can undergo a series of β-hydride eliminations and readditions with the metal migrating or 'walking' along the polymer chain. For example, β-hydride elimination from **25** produces the olefin hydride complex **26**. Readdition of hydrogen to the terminal methylene carbon will generate the secondary alkyl complex **28**. Trapping of **28** by ethylene to generate **29** followed by ethylene insertion will give a polymer with a methyl branch. Continuing β-hydrogen elimination and readdition from complex **28** will produce longer branches.

Chain transfer in these nickel and palladium systems occurs by exchange of the polymer-ended olefin of complex **26** with ethylene to generate the ethylene hydride complex **27**, which can initiate the growth of a new chain [35]. As noted above, this olefin exchange occurs via an associative displacement mechanism and is measurably slower as the size of the *ortho* or backbone substituents of the α-diimine ligand is increased. Brookhart and co-workers proposed that this slowing of associative olefin exchange was the key to forming high molecular weight polymer in these systems and that the structure of the α-diimine ligands is ideal for this purpose [35]. The α-diimine ligands bind to late metal d^8 square-planar complexes with the aryl rings lying roughly perpendicular to the square plane, which places the *ortho* substituents above and below the square plane [equation (9)]. Crystallographic data indicate that as the steric bulk of either the backbone substituents or the *ortho* substituents increases, the more perpendicular the aryl ring lies with respect to the square plane and the more effectively the *ortho* substituents block the axial sites [46, 53]. This blocking of the axial sites slows the approach of ethylene to the metal center and consequently slows the rate of chain transfer relative to chain propagation, allowing the formation of high molecular weight polymer [35]. In addition to these experimental results, further insights and proposals regarding the mechanism of polymerization were published in recent theoretical studies by the groups of Ziegler [54, 55], Morokuma [56] and Siegbahn [57, 58].

3.4 VARIATION OF POLYETHYLENE STRUCTURE AND PROPERTIES

One of the most remarkable features of the α-diimine catalysts is the ability to control polymer structure and properties through variations in catalyst structure and reaction conditions. For the palladium catalysts, Brookhart and co-workers reported that productivities and molecular weights are dependent on the α-diimine backbone substituents ($R = Me > An > H$) and on the *ortho* substituents of the *N*-aryl ring ($R' = {}^{i}Pr > Me$; Table 2), consistent with the proposed mechanism [35]. Owing to rapid rates of 'chain walking' relative to insertion, polyethylenes produced by the palladium catalysts are highly branched, typically exhibiting *ca* 100 methyl-ended branches per 1000 carbon atoms, independent of the reaction pressure (Table 2). However, the morphology of the polymer does vary with pressure with amorphous viscous oils produced at low pressures and amorphous rubbery solids at higher pressures.

According to McCord and co-workers, ^{13}C NMR analysis indicated that every length of short chain branch is present in the amorphous polyethylenes produced by the palladium catalysts with the branches becoming less prevalent as they increase in length [38]. However, this decrease is not monotonic as there are more even- than odd-numbered branches. These patterns were explained by the catalyst 'chain-walking' mechanism. In addition, the identification of *sec*-butyl-ended branches indicated that there are branches on branches in these new forms of polyethylene (Figure 3).

For the nickel-based catalysts, the extent of branching was reported by Brookhart and co-workers to be a function of ethylene concentration, reaction temperature and catalyst structure [35, 37]. The effect of pressure variation is illustrated in Table 3 for polyethylenes produced by the catalyst **23e**/MAO. As ethylene pressure (ethylene

Table 2 Branched polyethylene produced by palladium catalysts **20b** and **20d**[a]

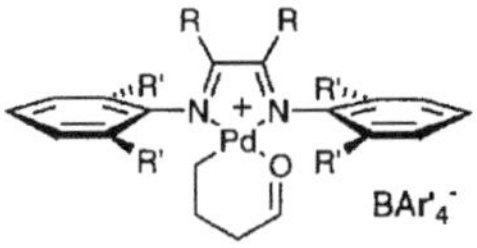

20b (R,R' = iPr)
20d (R,R' = Me)

R, R'	Pressure (atm)	Yield (total TO)	M_n ($\times 10^{-3}$)	M_w/M_n	Branches per 1000 °C
Me, iPr	2	8.80 g (30 200)	297^b	3.5^b	102
Me, iPr	11	13.70 g (48 700)	490^b	2.7^b	100
Me, iPr	29	8.10 g (28 500)	496^b	3.0^b	98
Me, Me	11	3.70 g (13 300)	$445/28^b$	—	106

[a] 0.01 mmol catalyst, 100 ml CH_2Cl_2, 25 °C.
[b] Bimodal distribution.

concentration) is increased, the number of branches per 1000 carbons atoms decreases significantly, whereas catalyst productivity and molecular weight are largely unaffected. The sensitivity of branching to ethylene pressure observed in these nickel systems provides additional experimental support for the proposed mechanism for olefin polymerization. The mechanism predicts that a competition exists between trapping with ethylene and 'chain walking'. At higher ethylene pressures trapping is favored, resulting in the formation of fewer branches.

The effect of temperature variation using the nickel catalyst **23b**/MAO is demonstrated in Table 4 [35, 37]. As the temperature increases, branching increases

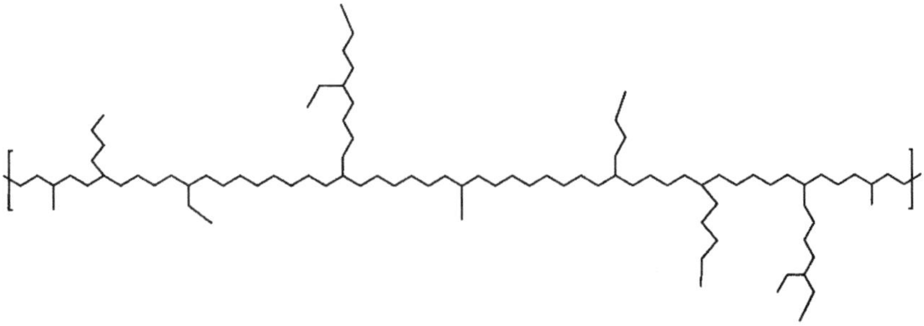

Figure 3 Representative structure of highly branched polyethylene produced by the palladium Versipol catalysts

Table 3   Effect of pressure variation on polyethylene properties and structure[a]

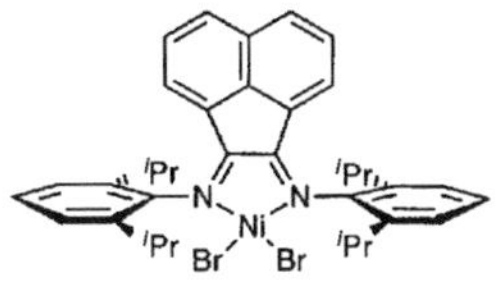

23e/MAO

Pressure (atm)	TOF $(\times 10^{-4}\,h^{-1})$	M_n	M_w/M_n	T_m (°C)	Branches per 1000 °C
1	19	170 000	2.0	46	65
15	180	470 000	2.1	118	30
30	230	490 000	2.2	123	20
42	190	510 000	2.2	126	5

[a]0.00083 mmol **23e**, ~100 equiv. MAO, 200 ml toluene, 10 min, 25 °C.

and the polymer molecular weight and melt transition temperature decrease. At 80 °C, a completely amorphous ethylene homopolymer ($T_m = -12\,°C$) was produced. This is explained by the fact that at higher reaction temperatures, unimolecular chain running is favored relative to bimolecular trapping and insertion, resulting in a more highly branched microstructure.

The influence of the α-diimine substituents on nickel-catalyzed polymerizations is summarized in Table 5 [35, 37]. A reduction in the steric bulk of either the backbone substituents or the *ortho* substituents of the aryl ring reduces the molecular weight and total number of branches for the resulting ethylene homopolymers. This is consistent with increased rates of associative olefin exchange for the less hindered

Table 4   Effect of temperature variation on polyethylene properties and structure[a]

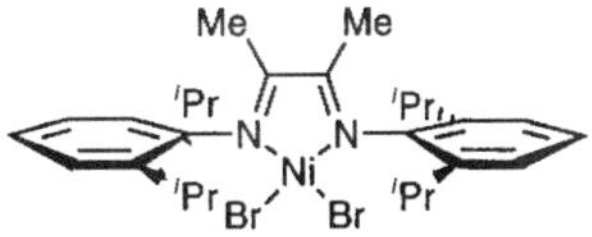

23b/MAO

Temperature (°C)	TOF $(\times 10^{-4}/h)$	M_n	M_w/M_n	T_m (°C)	Branches per 1000 C
25	77	850 000	2.3	99	30
50	56	260 000	2.8	50	67
65	68	180 000	2.5	24	80
80	9.0	150 000	1.8	−12	90

[a]~ 500 equiv. MAO, 200 ml toluene, 10–12 min, 25 °C, 0.0019 mmol **23b** (first entry) and 0.0024 mmol **23b** (other entries).

Table 5 Effect of catalyst tuning on polyethylene properties and structure[a]

23a/MAO (R = H, R' = ^{i}Pr)^i
23e/MAO (R = An, R' = ^{i}Pr)
23b/MAO (R = Me, R' = ^{i}Pr)^i
23d/MAO (R = Me, R' = Me)
23f/MAO (R = An, R' = H)

R, R'	Pressure (atm)	M_n	M_w/M_n	T_m (°C)	Branches per 1000 °C
H, ^{i}Pr	1	110 000	2.7	129	7
An, ^{i}Pr	1	650 000	2.4	112	24
Me, ^{i}Pr	1	520 000	1.6	109	48
Me, Me	1	170 000	2.6	115	20
An, H	15	Oligomers (85 % linear α-olefins)			
An, H	56	Oligomers (94 % linear α-olefins)			

[a] 100–200 ml toluene, 0 °C (first four entries), 35 °C (last two entries).

catalysts leading to chain transfer, as proposed above. A dramatic illustration of the effect of reduced steric bulk and increased chain transfer rates is the production of linear α-olefins, instead of high molecular weight polymer, by the sterically smaller N-phenyl catalyst **23f/MAO** with the selectivity for linear α-olefins increasing with increasing pressures (Table 5) [37, 59].

3.5 α-OLEFINS

Brookhart and co-workers reported that high molecular weight poly(α-olefins) with unusual microstructures can be prepared by these Ni(II) and Pd(II) catalysts with good activities [35–37]. The results in Table 6 illustrate the effect of catalyst structure on the properties and structure of polypropylene and are representative of the results obtained for other α-olefins. The amount of branching is dependent on the choice of metal and ligand with palladium catalysts such as **21b** and nickel

Table 6 Effect of catalyst structure on polypropylene properties and structure[a]

Compound	Time (h)	TOF (h^{-1})	M_n	M_w/M_n	T_g (°C)	Branches per 1000 °C
21b	16	—[b]	15 000	4.3	−43	213
23e/MMAO	1	3000	190 000	1.4	−24	272
23g/MMAO	1	1300	60 000	1.6	−55	159

[a] 1 atm propylene, 23 °C, 0.1 mmol **21b** and 0.017 mmol **23e** and **23g**.
[b] Total TO: 9300.

catalysts such as **23g** (one *ortho* *tert*-butyl substituent on the aryl ring) giving especially low levels of branching. For poly(α-olefins) synthesized by Ziegler–Natta and metallocene-based catalysts, 1,2-enchainment predominates to give a polymer with a branch on every other backbone carbon atom [equation (10)]. For example, atactic polypropylene with 1,2-enchainment has 333 branches per 1000 carbon atoms and a glass transition temperature of roughly $1\,^\circ$C [60]. Therefore, the low levels of branching and low glass transition temperatures obtained for the poly(α-olefins) made with the nickel and palladium α-diimine catalysts indicate that a significant amount of 2,1-insertion and subsequent 'chain-walking' to give 1,ω-enchainment and runs of polyethylene-like segments is occurring [equation (11)].

$$\underset{\substack{\\ \text{1,2-enchainment}}}{\text{M}\underset{2}{\overset{P}{\diagdown}}\!\!\diagup^{1} \;\;+\;\; n \;\; \overset{1\;\;2}{=\!=}\!\!\diagdown_R \;\;\xrightarrow{\;\text{1,2-insertion}\;}\;\; \text{M}\!\!\underset{1\;\;2}{\overset{R}{\diagup\!\!\diagdown}}^{P}_{n+1}} \tag{10}$$

$$\underset{\substack{\\ \\ \text{1, }\omega\text{-enchainment}}}{\text{M}\underset{2}{\overset{P}{\diagdown}}\!\!\diagup^{1}_{\substack{(CH_2)_n \\ | \\ CH_3 \\ \omega}}\;\;\xrightarrow{\;\text{2,1-insertion}\;}\;\;\xrightarrow{\;\text{chain migration}\;}\;\; \text{M}-\overset{\omega}{CH_2}-(CH_2)_{n+1}-\overset{1}{CH_2}-P} \tag{11}$$

The detailed ^{13}C NMR spectroscopic analysis of the microstructures of these poly(α-olefins) has been described in papers by McCord and co-workers [38, 39]. For a given α-olefin $H_2C{=}CH(CH_2)_nH$, their experiments support the occurrence of both 1,2- and 2,1-insertion and indicate that along with methyl and long chain branches, branches n carbons in length predominate. Branches of intermediate length are generally not observed. The ^{13}C NMR spectrum of poly(4-methyl-1-pentene) confirmed the ability of the catalyst to 'walk' past a tertiary carbon, and the spectrum of poly(propylene) established that at least three 1,3-enchainments of propylene can occur sequentially. Based on these NMR experiments, the following preliminary guidelines were given for α-olefin insertions, which apply only for the palladium catalysts: (i) both 1,2- and 2,-1-insertion of the α-olefin can occur; (ii) this insertion occurs exclusively into a primary palladium alkyl bond; and (iii) the metal center can 'walk' forward and backwards along the polymer chain and the newly added monomer, even past tertiary carbon atoms [38, 39].

3.6 LIVING POLYMERIZATIONS

Recently, Brookhart and co-workers described the use of α-diimine nickel catalysts for the living polymerization of α-olefins [36, 37]. Key factors for the living polymerization were the use of low temperatures ($-10\,^\circ$C) and low monomer concentrations (less than 1 M). For example, the polymerization of propylene by **23e**/MAO at $-10\,^{\circ}$C and 1 atm propylene resulted in the preparation of high molecular weight polypropylene with a narrow molecular weight distribution ($M_n = 160\,000$, $M_w/M_n = 1.13$). Further support for the living nature of the polymerization came from a plot showing a linear increase in the number-average molecular weight of the polypropylene with time and propylene conversion [36, 37]. Polymerizations of propylene at low temperatures, including the formation of syndiotactic polypropylene at $-78\,^{\circ}$C via a chain end control mechanism, have also been described by Pellecchia and Zambelli using catalyst **23a**/MAO [61].

For $1,\omega$-insertions of α-olefins, the length of the linear segments increases with the chain length of the α-olefin and results in higher melt transition temperatures. Illustrative examples include poly(1-octadecene) produced by **23e**/MMAO (45 branches per 1000 carbon atoms; $T_m = 56\,^{\circ}$C) and by **23g**/MMAO (33 branches per 1000 carbon atoms; $T_m = 78\,^{\circ}$C) [36, 37]. Brookhart and co-workers used this property together with the living nature of the nickel-catalyzed α-olefin polymerizations to prepare block copolymers with well defined architectures. For example, the synthesis of α-olefin **A–B–A** block copolymers where the semicrystalline **A** blocks are made up of poly(1-octadecene) and the **B** block is composed of a more highly branched amorphous random copolymer of propylene and 1-octadecene allowed the preparation of elastomeric polyolefins (**23e**/MMAO: $M_n = 253\,000$, $M_w/M_n = 1.17$, $T_m = -11$ and $40\,^{\circ}$C, $T_g = -38\,^{\circ}$C; **23g**/MMAO: $M_n = 112\,000$; $M_w/M_n = 1.43$, $T_m = 69$ and $0\,^{\circ}$C, $T_g = -52\,^{\circ}$C) [36, 37].

3.7 FUNCTIONALIZED OLEFINS

Brookhart and co-workers reported that the cationic palladium α-diimine catalysts are the first transition metal catalysts able to copolymerize ethylene and α-olefins with functionalized olefins such as acrylates to give high molecular weight polymers [42, 43]. To date, methyl vinyl ketone and methyl, *tert*-butyl and fluorinated octyl acrylate and other acrylate derivatives have been copolymerized with non-polar olefins by the palladium catalysts [42, 43]. In addition, even the copolymerization of acrylic acid has been reported [5].

Publications by Brookhart and co-workers have described the structure of the novel copolymers, and a representative structure of the ethylene–methyl acrylate (EMA) copolymers is shown in Figure 4 [38, 42–44]. Similarly to the ethylene homopolymers produced by the α-diimine palladium catalysts, the EMA copolymers are amorphous, highly branched materials with about 100 branches per 1000 carbon atoms. Glass transition temperatures typically range from -50 to $-70\,^{\circ}$C. A good match between the GPC analysis of the copolymers using both RI and UV detection indicates that they are indeed true copolymers with the ester groups randomly and evenly distributed throughout the polymer chain. Based on ^{1}H and ^{13}C NMR spectroscopy, most of the ester groups are located at the ends of branches. The

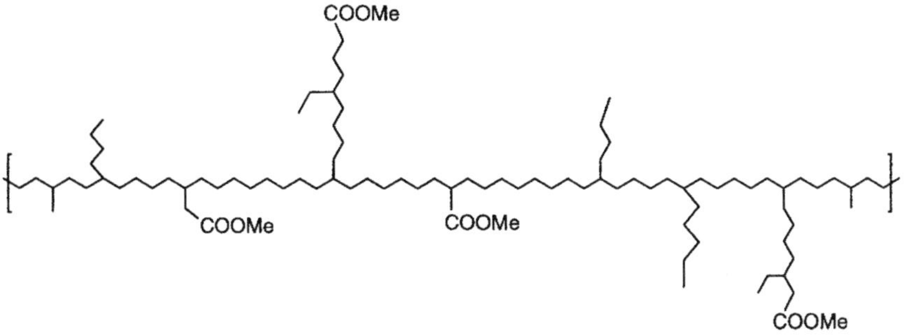

Figure 4 Representative structure of highly branched random copolymer of ethylene and methyl acrylate produced by the palladium Versipol catalysts

proportion of the ester groups directly attached to the backbone increases with increasing ethylene pressure.

The results of low-temperature NMR mechanistic studies of the copolymerization reaction were reported by Brookhart and co-workers and are illustrated in Scheme 5 [42, 43]. Key findings include the following: (i) acrylate insertion occurs predominantly in a 2,1 fashion and is followed by a series of β-hydride eliminations and readditions that result in the formation of a six-membered chelate complex, which is the catalyst resting state. This metal migration or 'chain-walking' accounts for the occurrence of the ester groups predominantly at the ends of branches. (ii) Strong binding of oxygen to palladium in the chelate complex makes chelate opening the turnover-limiting step of the copolymerization. (iii) Although acrylate insertion into the palladium alkyl bond is faster than ethylene insertion, the electrophilic palladium center preferentially binds the more electron-rich olefin ethylene.

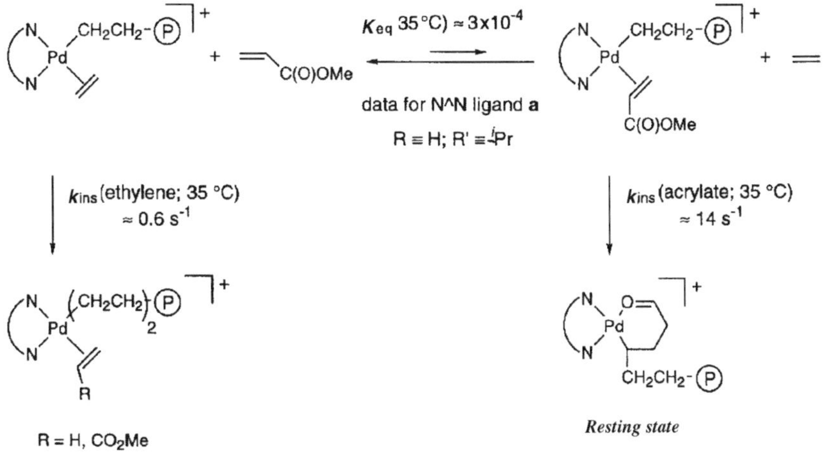

Scheme 5 Illustration of results from low-temperature NMR mechanistic studies of the copolymerization of ethylene and methyl acrylate [43].

Table 7 Copolymerization of ethylene and MA with **20b** (R = Me; R' = ^{i}Pr) and **20d** (R = Me; R' = Me)^a

Compound	E (atm)	MA (vol.%)	Yield (g)	MA incorporated (%)	Total TO	M_n	M_w/M_n	Branches per 1000 °C
20b	2	5	22.2	1	7800	88 000	1.8	103
20b	2	25	4.32	6	1400	26 000	1.6	103
20b	2	50	1.81	12	520	11 000	1.6	105
20b	6	50	11.2	4	3700	42 000	1.8	97
20d	6	50	2.3	14	630	7 000	2.1	116

^{a} 0.1 mmol **20b** and **20d**, total volume of CH$_2$Cl$_2$ and MA 100 ml, 35 °C, 18.5 h.

The findings of these low-temperature NMR studies are manifested in the preparative-scale copolymerizations of ethylene and MA as shown in Table 7 [42, 43], Entries 1–3 indicate that the fraction of acrylate incorporation is directly proportional to its concentration in solution. However, the net result of the poor relative binding of MA is that the molar percentage of MA incorporation in the copolymer is relatively low, even at high MA concentrations. As the percentage acrylate incorporation in the copolymer increases, the productivity decreases, owing to increased amounts of rate-retarding chelate formation. Increased acrylate incorporation can be achieved by decreasing the steric bulk of the diimine ligand or by incorporating more electron-donating substituents on the diimine. Both changes probably result in improved binding of MA to the palladium center. It should be noted that with increased acrylate incorporation, due to a decrease in the steric bulk of the ligand, concomitant decreases in copolymer molecular weights are also observed due to increased chain transfer (compare entries 4 and 5 in Table 7).

3.8 CYCLOPENTENE POLYMERIZATION

In addition to ethylene, α-olefins and functionalized olefins, the palladium and nickel complexes will catalyze the polymerization of acyclic and cyclic internal olefins. For example, McLain and co-workers reported that the polymerization of cyclopentene by both the nickel and palladium catalysts resulted in the formation of a new melt-processable polyolefin [40, 41] High molecular weight polycyclopentenes were produced with weight-average molecular weights as high as 251 000 and with broad melt transitions with end-of-melting points ranging from 241 to 330 °C. For comparison, all polycyclopentenes prepared with metallocene catalysts have very low molecular weights (M_n ≤ 2000). ^{13}C NMR studies and the results of hydro-oligomerization experiments indicated that the polycyclopentenes exhibit cis-1,3-enchainment and range in tacticity from atactic to partially isotactic. In addition, X-ray powder diffraction patterns indicated that the polycyclopentene produced by the α-diimine catalysts has a new crystalline form, which is different than that of the highly isotactic polycyclopentene produced by zirconium metallocene catalysts. Finally, a β-agostic resting state was observed for insertions of cyclopentene by both the nickel and palladium catalysts [equation (12)]. This is in contrast to the alkyl olefin complex resting state observed for the insertion of ethylene, α-olefins, and internal acyclic olefins.

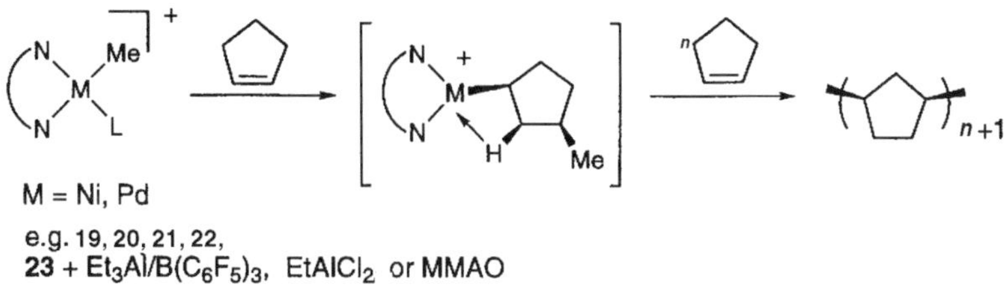

M = Ni, Pd

e.g. 19, 20, 21, 22,
23 + $Et_3Al/B(C_6F_5)_3$, $EtAlCl_2$ or MMAO

4 CONCLUSION

Many of the studies on nickel and palladium polymerization catalysts indicate the importance of a highly electrophilic, sterically hindered late metal center in producing high molecular weight polymer. The detailed studies by Brookhart and co-workers on the cationic α-diimine nickel and palladium catalysts provide a mechanistic rationale for these observations, and recent papers and patent applications indicate that a number of researchers are successfully applying these principles to develop new late metal polymerization catalysts [62, 63]. The functional group tolerance of late metal catalysts has been demonstrated, particularly in the polymerization of functionalized norbornenes by cationic nickel and palladium catalysts and in the copolymerization of ethylene and methyl acrylate by the α-diimine palladium catalysts. Finally, a number of unique polymers have been made using late metal polymerization catalysts, ranging from engineering polymers such as polycyclopentene and polynorbornene to elastomeric polymers of highly branched polyethylene and chain-straightened poly(α-olefins). The ability to prepare branched polyethylenes predictably from ethylene alone represents an opportunity to design ethylene homopolymers for target applications. Although industrial development is ongoing, the commercial impact of these polymers remains to be seen.

5 REFERENCES

1. Thayer, A. M., *Chem. Eng. News*, **73** (Sept. 11), 15 (1995).
2. Schumacher, J., in *Chemical Economics Handbook*, SRI International, Menlo Park, CA, 1994, p. 530.
3. Doak, K. W., in Mark, H. F. (Ed.), *Encyclopedia of Polymer Science and Engineering*, Wiley, New York, 1986, Vol. 6, p. 386.

4. Although some advances have been made in the homo- and copolymerization of functionalized olefins with early metals, these polymerizations are still limited to special substrates such as those in which the functional group is masked or placed in a remote position to the olefin: (a) Chung, T. C., *Macromolecules*, **21**, 865 (1988); (b) Chung, T. C. and Rhubright, D., *Macromolecules*, **26**, 3019 (1993); (c) Kesti, M. R., Coates, G. W. and Waymouth, R. M., *J. Am. Chem. Soc.*, **114**, 9679 (1992); (d) Aaltonen, P. and Lofgren, B., *Macromolecules*, **28**, 5353 (1995); (c) Galimberti, M., Giannini, U., Albizatti, E., Caldari, S. and Abis, L., *J. Mol. Catal*, **101**, 1 (1995).

5. Brookhart, M. S., Johnson, L. K., Killian, C. M., Arthur, S. D., Feldman, J., McCord, E. F., McLain, S. J., Kreutzer, K. A., Bennett, A. M. A., Coughlin, E. B., Ittel, S. D., Parthasarathy, A. and Tempel, D. J., *PCT Int. Appl.*, WO 96/23010 (1996).

6. (a) Brookhart, M., Volpe, A. F., Jr, Lincoln, D. M., Horvath, I. T. and Millar, J. M., *J. Am. Chem. Soc.*, **112**, 5634 (1990); (b) Schmidt, G. F. and Brookhart, M., *J. Am. Chem. Soc.*, **107**, 1443 (1985); (c) Brookhart, M., DeSimone, J. M., Grant, B. E. and Tanner, M. J., *Macromolecules*, **28**, 5378 (1995).

7. Wang, L. and Flood, T. C., *J. Am. Chem. Soc.*, **114**, 3169 (1992).

8. Timonen, S., Pakkanen, T. T. and Pakkanen, T. A., *J. Mol. Catal.*, **111**, 267 (1996).

9. (a) Keim, W., Behr, A., Limbacker, B. and Kruger, C., *Angew. Chem., Int. Ed. Engl.*, **22**, 503 (1983); (b) Keim, W., Behr, A. and Kraus, G., *J. Organomet. Chem.*, **251**, 377 (1983); (c) Peuckert, M. and Keim, W., *Organometallics*, **2**, 594 (1983); (d) Peuckert, M. and Keim, W., *J. Mol. Catal.*, **22**, 289 (1984); (e) Keim, W. and Schulz, R. P., *J. Mol. Cat.*, **92**, 21 (1994).

10. Flory, P. J., *J. Am. Chem. Soc.*, **62**, 1561 (1940).

11. (a) Schulz, G. V., *Z. Phys. Chem., Abt.*, **B30**, 379 (1935); (b) Schulz, G. V., *Z. Phys. Chem., Abt.*, **B43**, 25 (1939).

12. (a) Keim, W., Kowaldt, F. H., Goddard, R. and Kruger, C., *Angew. Chem., Int. Ed. Engl.*, **17**, 466 (1978); (b) Keim, W., *Makromol. Chem., Macromol. Symp.*, **66**, 225 (1993); (c) Hirose, K. and Keim, W., *J. Mol. Catal.*, **73**, 271 (1992); (d) Keim, W., *Angew. Chem., Int. Ed. Engl.*, **29**, 235 (1990).

13. Keim, W., Appel, R., Gruppe, S. and Knoch, F., *Angew. Chem., Int. Ed. Engl.*, **26**, 1012 (1987).

14. Klabunde, U. and Ittel, S. D., *J. Mol. Catal.*, **41**, 123 (1987).

15. (a) Starzewski, K. A. O., Witte, J., Reichert, K. H. and Vasiliou, G., in Kaminsky, W. and Sinn, H. (Eds), *Transition Metals and Organometallics as Catalysts for Olefin Polymerization*, Springer, Berlin, 1988, p. 349; (b) Starzewski, K. A. O. and Witte, J., *Angew. Chem., Int. Ed. Engl.*, **24**, 599 (1985); (c)) Starzewski, K. A. O. and Witte, J., *Angew. Chem., Int. Ed. Engl.*, **26**, 63 (1987).

16. (a) Desjardins, S. Y., Cavell, K. J., Jin, H., Skelton, B. W. and White, A. H., *J. Organomet. Chem.*, **515**, 233 (1996); (b) Desjardins, S. Y., Cavell, K. J., Hoare, J. L., Skelton, B. W., Sobolev, A. N., White, A. H. and Keim, W., *J. Organomet. Chem.*, **554**, 163 (1997).

17. (a) Wilke, G., *Angew. Chem., Int. Ed. Engl.*, **27**, 185 (1988); (b) Jolly, P. W. and Wilke, G., *The Organic Chemistry of Nickel*, Academic Press, New York, 1975, Vol. 2, Chapt. 1.

18. (a) Lippert, F., Hoehn, A. and Schauss, E., *PCT Int. Appl.*, WO 96 37522 (1996); (b) Lippert, F., Hoehn, A. and Schauss, E., *PCT Int. Appl.*, WO 96 37523 (1996); (c) Riehl, M. E., Girolami, G. S. and Wilson, S. R., in *Book of Abstracts, 203rd ACS National Meeting*, INOR 635, American Chemical Society, San Francisco, CA, April 1992.

19. Keim, W., Appel, R., Storeck, A., Kruger, C. and Goddard, R., *Angew. Chem., Int. Ed. Engl.*, **20**, 116 (1981).

20. (a) Schubbe, R., Angermund, K., Fink, G. and Goddard, R., *Makromol. Phys.*, **196**, 467 (1995); (b) Mohring, V. M. and Fink, G., *Angew. Chem., Int. Ed. Engl.*, **24**, 1001 (1985); (c) Fink, G., in Fontanille, M. and Guyot, A. (eds), *Recent Advances in Mechanistic and*

Synthetic Aspects of Polymerization, Reidel, Dordrecht, 1987, p. 515. (d) Fink, G., Mohring, V. M., Heinrichs, A., Denger, C., Schubbe, R. H. and Muhlenbrock, P. H., in Salamone, J. C. (Ed.), *Polymeric Materials Encyclopedia*, CRC Press, New York, 1996, vol. 6, p. 4720.
21. Deming, T. J. and Novak, B. M., *Macromolecules*, **26**, 7089 (1993).
22. (a) Ascenso, J. R., Dias, A. R., Gomes, P. T., Romao, C. C., Tkatchenko, I., Revillon, A. and Pham, Q.-T., *Macromolecules*, **29**, 4172 (1996); (b) Ascenso, J. R., Dias, A. R., Gomes, P. T., Romao, C. C., Pham, Q.-T., Neibecker, D. and Tkatchenko, I., *Macromolecules*, **22**, 998 (1989); (c) Ascenso, J. R., Carrondo, R. A. A. F. de C. T., Dias, A. R., Gomes, P. T., Piedade, M. F. M. and Romao, C. C., *Polyhedron*, **8**, 2449 (1989); (d) Ascenso, J., Dias, A. R., Gomes, P. T., Romao, C. C., Neibecker, D., Tkatchenko, I. and Revillon, A., *Makromol. Chem.*, **190**, 2773 (1989).
23. (a) Crossetti, G. L., Bormioli, C., Ripa, A., Giarrusso, A. and Porri, L., *Macromol. Rapid Commun.*, **18**, 801 (1997); (b) Longo, P., Grassi, A., Oliva, L. and Ammendola, P., *Makromol. Chem.*, **191**, 237 (1990).
24. (a) Goodall, B. L., Barnes, D. A., Benedikt, G. M., McIntosh, L. H. and Rhodes, L. F., *Polym. Mater. Sci. Eng.*, **76**, 56 (1997); (b) Goodall, B. L., Benedikt, G. M., McIntosh, L. H. and Barnes, D. A., *US Pat.*, 5 468 819 (1995); (c) Goodall, B. L., Benedikt, G. M., McIntosh, L. H. and Barnes, D. A., *US Pat.*, 5 569 730 (1996); (d) Goodall, B. L., Benedikt, G. M., McIntosh, L. H., Barnes, D. A. and Rhodes, L. F., *PCT Int. Appl.*, WO 95/14048 (1995).
25. (a) Mehler, C. and Risse, W., *Macromolecules*, **25**, 4226 (1992); (b) Melia, J., Connor, E., Rush, S., Breunig, S., Mehler, C. and Risse, W., *Macromol. Symp.*, **89**, 433 (1995); (c) Mathew, J. P., Reinmuth, A., Melia, J., Swords, N. and Risse, W., *Macromolecules*, **29**, 2755 (1996); (d) Seehof, N., Mehler, C., Breunig, S. and Risse, W., *J. Mol. Catal.*, **76**, 219 (1992); (e) Reinmuth, A., Mathew, J. P., Melia, J. and Risse, W., *Macromol. Rapid Commun.*, **17**, 173 (1996); (f) Mehler, C. and Risse, W., *Makromol. Chem., Rapid Commun.*, **12**, 255 (1991).
26. (a) Safir, A. L. and Novak, B. M., *Macromolecules*, **26**, 4072 (1993); (b) Safir, A. L. and Novak, B. M., *Macromolecules*, **28**, 5396 (1995).
27. Haselwander, T. F. A., Heitz, W. and Maskos, M., *Macromol. Rapid Commun.*, **18**, 689 (1997).
28. Sen, A. and Lai, T.-W., *Organometallics*, **1**, 415 (1982).
29. (a) Drent, E. and Budzelaar, P. H. M., *Chem. Rev.*, **96**, 663 (1996); (b) Sen, A., *Acc. Chem. Res.*, **26**, 303 (1993); (c) Rix, F. C., Brookhart, M. and White, P. S., *J. Am. Chem. Soc.*, **118**, 4746 (1996).
30. Moore, J. S., in Abel, E. W., Stone, F. G. A. and Wilkinson, G. (Eds); Hegedus, L. (Vol. Ed.), *Comprehensive Organometallic Chemistry II*, Pergamon Press, Oxford, 1995, Vol. 12, p. 1209.
31. (a) Horn, S., in Quirk, R. P. (Ed.), *Transition Metal Catalyzed Polymerizations*, Harwood, New York, 1983, Vol. 4, p. 257; (b) Hadjiandreou, P., Julemont, M. and Teyssie, P., *Macromolecules*, **17**, 2455 (1984); (c) Gin, D. L., Conticello, V. P. and Grubbs, R. H., *J. Am. Chem. Soc.*, **114**, 3167 (1992).
32. Pardy, R. B. A. and Tkatchenko, I., *J. Chem. Soc., Chem. Common.*, 49 (1981).
33. (a) Rush, S., Reinmuth, S. and Risse, W., *Macromolecules*, **30**, 7375 (1997); (b) Rush, S., Reinmuth, A., Risse, W., O'Brien, J., Ferro, D. R. and Tritto, I., *J. Am. Chem. Soc.*, **118**, 12230 (1996).
34. Hauptman, E., Sabo-Etienne, S., White, P. S., Brookhart, M., Garner, J. M., Fagan, P. J. and Calabrese, J. C., *J. Am. Chem. Soc.*, **116**, 8038 (1994).
35. Johnson, L. K., Killian, C. M. and Brookhart, M., *J. Am. Chem. Soc.*, **117**, 6414 (1995).

36. Killian, C. M., Tempel, D. J., Johnson, L. K. and Brookhart, M., *J. Am. Chem. Soc.*, **118**, 11664 (1996).
37. Killian, C. M., PhD Thesis, University of North Carolina–Chapel Hill, Chapel Hill, NC (1996).
38. McLain, S. J., McCord, E. F., Johnson, L. K., Ittel, S. D., Nelson, L. T. J., Arthur, S. D., Halfhill, M. J., Teasley, M. F., Tempel, D. J., Killian, C. and Brookhart, M. S., *Polym. Prepr., Am. Chem. Soc. Div. Polym. Chem.*, **38**, 772 (1997).
39. Nelson, L. T. J., McCord, E. F., Johnson, L. K., McLain, S. J., Ittel, S. D., Killian, C. M. and Brookhart, M., *Polym. Prepr., Am. Chem. Soc. Div. Polym. Chem.*, **38**, 133 (1997).
40. McLain, S. J., Feldman, J., McCord, E. F., Gardner, K. H., Teasley, M. F., Coughlin, E. B., Sweetman, K. J., Johnson, L. K. and Brookhart, M. *Polym. Mater. Sci. Eng.*, **76**, 20 (1997).
41. McLain, S. J., Feldman, J., McCord, E. F., Gardner, K. H., Teasley, M. F., Coughlin, E. B., Sweetman, K. J., Johnson, L. K. and Brookhart, M., *Macromolecules*, in press.
42. Johnson, L. K., Mecking, S., and Brookhart, M., *J. Am. Chem. Soc.*, **118**, 267 (1996).
43. Mecking, S., Johnson, L. K., Wang, L. and Brookhart, M., *J. Am. Chem. Soc.*, in press.
44. McLain, S. J., McCord, E. F., Arthur, S. D., Hauptman, E., Feldman, J., Nugent, W. A., Johnson, L. K., Mecking, S. and Brookhart, M., *Polym. Prepr., Am. Chem. Soc. Div. Polym. Chem.*, **76**, 246 (1997).
45. tom Dieck, H., Svoboda, M. and Grieser, T., *Z. Naturforsch., Teil B*, **36**, 832 (1981).
46. van Koten, G. and Vrieze, K., *Adv. Organomet. Chem.*, **21**, 151 (1982).
47. van Asselt, R., Elsevier, C. J., Smeets, W. J. J., Spek, A. L. and Benedix, R., *Recl. Trav. Chim. Pays-Bas*, **113**, 88 (1994).
48. (a) Kaminsky, W., *Nachr. Chem. Tech. Lab.*, **29**, 373 (1981); (b) Kaminsky, W., Miri, M., Sinn, H. and Woldt, R., *Macromol. Chem. Rapid Commun.*, **4**, 417 (1983).
49. (a) Beck W. and Sunkel, K., *Chem. Rev.*, **88**, 1405 (1988); (b) Strauss, S. H., *Chem. Rev.*, **93**, 927 (1993).
50. Brookhart, M., Grant, B. and Volpe, A. F., Jr, *Organometallics*, **11**, 3920 (1992).
51. The function of MAO in this system is believed to be similar to that reported for the reactivity of MAO with Group 4 metallocene catalysts: (a) Ref. 48; (b) Yoshida, T., Koga, N. and Morokuma, K., *Organometallics*, **14**, 746 (1995); (c) Sishta, C., Hathorn, R. M. and Marks, T. J., *J. Am. Chem. Soc.*, **114**, 1112 (1992); (d) Woo, T. K., Fan, L. and Ziegler, T., *Organometallics*, **13**, 2252 (1994).
52. Scientists at Union Carbide have described silica-supported chromium catalysts that polymerize ethylene to polyethylene with as many as 12 methyl branches per 1000 carbon atoms. The small amount of branching observed in the ethylene homopolymers prepared by these supported chromocene catalysts was attributed to a chain isomerization process: (a) Karol, F. J., Karapinka, G. L., Wu, C., Dow, A. W., Johnson, R. N. and Carrick, W. L., *J. Polym. Sci., Part A-1*, **10**, 2621 (1972); (b) Karol, F. J. and Johnson, R. N., *J. Polym. Sci., Polym. Chem. Ed.*, **13**, 1607 (1975).
53. Brookhart, M. and White, P. S., unpublished results.
54. Deng, L., Woo, T. K., Cavallo, L., Margl, P. M. and Ziegler, T., *J. Am. Chem. Soc.*, **119**, 6177 (1997).
55. Deng, L., Margl, P. and Ziegler, T., *J. Am. Chem. Soc.*, **119**, 1094 (1997).
56. Musaev, D. G., Froese, R. D. J., Svensson, M. and Morokuma, K., *J. Am. Chem. Soc.*, **119**, 367 (1997).
57. Siegbahn, P. E. M., Stromberg, S. and Zetterberg, K., *Organometallics*, **15**, 5542 (1996).
58. Stromberg, S., Zetterberg, K. and Siegbahn, P. E. M., *J. Chem. Soc., Dalton Trans.*, 4147 (1997).
59. Killian, C. M., Johnson, L. K. and Brookhart, M., *Organometallics*, **16**, 2005 (1997).

60. Measurement from DSC measurements on high molecular weight samples: Resconi, L. and Silvestri, R., in Salamone, J. C. (Ed.), *Polymeric Materials Encyclopedia*, CRC Press, New York, 1996, Vol. 9, p. 6609.
61. Pellecchia, C. and Zambelli, A., *Macromol. Rapid Commun.*, **17**, 333 (1996).
62. Feldman, J., McLain, S. J., Parthasarathy, A., Marshall, W. J., Calabrese, J. C. and Arthur, S. D., *Organometallics*, **16**, 1514 (1997).
63. Johnson, L. K., Feldman, J., Kreutzer, K. A., McLain, S. J., Bennett, A. M. A., Coughlin, E. B., Donald, D. S., Nelson, L. T. J., Parthasarathy, A., Shen, X., Tam, W., and Wang, Y., *PCT Int. Appl.*, WO 97 02298 (1997).

PART II

Ethylene Polymerization

12

Structure, Properties and Preparation of Polyolefins Produced by Single-site Catalyst Technology

S. P. CHUM, C. I. KAO AND G. W. KNIGHT†
Dow Chemical Co., Freeport, TX, USA

1 INTRODUCTION

Two major classes of single-site catalyst (SSC) technology were recently developed for the polymerization of ethylene and α-olefins. These two catalyst systems are the metallocene catalyst and the constrained geometry catalyst systems. The use of these catalyst technologies has allowed a very rapid development of olefin copolymers with a wide range of structures and related properties. This technology has initiated a major revolution for the polyolefin industry [1–4]. Several families of SSC technology-based polyolefin copolymers have been commercialized in the 1990s. These include polyolefin elastomers (e.g., ENGAGE from DuPont Dow Elastomers), polyolefin plastomers (e.g., AFFINITY from Dow Chemical; EXACT from Exxon Chemical), EPDM (NORDEL IP from DuPont Dow Elastomers), enhanced polyethylene (ELITE from Dow Chemical), gas-phase LLDPE (EXCEED from Exxon Chemical) and polypropylenes (ACHIEVE from Exxon Chemical). In addition to these commercial activities, several other single-site catalyst-related technologies

† Retired. Present address: 400 Cedar Creek Drive, Edmond, OK 73034, USA.

Metallocene-based Polyolefins   Edited by J. Scheirs and W. Kaminsky
© 2000 John Wiley & Sons Ltd

that allow the copolymerization of α-olefins with polar comonomers are also under industrial and academic development [5,6].

This chapter will focus on the solid-state structure and properties of homogeneous polyolefin copolymers made by copolymerization of ethylene and α-olefins (e.g. 1-butene, 1-hexene, 1-octene) using SSC technology. Unlike conventional heterogeneous linear low density polyethylenes (LLDPE) made by copolymerization of ethylene and α-olefins with Ziegler–Natta (Z–N) catalysts, homogeneous ethylene–α-olefin copolymers produced by SSC technology have narrow composition distributions (narrow molecular weight and comonomer distribution) and behave much more like ideal polymers. The polymerization kinetics and the resulting polymer and copolymer structures can be modeled. This significantly advances the fundamental understanding of the structure–property relationships. This capability allows polymer and materials scientists to use a molecular architecture approach to develop new products with exceptional speed. This chapter will cover the preparation, structure and property relationships of ethylene–α-olefin copolymers produced by SSC technology.

2 PREPARATION OF ETHYLENE–α-OLEFIN COPOLYMERS BY SSC TECHNOLOGY

Two major families of high-efficiency SSC, an unbridged biscyclopentadienyl single-site metallocene catalyst (MTC) (also known as Kaminsky catalyst) and a half sandwich, constrained geometry monocyclopentadienyl single-site catalyst [known as constrained geometry catalyst (CGC) under the trademark INSITE by Dow Chemical]. These two catalyst systems are illustrated in Figure 1. The Zr in the

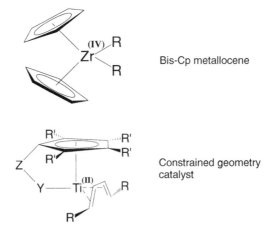

Figure 1 Structure of the unbridged bis-Cp metallocene catalyst and the constrained geometry catalyst

Zr-based MTC used by many SSC polymer producers has an oxidation state of +4 whereas the Ti in the Ti-based CGC system used by Dow Chemical has an oxidation state of +2.

SSC technology polyolefin copolymers can be produced by high-pressure, solution, gas-phase and slurry polymerization processes. Typical process conditions for making polyolefin copolymers in these processes are the following:

(i) *High-pressure process:* Two types of commercial reactors, stirred autoclave or a tube reactor, are used for the polymerization of ethylene and ethylene–α-olefin copolymers at high pressure. Ethylene and an α-olefin comonomer are usually compressed to at least 10 000 psi when fed into the reactor. The polymerization temperature usually exceeds 100 °C. Figure 2 is a schematic diagram of a typical high-pressure process.

(ii) *Solution process:* Stirred reactors are usually used for the polymerization of ethylene and ethylene–α-olefin copolymers in a solution phase. In most cases, C_6-C_8 hydrocarbons are used as the solvent. The reactors are operated at about 500 psi pressure and the polymerization is carried out at >60 °C.

(iii) *Gas-phase process:* In the gas-phase process, ethylene and an α-olefin comonomer (1-butene or 1-hexene) are polymerized in the solid state in a fluidized bed reactor into a powder form. The polymer powder is then converted into pellet

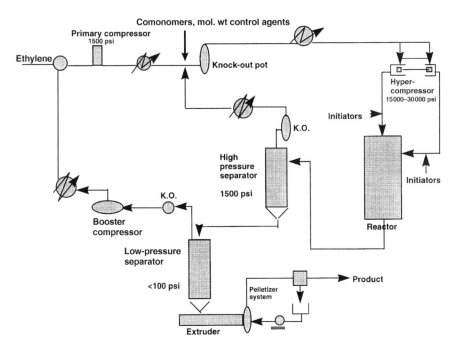

Figure 2   Schematic diagram of the high-pressure polyethylene polymerization process

form by an extrusion process. The reactor is usually set at >300 psi pressure and the polymerization process is generally carried out at $<90\,°C$.

(iv) *Slurry process:* In the slurry process, polymers are made in stirred reactors with an organic carrier liquid (C_4-C_6 hydrocarbons). The polymerization process temperature is generally $<90\,°C$ with a reactor pressure of <300 psi. Products made in this process are in powder form and can be converted into pellets using extrusion processes.

3 MOLECULAR STRUCTURE OF POLYETHYLENE COPOLYMERS MADE BY SINGLE-SITE CATALYST TECHNOLOGY

3.1 MOLECULAR WEIGHT DISTRIBUTION (MWD)

Homogeneous ethylene–α-olefin copolymers prepared by single-site catalyst technology tend to exhibit a narrower composition distribution than copolymers prepared by conventional Zeigler–Natta catalyst. Figure 3 compares gel permeation chromatography (GPC) traces of a conventional, heterogeneous ethylene–octene copolymer vs a homogeneous copolymer. Polymers with a narrow MWD in general have increased toughness and less 'solvent extractables'. On the other hand, the narrow MWD of a linear homogeneous copolymer also results in poor melt processability (low melt strength, high extruded back-pressure, high energy consumption during extrusion, etc.).

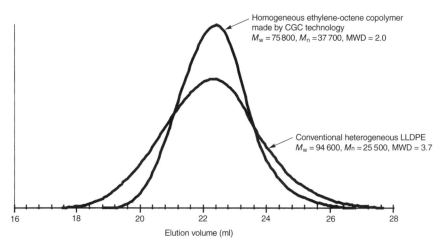

Figure 3 GPC molecular weight distribution comparison of two ethylene–octene copolymers having a density of $0.920\,g/cm^3$

3.2 DENSITY

The comonomer content in the copolymer has a profound effect on the properties of the polymer, including crystallinity, density and thermal and mechanical properties. The details of these structure–property relationships will be addressed later in this chapter. For reference purposes, the effect of comonomer content on the density and the crystallinity of ethylene–octene copolymers is illustrated in Figure 4.

3.3 COMONOMER DISTRIBUTION

Because of the significant effect of the α-olefin comonomer on the properties of the polymer, it is critical to understand and to measure the comonomer distribution in the polymer backbone. Intermolecular comonomer distributions can be measured by the temperature rising elution fractionation (TREF) technique [7]. As illustrated in Figure 5, the comonomer distribution for a homogeneous ethylene–octene copolymer prepared with the CGC catalyst is much narrower than that of a conventional LLDPE, which is a mixture of all different kinds of polymer molecules that have different levels of α-olefin comonomer incorporated in the polymer backbone.

The narrow TREF curve for the homogeneous copolymer signifies that the number of comonomer units per unit chain length between the homogeneous copolymer molecules is very similar, whereas the heterogeneous ones are not. However, although the intermolecular comonomer distribution for the homogeneous

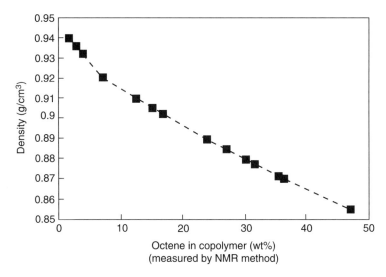

Figure 4 Density of homogeneous ethylene–octene copolymer vs octene content

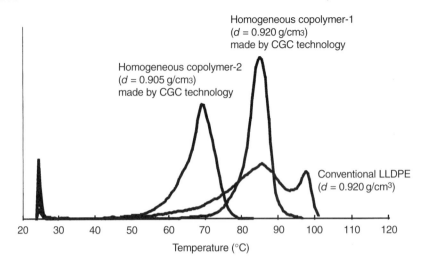

Figure 5 Comonomer distribution of ethylene–octene copolymers measured by TREF

polymer is very narrow, this does not mean that the intramolecular comonomer distribution is uniform. The narrow intermolecular comonomer distribution for the homogeneous polymer arises from the single-site nature of the MTC and CGC. The uniformity of the intramolecular comonomer distribution, however, is dictated by the reactivity ratio of the monomer and the comonomer (r_1/r_2) to the catalyst. From the polymer property point of view, the intramolecular comonomer distribution in homogeneous copolymers made by SSC technology can have a significant effect on the solid-state structure and properties of the polymer (e.g. thermal properties, dynamic mechanical properties). This will be discussed later in this chapter.

To address the issue of the effect for intramolecular comonomer distribution on polymer properties of a homogeneous polymer, the structural characteristics for CGC ethylene–octene copolymers were modeled using the Monte Carlo simulation [8]. Figure 6 illustrates the intramolecular comonomer distribution for three CGC copolymers, from 0.87 to $0.92\,\text{g/cm}^3$ density, in terms of ethylene block length between the hexyl branches within one polymer molecule. As illustrated in this figure, the intramolecular comonomer distribution for the higher density polymer $(0.92\,\text{g/cm}^3)$, in terms of ethylene block length distribution between the short branches that are formed from the α-olefin comonomers, is much broader than that of the lower density copolymers. This is due to the fact that the lower density copolymers have more comonomer units in the polymer backbone and, therefore, the short branches formed from the comonomer are more congested along the backbone. This results in a shorter ethylene block length between the short branches, and the

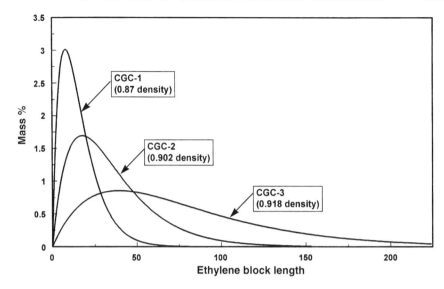

Figure 6 Model prediction for intramolecular comonomer distribution of three CGC ethylene–octene copolymers: ethylene block length distribution between comonomers

distribution is also narrower. The effect on the intramolecular comonomer distribution on polymer properties will be further discussed later in this chapter.

3.4 LONG-CHAIN BRANCHING

Homogeneous ethylene homopolymers and ethylene–α-olefin copolymers made by the CGC technology have one more, unique molecular structural feature not generally found in MTC technology polymers, viz. long-chain branching (LCB). Homopolymers and copolymers made by the CGC technology under certain process conditions contain a certain amount of long-chain branching. To summarize this, the molecular structure of three ethylene–α-olefin copolymers, conventional LLDPE made with a Z–N catalyst, homogeneous copolymers made by MTC and substantially linear homogeneous copolymers made by CGC, are illustrated schematically in Figure 7.

Polymer with LCBs made by CGC technology has many unique rheological properties [9,10]. A few of the most significant features for the CGC technology polymer are its melt fracture resistance and the control of shear thinning behavior, as illustrated in Figures 8 and 9, respectively.

Conventional LLDPE   Homogeneous copolymer   LCB substantially
                        by MTC (e.g. EXACT)   linear homogeneous
                                                   copolymer by CGC
                                                   (e.g. AFFINITY)

Figure 7   Molecular structure comparison of three different ethylene–α-olefin copolymers. EXACT is a trademark of Exxon Chemical and AFFINITY is a trademark of Dow Chemical

4 SOLID-STATE STRUCTURE AND MORPHOLOGY OF SSC TECHNOLOGY ETHYLENE HOMOPOLYMERS AND ETHYLENE–α-OLEFIN COPOLYMERS

Ethylene homopolymer and ethylene–α-olefin copolymers are semi-crystalline polymers. Three major factors that affect the crystal morphology and crystallinity of ethylene polymers in the solid state are (i) amount and size of α-olefin in the polymer, (ii) the molecular weight of the polymer and (iii) the crystallization conditions (crystallization temperature under isothermal conditions and/or cooling rate under non-isothermal crystallization conditions). A recent review on polymer crystals was written by Phillips [11].

The most generally accepted crystal morphology for ethylene polymers is lamellae formed from a folded chain conformation [12–14]. A computer model using the force field approach has been used to model the chain folding crystallization process for a model PE [15]. The results of the model are summarized in Figures 10 and 11, which describe the folded model of a polyethylene homopolymer and an ethylene–octene copolymer, respectively. As one can see in Figure 11, the existence of a short-chain branch from the octene comonomer significantly interrupts the folding process. Especially for copolymers made with higher molecular weight α-olefin comonomers (e.g. hexene, octene), the short branch in the copolymer backbone cannot incorporate into the crystal core and, therefore, is excluded from crystallizing with the rest of the chain [16,17]. For this reason, a major part of the regular chain close to the short branch was also not able to incorporate into the crystal core, as illustrated in Figure 11. This resulted in a significant lowering of the overall crystallinity and the crystal size of the polymer. This model is consistent with the experimental observations. From the polymer property point of view, the smaller

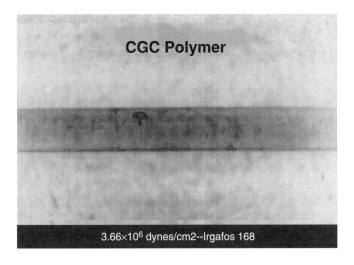

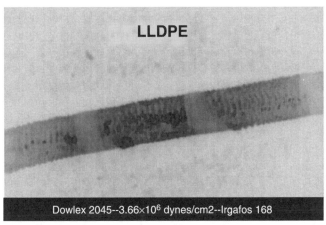

Figure 8 Extrudates of CGC polymer and LLDPE at $3.66 \times 10^6 \, \mathrm{dyn \, cm^{-2}}$ shear stress

and more uniform size distribution crystals significantly improve the optical properties and heat seal performance of the polymer.

For very low density SSC technology ethylene–α-olefin copolymers (which contain a high percentage of comonomer), the crystal morphology can be very different from the conventional lamellar model. For example, consider the morphological model of a 60 : 40 wt% ethylene–octene copolymer made by the Dow CGC technology. At this level of comonomer content the polymer exhibits approximately 10% crystallinity as measured by DSC and a density of approximately $0.87 \, \mathrm{g \, cm^{-3}}$. The intramolecular comonomer sequence distribution of the hexyl branch from the

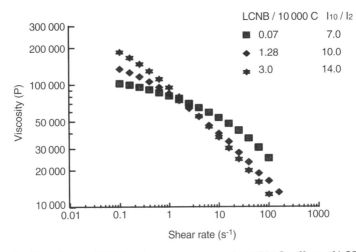

Figure 9 Rheology of CGC polymers measured at 190 °C: effect of LCB on shear thinning behavior of three 1 melt index (MI) polymers with various amounts of LCB. LCB 10 000 carbon atoms was estimated using a kinetic model

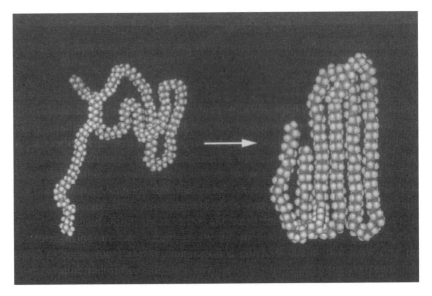

Figure 10 Lamellar crystal formed from chain folding of a PE molecule by molecular dynamic simulation

octene comonomer, calculated from the reactive ratio kinetic model (Figure 6) illustrates that more than 90% of the octene units are less than 50 ethylene units apart. The regular ethylene block length between the octene units is, therefore, much less than the minimum length to form one folded unit to form the thinnest possible

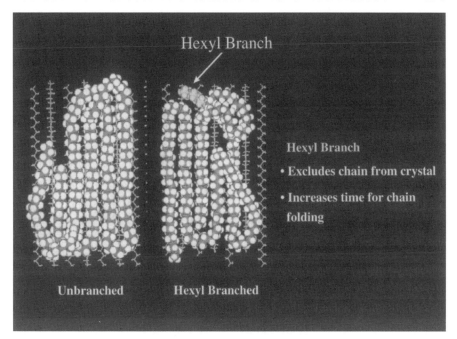

Figure 11 Lamellar crystal formed from an ethylene homopolymer and an ethylene–octene copolymer by molecular dynamic simulation

lamella (which is approximately $30\,\text{Å}$ thick). For a chain with this molecular structure, it is expected that the polymer has to crystallize in a crystalline form different from the conventional chain folded lamellar model.

A 'fringed micelle' model, as illustrated in Figure 12, had been proposed as a possible crystalline morphology for polyethylene by Hermann and Gerngoss [18] in 1930. Although this model was highly recommended by Flory [19], the actual structure was never observed in conventional ethylene–α-olefin copolymers in isolated form, even with advanced transmission electron microscopy (TEM), owing to the heterogeneity of the polymers. This is because heterogeneous copolymers always contain some ethylene chains that have very little comonomer and, therefore, can crystallize into the lamellar form via chain folding. However, because of the narrow composition distribution, fringed micelle morphology in a very low density homogeneous polymer can be clearly observed in TEM. Figure 13 is a TEM photo-micrograph of the crystal morphology of a higher density $(0.920\,\text{g/cm}^3)$, CGC technology ethylene–octene copolymer. At this density and octene level, the copolymer clearly shows a lamella morphology. Figure 14 is a TEM photo-micrograph of a very low density copolymer $(0.87\,\text{g/cm}^3)$. It clearly shows that the major crystalline structure of this very low density, low crystallinity copolymer exhibits 'spot-like' fringed micelle crystals.

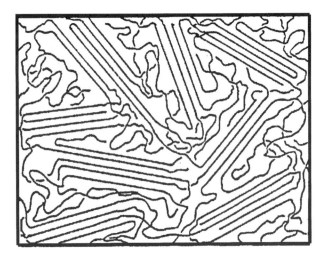

Figure 12   Fringed micelle model for very low density homogeneous ethylene–α-olefin copolymers

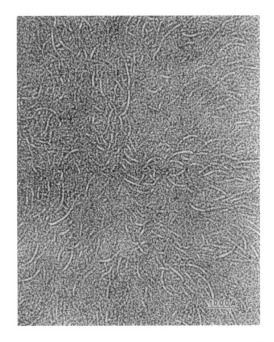

Figure 13   Transmission electron micrograph of a 0.920 g/cm³ ethylene–octene copolymer made by CGC technology

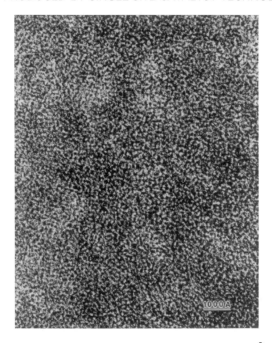

Figure 14 Transmission electron micrograph of a $0.87 \, g/cm^3$ ethylene–octene copolymer made by CGC technology

The mechanical properties of polymers with these two different types of crystal structures are expected to be different. Polymers with the lamellar-type crystal morphology are expected to have higher moduli and undergo yielding when deformed. Polymers with the fringed micelle crystals, however, are expected to be more like elastomers and do not have well defined yielding behavior. With these two different types of crystal structures, the SSC technology homogeneous ethylene–α-olefin copolymer can, therefore, be classified into four major domains [20] as illustrated in Figure 15. The thermal properties and the mechanical properties of these four major domains will be described in the following sections.

5 THERMAL AND DYNAMIC MECHANICAL PROPERTIES

5.1 MELTING BEHAVIOR OF SSC POLYMERS

The most popular method to study thermal properties (melting and crystallization) of semi-crystalline polymers is by differential scanning calorimetry (DSC). Much structural information can be uncovered by a careful interpretation of DSC thermo-grams generated under different conditions (cooling and heating at various rates, multiple cooling and heating conditions, etc.).

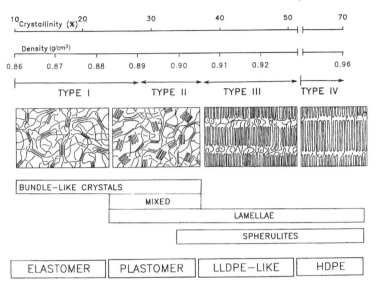

Figure 15 Classification of SSC technology ethylene–α-olefin copolymer based on crystal morphology

Owing to the narrow intermolecular comonomer distribution, SSC copolymers usually have a much narrower melting peak than their heterogeneous counterparts produced by multiple site Z–N catalysts. Figure 16 shows the thermal properties of two ethylene–octene copolymers, both at similar density, one made by conventional multiple site Z–N catalyst and the other by SSC catalyst technology.

The heterogeneous LLDPE copolymer always has a peak melting-point at around 120–130 °C at a broad density range (*ca* 0.89–0.95 g/cm^3). This is because the heterogeneous copolymer is a mixture and always contains some polymer chains that contain just a small amount of α-olefin. These chains can, therefore, crystallize to a rather large size crystal that has a melting-point in that temperature range. The peak melting-point of the homogeneous copolymer, however, drops accordingly as the density/crystallinity of the polymer is lowered. This characteristic has a very significant commercial value for heat-seal food packaging applications and is the subject of discussion in another chapter. Figure 17 summarizes the peak melting temperature of many heterogeneous and homogeneous copolymers across a broad range of polymer densities.

Many detailed studies of the thermal properties of SSC technology homogeneous ethylene-α–olefin copolymers have been performed by many different institutes [20,21]. The first publication showing the effect of increasing comonomer content on melting-point in homogeneous copolymers was a US Patent by Elston of DuPont in 1972 [22].

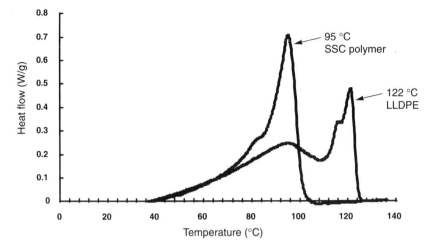

Figure 16 DSC comparison of two SSC and LLDPE ethylene–octene copolymers at 0.920 g/cm^3 density

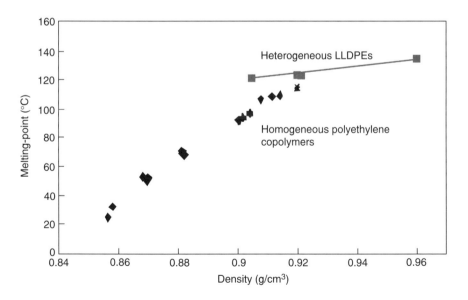

Figure 17 Peak melting temperatures of heterogeneous and homogeneous ethylene–octene copolymers vs polymer density

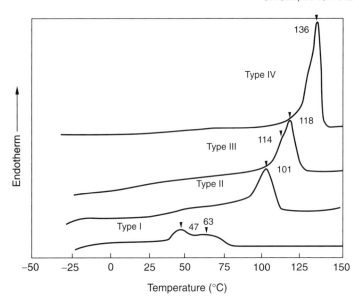

Figure 18 DSC thermograms of Type I, II, III and IV ethylene–octene homogeneous copolymers

Figure 18 shows the melting thermograms of four ethylene–octene copolymers from 0.87 to $0.96\,\mathrm{g/cm^3}$ density. The density range of these four samples represents all four types of polymer as discussed in the previous section (Type I, 0.870; Type II, 0.902; Type III, 0.918; and Type IV, $0.960\,\mathrm{g/cm^3}$). The samples were slowly cooled from the melt at $1\,°\mathrm{C/min}$, then the thermograms were obtained with a heating rate of $10\,°\mathrm{C/min}$. The peak melting temperature of the Type IV, $0.960\,\mathrm{g/cm^3}$ polymer was $138\,°\mathrm{C}$. With increasing comonomer content, the melting endotherm of Type I to Type III copolymers broadened and shifted to lower temperatures. However, the shape of the melting peaks of all the homogeneous copolymers still represents a single peak or a single peak with a shoulder, very different from that of the heterogeneous LLDPEs, as illustrated in Figure 16. The melting peak temperature for the Type III, $0.918\,\mathrm{g/cm^3}$ copolymer was $118\,°\mathrm{C}$ with a small shoulder at around $114\,°\mathrm{C}$; for the Type II, $0.902\,\mathrm{g/cm^3}$ copolymer it was $101\,°\mathrm{C}$. The Type I, $0.870\,\mathrm{g/cm^3}$ copolymer had a very broad melting peak at peak temperatures of 47 and $63\,°\mathrm{C}$. The broad melting range of the copolymers can be attributed to the intramolecular comonomer distribution. For example, the longer ethylene runs may form a larger lamella crystal with a higher melting-point and the shorter runs may form a small lamella crystal or even a fringed micelle crystal with a lower melting temperature.

5.2 DYNAMIC MECHANICAL PROPERTIES

Dynamic mechanical properties of these four types of homogeneous copolymers are illustrated in Figure 19. The γ-transition peaks of all four copolymers appear at

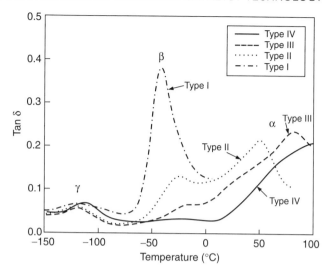

Figure 19 Dynamic mechanical properties of Type I to Type IV homogeneous ethylene–octene copolymers. Tan δ peaks were obtained at 1 Hz, cooled at 1 °C/min

around $-120\,°C$. In some publications the γ peak was designated as the true glass transition temperature of polyethylene. The β-transition peak for the Type I, $0.870\,g/cm^3$ copolymer appears at around $-40\,°C$, for the Type II, $0.902\,g/cm^{-3}$ copolymer it is about $-25\,°C$ and for the Type III, $0.918\,g/cm^3$ copolymer it is about $-20\,°C$. The Type IV, $0.960\,g/cm^3$ polymer has a broad β peak, also at around $-20\,°C$. In most cases the β-transition temperature can be related to the low temperature performance of ethylene-based polymers and copolymers. For example, below the β-transition temperature the polymer usually becomes brittle. Therefore, most PE users usually report the β-transition temperature as the practical glass transition temperature. The mechanical properties and deformation behavior of these four types of homogeneous polymers will be discussed further in the next section.

6 DEFORMATION BEHAVIOR AND MECHANICAL PROPERTIES OF HOMOGENEOUS ETHYLENE–α-OLEFIN COPOLYMERS MADE WITH SSC TECHNOLOGY

6.1 TENSILE PROPERTIES

The stress–strain curves of four materials representing Types I to IV are shown in Figure 20. The α-olefin comonomer content in the polymer backbone profoundly affected the response to deformation, as is evident by the broad spectrum of tensile properties. For copolymers with a low level of comonomer (higher density and

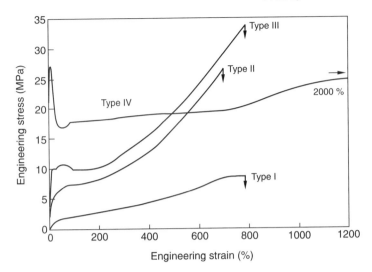

Figure 20   Engineering stress–strain properties of homogeneous ethylene–octene copolymers, measured at a strain rate of $0.450\,min^{-1}$ at room temperature

crystallinity, e.g. Types IV and III), the deformation had characteristics common to many semi-crystalline thermoplastics with localized yielding and cold drawing. For copolymers with higher levels of comonomer (lower density and crystallinity, e.g. Types II and I), the moduli were low and the deformation was essentially plastomeric and elastomeric. For this reason, Types II and I are being referred to 'polyolefin plastomers' (POP) and 'polyolefin elastomers' (POE), respectively.

Differences in yield behavior of the four types of materials are shown in the photographs in Figure 21. These pictures were taken at 150% engineering strain. As one can see, these polymers exhibit deformation behavior from necking (Type IV) to elastomeric, uniform deformation (Type I). The effect of increasing comonomer content on the tensile deformation behavior and the correlation between the structural classification and the large-scale deformation behavior have been studied and discussed in detail in many publications [23–25] and, therefore, will not be further discussed in this chapter.

6.2 MODULUS AND YIELD STRENGTH

Modulus and yield strength are important properties for ethylene-based polymers and copolymers for many commercial applications, such as packaging film and injection molded containers. The slope of the stress–strain curve at very low strain (less than 3–5% strain) is measured as the modulus of the polymer. Secant modulus at 1 or 2% strain and Young's modulus are the most commonly used ones.

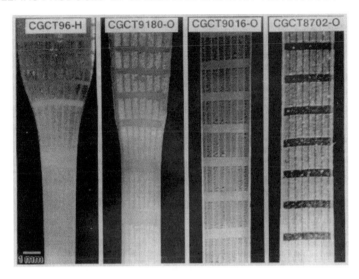

Figure 21 Photographs of profiles of deformation behavior of Type I to IV homogeneous ethylene–octene copolymers, taken at 150% engineering strain

The initial modulus is a complex characteristic of semi-crystalline polymers, including conventional PE and homogeneous ethylene–α-olefin copolymers. The moduli of these polymers in the solid state are strongly controlled by polymer crystallinity, molecular weight, polymer thermal history and fabrication conditions, which results in residual stress and orientation. Even with this complexity there are some rules of thumb which can be applied to the modulus. For example, it is generally true that the initial modulus increases with increasing density and crystallinity of the polymer. There is also a direct relationship between initial modulus and crystallite size. It has been shown [26] that two polyethylenes of the same degree of crystallinity can have very different moduli (up to 100% difference), which is related to a difference in crystallite size.

The moduli of a series of homogeneous ethylene–octene copolymers made by CGC technology, several homogeneous ethylene–butene copolymers made by MTC technology and several conventional LLDPE made with Z–N catalysts at a range of polymer densities and comonomer contents were measured for comparison purposes [27]. These polymers had similar molecular weights and the samples were prepared by compression molding under similar heating and cooling conditions to assure the same heat history on each of the samples. The data are summarized in Figure 22. It seems that, if the polymers have similar molecular weight and the polymer samples are prepared under similarly controlled conditions, the Young's moduli of these polymers are simply a function of the polymer density. However, it should be noted that it is very difficult to produce any ethylene copolymer having a density below $0.88 \, \text{g/cm}^3$ using Z–N catalysts. Therefore, data for Z–N heterogeneous LLDPE at

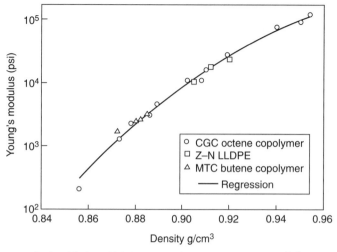

Log(modulus) = -130.0787 + 270.267(density) -134.832(density)^2, r^2 = 0.99

Figure 22   Young's modulus vs density of ethylene copolymers determined on compression molded plaques

below $0.88\,g/cm^3$ are not available for comparison purposes. Based on common knowledge of semi-crystalline polymers, the very low density heterogeneous polymer made using Z–N catalysts, if one can make it, may have a higher modulus than the homogeneous copolymer at similar densities. This is because the heterogeneous polymer, no matter how low the density is, will always have some polymer chains that have a very low level of comonomer incorporated. These polymer chains will crystallize to form larger size crystals and, therefore, may result in a higher modulus than the homogenous polymer.

6.3   ELASTIC PROPERTIES OF TYPE I POLYMERS; POLYOLEFIN ELASTOMERS (POE)

The Type I, high α-olefin content, very low density homogeneous ethylene–α-olefin copolymer has a fringed micelle crystal morphology and has a very different deformation behavior than the higher density copolymers that have lamella crystal structures. The small fringed micelle crystals dispersed in the soft, amorphous ethylene–α-olefin copolymer matrix act as tie points to anchor the amorphous chains during deformation and, therefore, result in an elastic deformation. This type of polymer is called polyolefin elastomer (POE) and is now commercially available from DuPont Dow Elastomers (ENGAGE). The load/unload property in a tensile deformation mode for a representative octene-based POE (ENGAGE EG-8100, $0.870\,g/cm^3$, 1 MI) is illustrated in Figure 23. The elastic recovery behavior of the POE sample is illustrated in Figure 24. The POE was first deformed in a tensile

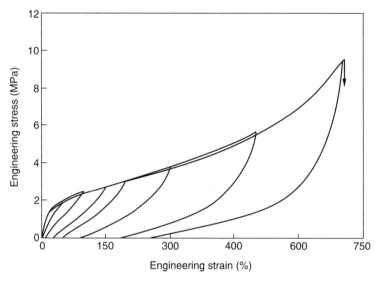

Figure 23 Load/unload behavior of a representative octene-based polyolefin elastomer (ENGAGE EG-8100, $0.870\,g/cm^3$ 1 MI). Strain rate, $2.25\,min^{-1}$; specimens were compression molded and cooled at $15\,°C\,min^{-1}$; the test was run at room temperature

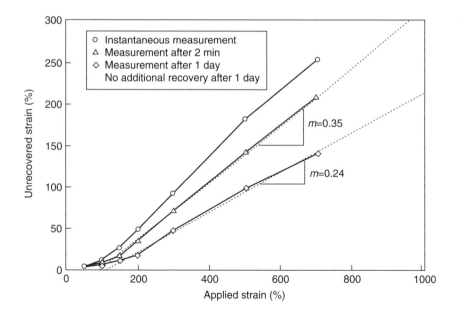

Figure 24 Elastic recovery behavior of POE ENGAGE EG-8100

deformation mode, then the stress was released, allowing the sample to recover for up to 24 h at room temperature. As one can see in Figure 24, the unrecovered strain of the POE sample, below 100% elongation, is very low. This means the sample is deforming in its elastic zone for up to 100% strain. However, the unrecovered strain became much higher beyond the 100% strain. This phenomenon can hypothetically be explained by a slip-linked theory in which the fringed micelle crystals are treated as slippage links. Beyond 100% strain, these crystals started to slip, which resulted in permanent deformation. A detailed discussion on the slip-linked model and the use of this model to predict the overall elastic properties of POE can be found in a recent paper by Baer and co-workers [25]. New applications for this class of elastic materials were developed commercially for wire and cable insulation, shoe soles, elastic fiber and films, foams etc., and are discussed in detail in another chapter.

7 TIE-MOLECULES IN ETHYLENE–α-OLEFIN COPOLYMERS MADE BY SSC TECHNOLOGY

Tie-molecules are polymer chains that link crystals together in a semi-crystalline polymer. Tie-molecules in the semi-crystalline polymer are critical for enhancing mechanical properties such as environmental stress crack resistance (ESCR), impact, tear and tensile strength. A schematic diagram of the tie-molecule structure in an ethylene–α-olefin semi-crystalline copolymer with a lamella morphology is illustrated in Figure 25. A main cause for rejection of a polyethylene chain from a crystal

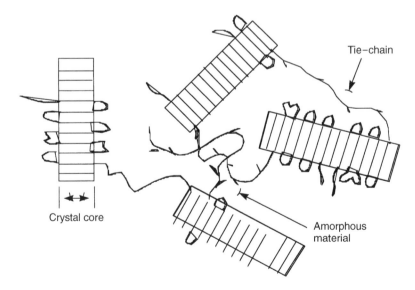

Figure 25   Schematic diagram of tie-molecule structure in an ethylene–α-olefin copolymer

is the presence of imperfections on the chain backbone, which are usually branch points formed by the α-olefin comonomer. Without these branch points the entire polymer chain can possibly be incorporated into the same lamella crystal and, thus, few tie-molecules can be formed. This will result in a polymer with very low mechanical strength.

In the past, the level of tie-molecules in a semi-crystalline polyolefin copolymer has been semi-quantitatively characterized using techniques such as infrared dichroism [28], measurement of the brittle fracture strength [29] and chain dimensions and topology [30,31]. Unlike the heterogeneous ethylene copolymer made by conventional Z–N catalyst technology, the SSC technology polymer has a homogeneous distribution of the α-olefin among the polymer chains. This allows all the polymer chains to crystallize similarly. If the copolymer has the optimized amount of α-olefin comonomer to enhance tie-molecule formation, all polymer chains can, therefore, crystallize to form tie-molecules. Therefore, the SSC technology homogeneous polyethylene copolymers, at the optimum density range, should form more tie-molecules overall than the conventional heterogeneous copolymers. This effect, plus the effect of branch size along the polymer backbone for tie-molecule formation, is modeled [31,32] and the results are illustrated in Figures 26 and 27. Figure 26 illustrates the optimum density range and the molecular weight effect on tie-molecule formation and Figure 27 illustrates the effect of branch size (size of α-olefin) on tie-molecule formation. Tear strengths of various ethylene α-olefin

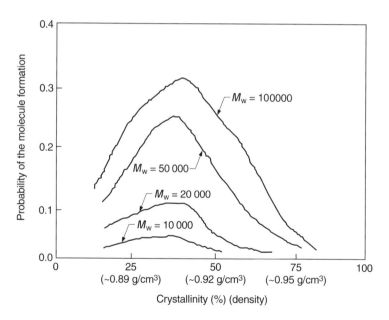

Figure 26 Probability of tie-molecule formation in an SSC technology ethylene–octene copolymer at various molecular weights [31].

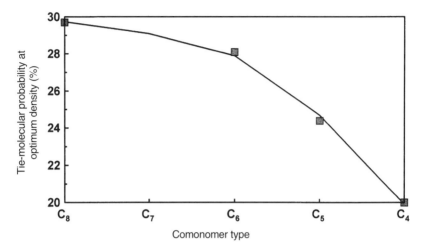

Figure 27 Probability of tie-molecule formation in SSC technology ethylene–α-olefin copolymers, with different types of α-olefin comonomer.

(propylene to octene) copolymers were measured [33] and the data are illustrated in Figure 28. As one can see, high α-olefin copolymers (octene and hexene) have much better tear strength than butene and propylene copolymers, with octene being the highest of all. The optimum tear strength for all polymers studied is found to be in the density range between 0.89 and 0.92 g/cm^{-2}. These experimental results are in good agreement with the model.

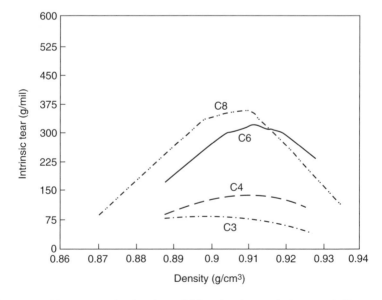

Figure 28 Tear strength of various SSC technology ethylene–α-olefin copolymers

8 CONCLUSION

The development and implementation of SCC technology have greatly expanded the product range of polyolefins, including lower density POE, POP and enhanced polyethylene. These homogeneous polymers have narrow molecular weight and composition distributions, resulting in enhanced mechanical, optical and heat-seal properties. Owing to the simple and predictable molecular structure of this class of polyolefin copolymers, building a model to predict properties for these polymers became feasible. This allows materials scientists to design products by a molecular architecture approach, using property and performance requirements as guidelines. With the advancement of the SSC technology for the production of polyolefin copolymers with well defined molecular structures, the authors expect that the polyolefin industry can take great advantage of this technology to design unique polymers and fulfil their customer needs in the future.

9 ACKNOWLEDGEMENTS

The authors express their sincere appreciation to Professor Eric Baer, Professor Anne Hiltner of Case Western Reserve University and Rajen Patel, Pradeep Jain of Dow Chemical. They provided many useful data for this chapter. The authors also want to remember the late Mr Gerry Lancaster, whose inspiration allowed them to develop and implement the SSC technology to the polyolefin industry with unprecedented speed.

10 REFERENCES

1. Swogger, K. W., in *Proceedings of the 1992 International Business Forum on Special Polyolefins, SPO '92*, 1992, p. 115.
2. Story, B. A. and Knight, G. W., in *Proceedings of Metcon '93*, Catalyst Consulting, Spring House, PA, 1993, p. 112.
3. Speed, C. S., Trudell, B. C., Mehta, A. K. and Stehling, F. C., in *Proceedings of Polyolefins VII International Conference*, 1991, p. 45.
4. Chum, S. P., Kao, C. I. and Knight, G. W., *Plast. Engi.*, **LI** (6), 21 (1995).
5. Johnson, L. K., Meeking S. and Brookhart, M., *J. Am. Chem. Soc.*, **118**, 267 (1996).
6. Johnson, L. K., Killian, C. M. and Brookhart, M., *J. Am. Chem. Soc.*, **117**, 6414 (1995).
7. Wild, L., Ryle, T., Knobeloch, D. and Peat, I., *J. Polym. Sci., Polym. Phys. Ed.*, **20**, 441 (1982).
8. Chum, S. P. and J. Ruiz, in *Proceedings of Aspen World Modeling Conference*, Boston, October 1997.
9. Lai., S. Y., Wilson, J. R., Chum, S. P., Knight, G. W. and Stevens, J. C., *US Pat.*, 5 272 236 (1993).
10. Knight, G. W. and Lai, S. Y., in *Proceedings of the Society of Plastics Engineers RETEC '93*, 1993.

11. Phillips, P. J., *Rep. Prog. Phys.*, **53**, 549 (1990).
12. Keller, A., *Philos. Mag.*, **2**, 1171 (1975).
13. Geil, P. H., *Polymer Single Crystals*, Wiley, New York, 1963.
14. Krimm. S. and Cheam, T. C., *Faraday Discuss. Chem. Soc.*, **68**, 244 (1979).
15. Chum, S. P., Knight, G. W., Ruiz, J. M and Phillips, P. J., *Macromolecules*, **27**, 656, (1994).
16. Mandlekern, L., Ergoz, E. and Fatou, J. G., *Macromolecules*, **15**, 147 (1972).
17. McFaddin, D. C., Russell, K. E. and Kelusky, E. C., *Polym. Commun.*, **29**, 258 (1988).
18. Hermann, K. and Gerngoss, O., *A. Phys. Chem.*, **10**, 371 (1930).
19. Flory, P. J., *J. Am. Chem. Soc.*, **84**, 2857 (1962).
20. Bensason, S., Minick, J., Moet, A., Chum, S., Hiltner, A. and Baer, E., *J. Polym. Sci., Part B: Polym. Phys.*, **34**, 1301 (1996).
21. Minick, J., Moet, A., Hiltner, A., Baer, E. and Chum, S., *J. Appl. Polym. Sci.*, **58**, 1371 (1995).
22. C. Elston, *US Pat.*, 3 645 992 (1972).
23. Bensason, S., Minick, J., Moet, A., Hiltner, A., Bear, E., Chum, S. and Sehanobish, K., in *Proceedings of the International SPE Annual Technical Conference ANTEC '95*, 1995, p. 2256.
24. Hwang, Y., Chum, S., Guerra, R. and Sehanobish, K., in *Proceedings of the International SPE Annual Technical Conference ANTEC '94*, 1994, p. 3414.
25. Bensason, S., Stepanov, E., Chum, S., Hiltner, A. and Baer, E., *Macromolecules*, **30**, 2436 (1997).
26. Popli, R. and Mandelkern, L., *J. Polym. Sci., Part B: Polym. Phys.*, **25**, 441 (1987).
27. Sehanobish, K., Patel, R., Croft, B., Chum, S. and Kao, C., *J. Appl. Polym. Sci.*, **51**, 887 (1994).
28. Lustiger, A. and Ishikawa, N., *J. Polym. Sci., Part B: Polym. Phys.*, **29**, 1047 (1991).
29. Brown, N. and Ward, I. M., *J. Mater. Sci.*, **18**, 1405 (1983).
30. Huang, Y. and Brown, N., *J. Polym. Sci., Part B: Polym. Phys.*, **29**, 129 (1991).
31. Hosoda, S and Uemura, A., *Polym. J.*, **24**, 939 (1992).
32. Patel, R., Sehanobish, K., Jain, P., Chum, S. and Knight, G., *J. Appl. Polym. Sci.*, **60**, 749 (1996).
33. Plumley, T., Knight, G. and Chum, S., *J. Plast. Film Sheeting*, **11**, 269 (1995).

13

New Developments in the Production of Metallocene LLDPE by High-pressure Polymerization

AKIRA AKIMOTO AND AKIHIRO YANO
Tosoh Corporation, Mie-ken, Japan

1 INTRODUCTION

The discovery of highly active metallocene catalysts by Kaminsky, Sinn and co-workers has stimulated the design of novel families of olefin homo- and copolymers [1]. The structures of metallocenes are closely linked to polymer microstructure, molecular mass and end groups. In stereo- and regioselective metallocene-catalyzed α-OLEFIN polymerization, steric control depends primarily on metallocene structure, especially metallocene symmetry. C_2 symmetric *ansa*-metallocenes produce isotactic poly(α-olefin)s, whereas C_s symmetric metallocenes produce syndiotactic poly(α-olefin)s [2].

In addition to the control of stereochemistry, control of comonomer incorporation represents a key feature of metallocene catalysts. A multiplicity of active centers is present at the catalyst surface in heterogeneous catalysts, which show different selectivities towards the comonomer. As a consequence, the copolymers produced by heterogeneous catalysts are inhomogeneous and can be separated into fractions having different compositions. On the other hand, the comonomer is randomly distributed over the polymer chain obtained with metallocene catalysts, which is typical of single-site catalysts. The amount of extractables is much lower than in

Metallocene-based Polyolefins, Edited by J. Scheirs and W. Kaminsky
© 2000 John Wiley & Sons Ltd

copolymers synthesized with Ziegler catalysts. Many studies concerning the effect of metallocene structures on the copolymerization of ethylene and α-olefins have been carried out [3]. Uozumi and Soga [4] reported that a syndiospecific metallocene was more effective for inserting an α-olefin into an ethylene copolymer than an isospecific metallocene or unbridged metallocene. Although these metallocene catalysts are highly active and show good copolymerization reactivity, the molecular weight of the copolymers obtained is not enough to meet the requirements of industrial processes. Moreover, the molecular weights of copolymers decrease with increasing polymerization temperature [5].

The main advantage of a high-pressure process is that no solvent is required and homogeneous catalysts are used without any modifications, but this process should be operated at high temperature to maintain a high productivity of linear low-density polyethylene (LLDPE), which indicates that designed metallocene catalysts which can produce high molecular weight LLDPE with high activity and good copolymerization reactivity are needed.

Since 1991, Exxon has launched the production of a range of ethylene–α-olefin copolymers named EXACT, using metallocene catalysts in a high-pressure process [6]. It was clearly shown that it is possible to use metallocene catalysts in a high-pressure process.

In this chapter, we explain the features of metallocenes in the following sections:

1. Ethylene polymerization with metallocene catalysts at low temperature.
2. Ethylene–α-olefin copolymerization with metallocene catalysts at low temperature.
3. Ethylene–α-olefin copolymerization with metallocene catalysts at high temperature.
4. Ethylene–α-olefin copolymerization with metallocene catalysts in a high-pressure process.

2 ETHYLENE POLYMERIZATION WITH METALLOCENE CATALYSTS AT LOW TEMPERATURE

2.1 METALLOCENE/MAO SYSTEMS

It is well known that zirconocene methylaluminoxane (MAO) catalysts show 10–100 times higher activity than conventional Ziegler catalysts for ethylene polymerization. Kaminsky et al. summarized ethylene polymerizations with many kinds of metallocene catalysts [7]. In general, zirconium catalysts are more active than titanium or hafnium catalysts. Of course, the activity is also influenced by the ligand structures and generally the introduction of electron-donating substituent groups to ligands gives rise to increased activity. Actually, for comparison of C_2 symmetric substituted bisindenylzirconium compounds, $[Me_2Si(2,4,7-Me_3Ind)_2]ZrCl_2$ which has an electron-donating substituent groups gives a highly active catalyst. On the other hand, C_s symmetric metallocenes show low activity for ethylene polymerization, although these catalysts can produce syndiotactic polypropylene with high activity [8]. Fink and co-workers explained that this was due to the low concentration of active centers based on a kinetic analysis of copolymerization [9]. Nevertheless, metallocene catalysts show high activity for ethylene polymerization and this is of great advantage.

The most significant problem with conventional metallocene catalysts is that the molecular weight of polyethylene decreases with increase in polymerization temperature [10]. For example, Table 1 shows the effect of polymerization temperature on the molecular weight of polyethylene with an $Et(Ind)_2ZrCl_2$-based catalyst. Actually, the M_w of polyethylene produced at low temperature is very high but the molecular weight decreases drastically with increasing polymerization temperature and at a polymerization temperature of 80 °C, the M_w of polyethylene is 46 000, which is very low for commercial use. This means that the design of metallocene compounds to produce high molecular weight polyethylene at high temperature is important in applying metallocene catalysts to a high-pressure process. It is well known that conventional Ziegler catalysts synthesize high molecular weight polyethylene, so chain termination agents such as hydrogen have to be applied. This indicates that with metallocene catalysts the situation is the opposite and the dependence of the molecular weight of polyethylene on metallocene structure is an important aspect when applying metallocene catalysts to a high-pressure process.

2.2 METALLOCENE/$B(C_6F_5)_4$ ANION SYSTEMS

Metallocene compounds must be activated by Lewis acid cocatalysts such as MAO, dehydroxylated Al_2O_3 and $MgCl_2$ to form active catalysts. Although MAO structurally remains incompletely characterized, it is currently by far the most effective cocatalyst, and MAO-containing catalysts have therefore been the most widely studied. The role of MAO is now generally thought to be to alkylate a metallocene

Table 1 Effect on polymerization temperature (T_p) on catalyst performance[a]

T_p (°C)	Activity [kg/(mmol Zr h)]	M_w ($\times 10^{-3}$)	M_n ($\times 10^{-3}$)	M_w/M_n
20	259	85.0	40.5	2.1
40	220	81.0	36.8	2.2
60	145	55.0	25.0	2.2
80	184	46.0	21.9	2.1

[a] Polymerization conditions: ethylene pressure 8 bar, polymerization time 60 min; catalyst, $Et(Ind)_2ZrCl_2$/MAO.

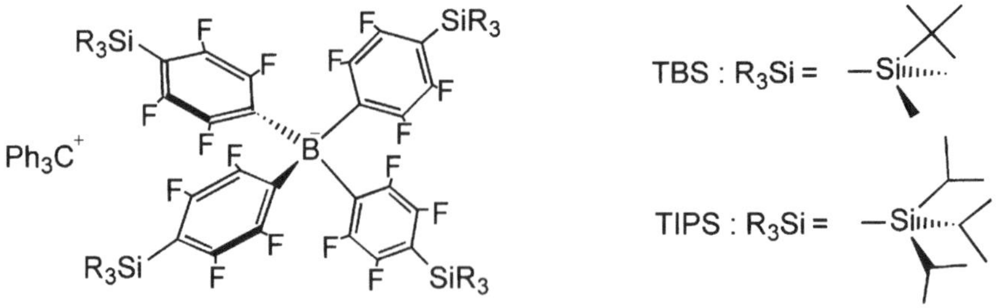

Figure 1   New activator compounds based on $B(C_6F_5)_4$ anion

and to activate the resulting complex by Lewis acid complexation. Based on these results, new activators such as $B(C_6F_5)_3$, $B(C_6F_5)_4$ anion and metallacarboranes have been studied [11]. Bochmann and Lancaster [12] reported that $(Me_3SiCp)_2MMe_2/Ph_3C \cdot B(C_6F_5)_4$ (M = Ti, Zr, Hf) gave new cationic catalysts, the ethylene polymerization activity increased in the order M = Ti < Hf < Zr, the activity of the Zr complex was comparable to that of Cp_2ZrCl_2/MAO system and there was a marked activity increase with temperature, with a corresponding decrease in polymer molecular weight. These results suggest that highly active cationic metallocene catalysts are closely connected with the nature of the cation–anion tight ion pairing. Recently, new activators [13] based on the $B(C_6F_5)_4$ anion have been synthesized as shown in Figure 1 and the real active site has been studied. Metallocene catalysts in conjunction with $Ph_3C \cdot B(C_6F_4TBS)_4$ or $Ph_3C \cdot B(C_6F_4TIPS)_4$ (TBS = *tert*-butyldimethylsilyl and TIPS = triisopropylsilyl) have polymerization activities roughly similar to that of the $B(C_6F_5)_4$ analog, but they afford thermally stable catalysts.

These results indicate that metallocene cation–anion ion pairs can be modified by the anion structure but catalytic features such as the catalytic activity and the molecular weight of polyethylene are mainly decided by the metallocene structure.

3   ETHYLENE–α-OLEFIN COPOLYMERIZATION WITH METALLOCENE CATALYSTS AT LOW TEMPERATURE

3.1   METALLOCENE/MAO SYSTEMS

Great interest has been focused on the copolymerization of ethylene and α-olefins with the soluble metallocene/MAO catalyst system because these catalyst systems show very high activity and can produce copolymers with a narrow chemical composition distribution as shown in Figure 2. Kaminsky and co-workers [14] first showed that Cp_2ZrMe_2/MAO is a good catalyst for the copolymerization of ethylene and propylene. The reactivity ratios of ethylene–propylene [15], ethy-

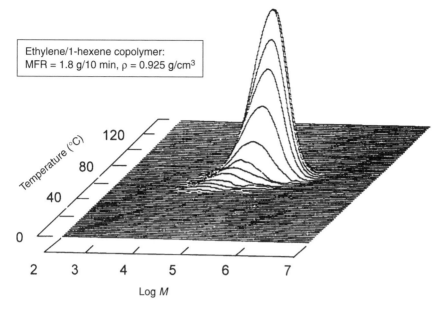

Figure 2 Bird's eye view of ethylene–1-hexene copolymer by cross-fractionation chromatography

lene–1-butene and ethylene–1-hexene [3b, 16] copolymerizations for several metallocene catalysts were reported as shown in Tables 2 and 3.

Zirconocenes having bulky ligands such as $(Me_5Cp)_2ZrCl_2$ have larger r_1 valves and the variation in r_1 value is consistent with the steric effects on coordination being more severe for propylene than ethylene. It may be said of these monomer reactivity ratios that the incorporation of propylene into the copolymer decreased in

Table 2 Copolymerization parameters for ethylene–propylene copolymerization

Metallocene	Temperature (°C)	r_1	r_2	r_1r_2	Ref.
$(Me_5Cp)_2ZrCl_2$	50	250	0.002	0.5	15a
Cp_2ZrCl_2	50	48.0	0.015	0.72	15a
$Et(Ind)_2ZrCl_2$	50	6.61	0.06	0.40	15b
$Et(Ind)_2ZrCl_2$	25	6.26	0.11	0.69	15b
$Et(Ind)_2ZrCl_2$	0	5.2	0.14	0.73	15c
$Et(Ind)_2ZrCl_2$	50	2.57	0.39	1.0	15d
$Et(Ind)_2ZrCl_2$	40	6.05	0.11	0.69	15e
$Me_2Si(Ind)_2ZrCl_2$	40	4.23	0.12	0.52	15e
$Et(H_4\text{-}Ind)_2ZrCl_2$	50	2.9	0.28	0.81	15d
$Me_2C(Cp)(Flu) ZrCl_2$	25	1.3	0.20	0.26	15c

Table 3 Copolymerization parameters for ethylene–1-butene or 1-hexene copolymerization

Metallocene	Temperature (°C)	α-Olefin	r_1	r_2	$r_1 r_2$	Ref.
Cp_2ZrCl_2	40	1-Butene	55	0.02		3b
	60		65	0.01		3b
	80		85	0.01		3b
	85		125	0.01		3b
$Et(Ind)_2ZrCl_2$	30	1-Butene	19.4	0.05	0.97	16b
	50		23.6	0.03	0.71	16b
	70		29.2	0.04	1.20	16b
$Et(Ind)_2HfCl_2$	30	1-Butene	5.4	0.17	0.92	16b
	50		6.6	0.10	0.66	16b
	70		6.8	0.21	1.43	16b
Cp_2ZrCl_2	40	1-Hexene	23.8	—	—	4
$Et(H_4\text{-}Ind)_2ZrCl_2$	40	1-Hexene	12.1	0.028	0.34	4
$Me_2C(Cp)(Flu)ZrCl_2$	40	1-Hexene	5.6	0.052	0.29	4
Cp_2ZrMe_2	20	1-Hexene	55	0.004		16a
	40		54	0.005		16a
	60		52	0.005		16a
	70		79	0.005		16a

the order syndiospecific catalyst > isospecific catalyst > aspecific catalyst. The numerical values of the product $r_1 r_2$ are close to 1 for C_2 precursors and < 1 for the C_s precursor. Generally, it may be said that an ethylene–propylene copolymer obtained with a isospecific catalyst is a statistically random copolymer but an ethylene–propylene copolymer obtained with a syndiospecific catalyst has an alternating character.

Et(Ind)$_2$HfCl$_2$/MAO catalyst shows a remarkably low r_1 value, indicating that Et(Ind)$_2$HfCl$_2$ is much more able to incorporate the comonomer into the copolymer than the zirconium analog. In this case, the $r_1 r_2$ values are about 1, which means that the insertion of 1-butene occurs randomly. The activity depends on the mole% of 1-butene, and with a low 1-butene content the activity of Et(Ind)$_2$HfCl$_2$/MAO catalyst is lower than that of the zirconium analog. The other advantage of the Et(Ind)$_2$HfCl$_2$/MAO catalyst is that it can produce higher molecular weight ethylene–propylene copolymer than the zirconium analog. Usually, the molecular weight of a copolymer decreases with increasing comonomer content, and a decrease in the molecular weight of ethylene–propylene copolymers with increasing propylene feed ratio using rac-Et(Ind)$_2$ZrCl$_2$/MAO [15c] or Cp$_2$ZrCl$_2$/MAO [17] has already been reported. For ethylene–propylene copolymerization with rac-Me$_2$Si(H$_4$Ind)$_2$ZrCl$_2$/MAO, a minimum molecular weight was observed at about 80% propylene in the copolymer [18]. Based on these results, the features of hafnium catalysts (such as good copolymerization reactivity, producing high molecular weight polymers) are a great advantage for applying them in a high-pressure process.

On the other hand, Fink and co-workers [9] reported that there was an increase in the molecular weight of copolymers and a polymerization rate enhancement at low 1-hexene concentration with $Me_2C(Cp)(Flu)ZrCl_2$/MAO, which was not observed with rac-$Me_2Si(Ind)_2ZrCl_2$/MAO. They speculated that the molecular weight enhancement effect caused by the strongly accelerated ethylene polymerization rate is greater than the molecular weight reducing effect caused by the β-H elimination reaction of the species [R–1-hexene–cat] or by the chain transfer reaction with 1-hexene.

The catalyst performance at high temperature is important in applying metallocene catalysts to a high-pressure process. From this point of view, it is less good that the reactivity ratio r_1 increases with increasing polymerization temperature. The r_1 value indicates how much faster an ethylene is inserted in the growing polymer chain than an α-olefin, when the last inserted monomer is an ethylene unit, so this means that the copolymerization reactivity is decreased with increasing polymerization temperature. On the other hand, conventional Ziegler catalysts [19] show opposite characteristics in the reactivity ratio r_1 as shown in Table 4. These results indicate that reactivity ratios of metallocene catalysts at high temperature must be understood in order to apply these catalysts to a high-pressure process.

3.2 $METALLOCENE$/$B(C_6F_5)_4$ ANION SYSTEMS

Metallocene cation–anion interaction is key factor in deciding the metallocene catalyst performance. From this point of view, four kinds of metallocene catalysts, Cp_2ZrCl_2, rac-$Et(Ind)_2ZrCl_2$, rac-$Me_2Si(Me_2Cp)ZrCl_2$ and $Me_2C(Cp)(Flu)ZrCl_2$, in combination with $Me_2PhNH·B(C_6F_5)_4$/$^iBu_3Al$ as cocatalyst were studied for the copolymerization of ethylene and 1-hexene. The copolymerization results and

Table 4 Copolymerization parameters for ethylene–α-olefin copolymerization

Ziegler catalyst	Temperature ($^\circ$C)	α-Olefin	r_1	r_2	Ref.
MgH_2/$TiCl_4$/Et_3Al	20	1-Octene	115	0.01	19a
	30		100	0.01	19a
	40		89	0.01	19a
	50		73	0.02	19a
	60		64	0.03	19a
Ziegler– Natta Catalyst	50	1-Butene	130	0.1	19b
	65		110	0.1	19b
	85		67	0.08	19b
	120		51	0.04	19b

Table 5 Copolymers synthesized at a 40 °C polymerization temperature[a]

Metallocene	Activity [kg/(mmol Zr h)]	M_w ($\times 10^{-4}$)	M_w/M_n	Methyl (No. per 1000 C)	T_m (°C)
Cp$_2$ZrCl$_2$	142	45.3	2.0	1.9	127
Et(Ind)$_2$ZrCl$_2$	220	7.3	1.9	5.7	120
Me$_2$Si(Me$_2$Cp)ZrCl$_2$	83	25.0	2.1	9.4	114
Me$_2$C(Cp)(Flu) ZrCl$_2$	47	13.0	2.3	17.5	104

[a] Polymerization conditions: ethylene pressure, 8 bar; 1-hexene, 15 ml; toluene, 450 ml; polymerization time, 60 min; metallocene/ Me$_2$PhNH·B(C$_6$F$_5$)$_4$/^{i}Bu$_3$Al = 1/2/250 µmol.

molecular structures of the copolymers obtained are summarized in Table 5. These four catalysts are highly active and the molecular weight distribution of the copolymers is *ca* 2, which is typical of copolymers synthesized with so-called single-site metallocene catalysts. The number of methyl branches formed in the copolymer increased and the melting-point of the copolymers decreased in the order Cp$_2$ZrCl$_2$, *rac*-Et(Ind)$_2$ZrCl$_2$, *rac*-Me$_2$Si(Me$_2$Cp)ZrCl$_2$, Me$_2$C(Cp)(Flu)ZrCl$_2$, which means that the incorporation of 1-hexene into the copolymer increased in

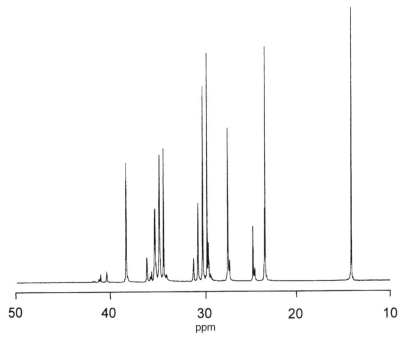

Figure 3 ^{13}C NMR spectrum of ethylene–1-hexene copolymer obtained with Me$_2$C(Cp)(Flu)ZrCl$_2$/Me$_2$PhNH·B(C$_6$F$_5$)$_4$/^{i}Bu$_3$Al catalyst.

Table 6 Monomer sequence distribution of ethylene–1-hexene copolymer[a]

[EE]	[EH] [HE]	[HH]	r_E	r_H	$r_E r_H$
46.6	50.0	3.4	8.1	0.03	0.25

[a] Polymerization conditions: polymerization temperature, 40 °C; ethylene pressure, 4 bar; 1-hexene, 200 ml catalyst, $Me_2C(Cp)(Flu)ZrCl_2/Me_2PhNH\cdot B(C_6F_5)_4/{}^iBu_3Al = 1/2/250\,\mu mol$. E = ethylene units; H = hexene units.

the same order. This tendency is the same as that for metallocene/MAO catalysts reported by Soga *et al.* [11] Figure 3 shows ^{13}C NMR spectrum of ethylene–1-hexene copolymer obtained with $Me_2C(Cp)(Flu)ZrCl_2/Me_2PhNH\cdot B(C_6F_5)_4/{}^iBu_3Al$ and Table 6 shows the monomer sequence distribution of ethylene–1-hexene copolymer indicated in Figure 3; $r_E = 5.93, r_H = 0.046$ and $r_E r_H = 0.27$ were calculated for this sample. These values are roughly the same as the reactivity ratios of ethylene–1-hexene copolymer obtained with $Me_2C(Cp)(Flu)ZrCl_2/MAO$ catalyst, which indicates that the copolymerization reactivity is mainly decided by the metallocene structure.

4 ETHYLENE–α-OLEFIN COPOLYMERIZATION WITH METALLOCENE CATALYSTS AT HIGH TEMPERATURE

4.1 METALLOCENE/B(C₆F₅)₄ ANION SYSTEMS

It is a key step for the design of the catalyst systems to understand the polymerization behavior of metallocene catalysts at high polymerization temperature. Conventional metallocenes activated with $B(C_6F_5)_4$ anion produce low molecular weight polyethylene with relatively good activity, as shown in Table 7. These results indicate that the breakthrough will occur when we find new catalysts which produce high molecular weight polyethylene without decreasing the catalytic activity. Ligand

Table 7 Results of ethylene polymerization with metallocene catalysts[a]

Metallocene	Amount (μmol)	Activity [kg/(mmol Zr)]	MFR (g/10 min)
Ind_2ZrCl_2	1.0	48	>500
$Et(Ind)_2ZrCl_2$	1.0	31	>500
$Me_2Si(Ind)_2ZrCl_2$	2.5	5	84

[a] Polymerization conditions: polymerization temperature, 150 °C; polymerization time, 20 min, ethylene pressure, 20 bar; solvent, $C_9–C_{13}$ hydrocarbon; catalyst, metallocene/$Me_2PhNH\cdot B(C_6F_5)_4/{}^iBu_3Al = 1/2/250\,\mu mol$.

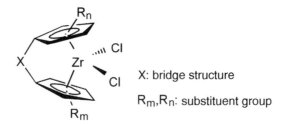

X: bridge structure

R_m, R_n: substituent group

Figure 4 Design of metallocene compound to produce high molecular weight ethylene–α-olefin copolymers.

Table 8 Results of ethylene–1-hexene copolymerization with metallocene catalysts[a]

Metallocene	1-Hexene (ml)	Activity [kg/(mmol Zr)]	M_w ($\times 10^{-4}$)	M_w/M_n
$Me_2C(Cp)(Flu)ZrCl_2$	20	6	3.2	2.5
$MePhC(Cp)(Flu)ZrCl_2$	20	32	3.6	2.0
$Ph_2C(Cp)(Flu) ZrCl_2$	20	172	10.0	2.0

[a] Polymerization conditions: polymerization temperature, 150 °C; polymerization time, 20 min; ethylene pressure, 20 bar; solvent, C_9–C_{13} hydrocarbon; catalyst, metallocene/Me_2PhNH·B$(C_6F_5)_4$/$^iBu_3$Al = 1/2/250 μmol.

structures are the most important factor in preparing metallocene catalysts, as shown in Figure 4. The production of high molecular weight polyethylene with high activity is achieved by selecting $Ph_2C(Cp)(Flu)ZrCl_2$ as the metallocene compound, as shown in Table 8 [20]. As already reported, the activity of this catalyst at low polymerization temperature is not very high, especially compared with bisindenyl metallocenes, but at high temperature the polymerization behavior changes. In this case, the bridge structure is a very important structure. By introduction of phenyl groups instead of methyl groups, the catalytic activity and molecular weight of polyethylene are drastically improved.

5   ETHYLENE–α-OLEFIN COPOLYMERIZATION WITH METALLOCENE CATALYSTS IN A HIGH-PRESSURE PROCESS

5.1   GENERAL FEATURES OF HIGH-PRESSURE POLYMERIZATION

The Production of LLDPE with Ziegler catalysts in a high-pressure process was developed by CdF Chimie in about 1970. The process flow is indicated in Figure 5. In a high-pressure process, polymerization is conducted at high temperature, which is important for maintaining high productivity, and the residence time is very short,

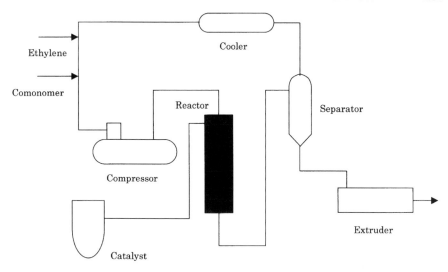

Figure 5 High-pressure process flow

so special catalyst performances are needed because of these severe polymerization conditions.

LLDPE have a great industrial potential and show a high growth rate, but the film properties of very low-density polyethylene (VLDPE) are inferior because the comonomer distribution is very wide owing to the multi-active sites of Ziegler catalysts. On the other hand, metallocene catalysts have a single active site and can produce LLDPE–VLDPE with a narrow molecular weight distribution and a narrow chemical composition distribution. As already shown, an improvement in molecular weight is needed in order to apply metallocene catalysts to a high-pressure process.

In 1991, EXXON started the commercialization of LLDPE to VLDPE using metallocene catalysts in a high-pressure process. Based on the EXXON patent [21], $(^nBuCp)_2ZrCl_2/MAO$ catalyst is used in the high-pressure process. Usually a high Al:Zr ratio of over 1000 is needed for sufficient catalytic performance, but in this case polymerizations are carried out at low Al:Zr ratios such as 12–100. The molecular weight of polyethylene is sufficient to meet the requirements of commercial use such as 100 000 g/mol. Ethylene–1-hexene copolymer which has a density of 0.90 g/cm^3 can be polymerized as shown in Table 9, indicating that this system can cover a range from HDPE to VLDPE. The conventional polymerization conditions are an ethylene pressure of over 500 bar and a polymerization temperature of 180 °C, which means a high ethylene concentration has a role in improving the molecular weight of polyethylene.

For comparison of metallocene compounds, the introduction of an *n*-butyl group to a cyclopentadienyl ring enhanced the catalytic activity and the molecular weight

Table 9 Ethylene–1-hexene copolymerization with metallocene/MAO in a high-pressure process

Metallocene	T_p (°C)^a	Al:Zr	C_6 (%)	P (bar)	M_w	Density (g/cm^3)
(^{n}BuCp)$_2$ZrCl$_2$	180	760	50	1000	57718	0.9244
Me$_2$Si(Cp)ZrCl$_2$	180	670	50	1000	11200	0.9361
Cp$_2$ZrCl$_2$	180	510	50	1000	37400	0.9312
(^{n}BuCp)$_2$ZrCl$_2$	170	185	66.7	2500	54100	0.9000
(^{n}BuCp)$_2$ZrCl$_2$	170	39	66.7	2500	54200	0.9028
(^{n}BuCp)$_2$ZrCl$_2$	170	16	66.7	2500	55200	0.9045

^{a} Polymerization temperature.

of copolymers was increased, but the introduction of a bridge structure (Me$_2$Si—) caused a decrease in the molecular weight of the copolymers.

Luft *et al.* [22] also studied ethylene polymerization phenomena with metallocene catalysts in a high-pressure process. The productivity and molecular weight of polyethylene decreased with increasing polymerization temperature. The activity at 80 °C is the highest with a value of 13 300 kg/g Zr and the activity at 220 °C is 3400 kg/g Zr. At temperatures above 200 °C, Mn was found in the range 36 000–30 000 g/mol. Simultaneously, the formation of low molecular weight wax was indicated. The formation of wax increases with increasing polymerization temperature. The polydispersities (M_w/M_n) are around 2 if the wax portion is excluded. For comparison of a heterogeneous supported catalyst, it is indicated that the productivity of a metallocene catalyst is four times higher than that of a heterogeneous catalyst under the same polymerization conditions, but the use of the heterogeneous catalyst system led to a much higher average molecular weight of the polymer. The most important difference between these two catalysts is the polydispersity. The polydispersity of the polymers obtained with a heterogeneous catalyst is distinctly higher and varied greatly with the polymerization temperature. On the other hand, the metallocene catalyst produced polymers with the same polydispersity (about 2). It is also clear from these experiments that it is most important to find a way to produce high molecular weight polymers in a high-pressure process.

Copolymerizations of ethylene and an α-olefin such as propylene, 1-butene, 1-hexene and 1-octene in a high-pressure process have also been studied [23]

Table 10 Reactivity ratios for copolymerization in a high-pressure process

Metallocene	Comonomer	r_1	r_2	Ref.
Modified silyl-bridged (H$_4$-Ind)ZrCl$_2$	Propylene	12.43	0.08	23a
	1-Butene	53.45	0.02	23b
	1-Hexene	62.7	0.02	23b
Et(Ind)$_2$ZrCl$_2$	1-Decene	80.02	0.01	26

and copolymerization reactivity ratios were determined using the Finemann and Ross method as shown in Table 10. The productivity decreased with increasing comonomer concentration in the feed. It is well known that the polymerization rate is enhanced by adding an α-olefin with metallocene catalysts [24]. However, it was reported that the addition of an α-olefin such as 1-hexene and 1-octene decreased the ethylene polymerization rate at a 95 °C polymerization temperature [25]. It may be said that this negative comonomer effect was seen when the polymer is soluble in the reaction medium. The same phenomenon probably occurs in these high-pressure ethylene–α-olefin polymerizations.

The molecular weight showed a steep decrease up to a comonomer concentration of 10–20 mol% in the feed, and then remained constant at higher comonomer concentrations. The molecular weight of ethylene–1-decene copolymer is lowest and that of ethylene–1-hexene copolymer is highest when these copolymers are compared at the same comonomer concentration in the feed. A chain transfer reaction from the polymer chain containing comonomer as the terminal unit is the main chain transfer reaction. These results indicate VLDPE production in a high-pressure process become more difficult than HDPE production, and designed metallocene catalysts are needed to produce high molecular weight VLDPE.

5.2 MODIFICATION OF METALLOCENE COMPOUND

The modification of metallocene catalysts for the production of LLDPE in a high-pressure process is focused on the design of metallocene compounds in synthesizing high molecular weight copolymers for commercial use. Bujadoux [26] carried out ethylene–1-butene copolymerization with various metallocene compounds in a high-pressure process and studied the relationships between metallocene structures and polymer structures such as molecular weight and polymer density. It was found that $Et(H_4\text{-Ind})_2ZrCl_2$ synthesized by the hydrogenation of $Et(Ind)_2ZrCl_2$ showed improved copolymerization reactivity and the ability to produce a high molecular weight copolymer. This tendency was observed between $Me_2C(Cp)(Flu)ZrCl_2$ and $Me_2C(Cp)(H_8\text{-Flu})ZrCl_2$. The replacement of one or two indenyl ligands of $Et(Ind)_2ZrCl_2$ by fluorenyl groups also enhances the molecular weight, whereas the density decreases only with the first substitution, as shown in Figure 6.

Copolymerization reactivity ratios $r_E = 26$ and $r_p = 0.22$ for $Et(Ind)_2ZrCl_2/MAO$ catalyst were calculated. This r_E value is higher than the value in low-temperature polymerization as shown in Table 2, which means that the copolymerization reactivity in a high-pressure process decreased especially compared with that of toluene slurry polymerization. The copolymerization reactivity of metallocene catalysts decreased with increase in polymerization temperature, so the decrease in copolymerization reactivity in a high-pressure process may be due to the effect of polymerization temperature.

For comparison of Zr and Hf compounds, $Et(Ind)_2HfCl_2$ and $Et(H_4\text{-Ind})_2HfCl_2$ produced copolymers with lower MI values and the density was lower than for the

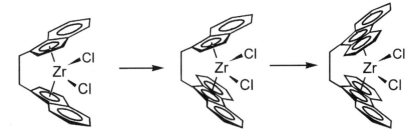

Figure 6 Replacement of indenyl rings by fluorenyls

zirconium analogs, which phenomena have already been observed for copolymerization at low temperature. It was indicated that the weak point is the activity, which is about five times lower than for the zirconium analogs.

Dow commercialized a new family of polyethylenes from INSITE technology in which $R^1R^2Si(R^xCp)(R^3N)MCl_2$/MAO catalysts (Figure 7) are used in the solution process [27]. These catalysts show good copolymerization reactivity and polyethylenes which have long-chain branches are produced under specific conditions by this technology [28]. The mechanism of the formation of a long chain branch was hypothesized as shown in Figure 8. The vinyl-terminated polymer chains may be considered as another comonomer in the polymerization. Recently, Soga [29] conducted the copolymerization of ethylene and polypropylene macromonomer ($M_n = 710$) with metallocene catalysts and showed that a CGC catalyst produced polyethylene with very long side-chains produced by the copolymerization between ethylene and propylene macromonomer having a terminal vinyl bond. EXXON [30] has carried out the copolymerization of ethylene and α-olefins with these types of catalysts in a high-pressure process. Copolymerization results achieved with these catalysts in a high-pressure process are given in Table 11. The effect of ligand structures and transition metal species on catalyst performance has been studied. MePhSi(Me$_4$Cp)(^{t}BuN)TiCl$_2$/MAO catalyst shows the highest activity of 61.7 kg/mmol catalyst at a 170 °C polymerization temperature. The molecular weight of the copolymers and the copolymerization reactivity are relatively high compared with the usual metallocene catalysts. It is indicated that the substituent group (R^3, R^x), Al:catalyst ratio and transition metal species (M) affect the catalyst performance. Actually, Bujadoux [26] carried out the copolymerization of ethylene and 1-butene

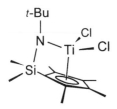

Figure 7 Representative constrained geometry catalysts

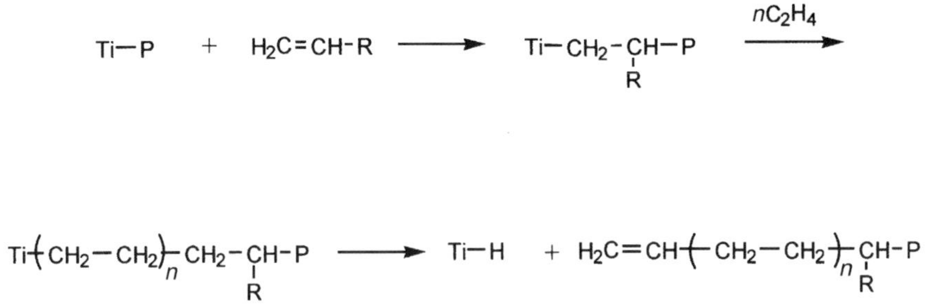

$$\text{Ti}-\text{P} \quad + \quad \text{H}_2\text{C}=\text{CH-R} \quad \longrightarrow \quad \text{Ti}-\text{CH}_2-\underset{\underset{\text{R}}{|}}{\text{CH}}-\text{P} \quad \overset{n\text{C}_2\text{H}_4}{\longrightarrow}$$

$$\text{Ti}\left(\text{CH}_2-\text{CH}_2\right)_{\!n}\!\text{CH}_2-\underset{\underset{\text{R}}{|}}{\text{CH}}-\text{P} \quad \longrightarrow \quad \text{Ti}-\text{H} \quad + \quad \text{H}_2\text{C}=\text{CH}\left(\text{CH}_2-\text{CH}_2\right)_{\!n}\!\underset{\underset{\text{R}}{|}}{\text{CH}}-\text{P}$$

Figure 8   Formation mechanism of a long-chain branch into polyethylenes

Table 11   Ethylene–1-butene copolymerization with various silylamido cyclopentadienyl compounds[a]

Metallocene	Al:TMC	T_p (°C)[b]	Activity (kg/mmol TMC)	M_w ($\times 10^{-4}$)	SCB[c] (per 1000 C)
MePhSi(Me$_4$Cp)(^{t}BuN)TiCl$_2$	1500	180	16.9	50.2	60.1
Me$_2$Si(Me$_4$Cp)(C$_6$H$_{11}$N)TiCl$_2$	1200	180	8.3	61.4	104.8
Me$_2$Si(Me$_4$Cp)(p-MeOC$_6$H$_4$N)TiCl$_2$	1300	180	9.3	65.0	55.5
Me$_2$Si(Me$_4$Cp)(PhN)TiCl$_2$	400	180	3.7	61.7	62.9
Me$_2$Si(Me$_4$Cp)(^{t}BuN)TiCl$_2$	600	180	8.8	50.8	69
Me$_2$Si(Me$_4$Cp)(C$_{12}$H$_{23}$N)TiCl$_2$	1400	180	10.7	72.6	110
MePhSi(Me$_4$Cp)(^{t}BuN)TiCl$_2$	1400	170	61.7	69.5	35.7
MePhSi(Me$_4$Cp)(^{t}BuN)ZrCl$_2$	—	180	3.2	31.9	46.6
MePhSi(Me$_4$Cp)(^{t}BuN)HfCl$_2$	—	180	9.3	40.8	36.9

[a] Polymerization conditions: pressure, 1300 bar; ethylene: 1-butene, 1.6.
[b] Polymerization temperature.
[c] Short chain branch.

with Me$_2$Si(Cp*)(^{t}BuN)TiCl$_2$/MAO catalyst and indicated that this catalyst showed lower activity than the metallocene catalysts.

5.3 B(C$_6$F$_5$)$_4$-ACTIVATED CATALYSTS

It is well known that metallocene/B(C$_6$F$_5$)$_4$ anion catalyst systems show high catalyst performance. EXXON applied these types of catalysts to a high-pressure process to produce LLDPE [31]. Me$_2$Si(H$_4$-Ind)ZrMe$_2$/Me$_2$PhNH·B(C$_6$F$_5$)$_4$ and Me$_2$Si(H$_4$-Ind)HfMe$_2$/B(C$_6$F$_5$)$_3$ were studied, as shown in Table 12. The advantage of these catalyst systems compared with MAO systems is their heat stability, so these catalysts show higher activity than MAO systems at high temperature. VLDPE copolymers of M_w 52 000 containing 19.4 w% 1-butene were produced at an activity of 60 kg/g activator with Me$_2$Si(H$_4$-Ind)HfMe$_2$/B(C$_6$F$_5$)$_3$.

Table 12 Ethylene–1-butene copolymerization with B compound-activated metallocene catalysts

TM/ activator[a]	Molar ratio	T_p (°C) (top/bottom)	$C_4:C_2$ wt ratio	Activity (kg/g activator)	Viscosity (cP)	C_4 (Wt%)
B/A	3.0	160/183	0.47	100	22510	15.1
B/A	3.0	180/199	0.47	150	5350	15.9
B/A	3.0	200/215	0.47	127	750	16.8
B/A	3.0	220/234	0.47	110	147	19.0
B/A	3.0	240/251	0.47	78	50	18.2
B/A	3.0	260/269	0.47	51	21	17.2
D/C	3.0	160/221	0.40	60		19.4

[a] A,$Me_2PhNH\cdot B(C_6F_5)_4$; B, $Me_2Si(H_4\text{-}Ind)_2ZrMe_2$; C, $B(C_6F_5)_3$; D, $Me_2Si(H_4\text{-}Ind)_2HfMe_2$.

Table 13 Results of ethylene–α-olefin copolymerization in a high-pressure process[a]

Catalyst	Comonomer	Activity (kg/mmol Zr)	Density (g/cm^3)	MFR (g/10 min)
$Ph_2C(Cp)(Flu)ZrCl_2$	1-Hexene	350	0.920	1.8
	1-Hexene	200	0.909	2.2
	1-Butene	800	0.883	3.1
	1-Butene	209	0.877	6.8
$Et(Ind)_2ZrCl_2$	1-Hexene	80	0.921	60.0

[a] Polymerization conditions: polymerization temperature, 150–190 °C; pressure, 900 bar; catalyst, metallocene/$Me_2PhNH\cdot B(C_6F_5)_4/^iBU_3Al = 1/2/250\,\mu mol$.

TOSOH studied the effect of metallocene structure on catalyst performance with $B(C_6F_5)_4$-activated catalysts and found that $Ph_2C(Cp)(Flu)ZrCl_2/Me_2PhNH\cdot B(C_6F_5)_4/^iBu_3Al$ produced high molecular weight polyethylene with good activity [20]. Representative copolymerization results are given in Table 13. For ethylene–1-butene copolymerization, high molecular weight VLDPE are synthesized with high activity at high temperature. $Et(Ind)_2ZrCl_2/Me_2PhNH\cdot B(C_6F_5)_4/^iBu_3Al$ produced ethylene–1-hexene copolymer with a higher MFR value (MFR = 60). This result is nearly the same as that with the MAO-activated catalyst obtained by Bujadoux [26], which indicated that the metallocene structure mainly decided the catalyst performance for copolymerization in a high-pressure process. The other important point is that this catalyst is heat stable and the activity at a 200 °C polymerization temperature was maintained at 800 kg/mmol Zr in a actual plant test as shown in Figure 9.

The catalyst performance was improved by introducing alkyl groups at the 2- and 7- positions in the fluorenyl ligand. Results for ethylene–α-olefin copolymerization at high temperature/low pressure and at high temperature/high pressure are summarized in Tables 14 and 15, respectively. These catalysts show the same activity as the

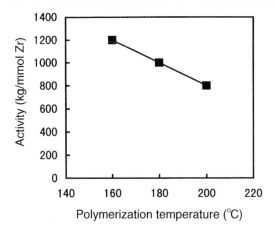

Figure 9 Effect of polymerization temperature on the catalytic activity for ethylene–1-hexene copolymerization with $Ph_2C(Cp)(Flu)ZrCl_2/Me_2PhNH \cdot B(C_6F_5)_4/^iBu_3Al$ catalyst

Table 14 Effect of substituent groups on catalyst performance[a]

Catalyst	Activity (kg/mmol Zr)	MFR (g/10 min)
$Ph_2C(Cp)(Flu)ZrCl_2$	148	4.1
$Ph_2C(Cp)(2,7-Me_2-Flu)ZrCl_2$	136	1.3
$Ph_2C(Cp)(2,7-^tBu_2-Flu)ZrCl_2$	144	1.3

[a] Polymerization conditions: polymerization temperature, 170 °C ethylene pressure, 20 bar; solvent; C_9–C_{13} hydrocarbon; catalyst, metallocene/$Me_2PhNH \cdot B(C_8F_5)_4{}^iBu_3Al = 1/2/250\,\mu mol$.

Table 15 Results of ethylene–α-olefin copolymerization in a high-pressure process[a]

Catalyst	Comonomer	Activity (kg/mmol Zr)	Density (g/cm³)	MFR (g/10 min)
$Ph_2C(Cp)(2,7-^tBu_2-Flu)ZrCl_2$	1-Hexene	380	0.918	1.7
$Ph_2C(Cp)(2,7-Me_2-Flu)ZrCl_2$	1-Hexene	320	0.918	1.6
$Ph_2C(Cp)(Flu)ZrCl_2$	1-Hexene	240	0.919	2.6
$Ph_2C(Cp)(2,7-^tBu_2-Flu)ZrCl_2$	1-Butene	660	0.886	2.8
$Ph_2C(Cp)(2, 7-Me_2-Flu)ZrCl_2$	1-Butene	480	0.882	3.5
$Ph_2C(Cp)(Flu)ZrCl_2$	1-Butene	800	0.881	3.8

[a] Polymerization conditions, polymerization temperature, 150–190 °C; pressure, 900 bar; catalyst, metallocene/$Me_2PhNH \cdot B(C_6F_5)_4/^iBu_3Al = 1/2/250\,\mu mol$.

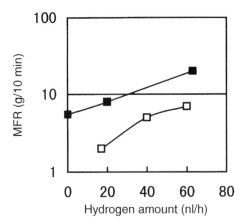

Figure 10 Molecular weight control with hydrogen. ■, metallocene, □, Ziegler catalyst.

$Ph_2C(Cp)(Flu)ZrCl_2$-based catalyst and produce higher molecular weight copolymers. This is attributed to the electron donation from the ligand to the metal. It is also indicated that hydrogen becomes a good chain transfer reagent in these catalysts. Kaminsky and Luker [32] studied the influence of hydrogen on the polymerization of ethylene and indicated that only traces of hydrogen were necessary to lower the molecular weight in a wide range, in contrast to most heterogeneous catalysts, and hydrogen lowered the activity of the catalyst in olefin polymerization. With this catalyst, the molecular weight of copolymer was controlled with hydrogen without a decrease in activity, as shown in Figure 10.

5.4 SUPPORTED METALLOCENE CATALYSTS

EXXON [33] prepared SiO_2-supported metallocene catalysts and conducted ethylene–1-butene copolymerizations in a high-pressure process. The polymerization reaction is carried out at $180\,^\circ C$ and a copolymer which has a density of $0.919\,g/ml$ and $M_w = 46\,900$ is obtained.

5.5 PROCESS

EXXON studied the pressure dependence of metallocene catalyst performance in a high-pressure process [34] and found that the productivity was increased for polymerization conducted at a 500 bar ethylene pressure, where phase separation occurred. The polymerization results are summarized in Table 16. $Me_2Si(C_{12}H_{23}N)(Cp^*)TiCl_2/MAO$ shows a higher activity for ethylene–α-olefin copolymerization at low pressure than at high pressure, and the decrease molecular

Table 16   Effect of ethylene pressure on catalyst performance in a high-pressure process[a]

Pressure (bar)	Temperature (°C)	Comonomer	Comonomer:C_2 ratio	Activity (kg/g TM)	MFR (g/10 min)	Density (g/cm^3)
1300	175	C_4	1.2	256.0	0.9045	1.6
600	170	C_4	1.2	194.9	0.9135	4.3
600	170	C_4	1.2	172.3	0.9126	3.3
500	180	C_4	1.2	89.6	0.9148	9.4
300	180	C_4	1.2	277.0	0.9125	15.4
200	162	C_4	1.2	273.8	0.9079	5.1
180	180	C_4	1.2	208.6	0.9072	34.0
220	152	C_4	2.3	91.3	0.8720	19.5
180	150	C_4	2.4	184.6	0.8897	7.1
1600	200	C_6	0.8	156.0	0.9153	15.5
180	180	C_6	0.8	202.3	0.9187	15.7

[a] Catalyst: $Me_2Si(H_4\text{-}Ind)ZrCl_2/MAO$.

weight with decrease in pressure is very small. This feature is a great advantage in cutting the operating costs of a high-pressure process.

Grunig and Luft indicated the formation of wax and the analysis of this fraction showed chain lengths of 50–60 carbon atoms and a low polydispersity similar to the low molecular weight polyethylene which was obtained with aluminium alkyls alone at high temperature and high pressure [35]. EXXON [36] also reported a polymerization reaction in the recycle line and the separator of a high-pressure process and the formation of wax was attributed to this reaction. A deactivation reagent such as water is used for polymerization with Ziegler catalysts, but water does not work as a deactivation reagent for MAO-activated metallocene catalysts because a new activator is formed by the reaction of MAO and water. New deactivation reagents have been reported for metallocene/MAO catalysts.

5.6   LANTHANOCENE-BASED CATALYSTS

Organolanthanides have recently been shown to be effective for olefin polymerization. The groups of Watson [37], Marks [38], Bercow [39] and Teuben [40] synthesized hydrides and alkyl lanthanocenes, scandocene and yttrocene, which are highly active in polymerization. Yasuda and co-workers reported the living polymerization of methyl methacrylate [41] and the block copolymerization of ethylene and alkyl methacrylates [42] using organolanthanide complexes.

Petit and co-workers [43] carried out ethylene polymerization with $Cp^*NdCl_2Li(OEt_2)_2/BuEtMg$ catalyst in a high-pressure pilot plant. High molecular weight polyethylene with good activity was obtained at a 160 °C polymerization temperature. Attempts at copolymerization with 1-butene have also been made. Polymers were obtained with good activity but the incorporation of 1-butene into

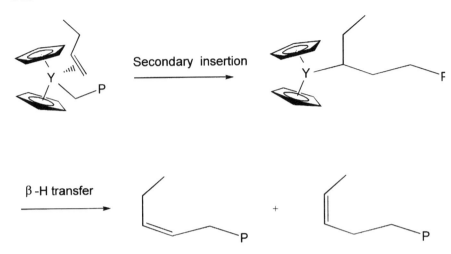

Figure 11 Chain transfer reactions in ethylene–1-butene copolymerization.

polymers did not occur. The melt flow index is drastically influenced by 1-butene concentration in these reactions, which means that 1-butene is a chain transfer reagent for ethylene polymerization with this catalyst system.

Other lanthanocene-based catalysts have also been studied. Although the neodymium catalyst is more active than Cp_2ZrCl_2/MAO, no copolymerization was observed. The less sterically crowded $(Cp_2YCl)_2$ isoelectronic system also did not give any copolymerization. Internal double bonds are present with the yttrium system, which indicates that 1-butene acts as a transfer reagent by β-H elimination from the growing polymer chain containing the secondary inserted 1-butene as terminal unit as shown in Figure 11. Further work should be done to extend the copolymerization of ethylene and α-olefins. Ligand modifications are currently being studied.

6 CONCLUSION

The main advantage of applying metallocene catalysts to a high-pressure process is that homogeneous catalysts can be used without any process changes and very low-density polyethylene is produced. Metallocene catalysts produce ethylene–α-olefin copolymers with a narrow molecular weight distribution and a narrow chemical composition distribution in a high-pressure process. The design of suitable metallocene ligands is needed for the synthesis of high molecular weight copolymers with high activity in a high-pressure process. Further work should be done to produce VLDPE with lower MFR.

7 REFERENCES

1. Sinn, H., Kaminsky, W., Vollmer, H. J. and Woldt, R. *Angew. Chem.,* **92**, 396 (1980).
2. (a) Ewen, J. A. *J. Am. Chem. Soc.,* **106**, 6355 (1984); (b) Kaminsky, W. Kulper, K. Brintzinger H. H. and Wild, F. R. W. P. *Angew. Chem., Int. Ed. Engl.,* **24**, 507 (1985); (c) Ewen, J. A., Jones, R. L. Razavi A. and Ferrar, J. D. *J. Am. Chem. Soc.,* **110**, 6255 (1988).
3. (a) Kaminsky W. and Miri, M. *J. Polym. Sci., Polym. Chem. Ed.,* **23**, 2151 (1985); (b) Kaminsky W. and Schlobohm, M. *Macromol. Chem. Macromol. Symp.,* **4**, 103 (1986); (c) Zambelli A. and Grassi, A. *Macromol. Chem., Rapid. Commun.,* **12**, 523 (1991).
4. Uozumi T. and Soga, K. *Macromol. Chem.,* **193**, 823 (1992).
5. Pietikainen P. and Seppälä, J. V. *Macromolecules,* **27**, 1325 (1994).
6. Montagna, A. A. and Floyd, J. C. in *MetCon '93,* Houston TX, May 26–28, 1993, Catalyst Consultants Inc., p. 171.
7. (a) Kaminsky, W., Engehausen, R., Zoumis, K., Spaleck W. and Rohrmann, J. *Macromol. Chem.,* **193**, 1643 (1992); (b) Kaminsky, W. *Macromol. Chem. Phys.,* **197**, 3907 (1996).
8. Herfert N. and Fink, G. *Macromol. Chem.,* **193**, 1359 (1992).
9. Herfert, N., Montag P. and Fink, G. *Macromol. Chem.,* **194**, 3167 (1993).
10. (a) Kaminsky, W., Kulper K. and Niedoba, S. *Macromol. Chem., Macromol. Symp.,* **3**, 377 (1986); (b) Eskelinen M. and Seppala, J. V. *Eur. Polym. J.,* **32**, 331 (1996).
11. (a) Jordan, R. F., Dasher W. E. and Echols, S. F. *J. Am. Chem. Soc.,* **108**, 1718 (1986); (b) Hlatky, G. G., Turner H. W. and Eckman, R. R. *J. Am. Chem. Soc.,* **111**, 2728 (1989); (c) Yang, X., Stern C. L. and Marks, T. J. *J. Am. Chem. Soc.,* **113**, 3623 (1991).
12. Bochmann M. and Lancaster, S. J. *J. Organomet. Chem.,* **434**, C1 (1992).
13. Yang, L. J. X., Stern C. and Marks, T. J. *Organometallics,* **13**, 3755 (1994); (b) Yang, L. J. X., Ishihara A. and Marks, T. J. *Organometallics,* **14**, 3135 (1995); (c) Chen, Y. X., Stern, C. L., Yang S. and Marks, T. J. *J. Am. Chem. Soc.,* **118**, 12451 (1996); (d) Chen, Y. X., Stern C. L. and Marks, T. J. *J. Am. Chem. Soc.,* **119**, 2582 (1997).
14. (a) Kaminsky W. and Miri, M. *J. Polym. Sci., Polym. Chem. Ed.,* **23**, 2151 (1985); (b) Kaminsky W. and Schlobohm, M. *Macromol. Chem., Macromol. Symp.,* **4**, 103 (1986).
15. (a) Ewen, J. A. in Keii, T. and Soga, K. (Eds), *Catalytic Polymerization of Olefins,* Kodansha, Tokyo, 1986, p. 271; (b) Drogemuller, H., Heiland, K. and Kaminsky, W. in Kaminsky W. and Sinn H. (Eds), *Transition Metals and Organometallics as Catalysts for Olefin Polymerization,* Springer, Berlin, 1988, p. 303; (c) Zambelli A. and Grassi, A. *Macromol. Chem., Rapid. Commun.,* **12**, 523 (1991); (d) Chien J. C. W. and He, D. *J. Polym. Sci., Part A: Polym. Chem.,* **29**, 1585 (1991); (e) Lehtinen C. and Lofgren, B. *Eur. Polym. J.,* **33**, 115 (1997).
16. (a) Kaminsky, W. in Keii, T. and Soga, K. (Eds), *Catalytic Polymerization of Olefins,* Kodansha, Tokyo 1986, p. 293; (b) Heiland K. and Kaminsky, W. *Macromol. Chem.,* **193**, 601 (1992).
17. Pietikainen P. and Seppälä, J. *Macromolecules,* **27**, 1325 (1994).
18. Sugano, T., Endo and J. Takahara, T. in *Science and Technology in Catalysis,* 1994, KODANSHA 1995, p. 37.
19. (a) Fink G. and Ojala, T. A. in Kaminsky, W. and Sinn, H. (Eds), *Transition Metals and Organometallics as Catalysts for Olefin Polymerization,* Springer, Berlin, 1988, p. 169; (b) Bohm, L. L. *J. Appl. Polym. Sci.,* **29**, 279 (1984).
20. Akimoto, A. in *Metallocene '95,* Brussels, Belgium, April 26–27, 1995, SCHOTLAND Business Research Inc., p. 439.
21. EXXON, Jpn. Pat., JP1-503788 and (1989).
22. Luft, G., Batarseh B. and Cropp, R. *Angew. Makromol. Chem.,* **212**, 157 (1993).
23. (a) Bergemann, C., Cropp R. and Luft, G. *J. Mol. Catal. A,* **102**, 1 (1995); (b) Bergemann, C., Cropp R. and Luft, G. *J. Mol. Catal. A,* **105**, 87 (1996).

24. Tsutsui T. and Kashiwa, T. *Polym. Commun.,* **29**, 180 (1988).
25. Koivumaki J. and Seppälä, J. *Macromolecules,* **26**, 5535 (1993).
26. Bujadoux, K. in *Metallocene '95,* 1995, p. 375.
27. Stevens, J. C. in *MetCon '93,* Houston TX, May 26–28, 1993 Catalyst Consultants Inc., p. 157.
28. Story, B. A. and Knight, G. W. in *MetCon '93,* Houston TX, May 26–28, 1993 Catalyst Consultants Inc., p. 111.
29. Soga, K. *Macromol. Symp.,* **101**, 281 (1996).
30. EXXON, Jpn. Pat., JP05-505593 (1993).
31. EXXON, PCT Int. Appl. WO93/25590 (1993).
32. Kaminsky, W. and Luker, H. *Macromol. Chem., Rapid Commun.,* **5**, 225 (1984).
33. EXXON, Jpn. Pat., JP03-501869 (1991).
34. EXXON, Jpn. Pat., JP07-501567 (1995).
35. Grunig H. and Luft, G. *Angew. Makromal. Chem.,* **142**, 161 (1986).
36. EXXON, Jpn. Pat., JP06-505046.
37. Watson P. L. and Parshall, G. W. *Acc. Chem. Res.,* **18**, 51 (1985).
38. Jeske, G., Lauke, H., Mauermann, H., Swepston, P. N., Schumann H. and Marks, T. J. *J. Am. Chem. Soc.,* **107**, 8091 (1985).
39. Burger, B. J., Thompson, M. E., Cotter W. P. and Bercow, J. E. *J. Am. Chem. Soc.,* **112**, 1566 (1990).
40. Den Haan, K. H., Wielstra, Y., Eshuis J. J. W. and Teuben, J. *J. Organomet. Chem.,* **323**, 181 (1987).
41. Yasuda, H., Yamamoto, H., Yokota, K., Miyake S. and Nakamura, A. *J. Am. Chem. Soc.,* **114**, 4908 (1992).
42. Yasuda, H., Furo, M., Yamamoto, H., Nakamura, A., Miyake S. and Kibino, N. *Macromolecules,* **25**, 5115 (1992).
43. (a) Olonde, X., Mortreux, A., Petit F. and Bujadoux, K. *J. Mol. Catal. A,* **82**, 75 (1993); (b) Pelletier, J. F., Mortreux, A., Petit, F., Olonde X. and Bujadoux, K. *Stud. Surf. Sci. Catal,* **89**, 249 (1994).

14

Metallocenes as Catalysts for the Copolymerization of Ethene with Propene and Dienes

MAURIZIO GALIMBERTI, FABRIZIO PIEMONTESI
AND OFELIA FUSCO
Montell Italia, G. Natta Research Center, Ferrara, Italy

1 INTRODUCTION

In recent years, research has been performed at Montell in the field of ethene copolymerization with propene and optionally a diene, promoted by metallocene-based catalytic systems (MBC) [1]. The aim of this chapter is to contribute to the scientific debate in the field [2, 3], focusing attention on the preparation of copolymers. Significant aspects of the copolymerization are discussed: the nature and behavior of catalytic systems, mechanism of copolymerization and microstructure of copolymers. Data are reported, working hypotheses for their interpretation are proposed and some conclusions are drawn. A glossary of terminology is presented at the end.

Copolymers with a chemical composition suitable for the preparation of elastomeric materials are in particular considered. The main product characteristics and process issues are discussed.

A comparison with the traditional catalysts for polyolefin elastomers, homogeneous vanadium-based and heterogeneous titanium-based, is presented.

Metallocene-based Polyolefins   Edited by J. Scheirs and W. Kaminsky
© 2000 John Wiley & Sons Ltd

2 CATALYSIS

2.1 A BREAKTHROUGH IN ZIEGLER–NATTA CATALYSIS: METALLOCENE/MAO CATALYTIC SYSTEMS

The interest in metallocenes, already used by Natta, Breslow and co-workers in the 1950s [4], dramatically increased after the discovery of Sinn, Kaminsky and co-workers, at the beginning of the 1980s, of a new cocatalyst, polymethylalumoxane (MAO) [5]. Metallocenes, combined with MAO, produced polyethene, low molecular mass atactic polypropene and ethene–propene (E/P) copolymers [6] with very high catalytic activity. The production of isotactic polyolefins with a racemic *ansa*-metallocene, reported by Ewen [7] and by Kaminsky *et al.* [8], revealed the potential importance of metallocenes as catalysts for polyolefins [5b, 9], also in industrial applications [10]. In a parallel way, many developments have been obtained in recent years in the field of ethene and 1-olefin copolymerization [1–3, 10].

2.2 SINGLE-CENTER CATALYSTS FOR POLYOLEFINS

The demonstrated potential of the metallocene/MAO system has paved the way to the discovery of many other organometallic complexes suitable for olefin polymerization. They are represented in Figure 1.

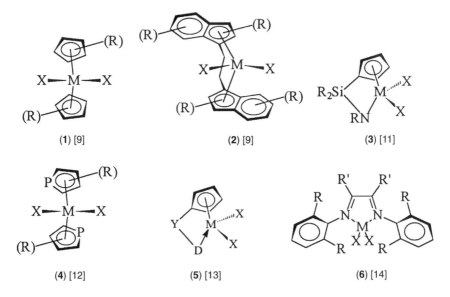

Figure 1   Organometallic complexes employed in olefin polymerization (references in brackets)

The catalytic systems based on these complexes are: (i) single-center [1c, 15], i.e. the catalytic centers responsible for the chain propagation have the same nature, (ii) soluble in most aliphatic and aromatic solvents and, nevertheless, active also in liquid monomers and in the gas phase and (iii) in many cases endowed with high catalytic activity. Furthermore, (iv) the organometallic complexes have a well defined chemical structure and (v) the π ligands remain coordinated to the transition metal atom during the course of the polymerization. The simultaneous occurrence of features (i)–(v) represents the real breakthrough in the field of Ziegler–Natta catalysis and, one can also say, in that of the insertion polymerization.

Nowadays, all these families of single-center catalysts are used for the preparation of polyolefins, to a different extent also on the industrial scale. As far as the performances of the single-center catalysts are concerned, it is possible to say that (i) only metallocenes are, at present, mature catalysts for the synthesis of stereoregular polyolefins and (ii) any of the single-center catalysts can be in principle suitable for preparing ethene-based homo- and copolymers. In fact, in this case the catalyst need not necessarily be stereospecific. This explains the wide application of complexes other than metallocenes e.g. **3** and **5** in Figure 1, for the preparation of ethene-based polymers, from HDPE (high-density polythene) to EP(D)M (elastomeric copolymers of ethene and propene with a diene). Nevertheless, this chapter will be focused on metallocenes, strictly interpreted from the organometallic point of view, apart from some comments made for the sake of comparison and specifically indicated.

2.3 COCATALYTIC SYSTEMS FOR METALLOCENE-BASED CATALYSIS

Cocatalysts alternative to MAO have appeared on the scene and have been proved able to promote the homo- and copolymerization of ethene and 1-olefins, first of all boron compounds [16]. Montell scientists have discovered and developed cocatalysts based on AlR_3 and H_2O [17]. They can be prepared and isolated as a tetralkylalumoxane or can be formed *in situ*, during the process, adopting a suitable $Al:H_2O$ ratio. Some examples of the precursors of these new cocatalysts are shown in Figure 2.

Tetraisobutylalumoxane (TIBAO) and tetraisooctylalumoxane (TIOAO) are in particular discussed in this chapter. In the scientific literature, the performances of different metallocenes are usually compared adopting MAO-based polymerization tests. As a consequence, any difference, for example, as far as the catalytic activity, is completely attributed to the structure of the metallocene. As has already been shown [1d, f], different aluminum-based cocatalysts give rise to different catalytic activities and an MAO-based system is not necessarily the most active one. This suggests the need for a new approach to the evaluation of the metallocene-based results. The performances of MBC are not only due to the different structures of the metallocenes, but should rather be attributed to the whole catalytic system, regarded as the combination of the metallocene and the cocatalyst. Later, the influence of the cocatalytic system on the copolymer microstructure will also be discussed.

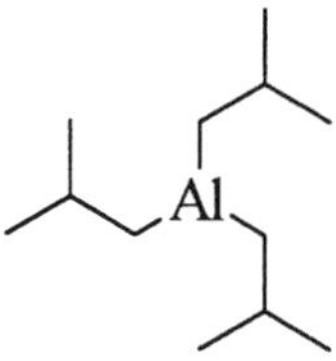

TIBA

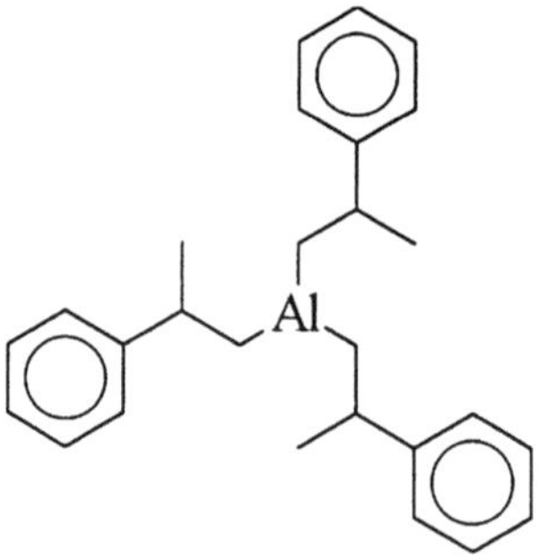

AlSty

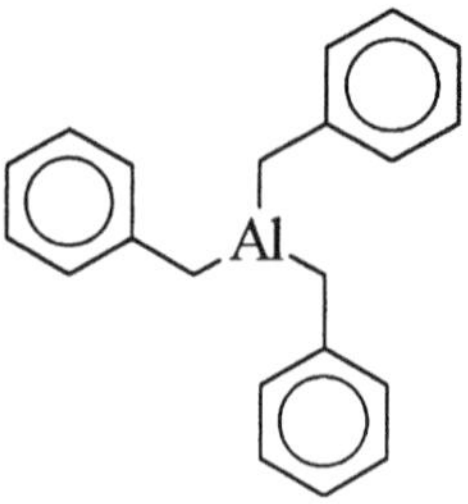

AlBz

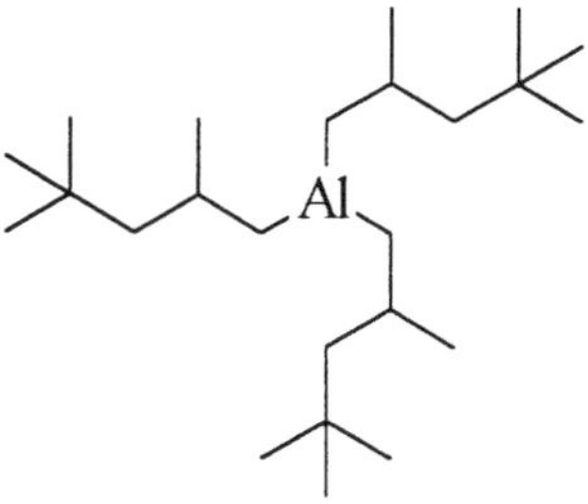

TIOA

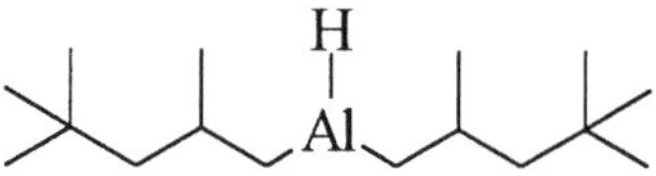

DIOAH

Figure 2 Aluminum alkyl compounds as cocatalyst precursors in metallocene-based polymerizations

2.4 CATALYTIC SYSTEM

It is worth attempting here (see Table 1) a summary of the main features of metallocene-based catalytic system, in comparison with the traditional titanium-and vanadium-based systems.

It is easy to draw an important conclusion. For the first time, with metallocenes, highly active single-center catalytic systems, with defined and tunable chemical structures, are available for the preparation of elastomeric copolymers and, more

Table 1 Main features of catalytic systems for ethene–propene (-diene) copolymerization

Catalyst feature	Catalytic system based on		
	Metallocenes	Vanadium	Titanium
Catalytic center	Single center	Single- or multi-center	Multi-center
Catalytic activity	High	Low	High
Chemical structure	Defined	Unknown	Unknown

Table 2 Catalytic activity (kg polymer/g metal) in ethene–propene copolymerizations as a function of catalytic system and polymerization process[a]

Process	T (°C)	Catalytic system based on		
		Metallocene	Vanadium	Titanium
Solution	50	600	80	–
Slurry	40	1000	100	500
Gas phase	50	500	80	150

[a] Polymerization time = 30 min for solution and slurry, 60 min for gas phase.

generally, of polyolefins. The presence of π ligands that remain coordinated to the transition metal atom during the polymerization allows one to steer the behavior of the copolymerization and the characteristics of the copolymers. The chance of designing the structure of the ligands has allowed the preparation of a large variety of metallocenes and, in turn, of copolymers. A correlation between the structure of the π ligands and the performances of the organometallic complexes can be attempted.

2.5 CATALYTIC ACTIVITY

It is not worth presenting tables full of data to try to compare the relative performances of different metallocenes. In fact, the relative catalytic activity depends on (i) the type of alumoxane, (ii) the Al:metal ratio, (iii) the polymerization temperature and (iv) the absolute and relative monomer concentration, to mention only the most important parameters. Upon changing any of them, even slightly, a new list of relative performances would need to be established.

However, a reliable order of magnitude can be established. In Table 2, the highest polymerization activities achieved at Montell with a metallocene-based catalytic system in the frame of different polymerization processes and, more in particular, adopting experimental conditions suitable for an industrial application are reported. A comparison with vanadium- and titanium-based catalysts is also indicated.

3 MOLECULAR PROPERTIES OF E/P COPOLYMERS FROM METALLOCENES

A set of molecular properties is the basic equipment of an E/P copolymer and contributes to determining its behavior in the elastomeric field. They are molecular mass, chemical composition and their intermolecular distribution. In Table 3, typical values obtained from MBC, and their ranges, are reported.

Table 3 Molecular properties of ethene–propene copolymers from MBC

Molecular property	Units	Minimum and maximum values[a]	Metallocene[b]
Molecular mass as intrinsic viscosity	dl/g	$\leqslant 1$ 20^c	$Me_2SiFlu_2ZrCl_2$[d] meso-EBDMIZrCl$_2$[d]
MMD[e]	–	2 10	Any Mixtures
CCD[f]	v coefficient	<1 15	Any Mixtures

[a] Obtained either in solution or in slurry processes, at 50 and 40 °C, respectively, for E/P copolymers having ethene content between 45 and 75 wt%.
[b] Metallocene that can be used to obtain the reported values.
[c] Obtained in slurry.
[d] For the nomenclature of metallocenes, see Glossary at the end of the text.
[e] MMD = Molecular mass distribution.
[f] CCD = chemical composition distribution.

As far as the molecular mass of the copolymers is concerned, the following comments can be made, based on results obtained with experimental conditions of industrial significance.

(i) Metallocenes are able to cover a broad range of molecular masses. However, whereas many candidates are available for obtaining low molecular mass and only a typical example among them is reported in Table 3, the preparation of high Mooney [18] grades is a problematic issue. meso-EBDMIZrCl$_2$ shows, in this respect, a peculiar behavior. The molecular mass of an E/P copolymer is indeed a critical parameter that brings about a dramatic selection among the metallocenes: as the molecular mass increases, the number of metallocenes available for their preparation decreases. A qualitative picture of the situation is shown in Figure 3.

(ii) For copolymers obtained with the same metallocene and under the same experimental conditions, the molecular mass depends on their chemical composition: the higher is the 1-olefin content, the lower is the molecular mass. This result, in agreement with what is traditionally observed with other catalytic systems, is presented in Figure 4 for copolymers prepared with metallocenes representative of different classes (see Scheme 1),

(iii) With the same metallocene and at the same temperature, higher molecular masses are obtained in slurries rather than in solution and this is essentially due to the higher concentration of monomers. Examples are given in Table 4.

Preliminary comments can be attempted in the direction of a correlation between the molecular mass of the copolymer and the structure of the metallocene, without any presumption of rationalization. Metallocenes can be organized in eight groups, as shown in Scheme 1, as a function of the presence of a bridge or substituents other than hydrogen on the π ligands and of the nature, aromatic or aliphatic, of the C_6

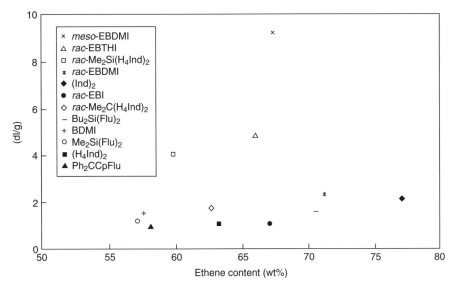

Figure 3 Intrinsic viscosity of ethene–propene copolymers as a function of ethene content and type of metallocene. Polymerization conditions: solution tests in hexane; $T = 50\,^\circ$C; $[E + P] = 3$–5 M; $[Zr] = 10^{-6}$ M; Al:Zr = 1000 mol/mol

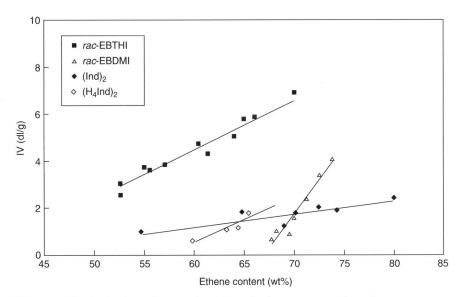

Figure 4 Intrinsic viscosity as a function of ethene content for ethene–propene copolymers obtained with different metallocenes. Polymerization conditions as in Figure 3

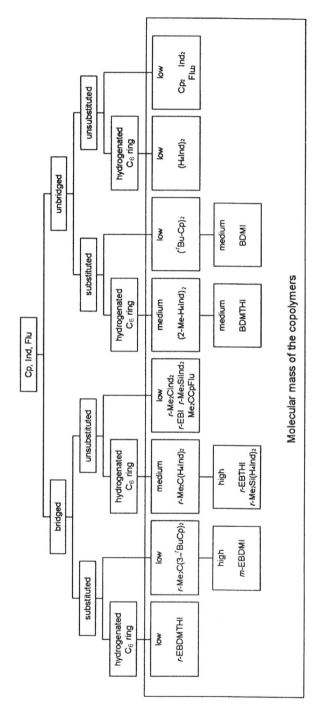

Scheme 1 The molecular mass of the copolymers is indicated as a function of $\bar{M}_w$ (g/mol) range: low = up to 150 000, medium = between 150 000 and 300 000, high = above 300 000. For experimental conditions, see Figure 3

Table 4   Molecular mass of ethene–propene copolymers from MBC as a function of polymerization process[a]

Metallocene (type)	Process (type)	$[E + P]_{polymerization}$ (mol/l)	$Ethene_{polymer}$ (wt %)	$IV_{polymer}$ (dl/g)
$(H_4Ind)_2ZrCl_2$	Solution[b]	5.1	63.0	1.1
	Slurry[c]	11	60.4	2.4
rac-EBTHIZrCl$_2$	Solution[b]	3.3	66.1	4.9
	Slurry[c]	11	66.5	5.9

[a]Polymerization conditions: $T = 50\,^\circ$C; alumoxane = TIOAO; Al:Zr = 1000.
[b]Solvent = hexane.
[c]Liquid monomers without any diluent.

ring optionally fused on the Cp group. In almost every group, there are metallocenes able to prepare copolymers with low molecular mass. Under the experimental conditions selected for this classification, this means a maximum $\bar{M}_w$ of about 150 000 g mol^{-1} for a copolymer with about 60 % by weight ethene content.

Unbridged metallocenes are not able to prepare high molecular mass elastomeric copolymers. To cover the latitude of the commercial grades, a metallocene has to present a bridge between the π ligands and should be either hydrogenated or substituted. In fact, the presence of a bridge is not sufficient to have a high intrinsic viscosity (IV). For example, syndiospecific metallocenes or the isopecific metallocenes of the type XInd$_2$ZrCl$_2$, where X = CH$_2$CH$_2$ or Me$_2$C or Me$_2$Si, give rise to low IV, not to mention the bisfluorenyl-type complexes. In the case of the mentioned isospecific metallocenes, the hydrogenation of the indenyl ligands brings about an increase of the molecular mass of the copolymers. An appropriate selection of the type and position of the substituents has to be performed to achieve the same result. For example, rac-Me$_2$C(3-^{t}BuCp)$_2$ZrCl$_2$ and rac-EBIZrCl$_2$ show similar behavior, preparing low molecular mass copolymers. rac-EBDMIZrCl$_2$ produces copolymers with higher molecular mass and, to underline the role played also by the isomeric nature of the complex, the meso isomer prepares copolymers with the highest molecular mass. It therefore seems that, at present, a fine interpretation can be reasonably approached only by correlating the copolymer characteristics with the structure of a particular metallocene. One must stress again that the molecular mass of a polyolefin elastomer is a limitation of the metallocene-based technology, particularly because it becomes very relevant when the EPDM grades, i.e. the important products on a commercial scale, have to be prepared. This hampers their introduction in commercial plants, taking also into account the ability of the traditional vanadium-based catalysts to prepare oil extended grades [18] with a high diene content.

In many publications the narrow molecular mass distribution of copolymers from MBC is reported: $\bar{M}_w/\bar{M}_n$ values close to 2 are repeatedly obtained. It is worth underlining here that we have obtained these results in both solution and suspension processes [1f]. In the gas phase similar values were also reproduced and higher values (up to 3.5) were observed upon manipulating the process conditions [1g].

Table 5 Chemical composition distribution of ethene–propene copolymers: v coefficient[a]

	Catalytic system based on		
Process	Metallocene	Vanadium	Titanium
Solution	$\leqslant 1\text{--}6$	1–12	–
Slurry	$2\text{--}20^{b}$	2–16	10–20

[a]For v coefficient, see Ref. 19. Average values detected on commercial samples (vanadium catalysts) or on laboratory-made samples (metallocenes and titanium catalysts) prepared with Montell technology.
[b]Copolymers obtained either with a mixture of metallocenes or with a bad process behavior.

The chemical composition distribution (CCD) of copolymers from MBC has been thoroughly investigated, essentially through fractionation techniques, also in comparison with products from traditional catalysis. A coefficient was defined (v coefficient) [19] to describe CCD. In Table 5, only a summary of the results is reported.

A general comment should be made. From the point of view of the intermolecular distribution of the molecular properties, MBC, on one side, are a completely different world from the titanium-based catalysts but, on the other side, do not introduce anything new with respect to the vanadium tradition. Furthermore, whereas the control of these distributions is a feature of vanadium technology, attempts are still in progress with metallocenes, essentially based on mixtures of metallocenes or cocatalysts [1f].

3.1 FROM ETHENE AND PROPENE TO THE BUILD-UP OF THE MACROMOLECULAR CHAIN

The process that leads a metallocene-based catalytic system, through interaction and cooperation with ethene and propene, to build up an E/P copolymer, was studied with attention focused on three main aspects: (i) the 'type of comonomers' that are present in the macromolecular chain, (ii) the relative reactivity of ethene and propene and (iii) the way in which the comonomers distribute themselves along the chain.

The approach adopted for the study of these aspects is able to account for all of them at the same time, correlating the relative reactivity of the comonomers, through the product of their reactivity ratios, with the microstructure of the macromolecule. The influence of both the components of the catalytic system, the metallocene and the alumoxane, was studied and is discussed as follows. Metallocenes were investigated using MAO as cocatalyst, whereas a metallocene, rac-EBTHIZrCl$_2$, was selected to compare the performances of MAO, TIOAO and TIBAO. Before starting a detailed discussion, it is worth summarizing the performances of MBC, reporting (in Table 6) the ranges of values detected for the main parameters of

Table 6 Ethene–propene copolymerizations with a metallocene-based catalytic system: main parameters of copolymerization and copolymer microstructure

Parameter	Units	Value[a]	Metallocene[b]
$r_1 r_2$	—	0.1	Me₂SiFlu₂ZrCl₂
			Most
		0.4–0.9	r-Me₂C(3-^{t}BuCp)₂ZrCl₂
			Me₂SiFlu₂ZrCl₂,
		>2	r-Me₂C(3-^{t}BuCp)₂ZrCl₂
			r-EBTHIZrCl₂
Statistical model of copolymerization	—	Markov 1st order	r-Me₂C(3-^{t}BuCp)₂ZrCl₂
		Markov 1st order	Cp₂ZrCl₂
		Markov 2nd order	Me₂CCpFluZrCl₂
Propene sequences:			
Tacticity	mm diads, mol%	100	Most
		30	r-EBDMIZrCl₂
Regioirregular units	mol%	0	
		0	
		5	

[a]Obtained either in solution or in slurry processes, for E/P copolymers having ethene content between 45 and 75 wt%.
[b]Metallocene that can be used to obtain the reported values.

copolymerization and copolymers. Some metallocenes are also indicated as an example, just to suggest a suitable candidate for obtaining the reported parameters.

3.2 COMONOMERS IN AN ETHENE-PROPENE MACROMOLECULAR CHAIN

In the case of an ethene–propene copolymer, five different comonomer units can be in principle present in the macromolecular chain. Besides ethene, four others generated by the different possible insertions of propene into the metal–carbon bond: 1,2 or 2,1, with *re* face or *si* face (see Scheme 2).

These aspects of the copolymer microstructure are particularly important in view of the introduction of metallocenes in the world of EP(D)M, still essentially vanadium based. In fact, it is well known that vanadium-based catalysts prepare the macromolecular chain of an E/P copolymer by combining ethene with all the possible propene insertions. The typical feature of a vanadium based copolymer is the presence, in a remarkable amount, of 2,1-inserted propene units [20a]. In Table 7, data are reported to characterize the microstructure of some commercial grades of vanadium-based EPM.

In the homopolymerization of propene, metallocenes are able to give rise to the comonomer units shown in Scheme 2. Furthermore, there are also metallocenes able to generate 3,1-inserted propene units [9c,21]. Although these results suggest the potential of metallocenes to generate a rich variety of copolymer microstructures, by combining all the possible comonomer insertions, yet the occurrence of any type of them in the macromolecular chain of an E/P copolymer has to be verified. A

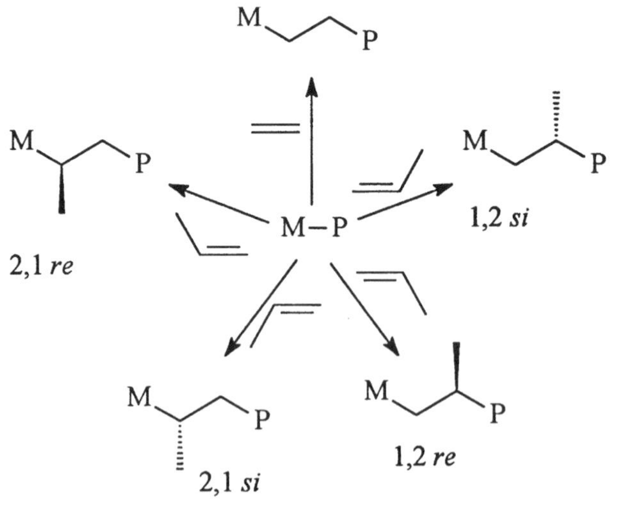

Scheme 2

Table 7 Microstructure of commercial grades of ethene–propene copolymers from vanadium-based catalysts

Commercial grade	C_3 (mol%)	$r_1 r_2$[a]	Propene sequences mm (mol%)	Regioirregularities[b] (mol%)
DUTRAL CO034[c]	21.5	0.52	n. d.	13.6
DUTRAL CO054[c]	27.7	0.66	37	11.0
Polysar 306[d]	19.8	0.68	n.d	13.2

[a]Estimated values obtained from ^{13}C NMR triad distribution neglecting the presence of regioirregularities.
[b]Values corresponding to the lower limits calculated according to Ref. 26.
[c]From EniChem Elastomeri.
[d]From Bayer.

thorough study was performed, moving from the analysis of the behavior of metallocenes in the homopolymerization of propene. By carefully examining the results reported in the patent and scientific literature, the following main conclusions can be drawn as the most useful ones to approach the study of the copolymers.

(i) Metallocenes are available for the preparation of highly isotactic [9c, 22] and syndiotactic [23] and also of essentially atactic polypropene [24], with an homopentad content up to about 99 % (mmmm) and 96 % (rrrr), respectively, for the stereoregular polyolefins. Furthermore, a different degree of stereo-regularity can be achieved by selecting the appropriate metallocene.

(ii) The concentration of propene plays a dramatic role in the control of the stereoregularity of a polypropene prepared with some isospecific metallocenes [25]. Moreover, the effect of the concentration depends on the metallocene.

(iii) At least on the basis of the available results, only isospecific metallocenes appear able to give rise to regioirregular insertions of propene [26]. Both aspecific and syndiospecific metallocenes are highly regiospecific.

A large number of metallocenes have been tested in E/P copolymerization, particular attention being paid to the poorly regiospecific, isospecific metallocenes. The propene concentration was varied in the range 0.1–11 mol/l, thus exploring conditions either suitable to decrease the stereoregularity of the propene sequence, as mentioned in item (ii) above, or typical of industrial practice. It is worth reminding here that the propene concentration in an industrial process for EPDM production is, in solution, at least $2 \, \text{mol} \, \text{l}^{-1}$.

Results are presented in Table 8, where some metallocenes identified as suitable for the preparation of different copolymer microstructures are also shown.

Copolymers with two levels of chemical composition are in particular character-ized. The lowest propene content is within or at the limit of the chemical composition suitable for the preparation of an elastomeric copolymer, while the highest content is that required to detect the presence of propene misinsertions, if any. The following comments can be made.

Table 8 Microstructures of ethene–propene copolymers obtained with different metallocenes at high and low monomer concentrations

Metallocene	P content in the copolymer (mol%)		r_1r_2[a]		PPP/P		Iso-index[b]		Regioirr.[c] (mol%)	
	High [P][d]	Low [P][d]	High [P][d]	Low [P][d]	High [P][d]	Low [P][d]	High [P][d]	Low [P][d]	High [P][d]	Low [P][d]
r-EBI	n.a.[e]	66.0	n.a.	0.52	n.a.	0.43	n.a.	81.8	n.a.	2.4
r-EBTHI	n.i.[e]	25.3	n.i.	0.41	n.i.	0.48	n.i.	100	n.i.	n.d.[e]
	64.5	77.7	1.1	0.93	0.51	0.63	100	62.3	1.7	1.2
	n.i.	27.5	n.i.	0.46	n.i.	0.09	n.i.	100	n.i.	n.d.
r-EBDMI	42.2	45.2	2.5	2.88	0.31	0.34	100	100	2.2	2.5
	19.9	25.9	2.4	3.07	0.12	0.19	100	100	1.1	n.d.
r-Me$_2$C(3-^{t}BuCp)$_2$	35.1	79.5	4.2	3.46	0.35	0.69	100	100	n.d.	1.4
	n.i.	14.9	n.i.	0.42	n.i.	0.07	n.i.	100	n.i.	n.d.
Me$_2$C(Cp)(Flu)	n.i.	74.3	n.i.	0.41	n.i.	0.49	n.i.	6	n.i.	n.d.
	n.i.	26.1	n.i.	n.i.	n.i.	0	n.i.	n.d.	n.i.	n.d.
Ind$_2$	26.6	n.i.	0.9	n.i.	0.15	n.i.	33	n.i.	n.d.	n.i.
(2MeInd)$_2$	n.i.	54.8	n.i.	0.5	n.i.	0.41	n.i.	13	n.i.	n.d.

[a] Calculated as described in Ref. 20.
[b] Calculated as iso-index $= (T_{\beta\beta})_{mm}/[(T_{\beta\beta})_{mm} + (T_{\beta\beta})_{mr+rr}]$, where $T_{\beta\beta} = (CH)_{PPP}$.
[c] Values corresponding to the lower limits calculated according to Ref. 26.
[d] Propene molar concentration in the polymerization bath: high $= 3$–4 M; low $= 0.15$–0.33 M.
[e] n.a. $=$ Not available; n.d. $=$ not detectable; n.i. $=$ not investigated.

(i) Metallocenes allow the preparation of 'limit microstructures', unknown in the past. At both high and low propene concentrations, aspecific metallocenes prepare E/P copolymers with atactic propene sequences, without any regioirregularity, and syndiospecific complexes produce highly stereo- and regioregular syndiotactic propene sequences. Although in the case of a syndiotactic placement of a 1-olefin a large number of successive units are in principle required to ascertain the stereoregularity of the sequence, nevertheless in E/P copolymers having the chemicals composition of an elastomeric product and an r_1r_2 value of about 0.4, prepared with the known syndiospecific metallocenes $X_2CCpFluZrCl_2$ (X = Me, Ph), the syndiotactic sequences can be clearly identified [27]. Most of the isospecific complexes prepare elastomeric E/P copolymers, with a propene content from about 15 to about 45 mol%, with highly stereo- and regioregular propene sequences. The occurrence of stereo- and regioerrors is a minor event, even in the presence of relatively long propene sequences. These metallocenes are thus able to synthesize, even at low propene concentrations, elastomeric ethene–propene bipolymers [1k], with regioregular propene units inserted, in a single macromolecular chain, with the same enantioface.

(ii) The data collected with the isospecific metallocenes seem to suggest an easier occurrence of a regioerror than a stereoerror.

(iii) *rac*-EBDMIZrCl$_2$ seems to show the highest ability to generate regioerrors, whereas *rac*-Me$_2$C(3-^{t}BuCp)$_2$ZrCl$_2$ seems to be the best candidate to have highly regular propene sequences [22c].

As a logical conclusion to what has been reported so far, one has to say that metallocenes give rise to macromolecules of ethene/propene copolymers with a higher degree of order with respect to the reference materials available on a commercial scale. This means that the microstructure of vanadium-based EP copolymers cannot be reproduced, as an aspecific, poorly regiospecific metallocene has not been identified.

3.3 RELATIVE REACTIVITY OF THE COMONOMERS, PRODUCT OF REACTIVITY RATIOS

As already mentioned, we have adopted an approach based on the statistical treatment of polymerization and characterization data that is able to account for the relative reactivity of the comonomers and for their distribution along the macromolecular chain. The method has been thoroughly described elsewhere [1k] and allows one to estimate the reactivity ratios of the comonomers for copolymerizations that can be described with different statistical models (see below). Scheme 3 shows the application of this method to a first-order Markovian copolymerization [28] and leads, in this case, to the determination of r_1 and r_2.

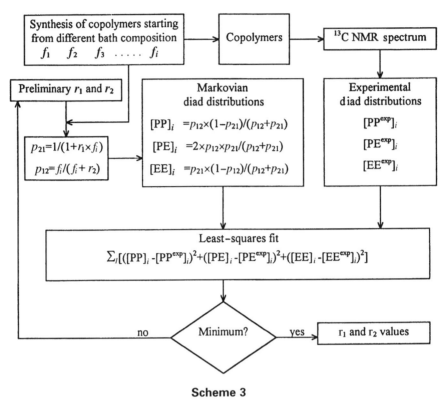

Scheme 3

Many values of r_1 and r_2 are available in the literature. Some are given in Table 9, with particular reference to the polymerization temperature and to the method applied for their calculation. The values derived by us with the method presented here, through both first- and second-order Markovian models, are shown in Table 10. Moreover, reactivity ratios reported in the literature obtained through a second-order Markovian model are also indicated.

Despite the considerable differences in polymerization and elaboration methods adopted by different authors, some general trends can be observed and some general comments can be attempted as follows.

(i) The product of reactivity ratios, $r_1 r_2$, is in most cases <1.
(ii) r_2 decreases dramatically as r_1 increases. This trend is clearly shown in Figure 5.
(iii) There is no apparent correlation between the stereospecificity or the symmetry of a metallocene and the r_1 and r_2 values.
(iv) There is an appreciable difference among the values obtained, often with different methods, by different authors.

Table 9 Reactivity ratios r_1 and r_2 for ethene–propene copolymerizations performed in the presence of MBC: literature data

325

Metallocene	T (°C)	r_1	r_2	r_1r_2	Ref.	Method[a]	Metallocene	T (°C)	r_1	r_2	r_1r_2	Ref.	Method[a]
Cp$_2$ZrCl$_2$	30	16	0.025	0.4	2f	1	Me$_2$SiCp$_2$ZrCl$_2$	50	24	0.029	0.7	2c	2
	40	16.5–25.9	0.03	0.81	3c	1	(Ind)$_2$ZrBz$_2$	40	25.4	–	–	3c	1
	40	16.0–16.9	0.029–0.033	0.49–0.53	2m	1	(Ind)$_2$ZrCl$_2$	40	13.7–29.8	0.09–0.18		3c	1
	50	48	0.015	0.72	2c	2		40	20	–	–	3c	1
Cp$_2$ZrMe$_2$	20	30	0.008	0.24	2f	1	r-EBIZrCl$_2$	0	5.2	0.14	0.73	2j	2
	60	31.5	0.005	0.25	2b	1		25	6.26	0.11	0.69	2e	2
Cp$_2$HfCl$_2$	50	20.6	0.074	1.52	2f	1		40	5.6–8.86	0.09–0.13	0.60–0.71	3c	1
Cp$_2$Ti=CH$_2$	50	24	0.0085	0.204	2c	1		50	6.61	0.06	0.4	2e	2
Cp$_2$TiCl$_2$	30	15.7	0.009	0.14	2f	1		50	2.6	0.39	1	2f	1
Cp$_2$TiMe$_2$	36	19	0.003	0.6	2a	2	r-Me$_2$(Si(Ind)$_2$ZrCl$_2$	20	1.31	0.36	0.47	2p	
Cp$_2$TiPh$_2$	40	–	–	0.29–0.34	2k	1	r-EBTHIZrCl$_2$	40	10.6–12.8	0.072–0.097	0.76–0.97	2m	1
	50	19.5	0.02	0.29	2c	2		50	2.9	0.28	0.81	2f	1
(MeCp)$_2$ZrCl$_2$	50	60	–	–	2c	2	r-Me$_2$Si(H$_4$Ind)$_2$ZrCl$_2$	20	12.4	0.08	0.99	3b	2 and 4
(^{n}Bu-Cp)$_2$ZrCl$_2$	40	19.6–79.6	–	–	3c	1	E(Flu)$_2$ZrCl$_2$	20	1.7	<0.01		2x	1
	40	27	–	–	3e	1	Me$_2$C(Flu)(Cp)ZrCl$_2$	25	1.3	0.2	0.26	2j	2
	80	19	0.005	0.095	3d	3		40	6.5–7.1	0.072–0.113	0.5–0.73	2m	1
(Me$_5$Cp)$_2$ZrCl$_2$	50	250	0.002	0.5	2c	2	Ph$_2$C(Flu)(Cp)ZrCl$_2$	80	3.8	0.1	0.38	3d	3
(Cp$_2$ZrCl)$_2$O	50	50	0.007	0.35	2c	1							

[a] Methods for the determination of r_1 and r_2: (1) from diad distribution and bath composition [26]; (2) from Finemann–Ross equation [30]; (3) from triad distribution and bath composition [2y]; (4) from Kissin–Boehm method [31].

Table 10 Reactivity ratios for ethene–propene copolymerization

Metallocene	$T(^\circ C)$	$r_1{}^a$	$r_2{}^a$	$r_1 r_2{}^a$	$r_{11}{}^b$	$r_{22}{}^b$	$r_{21}{}^b$	$r_{12}{}^b$
This work^c								
Me$_2$SiFlu$_2$ZrCl$_2$	0	1.861	0.023	0.044	–	–	–	–
Cp$_2$HfCl$_2$	50	13.481	0.041	0.548	–	–	–	–
Me$_2$CCpFluZrCl$_2$	25	1.786	0.233	0.415	–	–	–	–
r-EBDMIZrCl$_2$	50	36.533	0.065	2.390	–	–	–	
r-Me$_2$C(3-^{t}Bu-Cp)$_2$ZrCl$_2$	50	25.645	0.110	2.823	–	–	–	–
r-EBIZrCl$_2$	0	5.130	0.136	0.697	4.997	0.171	4.966	0.084
r-EBTHIZrCl$_2$	50	13.167	0.020	0.260	10.545	0.122	8.832	0.046
Literature data^d								
r-Me$_2$Si(Ind)$_2$ZrCl$_2$	25	–	–	–	4.1	0.153	3.9	0.065
Me$_2$CCpFluZrCl$_2$	25	–	–	–	3.4	0.270	2.2	0.153

^{a}First-order Markovian model.
^{b}Second-order Markovian model, $r_{11} = k_{111}/k_{112} = r_1$, $r_{22} = K_{222}/k_{221} = r_2$, $r_{21} = k_{211}/k_{212} = r_1$ and $r_{12} = k_{122}/k_{121} = r_2$.
^{c}Calculated with the method described in the text.
^{d}See Ref. 2n.

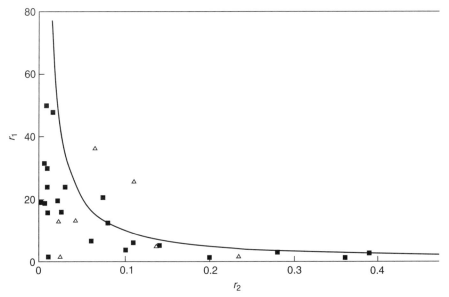

Figure 5 r_1 vs r_2 for metallocene-based copolymerizations: (■) literature data (Table 9); (△) this work (Table 10). The curve refers to $r_1 r_2 = 1$

(v) There is no clear evidence, based on the data presented, of a metallocene able to give rise to ethene–propene copolymerizations with $r_1 = 1/r_2$ ($r_1r_2 = 1$), i.e. ideal (random) copolymerizations.

These comments could be reasonably justified taking into account that the catalytic center arises from the cooperation between the π ligands, coordinated to the transition metal atom, and the growing chain. This concept of the catalytic center is at the basis of models and theories able to rationalize most of the results obtained in the field of 1-olefin polymerization and implies a fundamental role played by steric factors [29]. In the case of a copolymerization of ethene with propene (1-olefin), i.e., two comonomers very different from the steric point of view, a higher reactivity of the smaller monomer (ethene) has to be expected (see the values of r_1 and r_2 in Table 9 and 10). Moreover, the steric demand of the catalytic center depends on the nature of the growing chain, i.e. on the last inserted comonomer unit(s). This being the picture, it is reasonable to imagine that the relative reactivity of the comonomers depends on the last inserted unit(s), thus leading to a very low probability to the occurrence of random copolymerization.

The effect of the π ligands on the relative reactivity of ethene and propene can be examined, for example, comparing the behavior of *rac*-EBIZrCl$_2$ and *rac*-EBTHIZrCl$_2$ (see Table 10). The hydrogenation of the C$_6$ ring brings about an increased relative reactivity of ethene, as demonstrated by the higher r_1, r_{11} and r_{21} values.

Furthermore, two families of metallocene-based copolymerizations appear to be of particular interest, characterized by either a very low (<0.02) or high (>2) product of reactivity ratios. They are discussed in some detail below.

3.4 'ALTERNATE' DISTRIBUTION OF THE COMONOMERS

In 1995, we filed a patent to claim a new copolymer microstructure, characterized by an almost alternate distribution of the comonomers [32]. Long sequences of any of the comonomers were absent, in a broad range of chemical composition, and, in particular, the PP diad was less than 4 mol% for a P content of about 40 mol%. Accordingly, r_1r_2 values lower than 0.02 were determined. These results were obtained with Me$_2$SiFlu$_2$ZrCl$_2$, a metallocene originally applied at Montell for the preparation of high molecular mass atactic polypropene [24]. Some results have been reported since then and a detailed manuscript is in preparation. In Table 11, a summary of the most significant data is shown.

The statistical treatment of these data allowed us to describe the copolymerization with a first-order Markovian model. In Figure 6, the good agreement between the theoretical and experimental values can be seen for the diad distribution.

Chien and co-workers [2x] have also published on ethene–propene copolymerizations characterized by very low r_1r_2 values of about 0.01. The metallocene employed

Table 11 Ethene–propene copolymeriza-tion[a] with $Me_2SiFlu_2ZrCl_2$/MAO

f^b	F^c	$r_1r_2{}^d$
0.092	1.02	0.03
0.55	1.74	0.07

[a]Experimental conditions: $T = 0\,°C$; $p = 1.1\,atm$; solvent = toluene; Al:Zr = 1000.
[b]$f = [E]/[P]$ (mol/mol in the polymerization bath).
[c]$F = [E]/[P]$ (mol/mol in the copoymer).
[d]Calculated as described in Ref. 20.

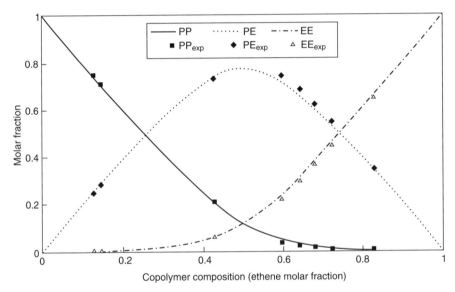

Figure 6 Fitting between theoretical (first-order Markovian model) and experi-mental diads for ethene–propene copolymerizations with $Me_2SiFlu_2ZrCl_2$/MAO

in these studies was $EtFlu_2ZrCl_2$ i.e. the homolog of the above-reported compound, with an ethene bridge in place of the dimethylsylilene.

As mentioned above, $X_2SiFlu_2ZrCl_2$ (X = Me, Bu) has been investigated with respect to atactic polypropene. By preparing random copolymers of propene with a minor amount of ethene, unusual low values of r_1r_2 were obtained. Moreover, low r_1r_2 values had been obtained earlier with the syndiospecific metallocene $Me_2C(Cp)(Flu)ZrCl_2$ [2j]. Recently, Waymouth reported on an 'alternating' co-polymerization promoted by $Me_2Si(3\text{-}Me\text{-}Cp)(Flu)ZrCl_2$ [33]. The common feature of all these metallocenes is the presence of at least a fluorenyl ligand. To investigate

Table 12 Metallocene for ethene–propene copolymers[a] with low $r_1 r_2$

Metallocene	$r_1 r_2$[b]
Flu$_2$ZrCl$_2$	0.77
CH$_2$Flu$_2$ZrCl$_2$	0.38
EBFluZrCl$_2$	0.07
EBFluHfCl$_2$	0.42
Me$_2$SiFlu$_2$ZrCl$_2$	0.07
Bu$_2$SiFlu$_2$ZrCl$_2$	0.07
Me$_2$C(Cp)(Flu)ZrCl$_2$	0.41
Me$_2$C(3-^{t}BuCp)(Flu)ZrCl$_2$	0.78
Me$_2$Si(H$_4$Flu)$_2$ZrCl$_2$	0.12
Me$_2$C(Cp)(H$_4$Flu)ZrCl$_2$	0.18

[a] Ethene content = 50–60 mol%.
[b] Calculated as described in Ref. 20.

the role played by this ligand, ethene–propene copolymerizations were performed with a series of metallocenes, having different steric surroundings of the zirconium atom but, as a common characteristic, the presence of a fluorenyl ligand or a close derivative. Results are given in Table 12.

It can be concluded that in most cases a low $r_1 r_2$ value was obtained. However, there are also clear examples, as in the case of copolymers from Flu$_2$ZrCl$_2$ or from Me$_2$C(3-^{t}BuCp)(Flu)ZrCl$_2$, of a distribution of the comonomers in the macromolecular chain definitely far from the alternate one. These findings perhaps indicate that a combination of both steric and electronic [34] factors is responsible for the alternate distribution of the comonomers. There is no doubt, in fact, that metallocenes with a structure similar to that of Me$_2$SiFlu$_2$ZrCl$_2$ or Me$_2$CCpFluZrCl$_2$, with a higher or lower steric demand but without fluorenyl groups, do not show any ability to distribute the comonomers in an alternate fashion. Let us recall here, for example, Me$_2$Si(Cp)$_2$ZrCl$_2$ and Me$_2$Si(Me$_4$Cp)$_2$ZrCl$_2$ that give rise to $r_1 r_2$ values of about 0.7 and 0.4, respectively. However, it also appears clearly that the absence of a bridge (—CH$_2$CH$_2$— or Me$_2$Si) leads a fluorenyl-containing metallocene to behave as other metallocenes similar only from the steric point of view, e.g. Ind$_2$ZrCl$_2$ or the hydrogenated homolog. Unfortunately, a clear picture of the electronic effect of the metallocene ligands is not available at present, in spite of the accurate works reported in the literature [35].

3.5 ETHENE–PROPENE COPOLYMERIZATIONS WITH A HIGH PRODUCT OF REACTIVITY RATIOS

We have reported on the ability of a metallocene-based catalytic system to promote ethene–propene copolymerizations with a high product of reactivity ratios [1a, c, d].

A representative metallocene, rac-Me$_2$C(3-^{t}BuCp)$_2$ZrCl$_2$, was in particular selected for its ability to generate copolymer microstructures that approach, as mentioned earlier, a *true* ethene–propene bipolymer [1k]. A thorough investigation of the copolymerization results has been reported elsewhere [1k]. Some of the main data are given in Table 13.

The statistical elaboration of these data allowed the copolymerization to be described with a first-order Markovian model. The good agreement between the theoretical curves and the experimental points is shown in Figure 7 for diad distribution.

It is interesting to underline that a high product of reactivity ratios does not originate from particularly high values of either r_1 or r_2 (see Table 10). The unusual event proposed for this metallocene is simultaneously high values of both the reactivity ratios, which means, on excluding the presence of a blend of macro-molecules having different chemical compositions [1k], the presence in the same macromolecule of long sequences of ethene and propene. In order to verify the role played by steric factors, metallocenes characterized by π ligands of different structure were employed and some candidates were found able to prepare E/P copolymers with high r_1r_2. Results are given in Table 14.

By examining the structure of the metallocenes, it appears that all of them are characterized by a highly steric-demanding surrounding of the transition metal atom. Before formulating any type of working hypothesis, it is worthwhile to consider the following comment. The formation of long sequences of both ethene and propene does not depend on different relative reactivities of the comonomers. In fact, a simple comparison of the metallocenes in Table 14 with those in Tables 9 and 10 shows that there are candidates presenting a much higher reactivity for ethene and much lower r_1r_2. This observation and the moderate catalytic activity of the complexes in Table 14 lead one to exclude that the catalyst can be in a starving situation for the most reactive comonomer, i.e. ethene, that could, in principle, favour the occurrence of successive insertions of the 1-olefin.

All these results suggest a role played by the steric factors and allow us to formulate the following working hypothesis. A metallocene presenting a surround-ing of the transition metal atom with a noticeable steric demand is able to promote

Table 13 Ethene–porpene copolymerization^a with rac-Me$_2$C(3-^{t}BuCp)$_2$ZrCl$_2$/MAO

f^b	F^c	$r_1r_2{}^d$
0.020	0.20	3.18
0.61	11.5	4.2

^{a} Experimental conditions: $T = 50\,°$C; $p = 1.1$ atm; solvent = toluene; Al:Zr = 1000.
$^b f = $ [E]/[P] (mol/mol in the polymerization bath).
$^c F = $ [E]/[P] (mol/mol in the copolymer).
^{d} Calculated as described in Ref. 20.

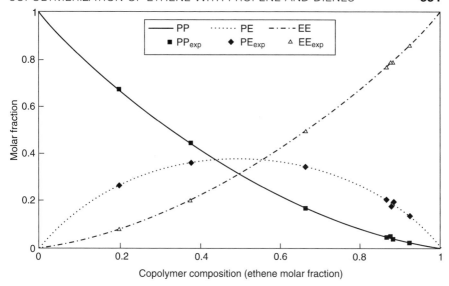

Figure 7 Fitting between theoretical (first-order Markovian model) and experimental diads for ethene–propene copolymerizations with rac-Me$_2$C(3-^{t}Bu-Cp)$_2$ZrCl$_2$/MAO

Table 14 Metallocenes for ethene–propene copolymers^a with high r_1r_2

Metallocene	$r_1r_2{}^b$
r-EBDMIZrCl$_2$	2.50
r-Me$_2$C(3-^{t}Bu-Ind)$_2$ZrCl$_2$	3.81
r-Me$_2$Si(2-Me,4-Ph-Ind)$_2$ZrCl$_2$	1.67

^{a} Ethene content = 50–60 mol%
^{b} Calculated as described in Ref. 20.

E/P copolymerizations with a high product of reactivity ratios. During the copolymerization, two different situations may occur, depending on the last-inserted comonomer unit. When ethene is the last-inserted comonomer, the activation energy for a further ethene insertion is lower, if not much lower, than that required for a propene insertion. This justifies the relatively high values of r_1, i.e. the long ethene sequences. When propene is the last-inserted comonomer, the steric demand is at such a level that the activation energy for an ethene insertion increases considerably, so that there is no longer a dramatic difference with respect to the activation energy required for a propene insertion [$\Delta(\Delta E)$ tends to zero]. This increases the probability of a propene insertion and, as a consequence also of a higher propene concentration in the polymerization bath, propene sequences are generated. This theory could hold also in the presence of the results obtained with 1-olefins other than propene. Preliminary results collected with 1-butene and 1-octene show that high $r_1 r_2$ values characterize these copolymerizations also.

3.6 STATISTICAL MODELS TO DESCRIBE THE COPOLYMERIZATIONS

Any of the copolymerizations discussed so far could be described with either a first- or a second-order Markovian model. The latter model was used to describe copolymerizations carried out with isospecific metallocenes of the type $XInd_2ZrCl_2$ and $X(H_4Ind)_2ZrCl_2$ (X = E, Me_2Si). A second-order Markovian model has been already reported by Fink and co-workers for metallocene-based copolymerizations, promoted by rac-$Me_2SiInd_2ZrCl_2$ and $Me_2CCpFluZrCl_2$ [2l,n]. However, a mechanicistic interpretation of this experimental finding is not yet available. Moreover, the elaborations that lead to the selection of the statistical models are based on assumptions whose validity has still to be verified, e.g. assuming the coefficient $a = 1$ in the expression $v_p = k[c^*][P]^a$.

As already discussed, a clear example of a random ethene–propene copolymerization has still to be identified. On the basis of the picture of the catalytic center proposed previously, this implies a fundamental role played by the growing chain, the type of comonomers being of great importance in determining the statistical law of the intramolecular distribution. Moving from ethene–propene to propene–butene copolymerization, an almost random distribution of the comonomers could be expected, whatever the catalytic system. First results collected with rac-$Me_2C(3$-$^tBuCp)_2ZrCl_2$, rac-EBTHIZrCl$_2$ and $Me_2CCpFluZrCl_2$ are in this direction.

3.7 EFFECT OF ALUMOXANE

The effect of different alumoxanes on the microstructure of ethene–propene copolymers has been investigated [1i]. Copolymerizations were performed with TIBAO, TIOAO and MAO as alumoxanes and with rac-EBTHIZrCl$_2$ as the metallocene. Detailed data and discussions are to be published. A summary of the results is given in Table 15. The main conclusions of this investigation are as follows: (i) TIBAO and TIOAO behave almost identically, (ii) a higher reactivity for propene could be, even if hardly, observed when MAO is used as a cocatalyst; however, the values obtained for the reactivity ratios with the three cocatalysts lie in a narrow range and one can conclude that (iii) they can be unequivocally correlated with the metallocene employed in copolymerization.

Table 15 Reactivity ratios^a for ethene–propene copolymerization^b with r-EBTHIZrCl$_2$/alumoxane

Alumoxane	r_{11}	r_{22}	r_{21}	r_{11}
TIBAO	15	0.07	15.7	0.025
TIOAO	14.7	0.092	19.3	0.021
MAO	10.5	0.12	8.8	0.046

^{a} Calculated with the method described in the text using a second-order Markovian model.
^{b} Experimental conditions: $T = 50\,^\circ$C; $p = 1.1$ atm; solvent = toluene; [Zr] $= 3.5 \times 10^{-5}$ M; Al:Zr $= 1000$.

3.8 RELATIVE REACTIVITY OF THE COMONOMERS

In this section, the relative reactivity of the comonomers is taken as the parameter that links their relative concentration in the copolymer and in the polymerization bath. It is expressed as follows:

$$R = (E/P)_{polymer}/(E/P)_{polym.\ bath}$$

This equation that has already been adopted to discuss metallocene-based copolymerizations [2y, 3b] and allows some issues related to the process for the production of the copolymers to be considered, as will be seen below.

It is easy to correlate the parameter R with f(E/P in the polymerization bath), through r_1 and r_2, by means of the following equation:

$$R = (r_1 f + 1)/(r_2 + f)$$

An analogous equation can be defined in the case of copolymerizations described by a second-order Markovian model [36].

From these equations, it appears that the dependence of R on f is not linear over the whole range of f. Curves that show the correlation of R and f for three metallocenes, particularly discussed above, are presented in Figure 8.

However, these graphs also show that, in an f range suitable for the preparation of an elastomeric copolymer, R does not change substantially. Hence an R value can be identified for any metallocene, upon defining the polymerization conditions. In Table 16, R values are reported for some typical metallocenes.

In Figure 9, the graph of F vs f, in the range of f suitable to have a constant R, suggests some interesting comments on the behavior of the copolymerization process.

MBC characterized by high R values allow the use of low ethene concentrations in the polymerization bath, thus implying a low polymerization pressure. These metallocenes are thus suitable to be applied in the frame of the existing technologies for EPDM production, characterized by low operating pressures. In fact, typical average values of R for vanadium-based catalysts are between 20 and 40. At the same time, a high R value brings about a large variation of the copolymer

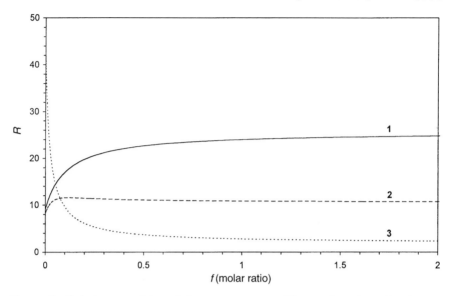

Figure 8 Relative reactivity of the comonomers (R) as a function of polymerization bath composition for three metallocenes: rac-Me$_2$C(3-^{t}Bu-Cp)$_2$ZrCl$_2$ (curve 1), rac-EBTHIZrCl$_2$ (curve 2) and Me$_2$SiFlu$_2$ZrCl$_2$ (curve 3)

composition for a small variation in the relative concentration of the comonomers in the polymerization bath. This could in particular affect the results obtained on the laboratory scale, through discontinuous or semi-continuous polymerizations, traditionally at the basis of any kind of scientific elaborations and discussions. On the other hand, MBC with low R values are less critical for the CCD of the copolymers but, on a larger scale, imply the availability of plants able to tolerate high pressures

Table 16 Relative reactivity of ethene and propene for different metallocenes^a

Metallocene	R
Me$_2$SiFlu$_2$ZrCl$_2$	5
r-EBIZrCl$_2$	15
r-EBTHIZrCl$_2$	16
r-Me$_2$C(3-^{t}Bu-Cp)$_2$ZrCl$_2$	20
BDMIZrCl$_2$	35
Ind$_2$ZrCl$_2$	45
(H$_4$Ind)$_2$ZrCl$_2$	50

^{a} Polymerization conditions: solvent = hexane; [E + P] = 3–4 M; $T = 50\,^\circ$C; [Zr] = 10^{-6} M; Al:Zr = 1000.

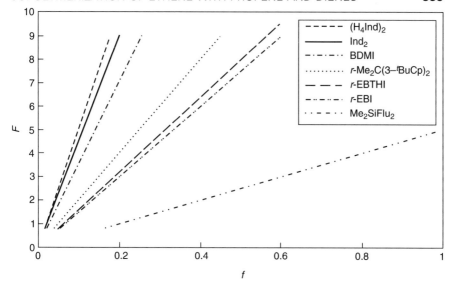

Figure 9 F vs f for a series of zirconocenes. R tends to a limit value $(= 1/r_2)$ for f tending to zero.

or the adoption of unusual polymerization conditions, e.g. a particularly low concentration of the comonomers.

3.9   COPOLYMERIZATION OF ETHENE AND PROPENE WITH A DIENE

The performances of metallocenes in the copolymerization of ethene and propene with a diene are discussed here with particular reference to the diene most commonly applied on the industrial scale, 5-ethylidene-2-norbornene (ENB). Some data and comments are reported also for other norbornyl-type dienes, such as 5-vinyl-2-norbornene (VNB) and dicyclopentadiene (DCPD).

The catalytic activity of a metallocene-based polymerization decreases in the presence of a diene [2b, w, x, z, 3a]. The extent of the decrease depends on the nature of the diene, as a exemplified in Table 17 for ethene–propene–diene copolymerizations promoted by EBTHIZrCl$_2$/TIOAO.

It appears that ENB and VNB behave similarly and have less effect on the catalyst productivity than DCPD. Table 17 shows also a critical aspect of the metallocene-based technology for EPDM, namely a relatively low reactivity of the diene. The ENB conversion was investigated by comparing different single-center catalysts with a traditional vanadium-based catalyst, as shown in Figure 10.

Table 17   Ethene–propene–(diene) copolymerization^a with r-EBTHIZrCl$_2$

| | | Copolymer | | | | |
Diene	Activity kg polymer/(g Zr h)	E (mol%)	Diene (mol)	IV (dl/g^{-1})	$R_E{}^b$	$R_{diene}{}^c$
None	500	68	–	3.2	16	–
ENB	170	71.9	1.3	2.1	16	8
VNB	150	66.4	1.2	2.3	18	8.5
DCPD	32	73.0	1.7	2.0	17.6	14

^{a}Polymerization conditions: solvent = hexane; [E + P] = 2.1 M; [diene] = 0.01 M; $T = 50\,^\circ$C; cocatalyst = TIOAO; Al:Zr = 300.
^{b}(E/P)$_{\text{in the polymer}}$/(E/P)$_{\text{in the polymerization bath}}$.
^{c}(diene/P)$_{\text{in the polymer}}$/(diene/P)$_{\text{in the polymerization bath}}$.

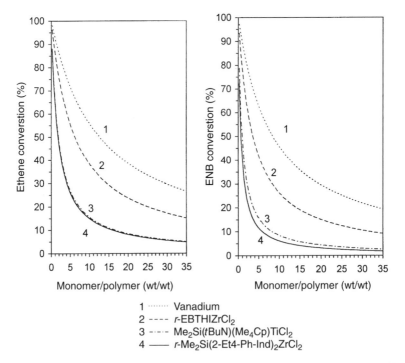

1 ⋯⋯ Vanadium
2 ---- r-EBTHIZrCl$_2$
3 -·--- Me$_2$Si(tBuN)(Me$_4$Cp)TiCl$_2$
4 ——— r-Me$_2$Si(2-Et4-Ph-Ind)$_2$ZrCl$_2$

Figure 10   Monomers conversion versus monomer:polymer ratio for different catalysts.

It is interesting to observe that a highly isospecific metallocene, that requires a surrounding of the transition metal atom highly demanding from the steric point of view, shows the lowest reactivity for ENB.

The selectivity of the metallocene-based catalytic system towards the two double bonds of the diene was carefully studied through the analysis of the microstructure of the copolymers and hydrooligomerization reactions [37]. The results show that with metallocenes belonging to different classes only the endocyclic double bond of ENB undergoes an insertion reaction, whereas with a metallocene presenting some steric demand, such as rac-EBTHIZrCl$_2$, both the double bonds of VNB seem to be inserted in the copolymer chain.

The presence of a diene in the polymerization bath brings about an appreciable reduction of the molecular mass of the copolymer. Data in Table 17 are reported only as an example.

No effect has either been observed by us or reported by other workers on the copolymer microstructure, which remains as described in previous sections.

4 POLYMERIZATION PROCESS FOR THE PRODUCTION OF METALLOCENE-BASED EPDM

Thanks to the low chlorine content, the absence of strong acidic sites and the high catalytic activity, metallocenes based catalytic systems allow us to

(i) prepare EPDM with low ash content, also through processes that do not require any washing or deashing step;
(ii) avoid the formation of gels or indispersible particles, to improve both product properties and process behavior; and
(iii) reduce, to a dramatic extent, corrosion problems in the production plant.

4.1 SOLUTION PROCESS

The catalysts that could bring about a real breakthrough for the solution process are those able to produce commercial grades of EPDM at very high polymerization temperatures. Among the advantages of a high process temperature are an increase in catalyst activity and process productivity and an improvement in the thermal balance and, in a nutshell, of the economic balance of the process. As discussed above, a limited choice is given by metallocenes and, perhaps, more suitable candidates should be selected in the families **3** and **5** in Figure 1. The former is already applied on an industrial scale.

4.2 SLURRY PROCESS

Metallocenes are able to reproduce the typical advantages of a slurry process for EPDM, from a higher catalytic activity and polymer concentration in the polymerization medium to a higher molecular mass of the copolymers. Moreover, they are the catalysts for a new generation of slurry processes, as they allow us (i) to obtain an improved process behavior over a wide range of copolymer characteristics and (ii) to prepare high-quality polymers, with low or very low ash content.

4.3 GAS-PHASE PROCESS

A new technology has been recently developed by Montell, named Multicatalyst Reactor Granule Technology (MRGT) [1f–h] and described in Scheme 4.

MRGT allows us to prepare a wide range of products, either crystalline or elastomeric, and blends of them and, in particular, polyolefin elastomers in the gas phase with a single-center catalyst.

Some comments on the performance of metallocenes in the gas phase can be made, based on the experience with MRGT:

(i) it is possible to support metallocenes on inert carriers, such as porous polyolefins, that do not necessarily imply a chemical bond with the catalytic

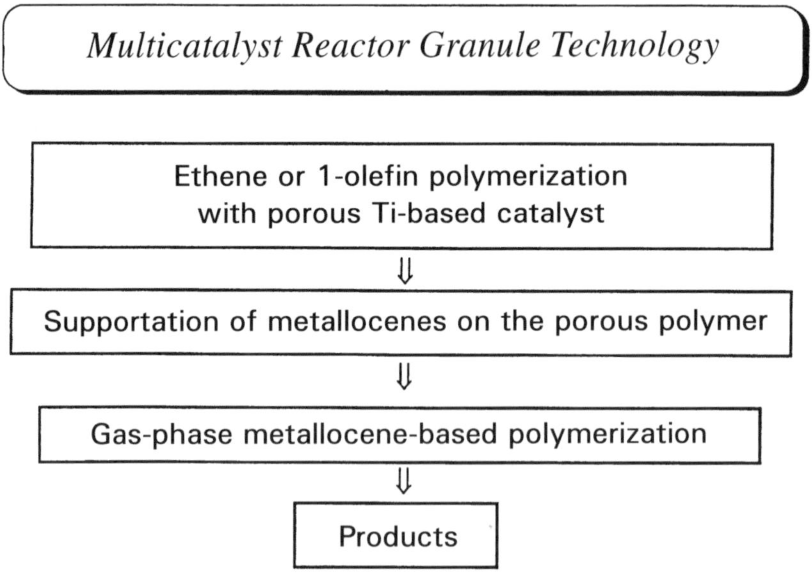

Multicatalyst Reactor Granule Technology

Ethene or 1-olefin polymerization
with porous Ti-based catalyst

⇓

Supportation of metallocenes on the porous polymer

⇓

Gas-phase metallocene-based polymerization

⇓

Products

Scheme 4

system but, nevertheless, achieve high catalytic performance and outstanding product morphology;

(ii) a high flexibility in the composition of the catalytic system can be exploited and very low Al:Zr ratios can be adopted;

(iii) it is worth stressing the low level of catalytic residues in a product coming from a solvent-free process, prepared with a short residence time;

(iv) the typical features of metallocene-based chemistry are reproduced in the gas phase and narrow molecular mass and chemical composition distributions are obtained: $\bar{M}_w/\bar{M}_n$ between 2.5 and 5 and a v coefficient close to 2 were observed;

(v) no limitations on the molecular mass of elastomers are suffered: IV, as high as 3–4 or even 6–7 were obtained with *rac*-EBTHIZrCl$_2$ and *meso*-EBDMIZrCl$_2$, respectively.

As a general comment on the performance of metallocenes in the frame of the different processes, the 'fingerprint' of the copolymers is the same, regardless of the polymerization process. To support this contention, in Figure 11 the triad distributions for three copolymers prepared with *rac*-EBTHIZrCl$_2$ in solution, slurry and the gas phase are reported.

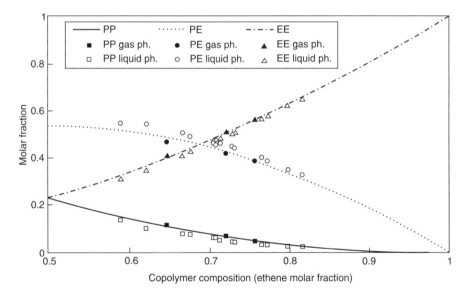

Figure 11 Diad distributions for ethene–propene copolymers with *rac*-EBTHIZrCl$_2$ obtained in liquid and gas phases

5 GLOSSARY OF METALLOCENES

Ligands	
Cp	cyclopentadienyl
I or Ind	indenyl
Flu	fluorenyl
DMI	4,7-dimethylindenyl
THI or H_4Ind	4,5,6,7-tetrahydroindenyl
B	bis
Isomers	
r- or *rac*-	racemic mixture
m- or *meso*-	*meso* form
Bridges	
M or CH_2	methylene
E or CH_2CH_2	ethylene
Me_2C	isopropylidene
Ph_2C	diphenylmethylene
Me_2Si	dimethylsilylene
Bu_2Si	dibutylsilylene

The bridge is in position 1 of an indenyl or position 9 of a fluorenyl group. Example: *r*-EBTHIZrCl$_2$ = racemic ethylenebis(4,5,6,7-tetrahydroindenyl)zirconium dichloride. Note: when only the π ligands are indicated, ZrCl$_2$ is omitted.

6 REFERENCES

1. (a) Galimberti, M., Martini, E., Sartori, F., Piemontesi, F. and Albizzati, E., in *Proceedings of MetCon '94*, Houston, TX, USA, May 25–27, 1994 (available from The Catalyst Group, P.O. Box 637, Spring House, PA 19477, USA); (b) Galimberti, M., Dall'Occo, T., Camurati, I., Sartori, F. and Piemontesi, F., in *Proceedings of MetCon '95*, Houston, TX, USA, May 17–19, 1995 [available as in (a)]; (c) Galimberti, M., Martini, E., Piemontesi, F., Sartori, F., Resconi, L. and Albizzati, E., *Macromol. Symp.*, **89**, 259 (1995); (d) Galimberti, M., Dall'Occo, T., Piemontesi, F., Camurati, I., Collina, G. and Battisti, M., in *Proceedings of MetCon '96*, Houston, TX, USA, June 12–13, 1996 [available as in (a)]; (e) Galimberti, M., Piemontesi, F., Fusco, O. and Camurati, I., communication at the Workshop on Recent Advances in Homogeneous Ziegler–Natta Polymerization, IIASS, Vietri sul Mare (SA), Italy, September 19–20, 1996; (f) Galimberti, M., Baruzzi, G., Camurati, I., Fusco, O., Piemontesi, F. and Vianello, M., in *Proceedings of Metallocenes Europe '97, The 3rd International Congress on Metallocenes Polymers*, Düsseldorf, April 8–9, 1997 (available from Schotland Business Research, Skillman, NJ 08558, USA); (g) Galimberti, M., Ferraro, A., Baruzzi, G., Sgarzi, P., Camurati, I., Piemontesi, F., Mingozzi, I. and Vianello, M., in *Proceedings of MetCon '97*, Houston, TX, USA, June 4–5, 1997 [available as in (a)]; (h) Ferraro, A., Galimberti, M., Baruzzi, G. and Di Diego, M., in *Proceedings of MetCon '97*, Houston TX, USA, June 4–5, 1997

[available as in (a)]; (k) Galimberti, M., Piemontesi, F., Fusco, O., Camurati, I. and Destro, M. *Macromolecules*, **31**, 3409 (1998); (l) Galimberti, M., Destro, M., Fusco, O., Piemontesi, F. and Camurati, I., in *Proceedings of XIII Convegno Italiano di Scienza e Tecnologia delle Macromolecole, AIM*, Genova, Italy, September 2–25, 1997 (available from AIM, c/o Dipartimento di Chimica e Chimica Industriale, 56126 Pisa, Italy).
2. (a) Busico, V., Mevo, L., Palumbo, G., Zambelli, A. and Tancredi, T. *Makromol. Chem.*, **184**, 2193 (1983); (b) Kaminsky, W. and Miri, M., *J. Polym. Sci., Polym. Chem. Ed.*, **23**, 2151 (1985); (c) Ewen, J. A., in Kei, T. and Soga, K. (Eds), *Catalytic Polymerization of Olefins* Kodansha, Tokyo, 1986, p. 271; (d) Kaminsky, W. and Schlobohm, M., *Makromol. Chem., Macromol. Symp.*, **4**, 103 (1986); (e) Drögemüller, H., Heiland, K. and Kaminsky, W., in Kaminsky, W. and Sinn, H. (Eds), *Transition Metals and Organometallics as Catalysts for Olefin Polymerization*, Springer, Berlin, 1988, p. 303; (f) Chien, J. C. W and He, D. J., *J. Polym. Sci. Part A: Polym. Chem.*, **29**, 1585 (1991); (g) Chien, J. C. W and He, D. J., *J. Polym. Sci. Part A: Polym. Chem.*, **29**, 1595 (1991); (h) Chien, J. C. W and He, D. J., *J. Polym. Sci. Part A: Polym. Chem.*, **29**, 1603 (1991); (i) Chien, J. C. W and He, D. J., *J. Polym. Sci. Part A: Polym. Chem.*, **29**, 1609 (1991); (j) Zambelli, A., Grassi, A., Galimberti, M., Mazzocchi, R. and Piemontesi, F., *Makromol. Chem., Rapid Commun.*, **12**, 523 (1991); (k) Soga, K., Park, J. R. and Shiono, T., *Polym. Commun.*, **32**, 310 (1991); (l) Herfert, N. and Fink, H., *Polym. Mater. Sci. Eng.*, **67**, 31 (1992); (m) Soga, K. and Uozomi, T., *Makromol. Chem.*, **193**, 823 (1992); (n) Herfert, N., Montag, P. and Fink, G., *Makromol. Chem.*, **194**, 3167 (1993); (o) Randall, J. C. and Rucker, S. P., *Macromolecules*, **27**, 2120 (1994); (p) Tsai, W. M., and Chien J. C. W., *J. Polym. Sci. Part A: Polym. Chem.*, **32**, 149 (1994); (q) Soga, K. and Kaminaka, M., *Macromol. Chem. Phys.*, **195**, 1369 (1994); (r) Soga, K., Uozumi, T., Saito, M. and Shiono, T., *Macromol. Chem. Phys.*, **195**, 1503 (1994); (s) Pietikainen, P. and Seppala, J. V., *Macromolecules*, **27**, 1325 (1994); (t) Koivumaki, J. and Seppala, J. V., *Eur. Polym. J.*, **30**, 1111 (1994); (u) Banzi, V., Angiolini, L., Caretti, D. and Carlini, C., *Angew. Makromol. Chem.*, **229**, 113 (1995); (v) Mitani, M., Ouchi, K., Hayakawa, M., Yamada, T. and Mukaiyawa, T., *Polym. Bull.*, **35**, 677 (1995); (w) Marques, M., Yu, Z, Rausch, M. D. and Chien, J. C. W., *J. Polym. Sci. Part A: Polym. Chem.*, **33**, 2787 (1995); (x) Yu, Z, Marques, M., Rausch, M. D. and Chien, J. C. W., *J. Polym. Sci. Part A: Polym. Chem.*, **33**, 2795 (1995); (y) Bergemann, C., Cropp, R. and Luft, G., *J. Mol. Catal. A: Chem.*, **102**, 1 (1995); (z) Yu, Z., Marques, M., Rausch, M. D. and Chien, J. C. W., *J. Polym. Sci. Part A: Polym. Chem.*, **33**, 979 (1995).
3. (a) Dolatkhani, M., Cramail, H. and Deffieux, A., *Macromol. Chem. Phys.*, **197**, 2481 (1996); (b) Bergmann, C., Cropp, R. and Luft, G., *J. Mol. Catal. A: Chem.*, **105**, 87 (1996); (c) Lehtinen, C. and Löfgren, B., *Eur. Polym. J.*, **33**, 115 (1997); (d) Karol, F. J., Kao, S.-C., Wasserman E. P. and Brady, R. C., *New J. Chem.*, **21**, 797 (1997); (e) Lehtinen, C., Stark, P. and Löfgren, B., *J. Polym. Sci. Part A: Polym. Chem.*, **35**, 307 (1997).
4. (a) Natta, G., Pino, P., Mazzanti, G. and Lanzo, R., *Chim. Ind. (Milan)*, **39**, 1032 (1957); (b) Breslow, D. S. and Newburg, N. R., *J. Am. Chem. Soc.*, **79**, 5072 (1957); (c) Natta, G., Pino, P., Mazzanti, G. and Giannini, U., *J. Am. Chem. Soc.*, **79**, 2975 (1957).
5. (a) Kaminsky W. and Hansen, H., *Ger. Pat.*, 3 240 383 (1982); (b) Sinn, H. and Kaminsky, W., *Adv. Organomet. Chem.*, **18**, 99 (1980).
6. Kaminsky, W., in Seymour, R. B. and Cheng, T. (Eds), *History of Polyolefins*, Reidel, Dordrecht, 1986 p. 257.
7. Ewen, J. A., *J. Am. Chem. Soc.*, **106**, 6355 (1984).
8. Kaminsky, W., Küpler, K., Brintzinger, H. H. and Wild, F. R. W. P., *Angew. Chem.*, **97**, 507, (1985); *Angew. Chem., Int. Ed. Eng.*, **24**, 507 (1985).
9. (a) Jordan, R. F., *Adv. Organomet. Chem.*, **32**, 325 (1991); (b) Gupta, V. K., Satish, S. and Bhardway, I. S. J., *Macromol. Sci. Rev. Macromol. Chem. Phys.*, **C34**, 439 (1994); (c) Brintzinger H. H., Fischer, D., Mülhaupt, R., Rieger, B. and Waymouth, R. M., *Angew.*

Chem., Int. Ed. Engl., **34**, 1143 (1995); (d) Huang J. and Rempel, G. L., *Prog. Polym. Sci.*, **20**, 459 (1995); (e) Bochmann, M., *J. Chem. Soc. Dalton Trans.*, 225 (1996).

10. See, for example, (a) *Proceedings of Metallocenes Europe '97, The 3rd International Congress on Metallocenes Polymers*, Düsseldorf, April 8–9, 1997; (b) *Proceedings of MetCon '97*, Houston, TX, USA, June 4–5, (1997); (c) *Metallocene Technology '97*, Chicago, IL, USA, June 16–17, 1997.

11. (a) Stevens, J. C., Timmers, F. J., Wilson, D. R., Schmidt, G. F., Nickias, P. N., Rosen, R. K., Knight, G. W. and Lai, S., *Eur. Pat.*, 416815A2 to Dow (1991); (b) Canich, J. A. M., *Eur. Pat.*, 420436A1, to Exxon (1991).

12. (a) Yamada, S., Sone, M., Hasegawa, S. and Yano, A. *Eur. Pat.*, 0570982A1, to Tosoh (1993); (b) Yamada, S. and Yano, A., *Eur. Pat.*, 0582268A2, to Tosoh (1994).

13. Van Beek, J. A. M., Van Doremaele, G. H. J., Gruter, G. J. M., Arts, H. J. and Eggels, G. H. M. R., *PCT Int. Pat. Appl.*, WO 96/13529, to DSM (1996).

14. (a) Johnson, L. K., Killian, C. M., Arthur, S. D., Feldman, J., McCord, E. F., McLain, S. J., Kreutzer, K. A., Bennet, M. A., Coughlin, E. B., Ittel, S. D., Parthasarathy, A., Tempel, D. J. and Brookhart, M. S., *PCT Int. Pat. Appl.*, WO 96/23010, to DuPont (1996); (b) Johnson, L. K., Feldman, J., Kreutzer, K. A., McLain, S. J., Bennet, M. A., Coughlin, E. B., Donald, D. S., Nelson, L. T. J., Parthasarathy, A., Shen, X., Tam, W. and Wang, Y., *PCT Int. Pat. Appl.*, WO 97/02298, to DuPont (1997).

15. Resconi, L., Fait, A., Piemontesi, F., Colonnesi, M., Rychlicki, H. and Ziegler, R., *Macromolecules*, **28**, 6667 (1995).

16. Hlatky, G. G., Turner, H. W. and Eckman, R. R., *J. Am. Chem. Soc.*, **111**, 2728 (1989); (b) Turner, H. W., *Eur. Pat. Appl.*, 277004 (1988); *Chem. Abstr.*, **110**, 58290a (1989); (c) Turner, H. W. and Hlatky, G. G., *Eur. Pat. Appl.*, 277003 (1988); *Chem. Abstr.*, **110**, 58290b (1989).

17. (a) Resconi, L., Giannini, U. and Albizzati, E., *Eur. Pat.*, 0384171, to Montell Technology (1990); (b) Resconi, L., Galimberti, M., Piemontesi, F., Guglielmi, F. and Albizzati, E., *Eur. Pat.*, 0575875, to Montell Technology (1993); (c) Dall'Occo, T., Galimberti, M., Resconi, L., Albizzati, E. and Pennini, G., *PCT Int. Pat. Appl.*, WO 96/02580, to Montell Technology (1996); (d) Galimberti, M., *PCT Int. Pat. Appl.*, WO 97/00897, to Montell Technology (1997).

18. Morton, M. (Ed.), *Rubber Technology*, R. E. Krieger Publishing, Malabar, FL, 2nd edn (1981).

19. The v coefficient is a relative dispersion index. It is obtained through copolymer fractionation and is given by the expression must 100 × (standard deviation/arithmetic mean). Substantially, the ethene content of a copolymer is given by the expression (average ethene content) ± $(v/100)$ × (average ethene content).

20. (a) Carman, C. J., Harrington, R. A. and Wilkes, C. E., *Macromolecules*, **10**, 536 (1977); (b) r_1r_2 calculated with the following expressions: (i) $r_1r_2 = 1 + F(\chi + 1) - (F + 1)(\chi + 1)^{1/2}$, where $\chi = [PPP + PPE]/[EPE]$, when E > P [20a]; (ii) $r_1r_2 = 1 + (1/F)(\chi + 1) - [(1/F) + 1](\chi + 1)^{1/2}$, where $\chi = [EEE + EEP] / [PEP]$, when P > E; in both (i) and (ii), $F = [E]/[P]$ (mol mol^{-1} in the copoymer).

21. (a) Grassi, A., Zambelli, A., Resconi, L., Albizzati, E. and Mazzocchi, R., *Macromolecules*, **21**, 15 (1988); (b) Soga, K., Shiono, T., Takemura, S. and Kaminsky, W., *Macromol. Chem., Rapid Commun.*, **8**, 305 (1987).

22. (a) Spaleck, W., Antberg, M., Rohrmann, J., Winter, A., Bachmann, B., Kiprof, P., Behm, J. and Herrmann, W. *Angew. Chem., Int. Ed. Engl.*, **31**, 1347 (1992); (b) Spaleck, W., Küber, F., Winter, A., Rohrmann, J., Bachmann, B., Antberg, M., Dolle, V. and Paulus, E., *Organometallics*, **13**, 954 (1994); (c) Resconi, L., Piemontesi, F., Nifant'ev, I., Ivchenko, P. and Albizzati, E., *PCT Int. Pat. Appl.*, WO 96/22995, to Montell Technology (1995).

23. Ewen, J. A., Jones, R. L., Razavi, A. and Ferrara, J. D., *J. Am. Chem. Soc.*, **110**, 6255 (1988).
24. (a) Resconi, L., Jones, R. L., Albizzati, E., Camurati, I., Piemontesi, F., Guglielmi, F. and Balbontin, G., *ACS Polym. Prepr.*, **35**, 663 (1994); (b) Resconi, L., Jones, R. L., Rheingold, A. L. and Yap, G. P. A., *Organometallics*, **15**, 998 (1996)
25. (a) Busico, V. and Cipullo, R., *J. Am. Chem. Soc.*, **116**, 9329 (1994); (b) Resconi, L., Fait, A., Piemontesi, F., Colonnesi, M., Rychlicki, H. and Ziegler, R., *Macromolecules*, **28**, 6667 (1995).
26. Randall, J. C., *Macromolecules*, **11**, 33 (1978).
27. Tritto, I., Fan, Z.-Q., Locatelli, P., Sacchi, M. C., Camurati, I. and Galimberti, M., *Macromolecules*, **28**, 3342 (1995).
28. Randall, J. C., *Polymer Sequence Determination. Carbon-13 NMR Method*, Academic Press, New York, 1977, p. 75.
29. (a) Guerra, G., Cavallo, L., Moscardi, G., Vacatello, M. and Corradini, P., *J. Am. Chem. Soc.*, **116**, 2988 (1994); (b) Guerra, G., Cavallo, L., Moscardi, G., Vacatello, M. and Corradini, P., *Macromolecules*, **29**, 4834 (1996), and references cited therein.
30. Finemann, M. and Ross, S. D., *J. Polym. Sci.*, **5**, 259 (1950).
31. Kissin, Y. V. and Beach, D. L., *J. Appl. Polym. Sci.*, **29**, 1171 (1984).
32. Galimberti, M., Resconi, L. and Albizzati, E., *Eur. Pat.*, 632066, to Montell Technology (1995).
33. Waymouth, R. M., presentation at the International Symposium on Methatesis and Related Chemistry (ISOM 12), St Augustine, FL, USA, July 13–18, 1997.
34. Resconi, L. and Piemontesi, F., communication at the Workshop on Recent Advances in Homogeneous Ziegler–Natta Polymerizaton, IIASS, Vietri sul Mare (SA), Italy, September 19–20, 1996.
35. Alameddin, N. G., Ryan, M. S., Eyler, J. R., Siedle, A. R. and Richardson, D. E., *Organometallics*, **14**, 5005 (1995).
36. Ham, G. E., in Ham, G. E. (Ed.), *Copolymerization*, Interscience, New York, 1964, p. 10.
37. Di Silvestro, G., Caronzolo, N., Galimberti, M. and Fusco, O., in *Proceedings of XIII Convegno Italiano di Scienza e Technologia delle Macromolecole*, AIM, Genova, Italy, September 21–25, 1997.

15

Copolymerization of Ethylene with Dienes Using Metallocene/MAO Catalysts

GEORGE J. JIANG
Chung-Yuan Christian University, Chung-Li, Taiwan

1 INTRODUCTION

Kaminsky, Sinn and co-workers [1,2] developed a metallocene/methylaluminoxane (MAO) catalyst system. Homogeneous metallocene catalytic systems such as Cp_2ZrCl_2/MAO and $Et(Ind)_2ZrCl_2$/MAO have received much attention recently [3,4]. These catalysts are much more effective than their heterogeneous counterparts for incorporation of a diene monomer into a polyethylene structure, and they reduce the difference in activity between comonomers and more easily control the composition of the copolymers. Especially the bridged $Et(Ind)_2ZrCl_2$/MAO catalytic system with strained geometry provides an unsaturated polypropylene with a narrow molecular weight distribution and controllable diene concentration. Also, the double bonds appear in the side-chain, and maintain the main chain unchanged. The results maintain the crystallinity of the polypropylene; the side-chain double bond is located in the amorphous phase, remaining available for further modifications.

The chemical modification of polyolefins, especially polyethylene and polypropylene, has been an area of increasing interest. Chung and co-workers [5,6] developed a novel method for the preparation of functionalized polyolefins by a borane approach. The method involves borane-containing intermediates which can be obtained by a direct or a post-polymerization process. In post-polymerization, hydroboration of unsaturated polyethylene prepared from ethylene and diene can be

Metallocene-based Polyolefins   Edited by J. Scheirs and W. Kaminsky
© 2000 John Wiley & Sons Ltd

used. Many functionized polyolefins and copolymers have been synthesized via a metallocene and borane procedure [7–9].

A borane-containing copolymer can serve as a macroinitiator for the free radical polymerization of methyl methacrylate, etc. [10–12]. Functional polyolefins can also be utilized for grafting with ε-caprolactone and ε-caprolactam. Various graft or block copolymers can also be prepared [8,9].

In recent years, graft or block copolymers have been found to be effective compatibilizers [13] which can significantly alter the morphology of a polymer blend by increasing the interfacial interaction and reducing the domain size. Despite the potential uses, the availability of graft or block copolymers is very limited, mainly owing to the difficulties in their preparation. The diene-containing monomers in ethylene copolymerization can easily be converted to graft or block copolymer with a polar segment [7–9]. Copolymerization of ethylene and dienes has been catalyzed by metallocene/MAO, thereby providing a novel approach for the preparation of a graft or block copolymer. This chapter provides an outline of recent results for the incorporation of a diene monomer into a polyethylene, following which a graft functional polyolefin or block copolymer can be prepared.

2 METALLOCENE CATALYSTS

2.1 METALLOCENE/MAO STRUCTURE

The catalytic systems used in the copolymerization of ethylene with dienes include Cp_2ZrCl_2/MAO, $Et(Ind)_2ZrCl_2$/MAO, $CpTpZrCl_2$/MAO and $Cp_2YCl(THF)$/ MAO. The basic structure of metallocene catalysts consists of a parallel (or having a certain angle) cyclopentadienyl (Cp) ring and a metal. The bonding between a Cp ring and a metal is π coordination. The central metal is bonded to chloride to produce an active site. The metals could be Group IIIB and IVB transition metals, such as Zr, Hf, Ti and Y. There can also be a substitution on the Cp ring, and the Cp ring can be substituted with a two or three fused ring system such as indenyl or fluorenyl. The substitution alters the symmetry of the metallocene structure to increase the selectivity. Also, the two-ring system can be linked by a carbon or silicon atom to form a bridged metallocene catalyst.

The cocatalyst is methylaluminoxane (MAO), obtained by the hydration of trimethylaluminum. The structure of MAO can be linear or cyclic. The structures of the metallocenes mentioned above and linear and cyclic MAO are illustrated

2.2 MECHANISM OF METALLOCENE CATALYTIC POLYMERIZATION

The mechanism coincides with the chain polymerization process in the polymerization of olefin monomers via metallocene catalysts. The chain polymerization process

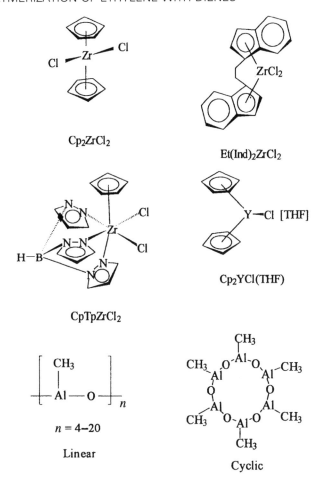

Cp$_2$ZrCl$_2$

Et(Ind)$_2$ZrCl$_2$

CpTpZrCl$_2$

Cp$_2$YCl(THF)

$n = 4$–20

Linear

Cyclic

includes chain initiation, propagation and termination reactions. The processes are shown in Schemes 1–3.

3 COPOLYMERIZATION OF ETHYLENE WITH DIENES

There are three types of copolymerization of ethylene with dienes. First, there are comonomers which are linear non-conjugated dienes, such as 1,4-hexadiene [7,14], 1,5-hexadiene [15] and 3-methyl-1,6-octadiene [16]. The second type of comonomer is a cyclic diene such as 5-ethylene-2-norbornene (ENB) [9,17]. Third, terpolymer-

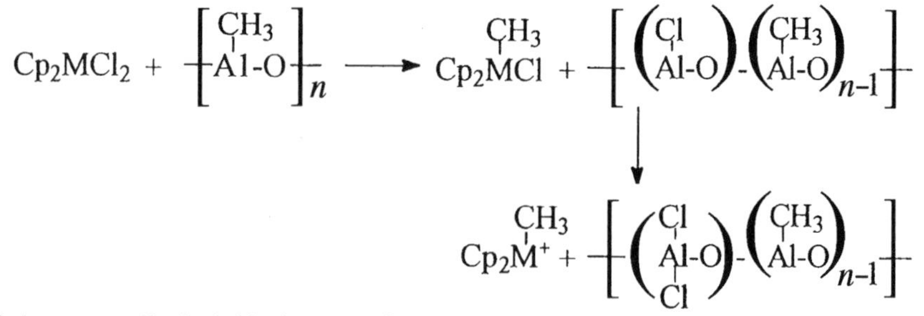

$$Cp_2MCl_2 + \left[\begin{array}{c} CH_3 \\ | \\ Al\text{-}O \end{array}\right]_n \longrightarrow Cp_2MCl + \left[\left(\begin{array}{c} Cl \\ | \\ Al\text{-}O \end{array}\right)\text{-}\left(\begin{array}{c} CH_3 \\ | \\ Al\text{-}O \end{array}\right)_{n-1}\right]$$

$$\downarrow$$

$$Cp_2M^+ + \left[\left(\begin{array}{c} Cl \\ | \\ Al\text{-}O \\ | \\ Cl \end{array}\right)\text{-}\left(\begin{array}{c} CH_3 \\ | \\ Al\text{-}O \end{array}\right)_{n-1}\right]$$

Scheme 1   Chain initiation reaction

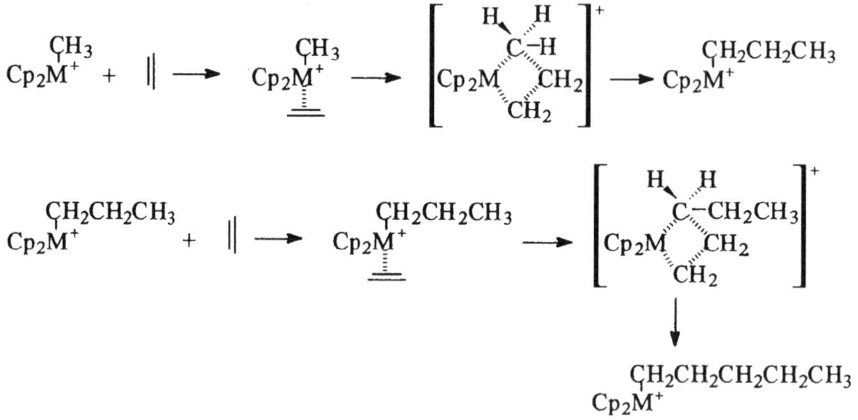

$$Cp_2M^+\!\!\!\overset{CH_3}{} + \; \| \longrightarrow Cp_2M^+\!\!\!\overset{CH_3}{} \longrightarrow \left[Cp_2M\overset{\displaystyle H\;\;H}{\underset{\displaystyle CH_2}{\overset{C\text{-}H}{\cdots CH_2}}}\right]^+ \longrightarrow Cp_2M^+\!\!\!\overset{CH_2CH_2CH_3}{}$$

$$Cp_2M^+\!\!\!\overset{CH_2CH_2CH_3}{} + \; \| \longrightarrow Cp_2M^+\!\!\!\overset{CH_2CH_2CH_3}{} \longrightarrow \left[Cp_2M\overset{\displaystyle H\;\;H}{\underset{\displaystyle CH_2}{\overset{C\text{-}CH_2CH_3}{\cdots CH_2}}}\right]^+$$

$$\downarrow$$

$$Cp_2M^+\!\!\!\overset{CH_2CH_2CH_2CH_2CH_3}{}$$

Scheme 2   Chain propagation reactions

izations of ethylene with α-olefins and ethylidene norbornene have been achieved [18–21].

3.1   HOMOPOLYMERIZATION OF ETHYLENE WITH METALLOCENE/MAO CATALYSTS

In the homopolymerization of ethylene, the activity varies with the ratio of the catalyst to the cocatalyst and with temperature. The activity increase dramatically with an [Al]/[Zr] ratio between 500 and 1000 but increases only very slightly beyond 1000. The variation of the activity in ethylene polymerization with temperature is shown in Figure 1. With the Cp_2ZrCl_2/MAO catalyst system, there is a minimum at 50 °C and a maximum at 90 °C. However, with the Cp_2ZrCl_2/MAO catalyst system, the minimum and maximum shift to 40 and 70 °C, respectively. The data are shown in Table 1. The activity with $CpTpZrCl_2$/MAO is much lower than that with the Cp_2ZrCl_2/MAO catalyst system. The molecular weight decreases with increase in

(1) Chain termination reaction of intramolecular β-hydrogen:

$$\left[\begin{array}{c} Cp_2MCH_2CHR \\ | \\ H \end{array} \right]^+ \longrightarrow [Cp_2MH]^+ + CH_2{=}CH(CH_2CH_2)_n$$

(2) Chain termination reaction of chain transfer to monomer:

$$\left[\begin{array}{c} Cp_2MCH_2CHR \\ | \\ CH_3 \end{array} \right]^+ + CH_3CH{=}CH_2 \longrightarrow [Cp_2MCH_2CH{=}CH_2]^+ + \begin{array}{c} CH_3CHR \\ | \\ CH_3 \end{array}$$

(3) Chain termination reaction of chain transfer to cocatalyst, MAO:

$$[Cp_2MR]^+ \left[\begin{array}{c} CH_3 \\ | \\ Al{-}O \end{array} \right] \longrightarrow [Cp_2MCH_3]^+ + \left[\begin{array}{c} R \\ | \\ Al{-}O \end{array} \right]$$

(4) Chain termination reaction of chain transfer to hydrogen molecule:

$$[Cp_2MR]^+ + H_2 \longrightarrow [Cp_2MH]^+ + RH$$

Scheme 3 Chain termination reactions

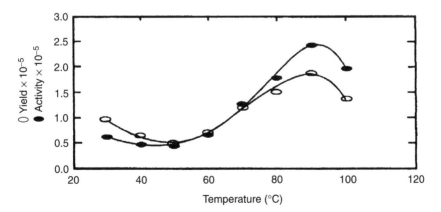

Figure 1 Variation of ($\bigcirc$) yield and ($\bullet$) activity in ethylene polymerization with temperature using Cp$_2$ZrCl$_2$/MAO

Table 1   Effects of ethylene polymerization as a function of temperature

Temperature (°C)	Activity[a] ×10^{-5}	Activity[b] ×10^{-4}	M_w[b] ×10^{-4}	M_n[b] ×10^{-4}	MWD
30	0.63	0.23	250	128	2.0
40	0.48	0.18	224	102	2.2
50	0.43	0.47	148	65	2.3
70	1.37	1.46	127	53	2.4
80	1.78	1.17	64	24	1.9
90	2.46	0.57	40	17	2.9

[a] $Cp_2ZrCl_2 = 1.0 \,\mu mol$.
[b] $CpTpZrCl_2 = 1.0 \,\mu mol$.
Activity $= kg\,PE/(mol\,Zr\,h[C_2H_4])$. MAO $= 1.0\,mmol$, toluene $= 150\,ml$, $P_{C_2H_4} = 10\,psi$, time $= 30\,min$.

temperature. In Figure 2, the GPC curves of polyethylene show two separated peaks. Integration of the two peaks to separately calculate the M_w and M_n gives a polydispersity of 2 for each peak, clearly suggesting that two active catalyst sites are present [3,22]. From the GPC curves it can be seen that the catalyst sites of the higher ratio (site II) of chain-transfer (C) to propagation-ratio (P) constants can gradually converge to a site type with a lower ratio (site I) of C/P with increase in temperature.

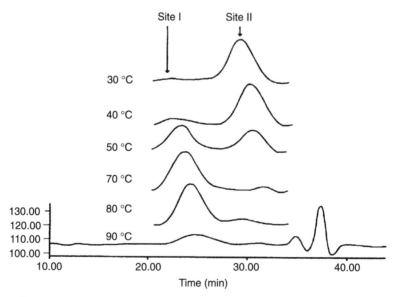

Figure 2   GPC curves of PE at temperatures differing from the reaction temperature ($CpTpZrCl_2$/MAO)

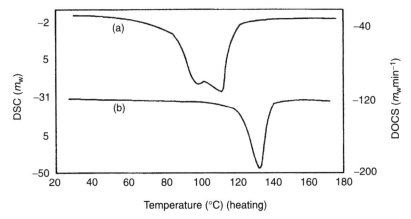

Figure 3 DSC curves for polyethylene prepared using metallocene/MAO catalyst (a) with $ZnEt_2$ and (b) without $ZnEt_2$

By use of the Haward kinetic and molecular weight method, the concentration of active sites $[C^*]$, polymerization rate constant (R), chain transfer constant (k_{tr}) and propagation rate constants (k_p) can be obtained, where $[C^*] = 0.7$ mol PE/mol Zr, $R = 92.02$ mol C_2H_4/(mol Zr s), $k_{tr} = 0.076\,s^{-1}$ and $k_p = 3078$ PE/s.

In addition to the metallocene/MAO catalysts, $ZnEt_2$ is added to the reaction mixture. The Lewis acid $ZnEt_2$ can act as a chain transfer reagent and linear low-density polyethylene (LLDPE) is obtained. The DSC curves for PE and LLDPE are shown in Figure 3. The T_m for HDPE is 134 °C and those for LLDPE are 99 and 109 °C, which are much lower than that the HDPE.

3.2 COPOLYMERIZATION OF POLY(ETHYLENE–co-1,4-HEXADIENE)

$$CH_2{=}CH_2 \;+\; CH_2{=}\underset{\underset{\underset{\underset{CH_3}{|}}{CH}}{\overset{\overset{CH_2}{|}}{\underset{|}{CH}}}{}\quad \xrightarrow[\;n\text{-heptane}\;]{\text{Metallocene/MAO}} \quad {-}(CH_2CH_2)_x{-}(CH_2\underset{\underset{\underset{\underset{CH_3}{|}}{CH}}{\overset{\overset{CH_2}{|}}{\underset{|}{CH}}}{}}{CH})_y{-}$$

The heterogeneous Ziegler–Natta catalyst performs very poorly in the incorporation of dienes in polyethylene owing to the large difference in activity between comonomers. However, homogeneous metallocene/MAO catalysts will perform much better in the incorporation of the diene in EPDM. The metallocene catalysts include Cp_2ZrCl_2 and $Et(Ind)_2ZrCl_2$ [7,14]. The activity of copolymerization increases with increase in the concentration of 1,4-hexadiene, as shown in Table 2.

Table 2 Copolymerization of ethylene with 1,4-hexadiene^a

[Diene]	Activity^b	Diene in copolymer (mol%)	G^c	F^c
0	125	0	0	0
0.061	104	0.20	3.40	0.0222
0.121	115	0.55	1.71	0.0162
0.242	151	0.63	0.85	0.0047
0.487	169	1.6	0.42	0.0029

^{a}Cp$_2$ZrCl$_2$ = 11.4 μmol, MAO = 11.4 mmol, [C$_2$H$_4$] = 0.208 mol/l, $P_{C_2H_4}$ = 30 psi, temperature = 30 °C, time = 20 min.
^{b}Activity = kg PE/(mol Zr h).
^{c}G and F values are defined in the Fineman–Ross equation: $G = -r_c + r_e F$ (see text).

The ^{1}H NMR and IR spectra are shown in Figure 4. The chemical shift at 5.5 ppm corresponds to the CH=CH group in the diene unit. The composition of the diene can be calculated from the spectrum. The FT-IR spectrum of unsaturated CH=CH absorption is also shown as 965 cm^{-1}. The activity of the catalyst increases with an increase in temperature and [Al]/[Zr] ratio.

The reactivity ratio can be obtained from the Fineman–Ross equation:

$$G = -r_c + r_e F$$

where r_c = comonomer reactivity ratio, r_c = ethylene reactivity ratio, $G =$ $[(m_e/m_c) - 1](M_e/M_c)/(m_e/m_c)$, $F = (M_e/M_c)^2/(m_e/m_c)$, $M_e = $ [C$_2$H$_4$], $M_c = $ [diene] and (m_e/m_c) = mol% of diene in copolymer. The reactivity ratio for ethylene is $r_e = 0.01$–0.324 and for 1,4-hexadiene $r_c = 120$–150; $r_e r_c = 1$–50.

3.3 COPOLYMERIZATION OF POLY(ETHYLENE–co-ETHYLIDENENORBORNENE)

$$CH_2=CH_2 \quad + \quad \text{(bicyclic diene)} \quad \xrightarrow{\text{metallocene/MAO}} \quad -(CH_2CH_2)_{\overline{x}}(CHCH)_{\overline{y}}$$

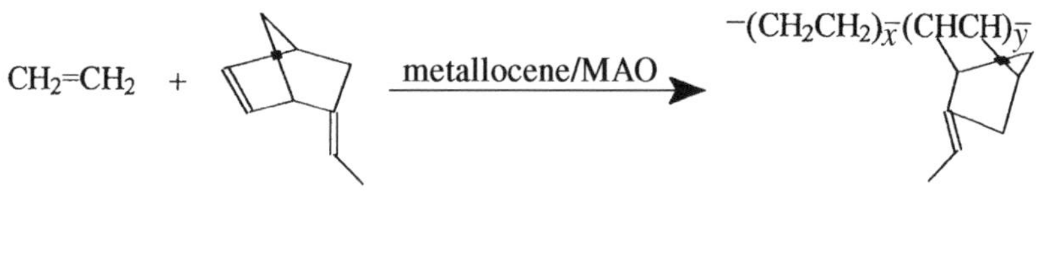

The copolymerization of ethylene and 5-ethylidene-2-norbornene (ENB) is catalyzed by Et(Ind)$_2$ZrCl$_2$ in the presence of the cocatalyst MAO (Al/Zr = 1000).

Figure 5 shows the ^{1}H NMR spectrum of a copolymer with an ENB ratio of 2.04 mol%. The olefinic protons can be observed from the peaks at 5.4 and 5.2 ppm, which are the absorptions of unsaturated hydrogen in the *exo-* and the *endo*-ethylidene groups in ENB, respectively. The mole ratio of *exo* to *endo* is about

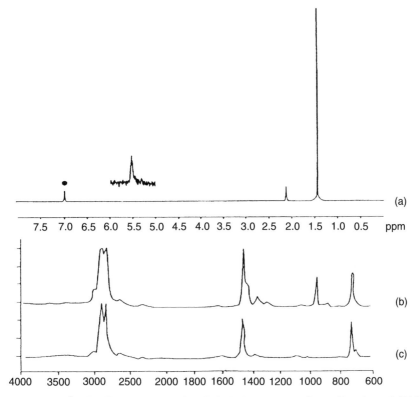

7.5 7.0 6.5 6.0 5.5 5.0 4.5 4.0 3.5 3.0 2.5 2.0 1.5 1.0 0.5   ppm

4000   3500   3000   2500   2000 1800 1600 1400 1200 1000 800   600

Figure 4   (a) ^{1}H NMR spectrum of poly(ethylene–co-1,4-hexadiene) and FT-IR spectra of (c) polyethylene and (b) poly(ethylene–co-1,4-hexadiene)

3:1, a ratio similar to that of the comonomer, ENB. The spectrum also shows a similar pattern. The results also show that the selectivity of the *exo* and *endo* forms in a copolymerization reaction with a metallocene catalyst is the same. The composition of ENB in copolymer was calculated from the system.

The activities of the catalyst under different conditions are shown in Table 3. The activity increases with an increase in ENB concentration and reaches a maximum. The results show a 'comonomer effect'; however, the molecular weight M_n (which can be calculated from $[\eta]^0$) decreases with an increase in the ENB mole ratio. This is reasonably due to the steric effect of the ENB moiety.

3.4   COPOLYMERIZATION OF ETHYLENE, α-OLEFINS AND DIENES

Ethylene, propylene and ethylidenenorbornene were terpolymerized by the bridged catalyst Et(Ind)$_2$ZrCl$_2$/MAO in solution [23]. The presence of ENB decreases the

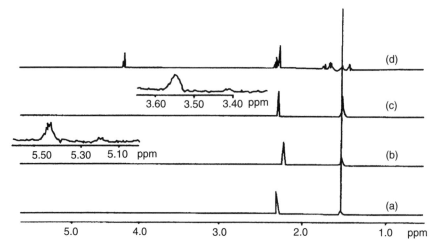

Figure 5 ¹H NMR spectra of functionalized PE copolymers: (a) pure PE; (b) PE– co-ENB; (c) PE–co-ENB-hydroxyl; (d) PE–g-PCL

Table 3 Copolymerization^a of ethylene and ENB using Et(Ind)$_2$ZrCl$_2$/MAO

ENB feed (mol/l) $\times$ 10^2	Yield (g)	Catalyst activity [kg/(mol h)]	ENB in PE^b (mol%)	$[\eta]^c$ (100 ml/g)
0.00	2.59	2303	0.00	1.94
2.50	2.75	2444	2.04	1.84
4.94	5.98	5315	3.90	1.08
9.93	3.28	2917	4.08	1.00

^{a}Ethylene 45 psi, toluene 150 ml, temperature 30 °C, time 30 min.
^{b}Calculated from NMR spectrum.
^{c}In trichlorobenzene at 105 °C.

rate of polymerization [23], especially at low E/P ratios [24]. The results are shown in Table 4. The molecular weight, M_n, also decreases with the addition of ENB; however, ENB incorporation is favorable at a lower E/P ratio.

EPDM terpolymer synthesis with a Cp$_2$ZrCl$_2$/MAO catalyst system [24] has also shown a high activity of 100–1000 kg EPDM/(mol Zr h); however, these activities are only high with an Et(Ind)$_2$ZrCl$_2$/MAO bridged catalyst. The reactivity ratio was found to be $r_1/r_2 \approx 0.3$ (subscripts 1 and 2 designate ethylene and propylene, respectively). The results show that the distributions of ethylene and propylene are nearly random since there is no detectable melting transition in the EPDM on DSC measurement. The narrow distribution ($M_w/M_n \approx 1.7$) implies that one active center is dominant. The bridged and unbridged catalysts are an isospecific and an aspecific metallocene catalyst, respectively, which are suitable catalysts for EPDM

Table 4   Activity of EPDM terpolymerization[a]

Ethylene (mol%)	Propylene (mol%)	ENB (g)	Activity [kg product/(mol Zr h)]	ENB (wt%)	M_n $\times 10^{-4}$
0.8	0.2	0	8100	0	9.52
0.8	0.2	0.09	4800	3.2	13.6
0.8	0.2	0.45	3800	16.4	12.0
0.4	0.6	0	2800	0	8.0
0.4	0.6	0.09	430	22.5	5.4

[a]$[Et(Ind)_2ZrCl_2] = 10^{-4}$ M, $[Al]/[Zr] = 2500$, temperature $= 50\,^\circ$C, $P_{C_2H_4} = 10$ psi.

copolymerization. However, the high copolymerization rate and the controllable molecular weight are still the most interesting research topics.

4 MODIFICATION OF UNSATURATED POLYETHYLENE

4.1 FUNCTIONALIZATION OF PE–co–HD AND PE-co-ENB

The chemical modification of polyolefins to enhance their chemical and physical properties is an area of great interest. The incorporation of a functional group in hydrocarbon polymers can provide the polymers with improved adhesion, dyeability, printability and compatibility. Furthermore, functional groups offer sites for initiating graft copolymers [7]. Recently, Chung and co-workers [29] investigated borane-containing copolymers and successfully converted the borane group to polar functionalities. The functionization reaction is summarized as follows:

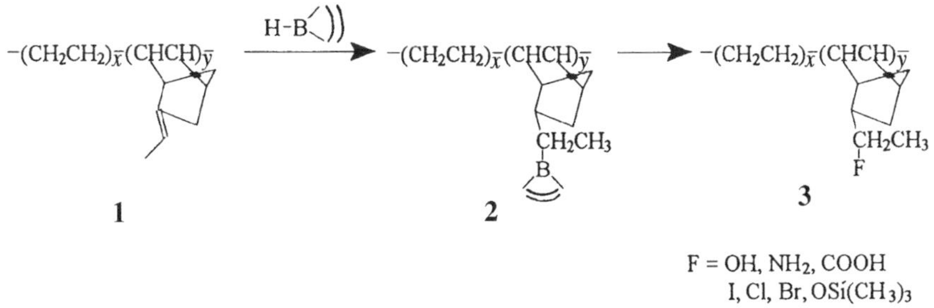

F = OH, NH₂, COOH
I, Cl, Br, OSi(CH₃)₃

From the borane-containing copolymer, a functionalized polyethylene having hydroxyl, amino, carboxyl, halide and silane groups is obtained. Apparently, the double bonds in the side-chain are located in the amorphous phase and are available for the reaction. A typical hydroxyl functionalization is described as follows. The pendent olefin of copolymer 1 can react with 9-borabicyclononane (9-BBN) to give the borane-containing copolymers **2**. To ensure a complete reaction of the internal double bonds, a higher reaction temperature of 60 °C is required for the hydroboration. The olefinic peaks in the ^{1}H NMR spectrum (Figure 5) disappear with a new signal appearing due to a hydroxylmethyl proton at 3.5 ppm.

On using the Cp_2ZrCl_2/MAO catalyst, the activity did not show a comonomer effect; however, for the Cp_2ZrCl_2/MAO catalyst system [8], the activity is as low as 50 kg product (mol h) at 20 °C. At a higher temperature (60 °C), the activity is 2100 kg product (mole h)$^{-1}$, which shows a comonomer effect. A diene incorporation of as high as 9.8 mol% was obtained. The incorporation of a high ratio may be due to the reaction of the double bond being promoted by the energy of the strained ring.

4.2 Graft Copolymerization

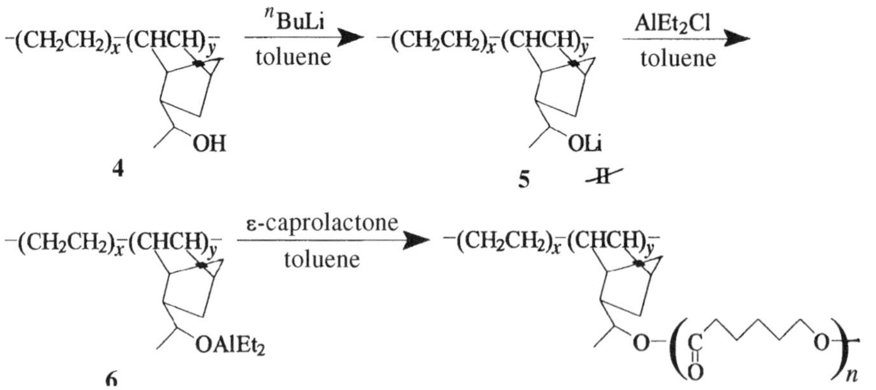

Although the ring-opening polymerization of ε-caprolcatone can be initiated by an initiator generated *in situ* via the reaction of the hydroxyl-containing copolymer **4** and BuLi, this process is complicated by intermolecular Claisen back-biting with the formation of macrocyclics. A cationic exchange with $AlEt_2Cl$ can solve this problem. Compound **6** initiates the ring-opening polymerization of ε-caprolactone. NMR signals show the characteristic peaks of the graft copolymer **7** at 4.06, 2.25, 1.65 and 1.35 ppm (Figure 5). The IR spectrum (Figure 6) shows an absorption at 1710 cm^{-1} from the carbonyl group of the caprolactone moiety.

5 PHYSICAL PROPERTIES AND APPLICATIONS

The metallocene/MAO catalysts reduce the difference in activity between ethylene and comonomers and are easier to use in controlling the composition of the copolymer; therefore, a higher percentage of comonomer in polyethylene can be

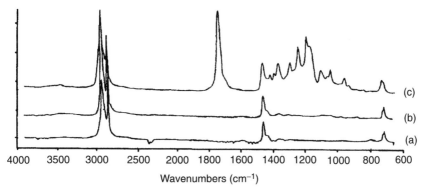

Figure 6 FT-IR spectra of functionalized PE copolymers: (a) PE–co-ENB; (b) PE–co-ENB-hydroxyl; (c) PE– g-PCL

obtained. The modified polyethylene can significantly alter the morphology of the polymer blend by increasing the interfacial interaction.

The unsaturated polyethylene can be prepared by copolymerization of ethylene with linear non-conjugated dienes and cyclodienes. The functionalized linear low-density polyethylene (LLDPE-f) can be obtained from PE–co-HD.

$$-(CH_2CH_2)_x-(CH_2CH)_y- \\ \quad\quad\quad\quad\quad CH_2 \\ \quad\quad\quad\quad\quad CH \\ \quad\quad\quad\quad\quad \| \\ \quad\quad\quad\quad\quad CH \\ \quad\quad\quad\quad\quad CH_3 \quad\longrightarrow\quad -(CH_2CH_2)_x-(CH_2CH)_y- \\ \quad\quad\quad\quad\quad CH_2 \\ \quad\quad\quad\quad\quad CH \\ \quad\quad\quad\quad\quad CH-F \\ \quad\quad\quad\quad\quad CH_3$$

$$F = H, OH, X \\ NH_2, COOH$$

The melting temperature (T_m) and crystallinity (X_c) of unsaturated copolymer **1** are dependent on the density of the side-chains. A higher density copolymer will have lower T_m and X_c. The T_m and X_c results show a similar molecular weight, and the side-chain density shows similar melting temperatures and crystallinities. There is no significant change in T_m and X_c after functionalization. The crystallization was observed by mixing of copolymers **1** and **2**. However, the polarity of the polyethylene has changed so significantly that the adhesion, dyeability, printability and compatibility with other polymers have improved. Also, the mechanical properties and the heat resistance can be maintained since the main chain is kept unchanged.

The DSC results from the reactions of PE–co-ENB and PE–co-ENB-g-PCL are shown in Tables 5 and 6. The copolymer of PE–co-ENB has a lower melting

Table 5 Summary of DSC data for PE–co-ENB[a]

Run No.	ENB in PE (mol%)	PE–co-ENB T_m(°C)	ΔH (J g^{-1})	PE–co-ENB X_c(%)[b]	T_5 %[c] (°C)
a$_2$-i	0.00	134.7	150.0	100.0	448.4
a$_2$-ii	2.04	116.4	61.0	40.1	342.4
a$_2$-iii	3.90	116.1, 129.8	46.4	31.0	329.7
a$_2$-iv	4.08	112.9, 129.3	41.6	27.7	329.2

[a]Ethylene = 45 psi, toluene = 150 ml, temperature = 30 °C, time = 30 min, Et(Ind)$_2$ZrCl$_2$ = 2.25 μmol, MAO = 2.25 mmol.
[b]X_c = [(ΔH copolymer)/(ΔH homopolymer)] × 100 %.
[c]$T_{5\%}$ is the decomposition temperature for a weight loss of 5 %.

Table 6 Summary of DSC data for PE– co-ENB-g-PCL

Reaction time (h)	PCL in graft (wt%)	PCL T_m(°C)	PCL ΔH(J/g)	PCL X_c (%)[a]	PE T_m (°C)	PE ΔH (J/g)	PE X_c (%)
3	18.7	–	–	–	120.7	86.9	71.2
6	22.1	52.4	6.3	39.6	121.3	81.8	70.0
9	32.9	54.2	13.4	56.7	121.4	70.1	69.6
12	37.2	53.6	17.4	65.2	121.0	66.7	70.8
24	51.8	55.3	28.4	76.4	121.1	49.5	69.1

[a]X_c(PCL) = [(ΔH graft-PCL/ΔH homo-PCL)(wt% of PCL)] × 100 %.

temperature of 112.9 °C compared with PE (134.7 °C). The lower T_m is consistent with the ΔH and X_c results. The crystallinity decreases by about 60 % when the mole ratio reaches about 2.04 % however, the tendency becomes much slower after that point. The lower percentage of crystallinity is due to the ENB moiety. At a high ENB concentration, the copolymers have two T_m, 116.1 and 129.8 °C, when the Et(Ind)$_2$ZrCl$_2$ catalyst is used (Figure 7). The multiple melting-points are possibly due to the two activity sites [25] or two different polymer phase formations. Two melting-points are also found in the copolymerization of ethylene with 5-ethylidene-2-norbornene when using the bridged Et(Ind)$_2$ZrCl$_2$ catalyst [26]; however, the Cp$_2$ZrCl$_2$/MAO catalyst does not show bimodal behavior at high ENB concentrations.

The DSC result from a graft copolymer of PCL shows the T_m of caprolactone at 52.4 °C and of PE at 121.3 °C. The T_m of the PE main chain remains unchanged, but the T_m of PCL in the side-chain increases with an increase in PCL. The ΔH and X_c values increase with wt% PCL on grafting much more than for the PE portion, since the PCL chain length increases with increase in wt% PCL. However, the PE chain length remains unchanged in the grafting process.

A polarized optical micrograph of a 50:50 PE–PVC blend is shown in Figure 8(a). The phase separation for two incompatible components can easily be seen; however,

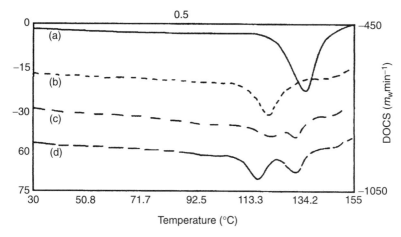

Figure 7 DSC curves for PE–co-ENB containing (a) 0.00 (b) 2.04 (c) 3.90 and (d) 4.08 mol% ENB, using Et(Ind)$_2$ZrCl$_2$/MAO

the dispersion is much smoother and the domain size is reduced dramatically after the addition of PE-g-PCL compatibilizer [Figure 8(b)]. The graft and diblock copolymers act as a compatibilizer.

EPDM terpolymers are commerically important in synthetic rubber. They have good resistance to oxygen, heat and chemical reagents and, because of these excellent properties, there are many applications, including in the automotive and housing industries.

EPDM can act as a compatibilizer for a polyolefin blend with natural rubber in different compositions to improve the hardness of the natural rubber. The crystalline polyolefin composition improves the impact resistance of the product, e.g. 60–95 parts of polyolefin blended with 5–40 parts of EPDM [26].

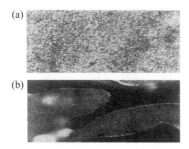

Figure 8 Polarized optical micrograph of polymer blends: (a) PE_PVC = 50:50(×250) and (b) PE_PVC_PE-g-PCL = 50:50:15(×100)

A copolymer containing of 62 mol% of ethylene, 33.1 mol% of 1-octene and 4.5mol% of 5-ethylidene-2-norbornene was prepared by use of a Cp_2ZrCl_2/MAO catalyst [27]. The molecular weight distribution was narrow and the T_g was $-67\,^\circ C$. This copolymer can be blended with polyolefins to prepare a vulcanized sheet exhibiting good tensile strength and low temperature resistance.

The Cp_2ZrCl_2/MAO catalyst exhibits a good ability for copolymerization and cyclopolymerization of ethylene and 1,5-hexadiene under a high pressure of 1500 bar. A predominance of *trans*-ring formation, as shown, was observed [28]. The cyclopolymer is a very good heat-resistant material.

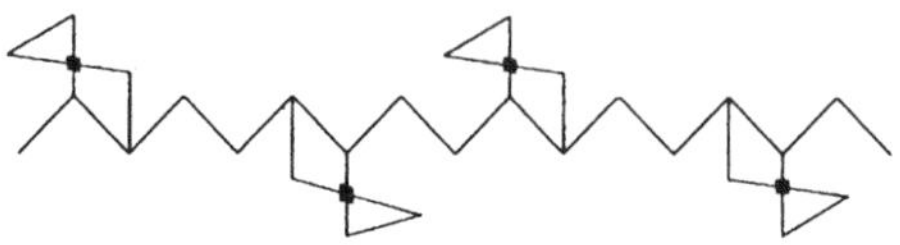

6 EXPERIMENTAL PROCEDURES

6.1 COPOLYMERIZATION OF ETHYLENE WITH DIENES

A typical copolymerization of ethylene with diene is illustrated by ethylidenenorbornene. The reaction was conducted in a pressure reactor equipped with a mechanical stirrer. A solution consisting of 0.45 g (3.7×10^{-3} mol) of ENB, 0.5 ml of MAO (30.6% in toluene) and 150 ml of toluene was placed in the reactor inside a dry-box. The reaction mixture was saturated with ethylene after the reactor had been removed from the dry-box. After the temperature had reached $30\,^\circ C$, a solution of 2.25×10^{-6} mol of $Et(Ind)_2ZrCl_2$ in 0.5 ml of toluene was added under ethylene pressure to catalyze the polymerization. After a predetermined reaction time the reaction was terminated by the addition of 50 ml of a dilute solution of HCl in methanol. The copolymer was isolated by filtration and washed with methanol before being dried in a vacuum oven.

6.2 FUNCTIONALIZATION OF UNSATURATED POLYETHYLENE COPOLYMER

A 0.5 g amount of copolymer was placed in a flask with 50 ml of THF and 0.218 g of 9-BBN. The reaction slurry was stirred at $60\,^\circ C$ for 6 h. Oxidation was implemented by the addition of 5 ml of NaOH solution (0.007 g NaOH in water) followed by dropwise addition of an H_2O_2 solution (0.2 ml of 30 % H_2O_2 in 10 ml of methanol). The slurry was stirred at $37\,^\circ C$ for 11 h. The product was filtered and dried in a vacuum oven at room temperature.

6.3 GRAFT COPOLYMERIZATION OF FUNCTIONALIZED POLYETHYLENE

The hydroxyl-containing polyethylene copolymer (1.86 mol% hydroxy group) was metallated by reacting a slurry of 0.50 g of polymer with an excess of ^{n}BuLi to form a lithium alkoxide by stirring for 12 h. The unreacted ^{n}BuLi was washed with toluene or n-hexane. The slurry was allowed to react for 12 h with 0.44 g of diethylalumnum chloride in 10 ml of fresh toluene. The aluminum alkoxide functionalized polymer was then placed in a slurry of 5.0 g of ε-caprolactone and 10 ml of toluene to start the graft reaction. After 4 h, the graft reaction was terminated by the addition of methanol. The product was fractionated by extraction with hot acetone in a Soxhlet apparatus. After removal of the acetone-soluble polycaprolactone, the acetone-insoluble graft copolymer PE–co–ENB-g-PCL was obtained.

7 CONCLUSIONS

The metallocene/MAO catalysts show high activity that systematically increases with increasing concentration of comonomers, especially for the Et(Ind)$_2$ZrCl$_2$/MAO catalyst. There is a 'comonomer effect' in the metallocene/ MAO catalysts. A high comonomer composition can be obtained.

The molecular weight decreases with increase in comonomer concentration with high steric hindrance, especially in 5-ethylidene-2-norbornene; however, the molecular weight distribution is still very narrow.

Varied functional polyolefins can easily be obtained in the side-chain, including OH, NH$_2$, COOH, OSi(CH$_3$)$_3$ and X, and retain the original properties.

With a homogeneous distribution of comonomers in most cases, especially for terpolymerization, no T_m can be detected in a DSC measurement.

Functional copolymers and graft copolymers have excellent properties, and many applications can be utilized.

8 REFERENCES

1. Sinn, H. and Kaminsky, W., *Adv. Organomet. Chem.*, **18**, 99 (1980).
2. Sinn, H., Kaminsky, W., Vollmer, H. J. and Woldt, R., *Angew. Chem., Int. Ed. Engl.*, **19**, 390 (1980).
3. Chien, J. C. W., *J. Polym. Sci., Polym. Chem.*, **28**, 15 (1990).
4. Rossi, A., *Macromolecules*, **28**, 739 (1995).
5. Chung, T. C., Raate, M., Berluche, E. and Schulg, D. N., *Macromolecules*, **21**, 1903 (1988).
6. Chung, T. C. and Rhubright, D., *Macromolecules*, **24**, 970 (1991).
7. Chung, T. C., Lu, H. L. and Li, C. L., *Macromolecules*, **27**, 7533 (1994).
8. Jiang, G. J., Lee, S. F. and Hsu, M. T., *Chung Yuan J.*, **26**, 1 (1997).

9. Jiang, G. J. and Wang, T., *J. Chin. Chem. Soc.*, **45**, 341 (1997).
10. Chung, T. C. and Jiang, G. J., *Macromolecules*, **25**, 4816 (1992).
11. Chung, T. C., Rhubright, D. and Jiang, G. J., *Macromolecules*, **26**, 3467 (1993).
12. Chung, T. C., Janvikul, W., Bernard, R. and Jiang, G. J., *Macromolecules*, **27**, 26 (1994).
13. Patta, S. and Lohse, D. J., *Polymeric Compatibilizers*, Hanser, Munich, 1996.
14. Jiang, G. J., Lai, K. J., Hwu, J. M., Wang, S. J. and Ting, C., *Polym. Prepr.*, **37**, 350 (1976).
15. Bergemann, C., Cropp, R. and Luft, G., *J. Mol. Catal. A: Chem.*, **116**, 317 (1997).
16. Kawaski, M., Okada, K., Toji, T., Tsutsui, T. and Aine, T., *Jpn. Kokai Tokyo Koho*, JP 09012801 (1997).
17. Wang, T., Lin, C. and Jiang, G. J., *Polym. Prepr.*, **37**, 641 (1995).
18. Kaminsky, W. and Miri, M., *J. Polym. Sci., Polym. Chem.*, **23**, 2151 (1991).
19. Chien, J. C. W. and He, D., *J. Polym. Sci., Polym. Chem.*, **29**, 1609 (1991).
20. Galimberto, M., Martini, E., Piemontesi, F., Sartori, F., Camurati, I., Resconi, L. and Albizzati, E., *Macromol. Symp.*, **89**, 259 (1995).
21. Kawada, T. and Mori, Y., *Jpn Kokai Tokyo Koho*, JP 01031292 (1997).
22. Eakelinen, M., *Eur. Polym. J.*, **32**, 331 (1996).
23. Kaminsky, K. and Mirc, M., *J. Polym. Sci., Polym Chem.*, **23**, 2151 (1985).
24. Chien, J. C. W. and He, D., *J. Polym. Sci., Polym. Chem.*, **29**, 1609 (1991).
25. Aaltonen, P. and Lofgren, B., *Macromolecules*, **28**, 5353 (1995).
26. Kawasaki, M., Okada, K., Tojo, T., Tsutsui, T. and Aine, T., *Jpn. Kokai Tokyo Koho*, JP 09012801 (1997).
27. Kawada, T. and Mori, Y., *Jpn. Kokai Tokkyo Koho*, JP 09031272 (1997).
28. Bergemann, C., Cropp, R. and Luft, G., *J. Mol. Catal. A: Chem.*, **116**, 317 (1997).
29. Chung, T. C., Lu, H. L. and Janvikul, W., *Polymer*, **38**, 1495 (1997).

16

Ethylene Polymerization Using a Metallocene Catalyst Anchored on Silica with a Spacer

DONG-HO LEE
Kyungpook National University, Taegu, Korea

SEOK KYUN NOH
Yeungnam University, Kyungsan, Korea

1 INTRODUCTION

Owing to Kaminsky's brilliant work, metallocene catalysts have become increasingly important as a potential new generation of Ziegler–Natta catalysts [1]. As an advantage over the conventional Ziegler–Natta catalysts, the metallocene/methylaluminoxane (MAO) systems combine high activity with the possibility of tailoring the polymer properties such as molecular weight, molecular weight distribution and stereochemical structure through an appropriate ligand design at the metal center [2]. Great efforts have been devoted to synthesizing new types of metallocene complexes to understand the correlation between the metallocene structure and the polymer properties.

As most of the existing polyolefin plants run in the slurry and gas phase with a heterogeneous Ziegler–Natta catalyst, a metallocene catalyst must be immobilized on a support not only to be applicable for the existing processes but also to control the morphology of the resulting polyolefins. For this reason, considerable studies have endeavoured to develop an efficient method to attach the metallocene to the support, thus preserving the advantages of the metallocene catalyst while improving

Metallocene-based Polyolefins   Edited by J. Scheirs and W. Kaminsky
© 2000 John Wiley & Sons Ltd

the limited practical application of metallocenes and the morphology of the polymers [3].

The first approach employed by Soga, Kaminsky and co-workers utilized the direct impregnation of a metallocene on a support [4]. More recently, Soga and co-workers [5] attempted an alternative strategy to prepare metallocenes chemically bonded to the support. The key aspect of this procedure is making use of a modified silica surface as a ligand attaching the heterogenized *ansa*-metallocene. Two typical beneficial aspects of a supported catalyst comparing with a homogeneous catalyst are the enhancement of the molecular weight of the generated polymer and the smaller amount of cocatalyst required in polymerization for optimum activity. On the other hand, a heterogenized metallocene leads to a decrease in the catalytic activity compared with that of the corresponding soluble catalyst, and the metal content of the immobilized metallocene is very low (about 0.7 wt%).

The above results led to the very important consequence that the characteristics of the active sites generated by a supported metallocene can be altered from those of the active sites by a homogeneous catalyst in some instances. To elucidate the cause of this transformation, it is necessary to probe the characteristics of the active sites formed at the surface of the heterogenized catalyst, which is extremely difficult owing to the lack of adequate analytical devices to provide insight into the structure of active sites. In order to minimize the uncertainties arising from the immobilization procedure, it would be worth trying to develop a simple but effective procedure to anchor the metallocene to a support.

Some interesting routes have been reported for preparing a modified silica surface which has cyclopentadiene units with a funtionalized silane incorporating an alkylidene spacer group [6]. A common feature of the reported processes is that most of the reactions were conducted in heterogeneous conditions, which indicates that the real structure and chemistry involved in the modification steps were hardly characterizable.

The main purpose of this chapter is to describe a new method to prepare an anchored metallocene catalyst having a spacer between the metallocene and the silica surface. The specific aims of the present work are (i) to probe the anchored metallocene resulting from the newly designed approach and (ii) to establish the differences between the new anchored metallocene and a heterogenized metallocene prepared by reported procedures.

2 NEW ANCHORED METALLOCENES

2.1 PREPARATION

In this section, a new strategy to accomplish the efficient attachment of a metallocene to a silica surface is presented on the basis of the philosophy that the supporting procedures involved are performed in as soluble, homogeneous and characterizable conditions as possible [7]. In order to reveal any differences in the catalyst properties among the immobilized metallocenes arising from the various anchoring methods, other supported metallocenes obtained by known routes were also investigated. Most of the experiments discussed here were performed with materials possessing a $CpIndZrCl_2$ fragment as a metallocene function and hexamethyltrisiloxane (TS) or pentamethylene (PM) unit as a spacer function.

Using the conditions of the 'newly designed approach' (route 1) conditions, the immobilized process is made up of two main steps [7]. The first step is the generation of monoanionic species having both a metallocene part as an active site and an ionic spacer part as a bonding site to the silica. The second step is transformation of the monoionic substance to the anchored metallocene through the contact of two bodies. In this new route, the important feature to be noted is that there is no inefficient metallation step in heterogeneous condition except this second step.

As shown in Scheme 1, the formation of the monoanionic species **3** is realized by the reaction of the dithallium salt **2** of the corresponding ligand, hexamethyltrisiloxanediylbis(cyclopentadienyl) (**1**) produced from 1,5-dichlorohexamethyltrisiloxane, with 1 equiv. of $IndZrCl_3$ in THF. The monoanion **3** can be separated as a greenish solid and is found to be soluble in diethyl ether and THF, but insoluble in hexane. Then **3** was contacted with tosylated silica (**4**) to obtain ZATS-1 (**5**) with a TS spacer.

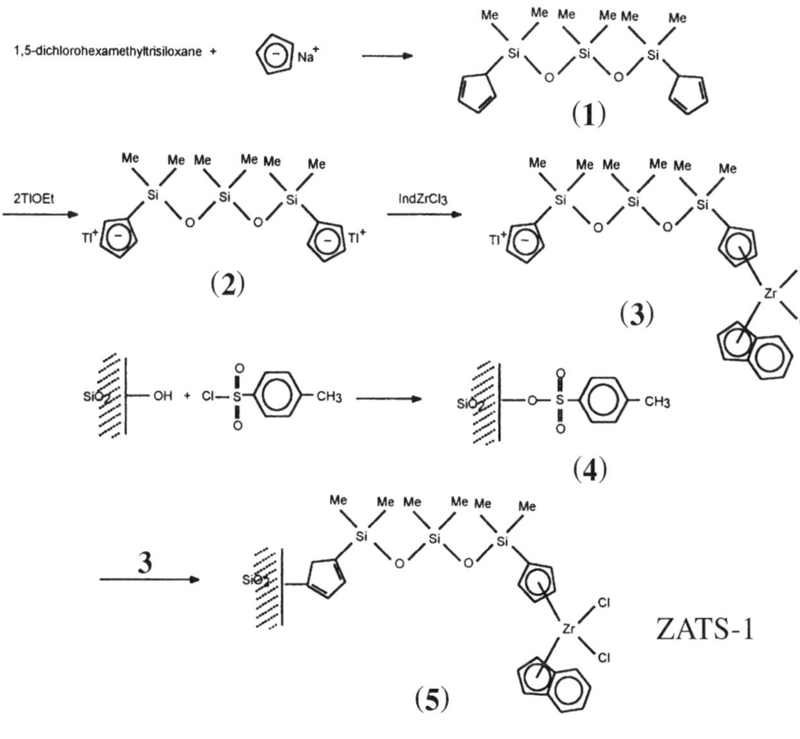

Scheme 1

By using 1,5-dibromopentane in place of 1,5-dichlorohexamethyltrisiloxane, ZAPM-1 with a PM spacer was prepared. In case of the TS spacer, the dithallium salt was produced quantitatively by treatment of the spacer ligand **1** with thallium ethoxide. On the other hand, the dilithium salt of the ligand with PM units was employed for convenience.

Owing to the sensitivity of monoionic species to moisture, a complete analysis could not be obtained. Instead, the formation of the monoionic intermediates can be confirmed indirectly by the chemical means: further reaction of **3** with metal chlorides such as IndHfCl$_3$ and TiCl$_4$ gave rise to the formation of new heterodinuclear metallocenes as follows [8]:

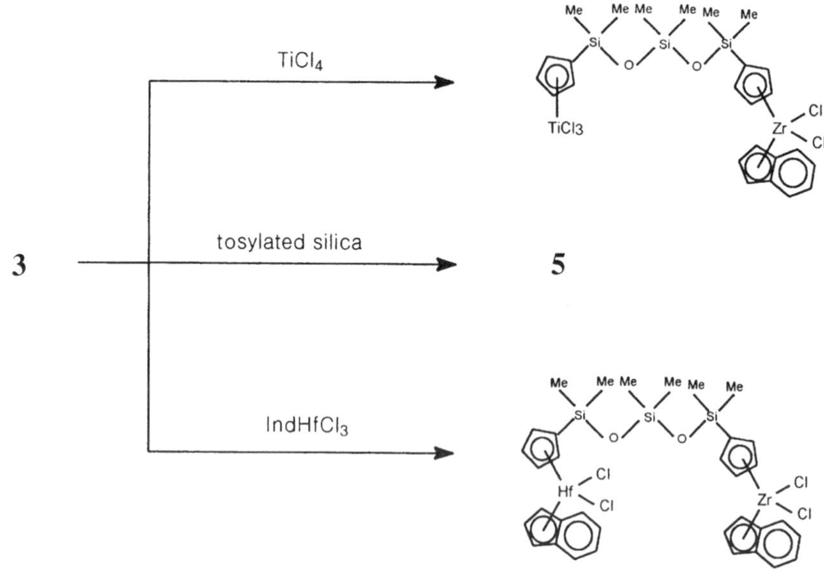

It turned out that these reactions provided not only indirect evidence for the existence and the reactivity of monoanionic species but also the versatility of the monoanion to result in a new dinuclear metallocene [9] that could be a useful catalyst for olefin polymerization. Immobilization, which is the only heterogeneous reaction, is simply achieved by mixing of the monoanionic materials with the tosylated silica which is pretreated with tosyl chloride.

In an effort to determine the differences associated with the immobilization process of the metallocene, two other anchoring procedures were separately examined.

The modified method (route 2) indicated in Scheme 2 is composed of two reactions under heterogeneous conditions. The anchored metallocene, ZATS-2 (**7**), would be formed by the metallation of IndZrCl$_3$ with the anchored anions (**6**) on the silica surface that were generated by the reaction between **2** and **4**. ZAPM-2 was obtained by using 1,5-dibromopentane instead of 1,5-dichlorohexamethyltrisiloxane.

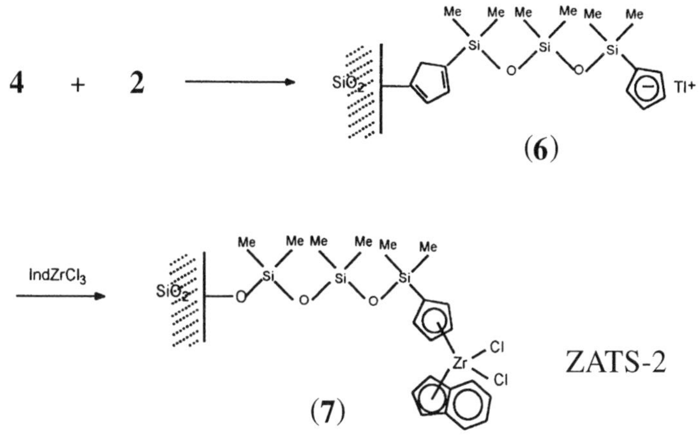

(6)

ZATS-2

(7)

Scheme 2

The other method (route 3, Scheme 3) is basically analogues to those reported by others [6]. All steps for the preparation of ZATS-3 (**12**) and ZAPM-3 are carried out under heterogeneous atmospheres.

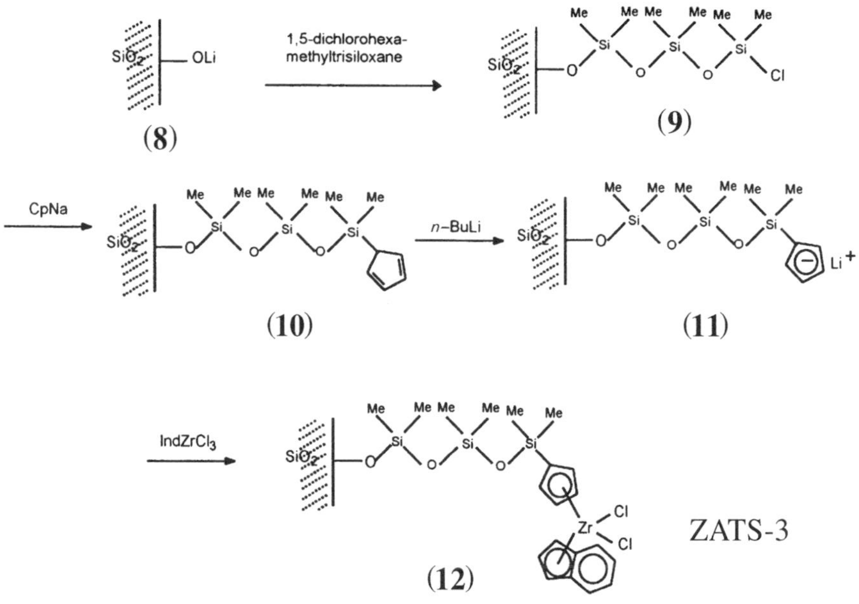

(8)   (9)

(10)   (11)

ZATS-3

(12)

Scheme 3

2.2 ZIRCONIUM CONTENT

The zirconium content of the prepared catalysts was determined by ICP, and the results revealed a significant difference among the supported metallocenes, as shown Table 1.

It was found that the zirconium content of anchored metallocene is strongly dependent upon the supporting method employed. The zirconium content of anchored catalysts possessing a TS linkage as a spacer unit decreased in the order ZATS-1 (3.4 wt%, route 1) > ZATS-2 (1.9 wt%, route 2) > ZATS-3 (0.6 wt%, route 3), which indicates that the metallocene anchored by the new approach is characterized by a much higher zirconium content than the two anchored metallocenes resulted from the other approaches. It is noteworthy that the immobilization method with more heterogeneous reaction steps leads to the formation of catalysts with a lower zirconium content.

In addition, it was observed that the anchored catalysts made from the same procedure exhibited almost identical zirconium contents regardless of the nature of the spacer. For example, ZATS-1 and ZAPM-1 from route 1 showed the zirconium contents of 3.4 and 3.5 wt%, respectively. Another feature to be noted is that 3.5 wt% of zirconium attached to the silica surface is the largest amount ever reported, as far as we know.

The key aspect of the new route 1 is the simplicity of the procedure. Furthermore, the reaction between the tosyl group on modified silica, which is an excellent leaving group, and the monoanionic species is expected to proceed very effectively to end up with the formation of the designed anchored metallocene. As a consequence, the catalyst obtained is expected to have only one structure of anchored zirconocene so that it could be considered as a real 'heterogeneous single-site' catalyst.

On the other hand, for the other two routes 2 and 3, some degree of flexibility can be permitted. Both procedures include a reaction step between the modified silica and the difunctionalized reactants, dianion species in route 2 and dihalide compounds in route 3. These reactants are able to make a bond not only with one end to form a desired catalyst precursor for metallation but also with both reactive

Table 1 Zirconium content of prepared catalyst

	Zr	
Catalyst	wt %	mmol (g cat)$^{-1}$
ZATS-1	3.4	0.39
ZAPM-1	3.5	0.39
ZATS-2	1.9	0.20
ZAPM-2	2.0	0.21
ZATS-3	0.6	0.07
ZAPM-3	0.7	0.08
$SiO_2/CpIndZrCl_2$	0.8	0.09

sites simultaneously on the silica surface to result in the creation of a 'dead or inert' site for further metallation reactions. In the metallation step, a 'dual' active site could be formed by double anchoring for both routes 2 and 3. In addition, in route 3 some additional active sites could be also produced by reaction of remaining—OLi sites and IndZrCl$_3$.

It is possible that under polymerization conditions a certain amount of the active complexes in the supported catalyst is leached into the solution, especially if more MAO cocatalyst or even alkylaluminum is added [10]. After ZATS-1 and ZAPM-1 had been treated with the modified methylaluminoxane (MMAO) cocatalyst in the absence of ethylene monomer for 1 h, the zirconium content of solid catalyst was measured again and it was found that the variation in zirconium content was negligible ($< 5\%$). In other word, the anchored zirconocenes are covalently bonded on silica by the new method so that the leaching is minimized.

3 POLYMERIZATION BEHAVIORS OF ANCHORED METALLOCENE

3.1 CATALYST TYPE

To investigate the effect of the anchoring type on the catalytic properties of prepared catalysts, the ethylene polymerizations were conducted with the anchored catalysts and MMAO cocatalyst in toluene, in comparison with the corresponding homogeneous metallocene CpIndZrCl$_2$ and the heterogeneous catalyst SiO$_2$/CpIndZrCl$_2$ obtained by direct impregnation of CpIndZrCl$_2$ with SiO$_2$. The experiments were carried out at 40 °C with an [Al]/[Zr] ratio of 5000 and the results are given in Table 2.

As expected, it was found the activity of the unsupported catalyst was much greater than that of the supported catalysts at 40 °C. However, the catalyst activity of CpIndZrCl$_2$ anchored on silica with TS or PM spacer (ZATS and ZAPM,

Table 2 Polymerization of ethylene initiated by various catalysts and MMAO at 40 °C^a

Catalyst	Activity^b	$M_{\mathrm{w}}(\times 10^{-3})$	$M_{\mathrm{w}}/M_{\mathrm{n}}$	$T_{\mathrm{m}}(°C)$
CpIndZrCl$_2$	2343	223	3.0	135
SiO$_2$/CpIndZrCl$_2$	130	301	3.9	134
ZATS-1	557	254	4.8	134
ZAPM-1	658	221	3.9	135
ZATS-2	284	204	4.2	135
ZAPM-2	368	196	3.9	134
ZATS-3	236	256	3.3	134
ZAPM-3	298	204	3.4	135

^{a}Polymerization conditions: [Zr] = (3.0–5.8) × 10^{-6}mol/l^{-1}, [Al]/[Zr] = 5000, 2 h,
^{b}Activity: kg PE (mol Zr h atm)$^{-1}$.

respectively) was much higher than that of $SiO_2/CpIndZrCl_2$. In addition, ZAPM exhibited higher activity than ZATS. This result is interpreted to mean that the polymethylene spacer increases the catalyst activity by electron donation enhancement to the metal center [11].

Among the anchored systems, the catalysts generated by the new route 1 exhibited 2–5 times larger activities than the catalysts made by the other two methods. Although the structures of the active sites for the prepared catalysts were expected to be identical, various active sites could be formed. ZATS-1 and ZAPM-1 have only a single kind of active site whereas a low-activity site was additionally present for ZATS-2 and ZAPM-2 owing to the double anchoring. In the case of ZATS-3 and ZAPM-3, the surface structure of the catalyst precursor was very complicated and some of the unreacted—OLi groups could be metallated by $IndZrCl_3$ to form less active sites.

In summary, the catalyst activity of the ZATS and ZAPM series decreased in the order ZAPM-1 > ZATS-1 ≫ ZAPM-2 > ZATS-2 ≈ ZAPM-3 > ZATS-3.

The weight-average molecular weight (M_w) of PE obtained with the catalysts used decreased in the order $SiO_2/CpIndZrCl_2$ > ZATS ⩾ ZAPM ⩾ unsupported $CpIndZrCl_2$. The polydispersity index (M_w/M_n) of PE produced using ZATS or ZAPM was slightly higher than that of PE produced using $SiO_2/CpIndZrCl_2$. The melting point (T_m) of PE was almost independent of the catalyst type, which indicated that the polymer produced is basically high-density PE.

3.2 POLYMERIZATION TEMPERATURE

To examine the effect of polymerization temperature on catalyst behavior, the polymerization of ethylene was also conducted with various catalysts at 70 °C and the experimental results are given in Table 3.

Comparison of the activities in Tables 2 and 3 revealed an even more important point. The activity of unsupported $CpIndZrCl_2$ is sharply reduced from 2343 at 40 °C to 1452 at 70 °C. On the other hand, the catalytic activities of the anchored

Table 3 Polymerization of ethylene initiated by various catalysts and MMAO at 70 °C[a]

Catalyst	Activity[b]	$M_w(\times 10^{-3})$	M_w/M_n	$T_m(°C)$
$CpIndZrCl_2$	1452	56	2.6	135
$SiO_2/CpIndZrCl_2$	162	102	3.4	135
ZATS-1	1026	104	3.4	134
ZAPM-1	1324	98	3.6	134
ZATS-2	621	99	3.5	135
ZAPM-2	864	101	3.4	134
ZATS-3	289	96	3.0	135
ZAPM-3	351	97	3.0	134

[a] Polymerization conditions: $[Zr] = (3.0-5.8) \times 10^{-6} mol/l^{-1}$, $[Al]/[Zr] = 5000$, 2 h.
[b] Activity: kg PE $(mol\, Zr\, h\, atm)^{-1}$.

systems in Table 3 are larger than those in Table 2. As a result, the anchored catalyst ZAPM-1 displays an activity almost identical with that of the unsupported, homogeneous catalyst in polymerization at 70 °C. These results illustrate an increased tolerance of the supported catalyst toward high temperature and strongly suggest that the supported catalyst examined maintained its thermal stability under the polymerization conditions. This low degree of deactivation might be due to the reduced freedom of active sites, which eventually protects the system from the formation of inactive dinuclear species by the approach of the active sites.

The higher activity characteristics of the immobilized catalyst prepared using the new procedure are likely to be more pronounced with polymerization at 70 °C. The observed superiority of the new supporting method over others can be further supporting evidence for the presence of a considerable amount of less active metal sites on the conventionally supported catalyst.

It is generally recognized that the M_w of a polymer is very sensitive to the polymerization temperature in the case of a soluble metallocene system. On this basis, it is not surprising to observe a drastic decrease in M_w from 223 000 at 40 °C to 56 000 at 70 °C for the soluble metallocene CpIndZrCl$_2$. This tendency can be effectively suppressed by immobilization of the homogeneous catalyst. As shown in Table 3, the M_w of PE from polymerization with the anchored catalysts at 40 °C is only double that of PE obtained at 70 °C. It should be noted that the anchoring methods employed do not have an appreciable influence on the M_w of the polymers formed.

3.3 COCATALYST AMOUNT

To study the effect of the amount of MMAO on the catalyst activity and M_w of PE, the polymerization of ethylene was conducted with various amounts of MMAO at 40 °C, and the results are summarized in Figures 1 and 2, respectively.

As can be seen in Figure 1, the catalyst activity increased with increasing amount of MMAO. At high mole ratio such as [Al]/[Zr] = 5000, ZATS-1 and ZAPM-1 became more active and the catalyst activity of ZAPM-1 was five times higher than that of SiO$_2$/CpIndZrCl$_2$. For the given range of MMAO amounts, the catalyst activity of ZAPM was slightly higher than that of ZATS.

In Figure 2, the M_w of PE decreases with increasing amount of MMAO owing to the chain transfer of MMAO [12]. The M_w of PE produced by ZATS or ZAPM was lower than that of PE obtained with SiO$_2$/CpIndZrCl$_2$, but higher than that obtained with unsupported CpIndZrCl$_2$. This result could be understood from the consideration that ZATS and ZAPM have a spacer between the silica surface and zirconocene, so that the zirconocene is away from the silica surface and exposed less to MMAO than unsupported free CpIndZrCl$_2$.

For a polymerization temperature of 70 °C, the effect of the amount of MMAO on catalyst activity and M_w of PE is shown in Figures 3 and 4, respectively.

With increasing amount of MMAO, the enhancement of the catalyst activity of ZATS and ZAPM was greater at 70 °C than that at 40 °C. At [Al]/[Zr] = 5000, the catalyst activity of ZAPM-1 was eight times higher than that of SiO$_2$/CpIndZrCl$_2$.

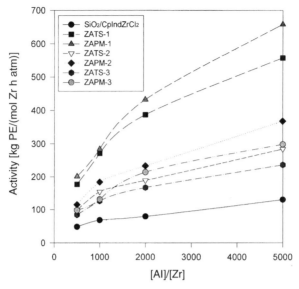

Figure 1 Effect of amount of MMAO on catalyst activity for ethylene polymerization with various supported catalysts. Polymerization conditions: $[Zr] = (3.0-5.8) \times 10^{-6}\,mol/l$, $40\,°C$, $2\,h$

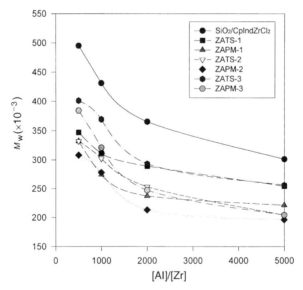

Figure 2 Effect of amount of MMAO on molecular weight of PE obtained with various supported catalysts. Polymerization conditions: $[Zr] = (3.0-5.8) \times 10^{-6}\,mol/l$, $40\,°C$, $2\,h$

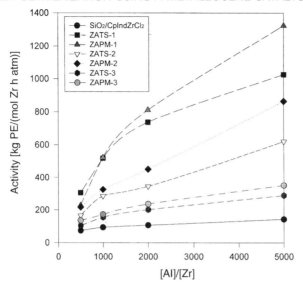

Figure 3 Effect of amount of MMAO on catalyst activity for ethylene polymerization with various supported catalysts. Polymerization conditions: [Zr] = $(3.0-5.8) \times 10^{-6}$ mol/l, 70 °C, 2 h

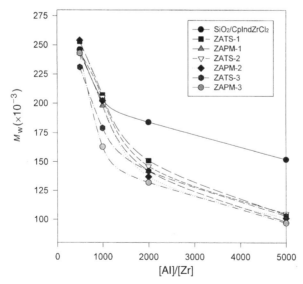

Figure 4 Effect of amount of MMAO on molecular weight of PE obtained with various supported catalysts. Polymerization conditions: [Zr] = $(3.0-5.8) \times 10^{-6}$ mol/l, 70 °C, 2 h

The catalyst activity of the silica-supported catalyst could be enhanced by introducing an appropriate spacer between the zirconocene and the silica surface. At 70 °C, the M_w of PE produced using ZATS and ZAPM was almost same and decreased sharply with increasing amount of MMAO.

3.4 MORPHOLOGY

The particle morphology of (A) ZATS-1 and (B) PE obtained with ZATS-1 and also of (C) PE obtained with CpIndZrCl$_2$ was observed by using SEM as shown in Figure 5.

Particles of both A and B had a spherical shape with a smooth surface. On the other hand, C was found to be irregular with a small particle size.

These results indicate that the replication phenomenon of morphology between catalyst particle and polymer particle is maintained even for the anchored catalyst.

3.5 SPACER LENGTH

The activity enhancement of an anchored metallocene may be due to less steric hindrance of the silica surface owing to the separation of the active site from the silica surface by the spacer. Therefore, CpIndZrCl$_2$ was anchored on silica with various lengths of —(CH$_2$)n— spacer by route 1 and the effect of the spacer length on the ethylene polymerization behavior was examined as shown in Table 4.

With increasing spacer length from trimethylene ($n = 3$) to pentamethylene ($n = 5$), the catalyst activity increased. In contrast, the activity decreased with further increase in spacer length. On the other hand, it should be noted that the Zr content of the prepared catalyst varied considerably from 3.2 wt% ($n = 3$) to 5.7 wt% ($n = 9$). The Zr content of the prepared catalyst might have some influence on the catalyst activity as an MgCl$_2$-supported TiCl$_4$ catalyst [13]. From this attempt, it was found that the Zr content of the anchored catalyst can increase up to ca 6 wt%.

From the above considerations, it is not possible so far to draw conclusions about the effect of spacer length on ethylene polymerization behavior. More detailed

Table 4 Polymerization behavior of CpIndZrCl$_2$ anchored on silica with various lengths of —(CH$_2$)$_n$— spacer[a]

n	Zr (wt %)	Activity[b]	$M_w(\times 10^{-3})$	M_w/M_n	$T_m(°C)$
3	3.2	450	382	3.2	135
5	3.5	658	221	3.9	135
7	4.9	626	412	3.5	135
9	5.7	395	586	3.8	135

[a] Polymerization condition: [Zr] = (2.0–4.0) × 10^{-6} mol μ^{-1}, [Al]/[Zr] = 5000, 40,°C, 1 h.
[b] Activity: kg PE (mol Zr h atm)$^{-1}$.

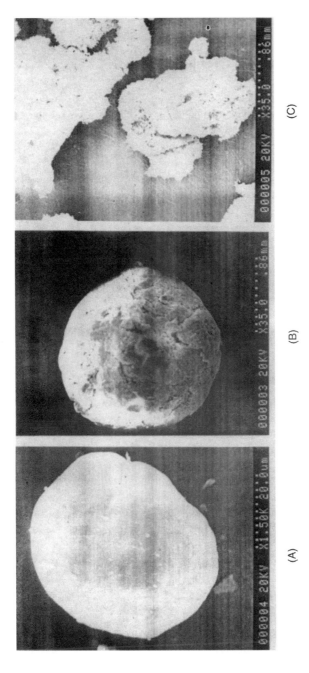

(A)   (B)   (C)

Figure 5 Particle morphology of (A) ZATS-1 and of PE obtained with (B) ZATS-1 and with (C) CpIndZrCl$_2$

studies are in progress on the effects of spacer length and Zr content of the anchored catalysts on ethylene polymerization behavior.

4 SUMMARY AND PROSPECTS

In order to accomplish efficient attachment of metallocenes to silica, a new strategy has been designed and investigated on the basis of the philosophy that the supporting procedures involved are performed in as soluble, homogeneous and characterizable conditions as possible. According to this idea, the synthesis of two supported catalysts holding different types of spacer ligands has been achieved by the new route 1. A further four catalysts have also been made based on other procedures (routes 2 and 3). Among the experimental studies which distinguish the anchored metallocenes from their preparation routes, the following points can be noted:

(i)   The zirconium content of the supported metallocenes is strongly dependent upon the supporting methods employed (route 1 > route 2 > route 3) and increases considerably up to 6 wt% with the new anchoring method of route 1, which has just one reaction step under heterogeneous conditions.
(ii) Among the anchored catalysts studied, the materials made by the new route 1 exhibited higher activity.
(iii) By route 1, it was possible to prepare a 'heterogeneous single-site' catalyst which might have only one type of active site.
(iv) The higher activity characteristics of the catalyst obtained via route 1 are even more pronounced at high polymerization temperatures.

It is believed that all of these outcomes demonstrate that the new anchoring procedure introduced here has definite advantages over other methods. It is certain, however, that many more efforts need be exerted in order to gain a better understanding and wider application of this method, although the new procedure has provided some important improvements. For example, the reaction between metallocenes having different types of spacer and metal centers and silica containing various groups on the surface should be examined to extend the advantages of the new method.

Although these tasks might not be easy, we still believe that the method employed in this work can be a good alternative to allow the rapid, practical application of heterogeneous metallocene catalysts.

5 REFERENCES

1. (a) Sinclair, K. B. and Wilson, R. B., *Chem. Ind. (London)*, 857 (1994); (b) Möhring, P. C. and Coville, N. J., *J. Organomet. Chem.*, **497**, 1 (1994); (c) Horton, A. D., *Trends Polym. Sci.*, **2**, 158 (1994); (d) Thaylor, A. M., *Chem. Eng. News*, **73**(37), 15 (1995); (e) Brintzinger, H. H., Fischer, D., Mülhaupt, R., Rieger, B. and Waymouth, R. M., *Angew. Chem., Int. Ed. Engl.*, **34**, 1143 (1995).

2. (a) Spaleck, W., Antberg, M., Dolle, V., Klein, R., Rohrman J. and Winter, A., *New J. Chem.*, **14**, 499 (1990); (b) Spaleck, W., Antberg, M., Rohrman, J., Winter, A., Bachmann, B., Kiplof, P., Behm, J. and Herrmann, W. A., *Angew. Chem., Int. Ed. Engl.*, **31**, 1347 (1992); (c) Stehling, U., Diebold, J., Kirstein, R., Roll, W., Brintzinger, H. H., Jungling, S., Mülhaupt, R. and Langhauser, F., *Organometallics*, **13**, 964 (1994); (d) Chacon, S. T., Coughlin, E. B., Henling, L. M. and Bercaw, J. E., *J. Organomet. Chem.*, **497**, 171 (1995); (e) Johnson, L. K., Mecking, S. and Brookhart, M., *J. Am. Chem. Soc.*, **118**, 267 (1996); (f) Leclerc, M. K. and Brintzinger, H. H., *J. Am. Chem. Soc.*, **118**, 9024 (1996); (g) Scollard, J. D. and McConvile, D. H., *J. Am. Chem. Soc.*, **118**, 10008 (1996); (h) Kravchenko, R., Masood, A. and Waymouth, R. M., *Organometallics*, **16**, 3635 (1997).
3. (a) Chien, J. C. W. and He, D., *J. Polym. Sci., Poly. Chem. Ed.*, **29**, 2603 (1991); (b) Holden, D. A., *Macromolecules*, **25**, 1780 (1992); (c) Bonni, F., Fraaijeand, V. and Fink, G., *J. Polym. Sci., Poly. Chem. Ed.*, **33**, 2393 (1995); (d) Ribeiro, M. R., Deffieux, A. and Portela, M. F., *Ind. Eng. Chem. Res.*, **36**, 1224 (1997).
4. (a) Kaminaka, M. and Soga, K., *Makromol. Chem. Rapid Commun.*, **12**, 367 (1991); (b) Kaminsky, W. and Renner, F., *Makromol. Rapid Commun.*, **14**, 239 (1994).
5. (a) Soga, K., Kim, H. J. and Shiono, T., *Macromol. Rapid Commun.*, **15**, 139 (1994); (b) Soga, K., Kim, H. J. and Shiono, T., *Macromol. Chem. Phys.*, **195**, 3347 (1994); (c) Kaminaka, M. and Soga, K., *Makromol. Rapid Commun.*, **15**, 593 (1994).
6. (a) Jutzi, P., Heidemann T. and Stammler H. G., *J. Organomet. Chem.*, **472**, 27 (1994); (b) van den Ancker Y. R. and Raston C. L., *Organometallics*, **14**, 584 (1995); (c) Gao H. and Angelici R. J., *J. Am. Chem. Soc.*, **119**, 6937 (1997)
7. Lee, D. H., Yoon, K. B. and Noh, S. K., *Macromol. Rapid Commun.*, **18**, 427 (1997).
8. Noh, S. K., Shim, J. Y., Lee, D. H. and Yoon, K. B., *Preprints of Asia Polymer Symposium*, Taegu, Korea, May 12–14, 1997, p. 150 (available from the Polymer Society of Korea, Hatchon Building, 831 Yeoksam-dong, Kangnam-ku, Seoul 135–792, Korea).
9. (a) Reddy, K. P. and Petersen, J. L., *Organometallics*, **11**, 665 (1992); (b) Cuenca, T., Flores, J. C., Gomez, R., Gomez-Sal, P., Parra-Hake, M. and Royo, P., *Inorg. Chem.*, **32**, 3608 (1993); (c) Ciruelos, S., Cuenca, T., Flores, J. C., Gomez, R., Gomez-Sal, P. and Royo, P., *Organometallics*, **12**, 944 (1993); (d) Jungling, S., Mülhaupt, R., and Plenio, H., *J. Organomet. Chem.*, **460**, 191 (1993); (e) Lee, D. H., Yoon, K. B., Lee, E. H., Noh, S. K., Byun, G. G. and Lee, C. S., *Macromol Rapid Commun.*, **16**, 265 (1995); (f) Lee, D. H., Yoon, K. B., Lee, Noh, S. K. and Kim, S. C., *Macromol. Rapid Commun.*, **17**, 639 (1996); (g) Noh, S. K., Byun, G. G., Lee, C. S., Lee, D. H., Yoon, K. B. and Kang, K. S., *J. Organomet. Chem.*, **518**, 1 (1996); (h) Ushioda, T., Reen, M. L. H., Haggitt, J. and Yan, X., *J. Organomet. Chem.*, **518**, 155 (1996); (i) Lee, D. H., Yoon, K. B., Noh, S. K. and Woo, S. S., *Macromol. Symp.*, **118**, 129 (1997); (j) Noh, S. K., Kim, S. C., Lee, D. H., Yoon, K. B. and Lee, H. B., *Bull. Korean Chem. Soc.*, **18**, 618 (1997); (k) Noh, S. K., Kim, S. C., Kim, J. H., Lee, D. H., Yoon, K. B., Lee, H. B., Lee, S. W. and Huh, W. S., *J. Polym. Sci., Poly. Chem. Ed.*, **35**, 3717 (1997).
10. Semikolenova N. V. and Zakharov V. A., *Macromol. Chem. Phys.*, **198**, 2889 (1997)
11. (a) Chien, J. C. W. and Razavi, A., *J. Polym. Sci., Part A: Polym. Chem.*, **26**, 2369 (1988); (b) Möhring, P. C. and Coville, N. J., *J. Mol. Catal.*, **77**, 41 (1992); (c) Möhring, P. C. and Coville, N. J., *J. Organomet. Chem.*, **479**, 1 (1994).
12. (a) Chien, J. C. W. and Razavi, A., *J. Polym. Sci., Poly. Chem. Ed.*, **26**, 2369 (1988); (b) Chien, J. C. W. and Wang, B. P., *J. Polym. Sci., Poly. Chem. Ed.*, **26**, 3089 (1988); (c) Chien, J. C. W. and Wang, B. P., *J. Polym. Sci., Poly. Chem. Ed.*, **28**, 15 (1990); (d) Resconni, L., Boossi, S. and Abis, L., *Macromolecules*, **23**, 4489 (1990); (e) Zambelli, A., Pellecchia, C., Oliva, L., Longo, P. and Grassi, A., *Macromol. Chem.*, **192**, 223 (1991).
13. Zakharov, V. A., Makhtarulin, S. I., Perkovets, D. V., Moroz, E. M., Mikenas, T. B. and Bukatov, D. in Keii, T. and Soga, K. (Eds), *Catalytic Polymerization of Olefins*, Kodansha, Tokyo, 1986, p. 71.

PART III

PROPYLENE POLYMERIZATION

17

Synthesis of Isotactic Polypropylene by Metallocene and Related Catalysts

KAZUO SOGA, TOSHIYA UOZUMI AND EIICHI KAJI
Japan Advanced Institute of Science and Technology, Ishikawa, Japan

1 INTRODUCTION

In the mid-1980s, Ewen [1] found that the Cp_2TiPh_2/MAO catalyst produces a stereoblock (atactic and isotactic) polypropylene at low temperatures by chain end control. When using a mixture of racemic and *meso*-$Et(Ind)_2TiCl_2$, he obtained a mixture of atactic and isotactic polypropylene having a microstructure in accordance with catalytic site control.

Since then, numerous metallocene compounds with C_2 symmetry have been synthesized for the production of highly isotactic polypropylene. Some metallocenes with C_1 symmetry are also known to produce isotactic polypropylene. In addition, a variety of supported-type metallocene catalysts have been developed from the viewpoint of industrial applications. This chapter describes recent developments in the field.

2 HOMOGENEOUS METALLOCENE CATALYSTS

2.1 MICROSTRUCTURE OF POLYPROPYLENE

The microstructure of polypropylene in terms of the enchainment of the monomer units and their configuration is determined by the regio- and stereospecificity of the monomer insertion.

Metallocene-based Polyolefins   Edited by J. Scheirs and W. Kaminsky
© 2000 John Wiley & Sons Ltd

Metallocenes favor consecutive primary insertions owing to their bent sandwich structure. However, secondary insertion also occurs to an extent determined by the structure of the metallocene, which causes an increased steric hindrance to the next primary insertion. The active center is blocked and, therefore, is regarded as a resting state of the catalyst [2]. The kinetic hindrance of chain propagation by another insertion favors chain termination and isomerization processes. One of the isomerization processes observed in metallocene-catalyzed propylene polymerization leads to the formation of 1,3-enchainment monomer units [3–6]. The mechanism, originally proposed to be of an elimination–isomerization–addition type, was recently discussed to involve transition metal mediated hydride shifts [7, 8] (Figure 1).

A single step of the polymerization is analogous to a diastereoselective synthesis. Thus, to achieve a certain level of chemical stereocontrol, chirality of the catalytically active species is necessary. In metallocene catalysis, chirality may be located at the transition metal itself, the ligand or the growing polymer chain, e.g. the terminal monomer unit. Therefore, two basic mechanisms of stereocontrol are possible [1, 9]: (a) catalytic (enantiomorphic) site control, which is connected with the chirality at the transition metal or the ligand, and (b) chain end control, which is caused by the chirality of the last inserted monomer unit. These two mechanisms cause different microstructures (Figure 2). In the case of catalytic site control, errors are corrected due to the regime of the catalytic site (Bernoullian statistics), whereas chain end-controlled propagation is not capable of doing so (Markovian statistics).

2.2 C$_2$ SYMMETRIC METALLOCENES

Systematic investigations of bis(indenyl)zirconocenes have clarified the effects of ligand modification on the catalyst performance. In metallocene polymerization, the main chain termination reaction is β-hydrogen transfer with the monomer [10, 11].

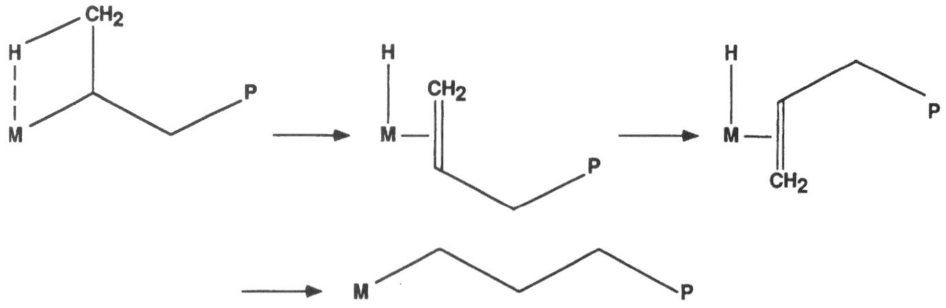

Figure 1 Elimination–isomerization–addition mechanism for the formation of a 1,3-enchained propylene unit [7].

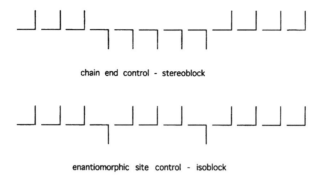

chain end control - stereoblock

enantiomorphic site control - isoblock

Figure 2 Microstructures of polypropylene resulting from different types of stereocontrol during insertion

The rate of β-hydride elimination is controlled by the catalyst employed, apparently with both structure and electronic contributions. Certain substitution patterns favor conformations of the chain which make β-hydride interaction with the active transition metal and consequently chain transfer difficult [12].

The β-hydride elimination is thus effectively suppressed by substituents in position 2 of the indenyl rings [10, 13, 14]. Substituents in position 4 also cause an increase in molecular weight by reducing 2,1-insertions which preferably result in chain termination by β-hydrogen elimination. In addition, electron-releasing substituents reduce the hydrogen abstraction owing to a decrease in the Lewis acidity on the active transition metal [15, 16]. As a result, the most active catalysts feature a methyl or ethyl group in position 2 and an aromatic group in position 4 of the indenyl rings. Spaleck et al. [17] have achieved a beautiful catalyst design utilizing these effects as functional increments. In Table 1 are shown typical results of propylene polymerization.

In the case of C_2 symmetric bridged bis(cyclopentadienyl)metallocenes of zirconium and hafnium, on the other hand, the key to high isotacticity seems to be the substituents in positions 2, 4, 3' and 5' [18].

An unbridged, chiral zirconocene producing an isotactic polymer was reported by Razavi and Atwood [19], i.e. MAO-activated (1-methylfluorenyl)$_2$ZrCl$_2$ gives isotactic polypropylene with a catalytic site-controlled pentad distribution ([mmmm] = 0.83 at 60 °C). The fluorenyl ligands in the complex are chirally disposed owing to their methyl substituents and their mutual rotation is strongly hindered. Therefore, the enantiomers are not interconverted during the growth of a polymer chain. Coates and Waymouth [20] also reported on an unbridged metallocene catalyst, (2-phenylindenyl)$_2$ZrCl$_2$/MAO. Isomerization of indenyl ligands between chiral and achiral coordination geometries during chain growth yields a highly stretchable atactic–isotactic stereoblock polypropylene with elastomeric properties (Figure 3).

Table 1 Polymerization of liquid polypropylene with zirconocene/methylalumoxane catalysts (70 °C, Zr:Al = 1 : 15 000) [17]. (Reprinted with permission from ref. 17. Copyright 1994 American Chemical Society)

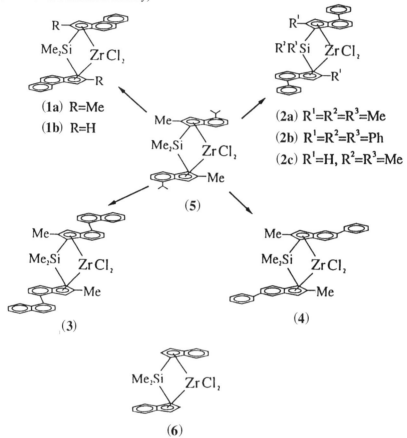

(1a) R=Me
(1b) R=H

(2a) $R^1=R^2=R^3=Me$
(2b) $R^1=R^2=R^3=Ph$
(2c) $R^1=H, R^2=R^3=Me$

(5)

(3)

(4)

(6)

Metallocene	Amount (mg)	Polymer yield (g)	Productivity [kg PP/(mmol Zr)h]	M_w (10^{-3} g/mol)	T_m (°C)	[mmmm] (%)
1a	2.88	2020	403	330	146	88.7
1b	2.74	1370	274	27	138	80.5
2a	2.20	2640	755	729	157	95.2
2b	2.40	1920	553	778	157	95.1
2c	3.00	240	48	42	148	86.5
3	2.18	2620	875	920	161	99.1
4	3.14	314	63	188	139	78.1
For comparison:						
5	2.09	956	190	36	138	81.7
6	2.80	1220	245	213	150	88.6

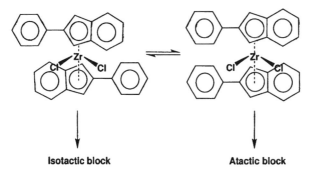

Figure 3 Reversible isomerization of an unbridged zirconocene catalyst between chiral and achiral geometries as found by Coates and Waymouth [20]. (Reprinted with permission from *Science*, **267**, 217 (1995). Copyright 1995 American Association for the Advancement of Science)

2.3 C_1 SYMMETRIC METALLOCENES

Variation of C_s symmetric metallocenes leads to C_1 symmetric compounds. If a methyl group is introduced at position 3 of the cyclopentadienyl ring, stereospecificity is disturbed at one of reaction sites. Every second insertion is thus random, and consequently a hemiisotactic polymer is produced [21–23]. If the methyl group is replaced by a *tert*-butyl group, stereoselectivity is inverted owing to enhanced steric hindrance, resulting in an isotactic polymer [24–29] (Figure 4).

A C_1 symmetric metallocene, *threo*-Me$_2$C(3-^{t}Bu-Cp)(3-^{t}Bu-Ind)ZrCl$_2$, has been claimed by Showa Denko, which affords highly isotactic polypropylene [30]. The isotactic polymerization with these C_1 symmetric metallocenes seems to proceed according to the mechanism proposed by Cossee using only one space for the coordination of an incoming monomer [31]. The mechanism is totally different from that proposed for typical C_2 symmetric metallocenes [32–35]. Typical results of propylene polymerization with C_1 symmetric metallocenes are given in Table 2.

When polymerization of propylene was conducted with typical C_2 symmetric metallocenes such as Et(Ind)$_2$ZrMe$_2$/Ph$_3$CB(C$_6$F$_5$)$_4$ and Me$_2$Si(Ind)$_2$ZrMe$_2$/Ph$_3$CB(C$_6$F$_5$)$_4$ in the presence of methyl methacrylate (MMA) or ethyl benzoate (EB), highly isotactic polypropylene with [mmmm] > 98 % and T_m = 160–161 °C was obtained at 0 °C. Even using an achiral zirconocene such as Cp$_2$ZrMe$_2$, isotactic polypropylene with T_m = 139 °C resulted. The zirconocene compounds might form C_1 symmetric complexes with these Lewis bases [36].

2.4 DIFFERENCE IN CATALYTIC ACTION BETWEEN C_2 AND C_1 SYMMETRIC METALLOCENES

In addition to the ligands of metallocene catalysts, some other factors are found to affect the stereoregularity of isotactic polypropylene. The polymerization tempera-

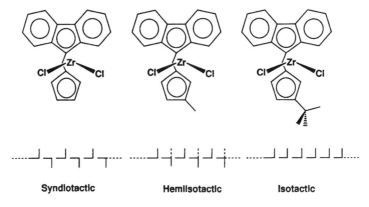

Figure 4 Relationship between catalyst structure and stereoselectivity as reported by Ewen and Elder [26, 28]

ture is believed to influence the rigidity of the metallocene frame which directly controls the stereospecifity [37]. When polymerization of propylene was conducted over the C_2 symmetric Et(2, 4, 7-Me$_3$Ind)$_2$ZrCl$_2$ (**I**) and Me$_2$Si(2-Me-4-Ph-Ind)$_2$ ZrCl$_2$ (**II**) combined with Ph$_3$CB(C$_6$F$_5$)$_4$, both the [mmmm] and T_m of resulting polymers increased markedly along with a decrease in the polymerization temperature from 30 to $-78\,°$C. Thus, isotactic polypropylene with $T_m = 168.9\,°$C was obtained from polymerization with catalyst I at $-78\,°$C. In contrast such a temperature dependence was hardly observed when the C_1 symmetric *threo*-Me$_2$C(3-^{t}Bu-Cp)(3-^{t}Bu-Ind)ZrCl$_2$ was employed [38].

Increased stereoerror frequencies at decreased olefin concentrations, observed by Busico and Cipullo [39] and Resconi *et al.* [40] were ascribed to an epimerization of the last inserted unit, which competes with olefin insertion. Direct evidence for stereoerror formation by chain-end isomerization was found by Leclerc and Brintzinger [41, 42] from a study of D-label distributions in isotactic polypropylene obtained from the polymerization of (*E*)- or (*Z*)-[D$_1$]-propylene with chiral *ansa*-zirconocene catalysts. The appearance of CH$_2$D instead of CH$_3$ groups in the mrrm stereoerror positions indicates that most of these errors arise from the isomerization. We have conducted propylene polymerization under reduced pressure (0.35–1.0 atm) using the C_1 symmetric *threo*-Me$_2$C(3-^{t}Bu-Cp)(3-^{t}Bu-Ind)ZrCl$_2$MAO and TiCl$_4$/MgCl$_2$/Al(^{i}Bu)$_3$ catalyst systems and investigated the effect of propylene pressure on the microstructure of resulting isotactic polypropylene. In the case of C_1 symmetric metallocene, the [mmmm] pentad fraction decreased moderately with a decrease in propylene pressure, but the magnitude of the decrease was much smaller than those reported by Busico and Cipullo [39] using typical C_2 symmetric metallocenes. On the other hand, the [mmmm] of isotactic polypropylene obtained with the conventional Ziegler–Natta catalyst remained almost unchanged in the pressure range examined. These results suggest that the catalytic action of C_1 symmetric metallocenes is close to that of Ziegler–Natta catalyst [43].

Table 2 Polymerization of propylene with C_1 symmetric metallocene/methylalumoxane catalysts [30]. (Reprinted with permission from ref. 30. Copyright 1995 American Chemical Society)

Metallocene	Polymerization temperature (°C)	Productivity [g PP/(mmol Zr)h]	$\overline{M}_w$ (10^{-3} g/mol)	$\overline{M}_w/\overline{M}_n$	[mm] (%)
Me$_2$C(3-^{t}Bu-Cp)(3-^{t}Bu-Ind)TiCl$_2$ (*threo*)	1	1950	34.1	2.6	99.6
Me$_2$C(3-^{t}Bu-Cp)(3-^{t}Bu-Ind)ZrCl$_2$ (*threo*)	1	620	105.0	2.3	99.6
	30	2000	38.0	2.4	99.2
	60	42000	9.0	2.7	99.2
Me$_2$C(3-^{t}Bu-Cp)(3-^{t}Bu-Ind)ZrCl$_2$ (*erythro*)	1	60	9.0	2.0	51.8
Me$_2$C(3-^{t}Bu-Cp)(3-^{t}Bu-Ind)HfCl$_2$ (*threo*)	1	30	39.0	3.3	99.5
	40	1400	11.9	2.7	n.d.
Me$_2$Si(3-^{t}Bu-Cp)(3-^{t}Bu-Ind)ZrCl$_2$ (*threo*)	1	110	28.0	3.5	98.6
	30	1900	10.2	2.3	n.d.
Me$_2$C(3-^{t}Bu-Cp)(3-Me-Ind)ZrCl$_2$ (*threo*)	1	1300	79.0	2.5	95.1
	40	92700	19.6	2.8	95.1

Al/Zr (Ti, Hf) = 2000; [metallocene] = $(2-4) \times 10^{-3}$ mmol/300 ml of toluene; [propylene] = 2 mol.

3 SUPPORTED-TYPE METALLOCENE CATALYSTS

Since current technology of polyolefin production is based on gas-phase and slurry processes, metallocene catalysts used as drop-in catalysts in existing plants have to be heterogenized. Carriers may be divided into two groups: (i) inorganics such as silica, alumina, zeolites and $MgCl_2$ [44–53] and (ii) organic materials such as cyclodextrins, starch and polymers [54–57].

The main preparatory routes reported in the literature for metallocene immobilization on these supports can be classified into the following three methodologies: (i) the first method involves immobilization of MAO on the support followed by reaction with a metallocene compound; a modified version of this method involves the replacement of MAO by an alkylaluminium; (ii) the second method involves direct impregnation of metallocene on the support which could be modified by previous treatment; (iii) the third method involves immobilization of aryl ligands on the support followed by addition of a metal salt such as zirconium halide. More common is heterogenization of the cocatalyst prior to mixing with the metallocene and activation by ordinary trialkylaluminum.

3.1 FIXING A METALLOCENE ON SUPPORTED MAO

Two approaches have been tried to prepare a supported methylaluminoxane: (i) synthesis of supported MAO by contacting trimethylaluminium with hydroxyl groups (starch, silica); and (ii) fixing MAO itself by reaction with hydroxyl groups of the carrier. In both cases, the resulting heterogeneous MAO is reacted with a metallocene dichloride followed by activation by trialkylaluminiums or an additional MAO. Metallocenes fixed on supported MAO exhibit similar behaviors in polymerization to their homogeneous analogues. Transfer of knowledge gathered in homogeneous processes is thus possible.

We have shown that $Et(IndH_4)_2ZrCl_2(SiO_2/MAO)$ associated with AlR_3 as cocatalyst induces the isotactic polymerization of propylene. Catalytic activity is dependent upon the cocatalyst and increases in the order $AlEt_3 < AlMe_3 < Ali(Bu)_3$ [45, 46].

We have also demonstrated that both the stability of catalysts and the morphology of polymer particles are markedly improved by using supported metallocenes. In addition, the shape of polymer particles can be easily controlled by conventional methods applied to Ziegler–Natta catalysts (prepolymerization). The lifetime of active species formed on the SiO_2 surface is also significantly enhanced [58].

3.2 ATTACHING A METALLOCENE TO A CARRIER

The procedure starts with a metallocene carrying an additional functionality either on the cyclopentadienyl rings or on the bridge, which may be used for the fixation.

Metallocenes can be fixed on silica (probably after modification of the hydroxyl groups) or other inorganic carriers as well as on polymeric materials.

Metallocenes may also be fixed on silica, alumina, $MgCl_2$, etc., by direct reactions between the two components. Marks [59] has demonstrated that reaction of dimethylmetallocenes with alumina results in the formation of an active catalyst. Kaminsky and Renner [60] obtain a highly isotactic polypropylene with $M_W = (6-8) \times 10^5$, [mm] $= 93-96\%$ and $T_m = 160-161\,^\circ C$ using an SiO_2-supported $Et(Ind)_2ZrCl_2$ catalyst combined with MAO. The catalyst activity can be enhanced by raising the polymerization temperature from 0 to 75 °C. Both the molecular weight and the melting temperature of the resulting polymer also increase with increase in temperature, in contrast to homogeneous catalysts.

We have prepared several kinds of supported metallocenes by direct impregnation [45, 61]. When Al_2O_3 and $MgCl_2$ were used as the carrier, the resulting zirconocene catalysts can be activated by ordinary alkylaluminums to give polypropylene with fairly high activity. In contrast, zirconocenes supported on SiO_2 combined with alkylaluminums are almost inactive. These results could be explained in terms of the acidity of carriers, i.e. carriers exhibiting strong Lewis acidity lead to active catalysts. A plausible model of the formation of an active species is illustrated in Figure 5 [46], taking the previous work of Marks and co-workers [62–64] into consideration.

3.3 SYNTHESIS OF A METALLOCENE ON A CARRIER

We have prepared several kinds of SiO_2-supported zirconocene catalysts, i.e. analogs of $Me_2Si(Ind)ZrCl_2$ (**1**), $Et(IndH_4)_2ZrCl_2$ (**2**) and $^iPr(Flu)(Cp)ZrCl_2$ (**3**,) etc., using the precursors shown in Figure 6 [65]. SiO_2 (Fuji Davison Co., No. 952, ca $300\,m^2/g$) was calcined at 200–900 °C. For example, the analog of **1** was prepared using precursor **A**, whereas those of **2** and **3** were prepared using precursor **B** according to the schemes shown in Figure 7.

Most of these catalysts are considered to contain a mixture of *meso* and racemic isomers. Therefore, atactic polypropylene produced with the *meso* isomer was removed by extracting with boiling heptane. Typical results obtained with catalysts **I, II** and **III** are shown in Table 3 [66].

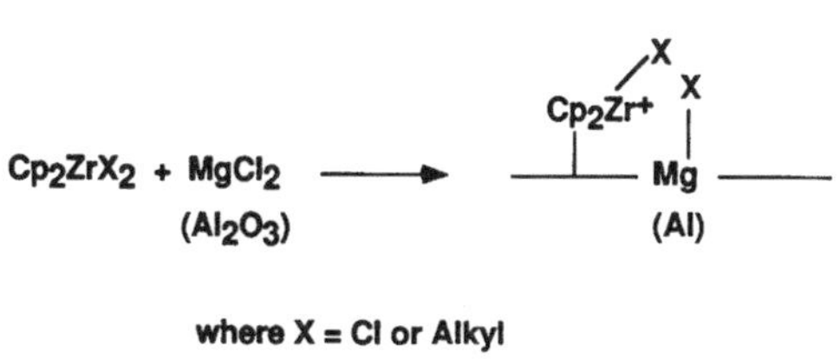

where X = Cl or Alkyl

Figure 5 A plausible model of catalyst activation

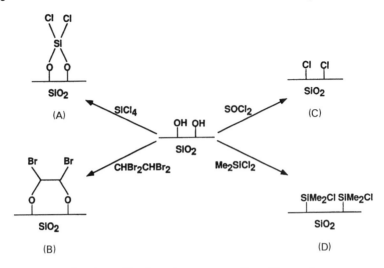

Figure 6 Modification of hydroxyl groups on the silica surface

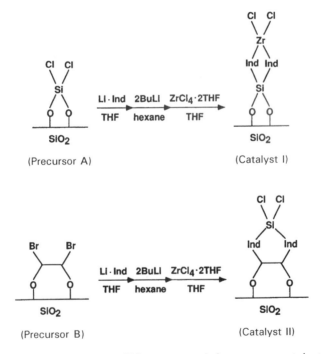

Figure 7 Synthetic routes to SiO_2-supported zirconocene catalysts

Table 3 Results of propylene polymerization with catalysts **I**, **II** and **III**[a]

Catalyst	Cocatalyst (mmol)	Yield (g)	$\overline{M}_w$ (10^{-4}g/mol)	T_m (°C)	I.I. (wt%)	[mmmm] (%)
Me$_2$Si(Ind)$_2$ZrCl$_2$ (**I**)	MAO (5)	0.37	3.0	142.1	32	83.3
Catalyst **I** (200 °C)[b]	MAO (3)	0.14	—	161.4	84	—
	Al(^{i}Bu)$_3$ (1)	0.22	—	157.7 162.7	63	97.5
(400 °C)	MAO (3)	0.33	—	156.1 162.3	67	94.3
	Al(^{i}Bu)$_3$ (1)	0.25	—	158.6 162.2	76	—
(900 °C)	MAO (3)	0.17	—	159.1	27	—
	Al(^{i}Bu)$_3$ (1)	0.30	—	158.7	26	94.5
Et(IndH$_4$)$_2$ZrCl$_2$	MAO (3)	2.07	0.3	111.0	—	71.0
Catalyst **II** (400 °C)	MAO (1)	0.47	—	149.9 160.0	46	93.0
	Al(^{i}Bu)$_3$ (1)	0.40	48.5	158.2	55	97.1
	AlEt$_3$ (1)	0.21	—	153.2 161.2	68	—
	AlMe$_3$ (1)	0.11	—	156.8 162.1	61	—
^{i}Pr(Cp)(Flu)ZrCl$_2$ (3)	MAO (3)	3.03	3.9	123.0	—	77(Syn.)[c]
Catalyst **III** (400 °C)	MAO (15)	0.60	27.0	162.0	25	95.0

[a] Homogeneous system: catalyst = 5×10^{-3} mmol, at 40 °C, 1 atm, for 1 h. Heterogenous system: catalyst = 100 mg at 40 °C for 20 h under pressure.
[b] Heating temperature of SiO$_2$.
[c] [rrrr].

As described above, the SiO$_2$-supported Et(IndH$_4$)$_2$ZrCl$_2$ catalyst prepared from the mixture of unmodified SiO$_2$ and Et(IndH$_4$)$_2$ZrCl$_2$ cannot be activated by ordinary trialkylaluminiums [68]. These catalysts were, however, found to be activated by trialkylaluminiums, especially by Al(^{i}Bu)$_3$. The catalyst isospecificity and the molecular weight of isotactic polypropylene increased considerably on rigidly immobilizing the zirconocene compounds on the chemically modified SiO$_2$ surface.

On the other hand, isotactic polypropylene obtained with catalysts **I** and **II**, which had been prepared using SiO$_2$ calcined at lower temperatures, displayed two melting-points, suggesting that there are different kinds of isospecific active species in those catalysts as illustrated in Figure 8 [69].

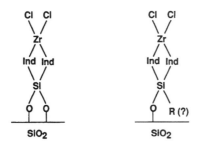

More rigidly immobilized style   **Less immobilized style**

Figure 8 Plausible structures of singly and doubly bound species

It is worthy of note that catalyst **III** gave highly isotactic polypropylene with a single T_m. The isospecific (racemic) active species in catalysts **I** and **II** are considered to possess C_2 symmetry, whereas such a structure cannot be expected in catalyst **III**. Accordingly, the catalyst isospecificity (chirality) might arise from another factor. A similar result was also reported by Kaminsky [70]. The surface structure of carrier might play a key role to determine the catalyst stereospecificity.

Different types of bridge to connect the metallocene ligands to the silica surface were also tested. Isotactic polypropylene obtained with the tin-bridged catalyst displayed a single T_m at 162 °C [69]. The higher reactivity of $SnCl_4$ toward the surface hydroxyl groups might be responsible for the formation of more uniform active species on SiO_2.

Some organic materials such as polystyrene beads have also been tested as a support [57].

3.4 NEW APPROACH FOR THE SYNTHESIS OF SUPPORTED-TYPE METALLOCENE AND RELATED CATALYSTS

Generally, the polymerization activities of supported-type metallocene catalysts prepared by the above methods for **2** and **3** are very low compared with the corresponding homogeneous systems. In addition, the stereospecificity of C_s symmetric zirconocenes drastically changes from syndiospecific to isospecific on supporting them on a solid surface [70]. These results might reflect the complicated surface structure of solid carriers.

We have employed polysiloxanes with simpler structures as the carrier. The results for propylene polymerization with typical polysiloxane-supported metallocene catalysts are shown in Table 4, where those obtained with the corresponding homogeneous systems are also attached for reference. Some of the catalysts proved to display very high activities. Plausible structures of aspecific and isospecific active species are illustrated in Figures 9 and 10, where the Zr–polymer bond is omitted for clarity [71].

We have also synthesized an isospecific heterogeneous zirconocene catalyst by supporting rac-$Cl_2Si(Ind)_2ZrCl_2$ on $MgCl_2$. Typical results for propylene polymerization with this catalyst are given in Table 5, together with a plausible structure of active species in Figure 11. It is not clear at present why the catalyst activity is much lower than that of the corresponding homogeneous catalyst [72].

It is recognized that the stability of active species depends substantially on the rigidity of the ligand framework. From such a viewpoint, $ansa$-metallocenes having phenyl substitutions on the silicon or carbon bridge, such as $Ph_2Si(Ind)_2ZrCl_2$ and $Ph_2C(Cp)(Flu)ZrCl_2$, have been developed. These catalysts were proved to show a high catalyst performance for ethylene polymerization [73]. More recently, we have synthesized a dinuclear $ansa$-zirconocene catalyst with high rigidity of the ligand framework, i.e. μ-$C_{12}H_8[PhSi(Ind)_2ZrCl_2]_2$. The dinuclear catalyst (composed of a mixture of racemic and $meso$-diastereomers) combined with MAO as cocatalyst

Table 4 Results of propylene polymerization with $(R_2SiO)_nZrCl_2$/MAO catalyst systems[a].

Catalyst	R	Al/Zr ratio	Productivity [kg f PP/(mol Zr) h]	I.I.[b] Soluble (wt%)	(C$_5$) Insoluble (wt%)	$\overline{M}_w$ (10^{-4}g/mol)	$\overline{M}_w/\overline{M}_n$	T_m (°C)	[mmmm] (%)
I	(Ind)$_2$	5.0×10^3	3.4×10^3	33.5	66.5	2.0	3.3	142.2	—
		1.0×10^4	2.8×10^3	40.1	59.1	2.2	2.9	142.5	87.0
II	(Flu)$_2$	5.0×10^3	5.9×10^3	—	—	18.5	2.0	n.d	mm = 15.2 mr = 47.9 rr = 36.9
IV	[(Me$_3$Ind)$_2$]	5.0×10^3	3.7×10^2	81.2	18.8	2.4	4.1	151.0	95.0
		1.0×10^4	1.3×10^2	75.5	24.5	2.1	4.7	150.9	—
V	(IndMe)	5.0×10^3	2.9×10^2	36.2	63.8	4.2	2.2	146.4	—
		1.0×10^4	2.0×10^2	42.0	58.0	3.8	1.9	145.8	94.4
VI	(FluMe)	5.0×10^3	3.0	54.2	45.8	4.2	Broad	148.5	—
		1.0×10^4	7.0	60.9	39.1	10.5	Broad	153.9	86.7
	Me$_2$Si(Ind)$_2$ZrCl$_2$[c]	5.0×10^3	2.2×10^4	—	—	8.0	1.7	142.0	89.0
	Me$_2$Si(Flu)$_2$ZrCl$_2$[d]	5.0×10^4	1.6×10^4	—	—	20.2	—	—	mm = 16.8 mr = 48.6 rr = 34.6

[a] Polymerization was conducted at 40 °C for 10 min in a 1 l glass autoclave containing 300 ml of toluene under 5 atm of propylene using 5.0×10^{-6} mol of Zr.
[b] I.I. = isotactic index = wt% of boiling pentane-insoluble fraction.
[c] Polymerization was conducted at 40 °C for 10 min in a 1 l glass autoclave containing 300 ml of toluene under 2 atm of propylene using 5.0×10^{-6} mol of Zr.
[d] From Ref.67.

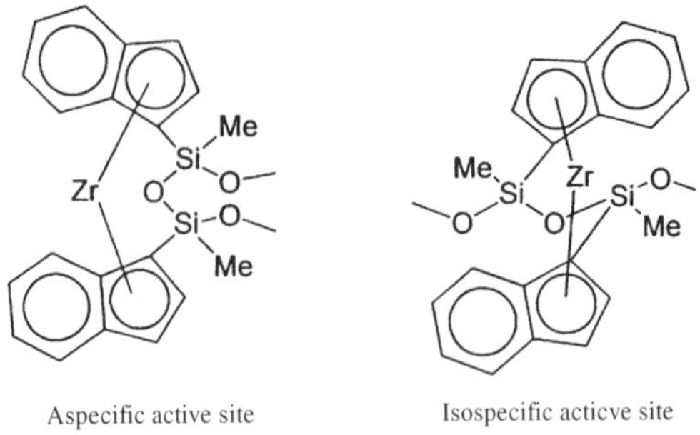

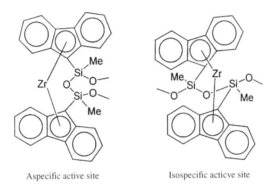

Aspecific active site   Isospecific acticve site

Figure 9 Plausible structures of active species in $(IndMeSiO)_nZrCl_2$

Aspecific active site   Isospecific acticve site

Figure 10 Plausible structures of active species in $(FluMeSiO)_nZrCl_2$

showed a much higher activity for ethylene polymerization than the mononuclear analog, $Ph_2Si(Ind)_2ZrCl_2$ [74]. It was also demonstrated that the active species formed in the dinuclear catalyst is thermally very stabilized. The dinuclear compound was separated into racemic and *meso*-diastereomers, and polymerizations of ethylene and propylene were conducted with them over a wide range of temperature.

Typical results for propylene polymerization together with the analytical results for the polymers produced are shown in Table 6 [75]. For propylene polymerization, the dinuclear catalyst showed less activity than the mononuclear catalyst. Thus, the dinuclear zirconocene catalyst may be called a typical ethylene-selective polymerization catalyst. The racemic and *meso* catalysts naturally yielded isotactic and atactic polypropylenes. However, the ^{13}C NMR spectrum of isotactic polypropylene did not display any peaks attributable to 2,1- and 1,3-insertions, suggesting that the regiospecificity of the racemic dinuclear catalyst is very high.

Recently, some zirconium β-diketonate complexes have been claimed as catalysts for olefin polymerizations, i.e. these complexes combined with MAO and AlR_nCl_{3-n} ($R = Me$, Et and ^iBu, $n = 2$, 3) give polyethylene and ethylene oligomer, respec-

Table 5 Results of propylene polymerization with MgCl$_2$-supported zirconocene catalyst systems

Catalyst	Cocatalyst	Al/Zr ratio	T_p (°C)	Productivity [kg PP/(mol Zr) h]	I.I. (C$_5$) (wt%)	$\overline{M}_w$ (10^{-4} g/mol)	$\overline{M}_w/\overline{M}_n$	T_m (°C)	[mmmm] (%)
MgCl$_2$-supported	MAO	1.0×10^4	40	3.3×10^2	85.6	2.4	2.7	143.7	—
	MAO	5.0×10^3	40	2.3×10^2	91.9	2.4	2.7	142.9	91.8
	[Ph$_3$C][B(C$_6$F$_5$)$_4$]	1.0	40	1.1×10^2	97.6	4.7	2.8	146.5	92.7
Me$_2$Si(Ind)$_2$ZrCl$_2$	MAO	5.0×10^3	40	3.8×10^4	—	11.9	1.8	142.0	89.0

Polymerization was conducted at 40 °C in a 100 ml stainless-steel autoclave containing 30 ml of toluene under 7 l of propylene using 5.0×10^{-7} mol of Zr.

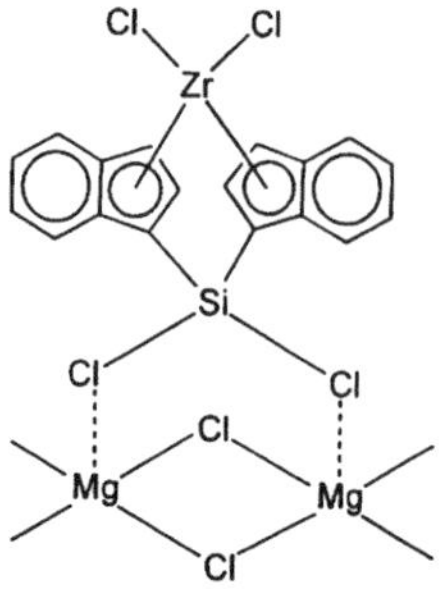

Figure 11   Plausible structure of active species formed on MgCl$_2$

Table 6   Results of propylene polymerization with the racemic and *meso* dinuclear catalysts[a]

Catalyst[b]	Al/Zr ratio	T_p (°C)	Productivity [kg PP/(mmol Zr) [Pr] h]	$\overline{M}_w$ (10^{-3}g/mol)	$\overline{M}_w/\overline{M}_n$	T_m (°C)	[mmmm] (%)
I	5000	40	0.22	42.2	2.4	140.6	92
-	10000	40	0.28	40.1	3.0	140.1	—
-	10000	60	0.36	29.0	2.9	133.2	—
II	5000	40	0.44	12.8	2.9	—	15
-	10000	60	0.75	4.7	4.3	—	—
III[c]	250	30	2.16	90	—	—	97

[a] Polymerization conditions: 1 l high-pressure glass reactor, propylene pressure = 2 atm, [Zr] = 2.0×10^{-6} mol/l, toluene = 295 ml, polymerization time = 1 h.
[b] I = racemic dinuclear catalyst; II = *meso* dinuclear catalyst; III = corresponding mononulear catalyst.
[c] Propylene pressure = 2.5 atm, [Zr] = 6.25×10^{-6} mol/l.

tively [76, 77]. The catalysts are, however, almost inactive for the polymerization of other α-olefins.

We have synthesized titanium analogs such as Ti(acetylacetonate)$_2$Cl$_2$, Ti(1-benzoylacetonato)$_2$Cl$_2$, Ti(2,2,6,6-tetramethyl-3,5-heptanedionato)$_2$Cl$_2$ and Ti(4,4,4-trifluoro-1-phenyl-1,3-butanedionato)$_2$Cl$_2$[Ti(BFA)$_2$Cl$_2$] and tested them in propylene polymerization. These complexes combined with MAO yielded only trace amounts of polypropylene. In contrast, when they were supported on MgCl$_2$, the resulting catalysts gave polypropylene in high yields even by using ordinary alkylaluminiums as cocatalyst. Among these titanium compounds, Ti(BFA)$_2$Cl$_2$ displayed the highest activity. The catalyst isospecificity was also markedly improved by the addition of a suitable Lewis base, especially di-iso-propyldi-methoxy-silane (DIPDMS). From the elemental analysis of the supported catalysts before and after being subjected to the cocatalyst, it was found that one of the β-diketone ligands is attached to the active Ti [78, 79]. A decrease in the loading amount of the titanium complex enhanced substantially both the catalyst activity and

Table 7 Results of propylene polymerization with $Ti(BFA)_2Cl_2/MgCl_2/TEA$ catalyst system

| Catalyst No. | Ti content (mmol/g) | [Si]/[Ti] (mol/mol) | Productivity [kg PP/(mol Ti)h][a] | | I.I.[b] (wt%) | $\overline{M}_n(10^{-4}$ g/mol) (C$_7$-insoluble) | $\overline{M}_w/\overline{M}_n$ (C$_7$-insoluble) | $T_m{}^c$ (°C) | [mmmm][d] (%) |
			C$_7$-insoluble (Iso.)	C$_7$-soluble (Ata.)					
1	0.017	0	114	378	23.4	3.2	3.6	158.1	88.9
		1.6	208	67	75.5	6.0	3.9	163.7	91.2
		3.1	113	18	85.7	8.4	3.7	164.9	93.2
		6.2	98	9	91.0	10.5	5.1	164.3	94.3
		12.4	62	7	90.8	15.1	4.4	165.7	94.9
2	0.011	0	173	473	26.7	3.8	2.9	159.9	86.6
		3.3	346	29	92.3	7.2	4.3	164.9	94.7
		6.5	201	10	95.4	7.0	5.0	166.2	95.2
		13.0	142	5	96.6	9.9	3.9	166.8	96.8
3	0.007	0	149	400	27.1	3.9	4.2	161.8	86.4
		3.3	401	172	70.0	6.2	4.1	163.9	
		6.5	429	33	92.9	11.1	3.7	166.2	94.8
		13.0	271	10	96.4	10.6	4.2	166.8	97.0
4	0.002	0	164	466	26.0	3.8	5.1	161.2	
		31.1	817	29	96.6	7.1	5.9	167.5	95.2
		62.5	416	10	97.7	10.4	4.5	168.0	98.3

Polymerization conditions: catalyst = $Ti(BFA)_2Cl_2/MgCl_2$, [Ti] = 0.01 mmol, [Al(TEA)]/[Ti] = 70, n-heptane = 100 ml. P = 1 atm, 40°C, 1 h.Si = di-iso-propyldimethoxysilane.
[a] Iso., isotactic part; Ata., atactic part.
[b] I.I. = isotactic index = wt% of boiling heptane-insoluble fraction.
[c] T_m: for boiling heptane-insoluble part.
[d] [mmmm] for boiling heptane-insoluble part.

isospecificity. Typical results for propylene polymerization over the catalysts with different Ti contents are shown in Table 7 [80].

The increase in the activity for isotactic polymerization upon the addition of DIPDMS becomes more prominent with a decrease in the Ti content (ca 5.0-fold for catalyst 4). Even in the ordinary $MgCl_2$-supported Ziegler–Natta catalysts containing much higher amounts of Ti, the addition of an external donor often causes an increase in the activity for isotactic polymerization [81–85], but such a drastic increase as observed here has not been reported previously.

Since we obtained a highly isotactic polypropylene with catalyst 4, the polymer (boiling heptane-insoluble fraction) was further extracted with boiling octane. The boiling octane-insoluble fraction (86 wt%) displayed [mmmm] and T_m as high as over 99 % and 169.3 °C, respectively.

4 REFERENCES

1. Ewen, J. A., *J. Am. Chem. Soc.,* **106**, 6355 (1984).
2. Busico, V., Cipullo, R. and Corradini, P., *Makromol. Chem., Rapid Commun.,* **14**, 97 (1993).
3. Grassi, A., Zambelli, A., Resconi, L., Albizzati, E. and Mazzocchi, R., *Macromolecules,* **21**, 617 (1989).
4. Soga, K., Shiono, T., Takemura, S. and Kaminsky, W., *Makromol. Chem., Rapid Commun.,* **8**, 305 (1987).
5. Cheng, H. N. and Ewen, J. A., *Makromol. Chem.,* **190**, 1931 (1989).
6. Rieger, B., Mu, X., Mallin, D. T., Rausch, M. D. and Chien, J. C. W., *Macromolecules,* **23**, 3559 (1990).
7. Schupfner, G. and Kaminsky, W., *J. Mol. Catal. A: Chem.,* **102**, 59 (1995).
8. Busico, V. and Cipullo, R., *J. Am. Chem. Soc.,* **117**, 1652 (1994).
9. Sheldon, R. A., Fueno, T., Tsuntsugu, T. and Kurukawa, J., *J. Polym. Sci., Part B,* **3**, 23 (1965).
10. Jüngling, S., Mülhaupt, R., Stehling, U., Brintzinger, H. H., Fischer, D. and Langhauser, F., *Macromol. Symp.,* **97**, 205 (1995).
11. Resconi, L., Fait, A., Piemontesi, F., Colonnesi, M., Rychlicki, H. and Ziegler, R., *Macromolecules,* **28**, 6667 (1995).
12. Krauledat, H. and Brintzinger, H. H., *Angew. Chem., Int. Ed. Engl.,* **29**, 1412 (1990).
13. Spaleck, W., Antberg, A., Rohrmann, J., Winter, A., Bachmann, B., Kiprof, P., Behn, J. and Herrmann, W. A., *Angew. Chem.,* **104**, 1373 (1992).
14. Stehling, U., Diebold, J., Kirsten, R., Röll, W., Brintzinger, H. H., Jüngling, S., Mülhaupt, R. and Langhauser, F., *Organometallics,* **13**, 964 (1994).
15. Brookhart, M. and Green, M. L. H., *J. Organomet. Chem.,* **250**, 395 (1983).
16. Brookhart, M., Green, M. L. H. and Wong, L., *Prog. Inorg. Chem.,* **36**, 1 (1988).
17. Spaleck, W., Kiiber, F., Winter, A., Rohrmann, J., Bachmann, B., Antberg, M., Dolle, V. and Psulus, E. F., *Organometallics,* **13**, 954 (1994).
18. Mise, Y., Miya, S. and Yamazaki, H., *Chem. Lett.,* 1853 (1989).
19. Razavi, A. and Atwood, J. L., *J. Am. Chem. Soc.,* **115**, 7529 (1993).
20. Coates, G. W. and Waymouth, R. M., *Science,* **267**, 217 (1995).
21. Farina, M., DiSilvestro, G. and Sozzani, P., *Prog. Polym. Sci.,* **16**, 219 (1991).
22. Farina, M. DiSilvestro, G. and Sozzani, P., *Macromolecules,* **26**, 946 (1993).

23. Herfert, N. and Fink, G., *Makromol. Chem., Macromol. Symp.,* **66**, 157 (1993).
24. Razavi, A., Nafplitotis, L., Vereecke, D., DenDauw, K., Atwood, L. J. and Thewald, U., *Polym. Prepr. Am. Chem. Soc.,* **32**, 469 (1991).
25. Razavi, A., Peters, L., Nafpliotis, L. and Atwood, J. L., in Soga, K. and Terano, T. (Eds), *Catalysts Design for Tailor-made Polyolefins,* Kodansha, Tokyo, 1994.
26. Ewen, J. A., *Macromol. Symp.,* **89**, 181 (1995).
27. Spaleck, W., Aulbach, M., Bachmann, B., Küber, F. and Winter, A., *Macromol. Symp.,* **89**, 237 (1995).
28. Ewen, J. A. and Elder, M. J., in Fink, G., Mülhaupt, R. and Brintzinger, H. H. (Eds), *Ziegler Catalysts,* Springer, Berlin, 1995, p. 99.
29. Razavi, A., Vereecke, D., Peters, L., DenDauw, K., Nafpliotis, L. and Atwood, J. L., in Fink, G., Mülhaupt, R. and Brintzinger, H. H. (Eds), *Ziegler Catalysts,* Springer, Berlin, 1995, p. 111.
30. Miyake, S., Okumura, Y. and Inazawa, S., *Macromolecules,* **28**, 3074 (1995).
31. Cossee, P., *Tetrahedron Lett.,* **12**, 17 (1960).
32. Pino, P., Cioni, P. and Wei, J., *J. Am. Chem. Soc.,* **109**, 6189 (1987).
33. Longo, P., Grassi, A., Pellecchia, C. and Zambelli, A., *Macromolecules,* **20**, 1015 (1987).
34. Castaonguay, L. A. and Rappe, A. K., *J. Am. Chem. Soc.,* **114**, 5832 (1992).
35. Hart, J. R. and Rappe, A. K., *J. Am. Chem. Soc.,* **115**, 6159 (1993).
36. Deng, H. and Soga, K., *Polym. Bull.,* **37**, 29 (1996).
37. Rieger, B., Mu, X., Mallin, D. T., Rausch, M. and Chien, J. C. W., *Macromolecules,* **23**, 3559 (1990).
38. Deng, H., Winkelbach, H., Taeji, K., Kaminsky, W. and Soga, K., *Macromolecules,* **29**, 6376 (1996).
39. Busico, V. and Cipullo, R., *J. Am. Chem. Soc.,* **116**, 9329 (1994).
40. Resconi, L., Fait, A., Piemontesi, F., Colonnesi, M., Rychlicki, H. and Ziegler, R., *Macromolecules,* **28**, 6667 (1995).
41. Leclerc, M. K. and Brintzinger, H. H., *J. Am. Chem. Soc.,* **117**, 1651 (1995).
42. Leclerc, M. K. and Brintzinger, H. H., *J. Am. Chem. Soc.,* **118**, 9024 (1996).
43. Schneider, M. J., Kaji, E., Uozumi, T. and Soga, K., *Macromol. Chem. Phys.,* **198**, 2899 (1997).
44. Kaminaka, M. and Soga, K., *Makromol. Chem. Rapid Commun.,* **12**, 367 (1991).
45. Soga, K. and Kaminaka, M., *Makromol. Chem. Rapid Commun.,* **13**, 221 (1992).
46. Soga, K. and Kaminaka, M., *Makromol. Chem.,* **194**, 1745 (1993).
47. Soga, K., Kaminaka, M. and Shiono, T., in *Proceedings of the Worldwide Metallocene Conference MetCon'93,* Catalyst Consultants, Houston, TX, 1993.
48. Soga, K., Kim, H. J. and Shiono, T., *Makromol. Chem. Rapid Commun.,* **15**, 139 (1994).
49. Chien, J. C. W. and He, D., *J. Polym. Sci. Part A, Polym. Chem.,* **29**, 1603 (1991).
50. Collins, S., Kelly, W. M. and Holden, D. A., *Macromolecules,* **25**, 1780 (1992).
51. Janiak, C. and Reiger, B., *Angew. Makromol. Chem.,* **215**, 47 (1994).
52. Ismayel, A., Sanchez, G., Arribas, G. and Ciardelli, F., *Mater. Eng.,* **4**, 267 (1993).
53. Woo, S. I., Koo, Y. S. and Han, T. K., *Macromol. Rapid Commun.,* **16**, 489 (1995).
54. Lee, D. H. and Yoon, K. B., *Macromol. Rapid Commun.,* **15**, 841 (1994).
55. Lee, D. H. and Yoon, K. B., *Macromol. Symp.,* **97**, 185 (1995).
56. Kaminsky, W., in Quirk, R. P. (Ed.), *Transition Metal Catalyzed Polymerizations, Alkenes and Dienes,* Cambridge University Press, Cambridge, 1988, p. 225.
57. Nishida, H., Uozumi, T., Arai, T. and Soga, K., *Macromol. Rapid Commun.,* **16**, 821 (1995).
58. Soga, K. and Kaminaka, M., *Macromol. Rapid Commun.,* **15**, 593 (1994).
59. Marks, T. J., *Acc. Chem. Res.,* **25**, 57 (1992).
60. Kaminsky, W. and Renner, F., *Macromol. Chem., Rapid Commun.,* **14**, 239 (1993).

61. Kaminaka, M. and Soga, K., *Polymer,* **33**, 1105 (1992).
62. He, M.-Y., Xiong, G., Toscano, P. J., Burwell, R. L., Jr, and Marks, T. J., *J. Am. Chem. Soc.,* **107**, 641 (1985).
63. Hedden, D. and Marks, T. J., *J. Am. Chem. Soc.,* **110**, 1647 (1988).
64. Finch, W. C., Gillespie, R. D., Hedden, D. and Marks, T. J., *J. Am. Chem. Soc.,* **112**, 622 (1990).
65. Soga, K., Arai, T., Hoang, B. T. and Uozumi, T., *Macromol. Rapid Commun.,* **16**, 905 (1995).
66. Soga, K., Kim, H. J. and Shiono, T., *Macromol. Chem. Phys.,* **195**, 1503 (1994).
67. Recconi, L., Jones, R. L., Rheingold, A. L., and Yap, G. P. A., *Organometallics,* **15**, 998 (1996).
68. Soga, K. and Kaminaka, M., *Macromol. Chem. Phys.,* **195**, 1369 (1994).
69. Soga, K., Arai, T., Nozawa, H. and Uozumi, T., *Macromol. Symp.,* **97**, 53 (1995).
70. Kaminsky, W., *Makromol. Symp.,* **89**, 203 (1995).
71. Arai, T., Hoang, B. T., Uozumi, T. and Soga, K., *J. Polym. Sci. Part A, Polym. Chem.,* in press.
72. Soga, K., Arai, T. and Uozumi, T., *Polymer,* **38**, 4993 (1197).
73. Akimoto, A., in *Proceedings of Metalloene '95, Brussels,* Schotland Business Research Inc., 1995, p. 439.
74. Soga, K., Ban, H. T. and Uozumi, T., *J. Mol. Catal.,* in press.
75. Ban, H. T., Uozumi, T. and Soga, K., *J. Polym. Sci. Part A, Polym. Chem.,* submitted for publication.
76. Janiak, C., Scharmann, T. G. and Lange, K. C. H., *Macromol. Rapid Commun.,* **15**, 655 (1994).
77. Oouchi, K., Mitani, M., Hayakawa, M., Yamada, T. and Mukaiyama, T., *Macromol. Chem. Phys.,* **197**, 1545 (1996).
78. Soga, K., Kaji, E. and Uozumi, T., *J. Polym. Sci. Part A, Polym. Chem.,* in press.
79. Soga, K., Uozumi, T., Arai, T., Ban, H. T. and Kaji, E., in *Proceedings of MetCon'97, Houston,* The Catalyst Group, 1997.
80. Kaji, E., Uozumi, T., Jin, J., Sano, T., and Soga, K., *Macromol. Chem. Phys.,* in press.
81. Kashiwa, N., *Polym. Bull.,* **12**, 99 (1984).
82. Kashiwa, N., Yoshitake, J. and Toyota, A., *Polym. Bull.,* **19**, 333 (1988).
83. Spitz, R., Bobichon, C. and Guyot, A., *Makromol. Chem.,* **190**, 707 (1989).
84. Barbe, P. C., Noristi, L. and Baruzzi, G., *Makromol. Chem.,* **193**, 229 (1992).
85. Sacchi, M. C., Tritto, I., San, C., Mendichi, R. and Noristi, L., *Macromolecules,* **24**, 682 (1991).

18

Synthesis and Properties of Metallocene Catalysts for Isotactic Polypropylene Production

WALTER SPALECK
Hoechst AG, Frankfurt am Main, Germany[†]

1 INTRODUCTION

Isotactic propylene polymerization by a metallocene catalyst was first described in 1984 [1, 2]. Six years later metallocene structures with technically relevant polymerization performances were developed [3] and 1995 was the year of the first production plant trials with supported versions of those metallocenes. Two years later several companies started regular production of metallocene-catalyzed isotactic polypropylenes (m-iPPs) (for a definition of 'metallocene isotactic polypropylene', see Chapter 19). The quantities produced are still small, in 1997 clearly below 150 000 t, i.e. less than 1 % of the world's isotactic polypropylene (iPP) production.

Metallocene iPP catalysts have also raised considerable interest in the scientific world. So far more than 400 papers have been published, a large portion of them on mechanistic topics. This literature has already been reviewed several times, also just recently (e.g. [4–7]). In this chapter, theoretical and mechanistical aspects are therefore only briefly discussed; the focus is on the synthesis and behavior of practical catalysts which can be used in technical polymerization processes.

[†] Current address: Spaleck Research, Elsa-Brändström-Str. 32, D-28359 Bremen, Germany

Metallocene-based Polyolefins   Edited by J. Scheirs and W. Kaminsky
© 2000 John Wiley & Sons Ltd

2 BASIC MECHANISTICAL ASPECTS

Metallocenes are usually synthesized as dichloro complexes of the type $Cp^*_2MCl_2$ (M = transition metal, e.g. Ti, Zr, Hf, and Cp^* = cyclopentadienyl, substituted or unsubstituted) which are only catalyst precursors. This precursor must be reacted with an activator, e.g. methylaluminoxane (MAO). The reaction generates the active catalytic complex by abstracting both chlorine atoms, putting a methyl group on the transition metal and thus creating a cationic species with a free coordination position (for the monomer) and a metal–carbon bond. Insertion of a propylene molecule into this primary metal carbon bond then initiates chain growth, which continues by repeated insertion into the bond between the catalyst's active metal center and the first carbon atom of the growing chain. Breaking of this bond by several reactions (see below) terminates chain growth, which may start again by a new chain initiation.

Formation of an isotactic polypropylene chain (Figure 1) needs the mode of insertion to be exclusively both 1,2-regiospecific (i.e. monomer enchainment in head-to-tail-fashion), and (iso-) stereospecific (i.e. putting all methyl-bearing carbons into the same configuration). Regiospecificity and stereospecificity of the insertion are controlled by the ligand framework of the active complex, i.e. the coordination position of the incoming monomer leading to the 'right' insertion mode is energetically favored over all other possible insertion pathways by the steric influence of the ligand framework [4,8–10].[†] This steric control is believed to be 'indirect', i.e. the incoming monomer is positioned by the end of the growing chain which itself is positioned by the metallocene ligands. Insertion of monomer by a disfavoured pathway ('insertion mistake') will produce a chain irregularity. The types of irregularities found in metallocene iPP are depicted in Figure 2. As shown there, these are isolated, i.e. after a single mistake the chain is continued as before. 1,2-Stereoerrors [Figure 2(b)] may typically occur in the range 0.5–5 % of all insertions and 2,1-Regioerrors [Figure 2(c)] and the so-called 1,3-enchainments [Figure 2(d)] clearly below 1 %.[‡] The macroscopic consequence of chain irregula-

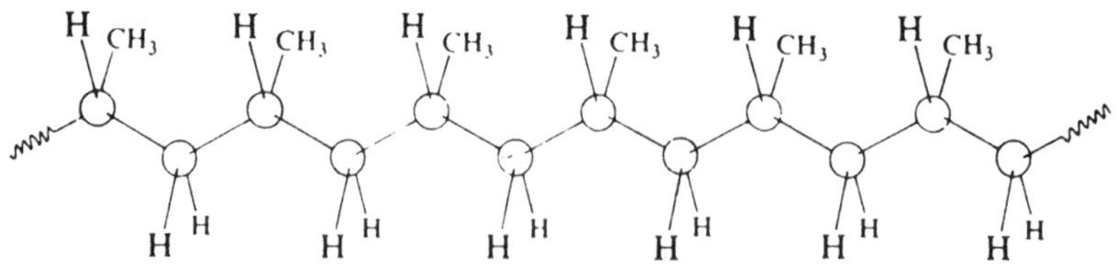

Figure 1 Isotactic polypropylene

---

† Electronic influences seem to play the minor role in most cases. For their discussion, see Refs 9 and 11.
‡ There are few exceptions from this general observation, see e.g. Refs 12 and 13. Stereospecificity and regioselectivity are related to each other, according to metallocene structure [14].

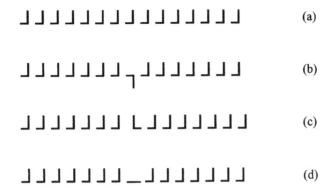

Figure 2 Chain irregularities in metallocene isotactic polypropylene: (a) perfectly isotactic chain (for comparison); (b) 1,2-stereoerror; (c) 2,1-regioerror; (d) 1,3-enchainment

rities is a reduction of the iPP's melting-point by hampering the crystallization of the chains. In this respect, mistakes of type 'c' and 'd' are more effective than 'b'.[†]

Several metallocene structure types have turned out to be suitable for isotactic propylene polymerization (see Section 3.1 and Figure 3); these include bridged structures of C_2 and C_1 symmetry (C_2, Figure 3, species **1–4**; C_1, Figure 3, species **5** and **6**) and even non-bridged metallocenes [14]. Bridged bisindenyl structures such as **1** and **2** (Figure 3) have been investigated most thoroughly. To illustrate our mechanistic discussion Figure 4 shows schematic representations of the active complex of **2** with the end of the growing chain and the incoming propylene coordinated in different insertion positions.

Not only the frequency of misinsertions and thus the melting-point of the polymer is controlled by the ligand framework, but also the frequency of chain termination determining the polymer's molecular weight. The chain can be terminated by elimination of the β-hydrogen at the chain end (see Fig. 4) with parallel cleavage of the metal–carbon bond; thus a double bond at the chain end is formed and the hydrogen eliminated is transfered to the active metal.[‡] The ligand framework may, by positioning the chain and thus the β-hydrogen, more or less disfavor β-hydrogen transfer and thus more or less enhance the molecular weight of the polymer produced.

Influences of the metallocene's ligand framework on the overall activity of the system are also discussed. These may operate via control of deactivation pathways (e.g. formation of a 'dormant' center by regiomisinsertion which will hamper a

---

† Conventional iPP homopolymer has a melting-point of 160–165 °C in spite of containing a similar amount of chain irregularities (only 1,2-stereoerrors); there is no melting-point depression because the irregularities are concentrated in the low molecular weight fraction of the polymer.
‡ Other chain termination mechanisms occur [15], to which similar steric argumentations may apply.

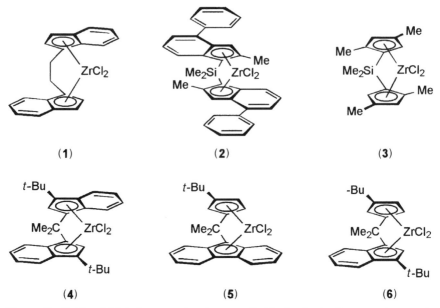

(1)   **(2)**   **(3)**

(4)   **(5)**   **(6)**

Figure 3 Bridged Metallocene structures suitable for isotactic propylene polymerization: Species **1** and **2**, bisindenyl (type A); species **3**, substituted biscyclopentadienyl (type B); species **4**, bisindenyl with large substitutions in 3, 3'-
positions (type C); species **5**, fluorenyl cyclopentadienyl with large substituent in
Cp-3 (type D); species **6**, indenyl cyclopentadienyl with large substituents in 3, 3'-
positions (type E)

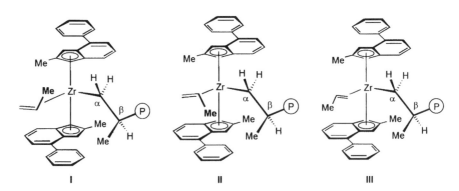

I   II   III

Figure 4 Coordination positions of propylene monomer in active complex of
metallocene **2** (bridge of metallocene omitted): positions I and II lead to stereoregular and stereoirregular 1,2-insertion and position III to a regioirregular 2,1-
insertion

consecuting insertion step [16]) or direct stabilization of the polymerizing species in its transition state (e.g. by agostic interaction of one α-hydrogen of the chain end with the transition metal; see Figure 4 and Ref. 4).

In addition to the elementary steps already discussed (initiation, insertion, termination) chain migration [17], epimerization [18] and exchange [19] may play a role if certain metallocene structures are used and selected polymerization conditions applied.

3 CATALYST COMPONENTS

3.1 METALLOCENES

Technical and economic factors decide on the suitability of a certain metallocene structure for industrial iPP production. Modern iPP processes normally run at temperatures of $\geqslant 60\,^\circ\text{C}$ and high propylene concentrations (liquid propylene or > 20 bar propylene in the gas phase). Under these conditions a suitable system must at least be able to make a homopolymer of $M_\text{w} = 80\,000\,\text{g/mol}$ and melting-point $140\,^\circ\text{C}$. Higher values, especially for M_w are of course desirable. For acceptable catalyst costs the activity of the system should exceed $A = 100\,\text{kg PP/(mmol}$ transition metal h).

The very first structures found to polymerize propylene isospecifically, such as the ethylene-bridged bisindenyl system **1**[†] (see Fig. 3), clearly did not fulfil these critcria (see entry 1 in Table 1). To optimize the performance, far more than 100 structure variations were synthesized (for a review, see Refs 11 and 12). Table 1 demonstrates the stepwise progress (entries 2–7). Replacement of zirconium by hafnium boosted M_w, but with unacceptable loss of activity [20]. Introduction of silicon as a bridge and alkyl substituents into the 2-positions of the indenyl rings [21] pushed M_w and the m.p. above the critical level, and appropriate aryl substitution of the six-membered rings led to systems (**2** in Fig. 3 and **8** in Fig. 5, entries 5–7 in Table 1) [11,22] which matched and exceeded conventional catalysts (very high activities and stereospecifities, whole commercial range of M_w, i.e. ca 80–500 000 g/mol is covered). Further variants of their basic design did not result in significant enhancement of performance [23–28].

A very early competing design was chiral bridged biscyclopentadienyl systems (**3** in Fig. 3, entries 8 and 9 in Table 1) [29]. Although their performance data are just borderline, their commercialization has been announced [30]. Possibly their optimization is based on special cocatalysts or supporting techniques [31,32].

[†] So far this system has been used as a model system in the majority of all studies published. Unfortunately, the properties of the system are in some respects completely different from those of the optimized structures developed in the meantime. Hence the results of several studies should not be generalized.

Table 1 Metallocene structure and catalyst performance in homogeneous propylene polymerizations

Entry No.	Type[a]	Metallocene structure[b]	Polymerization conditions	Activity [kg PP/ (mmol Zr) h]	M_w (kg/mol)	m.p. (°C)	Ref.
1	A	rac-C$_2$H$_4$(Ind)$_2$ZrCl$_2$ (**1**)	C$_3$(1), 70°C, 60 min, MAO, Zr:Al - 1: 15 000	188	24	132	11
2	A	rac-C$_2$H$_4$(Ind)$_2$HfCl$_2$	As entry 1, but Hf: Al - 1: 5000	10	170	134	20
3	A	rac-Me$_2$Si(Ind)$_2$ZrCl$_2$	As entry 1	190	36	137	11
4	A	rac-Me$_2$Si(2-MeInd)$_2$ZrCl$_2$ (**7**)	As entry 1	250	195	145	11, 92
5	A	rac-Me$_2$Si(2-Me-4,5-benzoInd)$_2$ZrCl$_2$ (**8**)	As entry 1	403	330	146	11
6	A	rac-Me$_2$Si(2-Me-4-PhInd)$_2$ZrCl$_2$ (**2**)	As entry 1	755	729	157	11
7	A	rac-Me$_2$Si(2-Me-4-NaphthInd)$_2$ZrCl$_2$	As entry 1	875	920	161	11
8	B	rac-Me$_2$Si(2,4-Me$_2$Cp)$_2$ZrCl$_2$ (**3**)	C$_3$(1), 70°C, 60 min, MAO, Zr: Al - 1: 4000	98	35	150	92
9	B	rac-Me$_2$Si(2,3,5-Me$_3$Cp)$_2$ZrCl$_2$	C$_3$(1), 50°C, MAO, Zr: Al - 1: 15 000	207	185	161	30
10	C	rac-Me$_2$C(3-Me$_3$SiInd)$_2$ZrCl$_2$	C$_3$(1), 70°C, 60 min, MMAO, Zr: Al - 1: 1500	26	45[c]	126	33
11	C	rac-Me$_2$C(3-^{t}BuInd)$_2$ZrCl$_2$ (**4**)	As entry 10, Zr: Al - 1: 8000	110	30[c]	140	33
12	D	Me$_2$CFlu(3-^{t}BuCp)ZrCl$_2$ (**5**)	C$_3$(1), 60°C, 60 min, MAO, Zr: Al = 1: 3600	20	62	127	17
13	D	Me$_2$SiFlu(3-^{t}BuCp)ZrCl$_2$	As entry 1	55	59	145	17, 93
14	E	threo-Me$_2$C(3-^{t}BuInd)(3-^{t}BuInd)$_2$ZrCl$_2$ (**6**)	Toluene, C$_3$ 6 mol/l, 60°C, MAO, Zr: Al - 1: 2000	42	9	159	34

[a] Type according to Figure 3.
[b] Numbering of species according to Figure 3 and 5.
[c] Estimated from viscosity numbers.

New designs C [33], D [17] and E [34] (see Figure 3 and Table 1) seem to be further away from technically relevant performance levels. They offer some interesting features, however, such as excellent regiospecifity and stereospecifity (C, E) or few-step synthesis (D), which may inspire optimization efforts.

3.2 COCATALYSTS

Methyl aluminoxane, the original cocatalyst of metallocene catalysts, is synthesized by incomplete and controlled hydrolysis of trimethylaluminum [6]. Considerable research efforts have been invested in the elucidation of its chemical nature, with limited success: It has been proven to be a mixture of oligomers of the $-O-Al(CH_3)-$ unit, most probably with cage-like structures, containing associated trimethylaluminum, and forming equilibria that are very sensitive to external influences [6,7,35]. MAO's cocatalyst function includes the ability to form a zirconocenium cation of type $Cp^*_2Zr(CH_3)^+$ from a neutral dichloro-or dialkylmetallocene of the type $Cp^*_2ZrCl_2$ or $Cp^*_2ZrR_2$ by parallel or consecutive halogen/alkyl abstraction and methylation, and to stabilize this cation, presumably by a voluminous non-coordinating anion, without preventing monomer coordination and polymerization. In solution, the molar ratio of aluminum in MAO to transition metal in metallocene must be at least $Al:Zr = 250:1$ to constitute an active catalyst; by increasing this ratio the activity observed will rise by a factor of 10 or even much more, reaching a maximum and then slightly decreasing with even higher MAO excess [36]. The curves look different for different metallocenes, monomers and polymerization conditions, a phenomenon which has inspired a lot of research and caused a lot of confusion. As metallocenes and MAO are expensive chemicals, a maximum of metallocene activity must be achieved with moderate $Al:Zr$ ratios. As a rule of thumb, $Al:Zr = 500:1$ is an upper limit for technical iPP catalysts. Attempts to develop chemically similar alternatives to neat methylaluminoxanes include MAOs with a certain content of alkyl groups other than methyl (e.g. isobutyl [37]) and reaction products of certain alkylaluminum compounds with water [38] or methylboronic acid [39], but at the moment their potential seems to be limited.

The obvious disadvantages of aluminoxane cocatalysts also inspired the search for alternatives. The most promising result of this effort was the discovery that perfluorinated tetraaryl- or triarylboron compounds could replace MAO without substantial losses in activity and stereospecifity of the metallocene systems [4–7].

Unlike MAO, they can be used in amounts equimolar to the metallocene, and their structure is fully analyzable and clearly defined. Active metallocene iPP catalysts with such defined arylboron anions have been generated by different routes, as follows.

Species of type $(Cp^*_2)ZrR^+B(C_6F_5)_4^-$ can be made from $(Cp^*_2)ZrR_2$ either (a) by protolysis with ammonium salts such as $(NHMe_2Ph)^+B(C_6F_5)_4^-$ [40] or (b) via abstraction of a methyl group by a strong Lewis acid such as $Ph_3C^+B(C_6F_5)_4^-$ [41]; reaction of $(Cp^*_2)ZrR_2$ with the neutral strong Lewis acid $(C_6F_5)_3B$ (c) generates

$(Cp*_2)ZrR^+RB(C_6F_5)_3^-$ [40]. $(Cp*_2)ZrR_2$ may be made by alkylation of $(Cp*_2)ZrCl_2$ with methyllithium or trimethylaluminum prior to reaction with the boron reagent or with a one-step procedure using trialkylaluminums such as TIBAL or TEAL as alkylating agent [32,41,42]. The activated species appear to be more sensitive to catalyst poisoning than analogous MAO systems; thus TEAL, TIBAL or TMAL are used as scavengers; they eventually form adducts with the active complex [4]. Table 2 shows homogeneous polymerization examples with boron activators (entries 1–3) and MAO modifications (entries 4 and 5) under near-technical conditions; they may be compared with the data for MAO-activated systems in Table 1.

In analogy to the work described, other (simple) Lewis acids such as $MgCl_2$ [40] or $(CH_3)_2AlF$ [43] or voluminous anions such as heteropoly acids [44] or tetraarylaluminates [45] have been used as activators, usually showing poorer performance; see Table 2 (entries 6 and 7).

3.3 SUPPORTS

The reactor and work-up section designs of industrial iPP production plants require a defined morphology of the polypropylene product: it should be a powder of regularly shaped particles with bulk densities $> 0.4 g/cm^3$, a narrow particle size distribution, an average particle diameter of at least 300 μm and a low ($< 2 wt\%$) content of fine particles. In most cases a homogeneous metallocene catalyst will not produce such morphologies; thus it has to be heterogenized by supporting it on an inert particle of proper size and shape. As is known from conventional catalysts, the polymer particle will be a magnified replica of the catalyst particle if the catalyst is designed well and the polymerization run in a suitable manner [15]. Further purposes of supporting metallocenes are prevention of reactor fouling and effective use of catalyst components (e.g. by reducing MAO consumption).

Many different concepts have been tested; they can be differentiated by choice of support material and method of support. Support materials include classical support materials known in polyolefin technologies such as silica (see below) or alumina [46], other inorganics such as zeolites [47], inert [48] or reactive [49] polymers and more exotic materials [50]. Support can be achieved in three alternative ways: (i) the active catalyst species is fixed to the support particle only by adsorption (Van der Waals forces), e.g. by impregnating porous polypropylene with metallocene/MAO solution (see below; [48]); (ii) the metallocene is 'directly' fixed to the support particle by a covalent or ionic bond, introduced by either reaction of an auxiliary substituent on the metallocene ligand framework [51] or complete synthesis of the metallocene structure on the support [52]; or (iii) the active metallocene cation is fixed to the support particle via the cocatalyst anion which is covalently bonded to the support. Theoretically none of these approaches can be ruled out as not practical. The result of screening programs at a certain stage of industrial research efforts is

Table 2 Alternative cocatalysts: homogeneous propylene polymerizations with isospecific metallocenes and non-MAO cocatalysts

Entry No.	Catalyst preparation	Polymerization conditions	Activity [kg PP/(mmol Zr) h]	Ref.
1	$[PhMe_2NH]^+[B(C_6F_5)_4]^-$ + $C_2H_4Ind_2ZrMe_2$	$C_3(l)$, 70 °C, 60 min, 0.5 mmol TEAL/l	113	40
2	$[Ph_3C]^+[B(C_6F_5)_4]^-$ + $C_2H_4Ind_2ZrMe_2$	$C_3(l)$, 70 °C, 60 min, 0.5 mmol TEAL/l	149	40
3	$(C_6F_5)_3B$ + $C_2H_4Ind_2ZrMe_2$	$C_3(l)$, 70 °C, 30 min, 0.5 mmol TEAL/l	27	40
4	$R_1R_2AlO(Me)BOAlR_1R_2(\mathbf{9})$+ $Me_2Si(IndH_4)_2ZrCl_2$	Toluene, 7 bar C_3, 120 min, 4 mmol **9**/l 0.6 mmol TEAL/l	6.2	39
5	1-Alkene-modified $(^iBu)_2AlH$ + H_2O^a + $Me_2Si(2\text{-}MeInd)_2ZrCl_2$	$C_3(l)$, 70 °C, 60 min, Al: Zr - 4000:1	74	38
6	$[Ph_3C]^+[Al(C_6F_5)_4]^-$ + $[C_2H_4Ind_2ZrCl_2$ + TEAL]	$C_3(l)$, 60 °C, 60 min, 0.6 mmol TEAL/l	7.5	45
7	$MgCl_2$ + $C_2H_4Ind_2ZrMe_2$	$C_3(l)$, 70 °C, 30 min, 0.5 mmol TEAL/l	3.5	40

[a]Defined amount of water is added in the polymerization reactor.

presented in the next section; it is type (iii) support with silica as support material. It may not be the only or final answer; further perspectives are discussed in Section 5.

4 PRACTICAL CATALYSTS

4.1 INTRODUCTION

Since 1995, technical metallocene iPP catalysts have been introduced into existing polymerization processes such as gas-phase and liquid propylene bulk polymerization (see Section 1). This was done in a 'drop-in' fashion, i.e. major changes either in polymerization equipment or in polymerization conditions were not necessary when switching the plant from a conventional catalyst to a metallocene system. The companies pioneering this new technology have also publicly revealed [53–55] what the basic ingredients of their catalysts are: selected bridged bisindenylzirconocenes with certain substitution patterns, methylalumoxane (MAO) and silica constitute a solid fine-particle catalyst which is used with an aluminum trialkyl such as triethylaluminum (TEAL) as cocatalyst. Thus the information sources for the following description of the synthesis, composition and behavior of those catalyst systems are mainly recent patents and technical publications of these companies, supplemented with some data from other companies' research groups and several scientific publications directly related to the topic.

4.2 SYNTHESIS AND BASIC PERFORMANCE

Three metallocene structures are most frequently cited within the catalyst patents: *rac*-dimethylsilylbis(2-methylindenyl)zirconium dichloride (**7**) [21], *rac*-dimethylsilylbis(2-methyl-4,5-benzoindenyl)zirconium dichloride (**8**) [11,22] (see Figure 5) and *rac*-dimethylsilylbis(2-methyl-4-phenylindenyl)zirconium dichloride (**2**) [11] (see Figure 3). Figure 6 shows the synthesis of **2** as an example. The five-step sequence can be divided into three segments: synthesis of a substituted indenyl structure (steps A–C), introduction of the bridge (step D) and formation of the complex (step E), with separation of the desired *rac* isomer (see Section 5). The

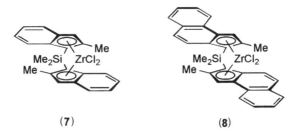

(**7**)   (**8**)

Figure 5 Commercially relevant metallocene structures most often cited

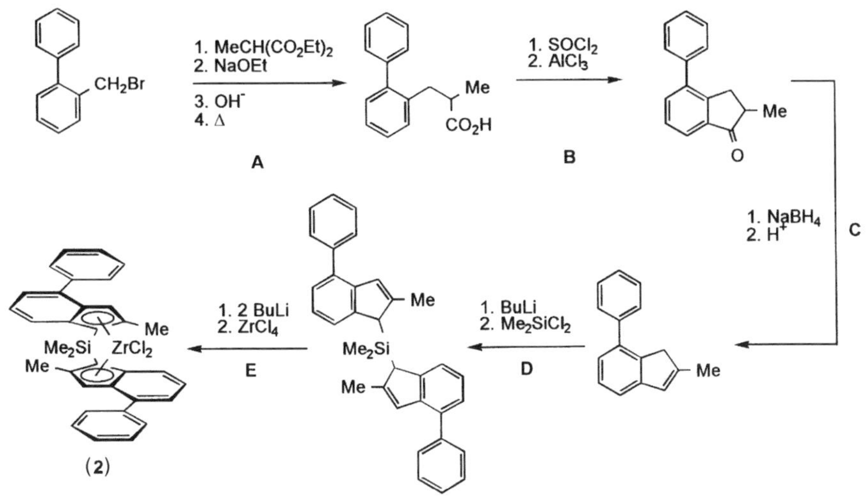

Figure 6 Synthesis route for metallocene **2**

overall yield of the sequence in the original procedure is only 17%, but has considerable potential for optimization [56–58].

The MAO qualities used for catalyst preparations are commercially available solutions in toluene. They are applied as purchased; in individual cases patents recommend the use of only visibly clear, gel-free MAO solutions [59] or to distil off the toluene and redissolve the solid MAO residue [60].

Like MAO, the silica support materials used are standard industrial qualities. They usually consist of spherical porous particles with average diameters between 20 and 80 μm. In all cases the silica is dried before use, applying temperatures between 140 and 800 °C and reduced pressure for several hours. Heating at 100 °C removes adsorbed and bulk water from the silica; at higher temperatures silanol (i.e. Si—OH) hydroxyl groups of adjacent silicons will condense into water molecules, forming siloxane groups (Si—O—Si). According to Ref. 61, the concentration of OH groups on the silica is reduced from 5.8 to 1 OH/nm^2 on going from 100 to 900 °C. Regardless of these facts, the exact choice of the drying temperature within the range 140–800°C is not very critical for final catalyst performance (see Table 3).

The catalyst formation procedures described are very simple and usually have two basic steps. Either subtype (a), where (1) MAO is reacted with silica, and (2) the product is further reacted with a solution of the metallocene in a neutral solvent [60] or in MAO–toluene solution [61]; or subtype (b), where (1) MAO is reacted with the metallocene and (2) this solution combined with the silica [62]. Within the subtypes further variants may include filtration, washing and drying steps and the use of aliphatic or aromatic hydrocarbons for solution, dilution and suspension of reaction components. As additional operations, the pretreatment of the silica with an

Table 3 Supported catalysts of type metallocene/MAO/silica: polymerization performance of preparation subtypes (see Section 4.1) under near technical conditions

	Catalyst preparation		Polymerization conditions and results					
Metallocene	Support method	Zr (wt%) Zr:Al	Conditions	Activity/cat. [kg PP/ (g solid cat.) h]	Activity/Zr [kg PP/ (mmol Zr) h]	M_w (kg/mol)	M.p. (°C)	Ref.
7	Subtype a (+ bisphenol A)	0.19 1:400	Liquid C_3, 70°C, TIBAL/Zr 600:1	2.9	139	280	149	65
7	Subtype b, ('pore-filling')	0.12 1:180	Liquid C_3, 65°C, TEAL 0.4 mmol/l	2.9	218	n.m.	n.m.	63
7	Homogeneous (comparison)	1:15000	Liquid C_3, 70°C	-	250	195	146	11, 92
8	Subtype a	0.08[a] 1:900[a]	Gas phase, 24 bar, C_3/C_4; 60°C, TIBAL 1.5 mmol/l	2.0	200	190[b]	142[b]	62
8	Subtype b ('Gel-free MAO')	0.19 1:570	Liquid C_3, 65°C, TEAL 1.5 mmol/l	2.6	125	170	144	69
8	Homogeneous (comparison)	1:15000	Liquid C_3, 70°C	-	403	330	146	11
2	Subtype a, (MAO dried and redissolved)	0.09 1:500	Liquid C_3, 70°C, TIBAL 1.5 mmol/l	2.1	213	600[c]	149	60
2	Subtype b	0.17 1:255	Liquid C_3, 65°C, TEAL 1.5 mmol/l	1.4	76	281	150	69
2	Homogeneous (comparison)	1:15000	Liquid C_3, 70°C	-	755	729	157	11

[a]Estimated from preparation data.
[b]Polymer contains 0.6 mol% C_4.
[c]Estimated from η or MFR value.

aluminum trialkyl [64], modification of MAO with bisphenol A [65] and final treatment of the catalyst with antifouling agents [63] or its prepolymerization with ethylene [66] or propylene [60] are mentioned in individual cases. The most elegant and economic supporting procedure seems to be the 'pore-filling-method' [63]: dry silica is treated with an amount of metallocene/MAO solution, the volume of which is equal to or slightly larger than the pore volume of the silica. The catalyst is finished by drying without any filtering or washing operations.

The finished solid catalyst particles retain the shape of the original silica particle [67]; the catalyst contains 0.1–0.4 wt% Zr (i.e. 0.6–2.5 ,wt% metallocene, a transition metal content a factor of 10 lower than in conventional Ti catalysts [15]) and 10–25 wt% Al (i.e. ca 20–50 wt% MAO), resulting in Zr : Al ratios from 1 : 150 to 1 : 900 (see Table 3).

According to Ref. 68, the chemical reactions forming the active supported catalyst are (i) reaction of MAO with the silanol groups of the silica, fixing the MAO to the silica by covalent Si—O—Al bridges, and (ii) reaction of the metallocene dichloride with the so-fixed MAO forming the active complex, an ion pair of zirconocene cation and a weakly coordinating MAO anion, as depicted in Figure 7. These ideas have been basically confirmed by IR spectroscopic examinations [67].

Table 3 shows analyses of typical catalyst samples prepared by one of the procedures described and their polymerization performance under near-technical conditions. Polymerization results of the respective homogeneous systems (maximum activities at very high Al : Zr ratios) are included for comparison. The preparation's activities per mmol Zr are between 10 and 80 % of the homogeneous maximum values. Polypropylene yields per gram of solid catalyst are between 1.5 and 3 kg and thus only 10–20 % of the comparable yields of conventional fourth-generation catalysts (e.g. Ti/MgCl$_2$/DiBP type, see Ref. 15). Comparing the homogeneous and the heterogeneous cases, differences in the molecular weights

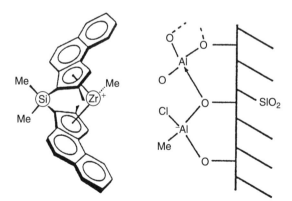

Figure 7   Active complex of metallocene **8**/MAO supported on silica. Reproduced from Ref. 53 by permission of Schotland Business Research

and melting-points of the products are also substantial. The deviations of molecular weight may be due to chain transfer (see above) by the aluminum cocatalyst or by MAO; metallocene catalysts are very sensitive to this reaction, as will be discussed below. In many cases, molecular weight distributions produced by supported metallocene catalysts are slightly broader ($M_w/M_n = 3.0–4.5$) than with their homogeneous counterparts. Lowering of the polymer melting-point by supporting the metallocene is a marked effect only in the case of metallocene **2**. The phenomenon, which was studied [67] under somewhat different conditions (slurry polymerization in toluene, $P_{C_3} = 2\,\text{bar}$, $T_{poly} = 20–70\,^\circ\text{C}$), is mainly due to a reduction of the metallocene's regioselectivity by the support. For example, at $T_{poly} = 50\,^\circ\text{C}$, the supported metallocene produced iPP with 0.6 % m-2, 1-misinsertions, melting-point 148.2 °C, while the homogeneous system gave iPP with melting-point 159.8 °C and only 0.1 % m-2,1-misinsertions. A similar reduction of the melting-point of iPP by reduction of the system's regiospecifity is also observed on comparing the homogeneous performance of **2** with its binuclear analog [13]. It can thus be interpreted as a result of the reduced mobility of the metallocene cation as a whole.

4.3 INFLUENCE OF VARIABLE PARAMETERS ON THE POLYMERIZATION RESULT

4.3.1 Catalyst and cocatalyst

A systematic study of the influence of the catalyst composition on the polymerization behavior has been published only in the case of the model metallocene rac-Me$_2$SiInd$_2$ZrCl$_2$ [67,70], which was investigated at $T_{poly} = 40\,^\circ\text{C}$. According to these studies the minimum limit of MAO loading for obtaining an active catalyst is between 10 and 20 wt% Al; above this a directly comparable activity of the homogeneous system (Al : Zr = 300 in both systems) may be reached by the heterogeneous system also.

The choice of the aluminum alkyl for cocatalyst is fairly critical, as demonstrated by Table 4 [65,71]. It may change the activity of a certain supported catalyst by a

Table 4 Influence of aluminum alkyl on polymerization performance of supported catalyst system Me(Me$_3$Si)Si(2-Me, 4-PhInd)$_2$ZrCl$_2$/MAO/silica. Reproduced from Ref. 65 by permission of Schotland Business Research

Aluminum alkyl	kg PP/(g cat) · 2 h	M_w (kg/mol)	M_w/M_n	T_m (°C)
TMAL	1.0	271	4.8	152
TEAL	4.0	369	2.8	152
TIBAL	1.6	362	4.5	153
TNHAL	2.1	551	5.8	152
DEAC	0.1	72	5.6	147
DEA(OEt)	0.7	323	6.9	152

Conditions: 70 °C; Al (alkyl):Zr mole ratio = 400; 0.37 mol% hydrogen.

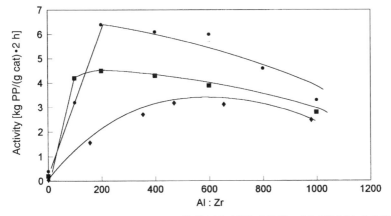

• Me$_2$Si(2-Me,4-PhInd)$_2$ZrCl$_2$ ■ Me(Me$_3$Si)Si(2-Me,4-PhInd)$_2$ZrCl$_2$ ♦ Me$_2$Si(2-MeInd)$_2$ZrCl$_2$

Figure 8 Influence of the cocatalyst concentration on the catalyst activity (three different catalysts of type metallocene/MAO/silica, TEAL as cocatalyst). Reproduced from Ref. 65 by permission of Schotland Business Research

factor of >10 and the molecular weight of the polymer by a factor of >2. These observations seem to indicate that the role of the aluminum alkyl is more than just being a scavenger of catalyst poisons. The influence of cocatalyst concentration on catalyst behavior resembles observations with conventional catalysts [15]. Starting at very low values, the catalyst activity reaches a maximum at a certain cocatalyst concentration and then decays with concentrations further enhanced (Figure 8). The cocatalyst concentration also influences the kinetic profile of the polymerization startup (Figure 9) [65,70].

4.3.2 Polymerization temperature, monomer concentration and polymerization time

As optimum use of reactor and cooling equipment is necessary, modern iPP processes are run within very narrow ranges of temperature ($T_{poly} = 60-70\,^\circ$C), monomer concentration (liquid propylene in bulk polymerization, 20–30bar in the gas phase) and (mean) polymerization time (process-dependent, e.g. in the Spheripol process: 90–150 min). The examinations published [67,70,71] deal with wider ranges of temperatures and lower monomer concentrations than in the industrial processes and should thus be interpreted with caution.

Table 5 [71] demonstrates the effects of temperature change on several parameters with both homogeneous and supported **8** (see Figure 5) as catalyst. According to Ref. 67, the Arrhenius plot of activity for both homogeneous and heterogenized **8** in the temperature range 20–70$\,^\circ$C is linear, but has a change of slope at about 40$\,^\circ$C, resulting in two activation energies. In the heterogeneous case they are

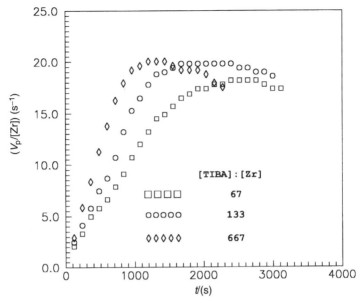

Figure 9 Influence of the TIBAL:Zr ratio on the experimental rate–time curves of propylene polymerizations with *rac*-Mc$_2$SiInd$_2$ZrCl$_2$/MAO/SiO$_2$ (catalyst, 17 wt% Al, Al : Zr = 284; polymerization, toluene,[Zr] = 3 × 10^{-5} mol/l, 40 °C, 2 bar). Reproduced from Ref. 67, 70 by permission of G. Fink, Max-Planck-Institut für Kohlenforschung

Table 5 Influence of polymerization temperature for both supported and homogeneous catalysts from metallocene **8** [From Ref. 71]

Catalyst[a]	T_{poly} (°C)	Activity[b]	M.p. (°C)	M_n (kg/mol)
8 + MAO	20	6	161.1	318
8 + MAO	40	30	155.4	138
8 + MAO	60	110	149.7	74
8/MAO/SiO$_2$ + TIBAL	40	1.1	143.9	76
8/MAO/SiO$_2$ + TIBAL	50	2.9	151.2	99
8/MAO/SiO$_2$ + TIBAL	60	18.0	151.1	83

[a]Entries 1–3: homogeneous catalyst, [Zr] = 1 μmol/l, [Al] = 20 mmol/l. Entries 4–6: supported catalyst, 0.07 wt% Zr, 14.7 wt% Al, cocatalyst TIBAL, 10 mmol/l.
[b]Activities in kg PP(mmol Zr) h (mol/l C$_3$)h (mol/l C$_3$) Entries 1–3: maximum activities, [C$_3$] = 0.91 mol/l. Entries 4–6: activity averaged over 3 h, [C$_3$] = 0.72 mol/l.

$91 \pm 4 \, \text{kJ mol}^{-1} (20-40\,^\circ\text{C})$ and $38 \pm 2 \, \text{kJ mol}^{-1} (40-70\,^\circ\text{C})$. The difference might be due to diffusion phenomena.

Comparable examinations of the influence of monomer concentration have been published for homogeneous systems only [72,73] (see also Ref. 15); partly reaction rate orders higher than one are found. For the heterogeneous case, comparison of data found in Refs 71, 74 and 62 is instructive. Thus, at $T_{\text{poly}} = 60-65\,^\circ\text{C}$, similar supported versions of K.403 have activities of ca 25 kg PP/(mmol Zr h) at 3 bar propylene pressure (slurry polymerization) and ca 200 kg PP/(mmol Zr h) at 26 bar (gas-phase polymerization), the polymer molecular weight being doubled from slurry to gas-phase conditions.

Typical curves of polymerization activity as a function of time are represented in Figure 9 [67,70]. After an induction period, maximum activity is reached; the decline from this maximum might be faster at higher temperatures and propylene concentrations.

4.3.3 Hydrogen concentration

In the industrial use of supported metallocene catalysts, hydrogen is used as effective molecular weight modifier which terminates chain propagation by transfer of the growing chain. As with several conventional catalysts [15], hydrogen has an activating effect. The magnitude of the effect is different from one metallocene to another. Taking supported **2** as an example [75], the catalyst activity is doubled on applying 0.4 bar of hydrogen; the molecular weight of the polymer decreases by a factor of 4, but still stays in the range of commercial polypropylene grades. The causes of activation by hydrogen and its effects on regiospecificity of the catalyst have been examined for several metallocenes in homogeneous polymerizations, but no simple and uniform explanation pattern was found [76].

Generally, supported metallocene catalysts are much more sensitive to hydrogen M_w regulation than modern conventional catalysts, as exemplified by Figure 10 [53]. The difference has practical consequences for the introduction of metallocene catalysts into production lines normally run with conventional catalysts; usually the equipment for hydrogen dosage and measurement has to be changed.

4.3.4 Copolymerization

Copolymers of propylene with ethylene (either random copolymers or so-called block copolymers) make up more than 30 % of the isotactic polypropylene sold worldwide. Propylene–ethylene copolymerizations with several metallocene systems, either homogeneous or supported, have been examined fairly often (for a review, see Ref. 4). The copolymerization parameters turned out to be strongly dependent on metallocene structure (see Table 6) [77,78]. The metallocenes

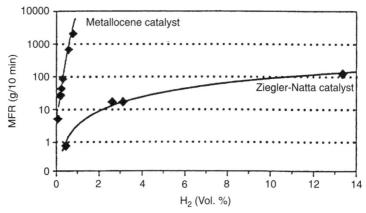

Figure 10 Hydrogen regulation of MFR value: comparison of supported metallocene **8** with conventional Ziegler–Natta catalyst in gas-phase polymerization of propylene. Reproduced from Ref. 53 by permission of Schotland Business Research

Table 6 Reactivity ratios from ethylene–propylene copolymerizations with three metallocene/MAO systems and two conventional Ti/MgCl₂ catalysts [77, 78]

Catalyst	r_E	r_P	$r_E r_P$
Ti/MgCl$_2$ (A)	8.8	0.207	1.82
Ti/MgCl$_2$ (B)	53.0	0.109	5.78
rac-C$_2$H$_4$Ind$_2$ZrCl$_2$ (**1**)	6.61	0.06	0.4
rac-Me$_2$SiInd$_2$ZrCl$_2$	5.3	0.13	0.7
rac-Me$_2$Si(2-Me, 4-PhInd)$_2$ZrCl$_2$ (**2**)	1.6	1.18	1.9

discussed here show a *poorer* ethylene incorporation than many conventional iPP catalysts [53]. Ethylene also operates as a chain transfer agent, lowering the molecular weight of the copolymer when compared with the M_w of an iPP homopolymer produced under same conditions. At high ethylene concentrations this molecular weight reduction might be more than 50 %. These observations have led to the development of special metallocenes suitable for such copolymerizations [79]. On the other hand, the metallocene catalysts discussed in this section show a much better incorporation of higher 1-olefins such as 1-hexene and 1-octene than conventional catalysts, which makes propylene–1-hexene copolymers, for example, commercially feasible [80]. 1-Octene as comonomer has been demonstrated to enhance catalyst activity and polymer molecular weight in propylene slurry polymerizations with supported **8** [71]. Such effects, however, have also been observed with conventional catalysts [15].

Figure 11   Scanning electron micrograph of a polypropylene particle obtained by a supported metallocene catalyst. Reproduced from Ref. 70 by permission of G. Fink, Max-Planck-Institut für Kohlenforschung

4.4 PARTICLE MORPHOLOGY

The polymer particle morphologies produced by supported metallocene catalysts under industrial conditions [81] are similar to those of conventional products: bulk densities above $0.4\,g/cm^3$, narrow particle size distributions, high average particle sizes ($> 500\,\mu m$) and low content of fines (less than $1\,wt\% < 125\,\mu m$). Also, the well known replica phenomenon has been demonstrated with metallocene iPP catalysts: both the individual shape of the particle and the shape of the particle size distribution curve are reproduced within the sequence support particle–supported catalyst particle–polymer particle [65]. A typical polymer particle is shown in Figure 11. With a selected kind of supported metallocene catalyst, the particle growth has been investigated and mathematically modeled [70,82]. In this case it turned out that the multigrain model widely used to explain particle growth with conventional catalysts [15] does not apply, and a different model has been proposed.

5 FURTHER DEVELOPMENTS

Although the supported catalysts described in Section 3 are now used in industrial processes, they have obvious shortcomings. Hence they need further optimization and single catalyst components or even the whole catalyst system might be replaced by something new.

Alternative designs of the metallocene component have been discussed in Section 3.1; whether optimizations of these will become technically important cannot be predicted. If not, progress concerning the metallocene component will be just synthesis optimization. The selectivity of the complex formation step of the synthesis has drawn most attention. Metallocene formation by reaction of the lithiated bridged bisindenyl ligand with zirconium tetrachloride usually results in a 1 : 1 mixture of the *rac*- and *meso*-isomers of the complex (see Figure 12). As only the *rac*-isomer is a stereospecific iPP catalyst, the yield of the complex formation step is only 50 % or below. Methods to enhance selectivity in favor of the *rac*-isomer have been studied with model systems; they include photochemical interconversions of given 1 : 1 *rac–meso* mixtures [83], but also direct selective synthesis. For example, in the case of the model metallocene *rac*-Me$_2$SiInd$_2$ZrR$_2$, a 93 % *rac* product is obtained via complex formation by amine elimination, i.e. reaction of the neutral bridged bisindenyl ligand with an zirconium tetraamide [84].

The basic problem with the now-used cocatalyst MAO has already been mentioned: it must be applied in an Al : Zr molar ratio of at least about 150 : 1, i.e. a weight ratio of MAO to metallocene of about 15 : 1, which means limited cost effectiveness and tight 'design limits' for a supported catalyst. This is different to triaryl- or tetraarylboron compounds as cocatalyst, where B : Zr molar ratios of near 1 are possible, which are equivalent to tetraarylboron : metallocene weight ratios of 1.5 : 1 to 1 : 1. In homogeneous polymerization, metallocenes activated by roughly equimolar amounts of tetraarylboron activator show performances similar to those of their MAO-activated analogs (see above). The few published examples of supported boron-activated species, however [85–87], still exhibit poor activities. Considerable progress in this field has been announced [81], but up to now has not been substantiated by data. Concerning activity in kg PP/g solid catalyst, metallocene systems must advance by a factor of 10 to match the level of modern conventional catalysts. The content of inorganic residues in the polymer, which is a function of the so-defined activity, would then decrease correspondingly. A different approach to reduce this content is the use of isotactic polypropylene as a support itself. This iPP might be a special microporous material [48], a standard iPP powder [88] or an *in*

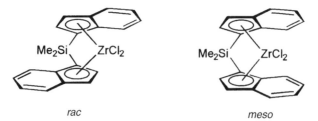

rac   *meso*

Figure 12   *rac*- and *meso*-isomers of metallocene Me$_2$SiInd$_2$ZrCl$_2$

situ prepared iPP support [89]. The support method may also influence the choice of the metallocene component. Bringing metallocene on to a support without the mediation of a cocatalyst has been found to 'tune up' the performance of very simple metallocene structures [90]: thus, so-supported $C_2H_4(Ind)_2ZrCl_2$ (activated with MAO) at $50\,^\circ C$ made iPP with m.p. $= 159\,^\circ C$ and $M_v = 518\,000\,g/mol$, which means increases in m.p. of $25\,^\circ C$ and in M_v of a factor of 20 compared with the product of the analogous homogenous system (see entry 1 in Table 1). The activity of the supported system, however, is extremely poor, hence the method needs dramatic optimization to gain practical value.

As described in section 2, the large variety of known isospecific metallocenes will make different polypropylene products under same polymerization conditions. These differences may include the melting-point,the molecular weight and in the case of copolymers, the comonomer content of the product and the comonomer distribution within the polymer chain. Combinations of two or more different metallocenes on one support will thus make a product which is a molecular blend of the single metallocenes' products. 'Designed' iPP products made with such metallocene multicenter catalysts have already been described, mostly without detailed data on the catalysts [54,91]. These catalysts will be the most important next development step in metallocene iPP technology.

6 REFERENCES

1. Ewen, J. A., *J. Am. Chem. Soc.*, **106**, 6355 (1984).
2. Kaminsky, W., Külper, K., Buschermöhle, M. and Lüker, H., *Eur. Pat.*, 185 918, to Hoechst (1984).
3. Winter, A., Antberg, A., Spaleck, W., Rohrmann, J. and Dolle, V., *US Pat.*, 5 145 819, to Hoechst (1990).
4. Brintzinger, H.-H., Fischer, D., Mülhaupt, R., Rieger, B. and Waymouth, R. M., *Angew. Chem., Int. Ed. Engl.*, **34**, 1143 (1995).
5. Bochmann, M., *J. Chem. Soc., Dalton Trans.*, 255 (1996).
6. Kaminsky, W. and Arndt, M., *Adv. Polym. Sci.*, **127**, 143 (1997).
7. Mashima, K., Nakayama, Y. and Nakamura, A., *Adv. Polym. Sci.*, **133**, 1 (1997).
8. Guerra, G., Corradini, P., Cavallo, L. and Vacatello, M., *Macromol. Symp.*, **89**, 307 (1995).
9. Razavi, A., Peters, L. and Nafpliotis, L., *J. Mol. Catal. A: Chem.*, **115**, 129 (1997).
10. Van der Leek, Y., Angermund, K., Reffke, M., Kleinschmidt, R., Goretzki, R. and Fink, G., *Chem. Eur. J.*, **3**, 585 (1997).
11. Spaleck, W., Küber, F., Winter, A., Rohrmann, J., Bachmann, B., Antberg, A., Dolle, V. and Paulus, E. F., *Organometallics*, **13**, 954 (1994).
12. Spaleck, W., Antberg, M., Aulbach, M., Bachmann, B., Dolle, V., Haftka, S., Küber, F., Rohrmann, J. and Winter, A., in Fink, G., Mülhaupt, R. and Brintzinger, H.-H. (Eds), *Ziegler Catalysts*, Springer, Berlin 1995, p. 83.
13. Spaleck, W., Küber, F., Bachmann, B., Fritze, C. and Winter, A., *J. Mol. Catal. A: Chem.*, **128**, 279 (1998).

14. Razavi, A., Vereecke, D., Peters, L., Den Dauw, K., Nafpliotis, L. and Atwood, J. L., in Fink, G., Mülhaupt, R. and Brintzinger, H.-H. (Eds), *Ziegler Catalysts*, Springer, Berlin, 1995, p. 111.
15. Albizzati, E., Giannini, U., Collina, G., Noristi, L. and Resconi, L., in Moore, E. P., Jr (Ed), *Polypropylene Handbook*, Carl Hanser Munich, 1996, p. 11.
16. Busico, V., Corradini, P. and Cipullo, R., *Makromol. Chem., Rapid Commun.*, **14**, 97 (1993).
17. Ewen, J. A. and Elder, M. J. in Fink, G., Mülhaupt, R. and Brintzinger, H.-H. (Eds), *Ziegler Catalysts*, Springer Berlin, 1995, p. 99.
18. Busico, V., Brita, D., Caporaso, L., Cipullo, R. and Vacatello, M., *Macromolecules* **30**, 3971 (1997).
19. Song, W., Yu, Z. and Chien, J. C. W., *J. Organomet. Chem.* **512**, 131 (1996).
20. Spaleck, W., Antberg, M., Dolle, V., Klein, R., Rohrmann, J. and Winter, A., *New J. Chem.*, **14**, 499 (1990)
21. Spaleck, W., Antberg, M., Rohrmann, J., Winter, A., Bachmann, B., Kiprof, P., Behm, J. and Herrmann, W. A., *Angew. Chem., Ind. Ed. Engl.*, **31**, 1347 (1992).
22. Stehling, U., Diebold, J., Kirsten, R., Röll, W., Brintzinger, H.-H., Jüngling, S., Mühlhaupt, R. and Langhauser, F., *Organometallics*, **13**, 954 (1994).
23. Imuta, J., Saito, J., Ueda, T., Kiso, Y., Mizuno, A. and Kawasaki, M., *Eur. Pat.*, 629 631, to Mitsui Petrochemical Industries (1993).
24. Fukuoka, D., Tashiro, T., Kawaai, K., Saito, J., Ueda, T., Kiso, Y., Mizuno, A., Kawasaki, M., Itoh, M., Imuta, J., Fujita, T., Nitabaru, M., Yoshida, M. and Hashimoto, M., *Eur. Pat.*, 629 632, to Mitsui Petrochemical Industries (1993).
25. Imuta, J., Fukuoka, D., Yoshida, M., Saito, J., Fujita, T., Tashiro, T., Kawaai, K., Ueda, T. and Kiso, Y., *Eur. Pat.*, 653 433, to Mitsui Petrochemical Industries (1993).
26. Schneider, N., Huttenloch, M. E., Stehling, U., Kirsten, R., Schaper, F. and Brintzinger, H.-H., *Organometallics*, **16**, 3413 (1997).
27. H. Uchino, J. Endo, T. Takahama, T. Sugano, K. Katoh, N. Iwama and E. Taniyama, *Eur. Pat.*, 693 502, to Mitsubishi Chemical (1994).
28. T. Sugano, H. Uchino, K. Imaeda, E. Taniyama and N. Iwama, *Eur. Pat.*, 697 418, to Mitsubishi Chemical (1994).
29. Mise, T., Miya, S. and Yamazaki, H., *Chem. Lett.* 1853 (1989).
30. Ushioda, T., Fujita, H. and Saito, J., *Proc. SPO '97*, 1997, p. 101, (available from Schotland Business Research, 16 Duncan Lane, Skillman, NJ 08558, USA).
31. Soga, K. and Kaminaka, M., *Macromol. Rapid Commun.*, **15**, 593 (1994).
32. Cohen, S. A., *Int. Pat.*, WO 97/14727, to Amoco (1995).
33. Resconi, L., Piemontesi, F., Nifant'ev, I. E. and Ivchenko, P. V., *Int. Pat.*, WO 96/22995, to Montell Technology (1995).
34. Miyake, S., Okumura, Y. and Inazawa, S., *Macromolecules*, **28**, 3074 (1995).
35. Mason, M. R., Smith, J. M., Bott, S. G. and Barron, A. R., *J. Am. Chem. Soc.*, **115**, 4971 (1993).
36. Fink, G., Herfert, N. and Montag, P., in Fink, G., Mülhaupt, R. and Brintzinger, H.-H. (Eds), *Ziegler Catalysts*, Springer, Berlin, 1995, p. 159.
37. Kurokawa, H. and Sugano, T., *Macromol. Symp.*, **97**, 143 (1995).
38. Dall'Occo, T., Galimberti, M., Resconi, L., Albizzati, E. and Pennini, G., *Int. Pat.*, WO 96/02580, to Montell Technology (1994).
39. Takahama, T., *Eur. Pat.*, 601 830, to Mitsubishi Petrochemical (1993).
40. Ewen, J. A. and Elder, M. J., *Makromol. Chem., Macromol. Symp.*, **66**, 179 (1993).
41. Chien, J. C. W. and Tsai, W.-M., *Makromol. Chem., Macromol. Symp.*, **66**, 141 (1993).
42. Deng, H., Winkelbach, H., Taeji, T., Kaminsky, W. and Soga, K., *Macromolecules*, **29**, 6371 (1996).

43. Zambelli, A., Longo, P. and Grassi, A., *Macromolecules*, **22**, 2186 (1989).
44. Soga, K., Uozumi, T. and Kishi, N., *Macromol. Rapid Commun.*, **16**, 793 (1995).
45. Elder, M. J. and Ewen, J. A., *Eur. Pat.*, 573 403, to Fina Technology (1992).
46. Inahara, K. and Yano, A., *Eur. Pat.* 553 941, to Tosoh (1991).
47. Ko, Y. S., Han, T. K., Park, J. W. and Woo, S. I., *Macromol. Rapid Commun.*, **17**, 749 (1996).
48. Sugano, T., Fujita, T. and Kitagawa, K., *Eur. Pat.*, 598 543 to Mitsubishi Chemical (1992).
49. Albizatti, E., Resconi, L., Dall'Oco, T. and Piemontesi, F., *Eur. Pat.*, 633 272, to Spherilene (1993).
50. Ribeiro, M. R., Deffieux, A. and Portela, M. F., *Ind. Eng. Chem.*, **36**, 1224 (1997).
51. Langhauser, F., Fischer, D., Kerth, J., Schweier, G., Barsties, E., Brintzinger, H.-H., Schaible, S. and Roell, W., *Eur. Pat.*, 670 336, to BASF (1994).
52. Soga, K., *Macromol. Symp.*, **89**, 249 (1995).
53. Hingmann, R., *Proc. Metallocenes Asia '97*, 1997, p. 53 (for availability, see Ref. 30).
54. Winter, A., *Proc. SPO '97*, 1997 (for availability, see Ref. 30).
55. Brekner, M. J., *Proc. Metallocenes '96*, 1996, p. 153 (for availability, see Ref. 30).
56. Weisse, L., Rohrmann, J., Küber, F. and Strutz, H., *US Pat.*, 5 329 050, to Hoechst (1992).
57. Aulbach, M. and Küber, F., *Eur. Pat.*, 704 454, to Hoechst (1994).
58. Kaufmann, W., Wisser, T., Streb, J., Rink, T., Zenk, R., Riedel, M. and Cabrera, I., *Eur Pat.*, 780 396, to Hoechst (1995).
59. Burkhardt, T., *Int. Pat.*, WO 95/18809, to Exxon Chemical Patents (1994).
60. Tsutsui, T., Yoshitsugu, K., Ohgizawa, M., Imuta, J. and Matsumoto, T., *Eur Pat.*, 697 419, to Mitsui Petrochemical Industries (1994).
61. Taylor, J. A. G. and Hockey, J. A., *J. Phys. Chem.*, **70**, 2169 (1966).
62. Fischer, D., Langhauser, F., Kersting, M., Hingmann, R. and Schweier, G., *Int. Pat.*, WO 97/10272, to BASF (1995).
63. Speca, A. N., Brinen, J., Vaughn, G. A., Brant, P. and Burkhardt, T. J., *Int. Pat.*, WO 96/00243, to Exxon Chemical Patents (1994).
64. Fritze, C., Bachmann, B. and Küber, F., *Int. Pat.*, WO 97/11775 to Hoechst (1995).
65. Ernst, D., Reussner, J., Denifl, P. and Neisl, W., *Proc. SPO '97*, 1997, p. 87 (for availability, see Ref. 30).
66. Burkhardt, T., Brinen, J., Hlatky, G. G., Spaleck, W. and Winter, A., *Int. Pat.*, WO 94/28034, to Exxon Chemical Patents and Hoechst (1993).
67. Fraaije, V., Dissertation, Düsseldorf (1995).
68. Chien, J. C. W. and He, D., *J. Polym. Sci., Part A: Polym. Chem.*, **29**, 1603 (1991).
69. Burkhardt, T., and Brandley, W., *Int. Pat.*, WO 97/05178, to Exxon Chemical Patents and Hoechst (1995).
70. Bonini, F., Fraaije, V. and Fink, G., *J. Polym. Sci, Part A: Polym. Chem.*, **33**, 2393 (1995).
71. Jüngling, S., Koltzenburg, S., and Mülhaupt, R., *J. Polym. Sci., Part A: Polym. Chem.*, **35**, 1 (1997).
72. Jüngling, S., Mülhaupt, R., Stehling, U., Brintzinger, H.-H., Fischer, D. and Langhauser, F., *J. Polym. Sci., Part A.: Polym. Chem.*, **33**, 1305 (1995).
73. Herfert, N. and Fink, G., *Makromol. Chem.*, **193**, 1359 (1992).
74. Hungenberg, K. D., Kerth, J., Langhauser, F., Marczinke, B. and Schlund, R. in Fink, G., Mülhaupt, R. and Brintzinger, H.-H., (Eds), *Ziegler Catalysts*, Springer, Berlin, 1995, p. 363.
75. Aulbach, M., Küber, F., Bachmann, B., Spaleck, W., and Winter, A., *Proc. Metallocenes '95*, 1995, p. 311 (for availability, see Ref. 30).
76. Sacchi, M. C. and Carvill, A., *Proc. MetCon '97*, 1997 (available from Catalyst Consultants, P. O. Box 637, Spring House, PA 19477, USA).

77. Van Os, G. and Pentz, H., *Proc. Metallocenes '95*, 1995, p. 520 (for availability, see Ref. 30).
78. Herfert, N., Dissertation, Düsseldorf 1992
79. Winter, A., Bachmann, B. and Küber, F., *Ger. Pat.*, DE 19 544 828, to Hoechst, (1995).
80. Mc Alpin, J. J. and Stahl, G. A., *Proc. MetCon '94*, (1994) (for availability, see Ref. 76).
81. Fischer, D., Bidell, W., Grasmeder, J., Hingmann, R., Jones, P., Kersting, M., Langhauser, F., Marczinke, B., Moll, U., Rauschenberger, V., Süling, C. and Popham, N., *Proc. SPO '97*, 1997 (for availability, see Ref. 30).
82. Steinmetz, B., Tesche, B., Przybyla, C., Zechlin, J. and Fink, G., *Acta Polym.*, **48**, 392 (1997).
83. Schmidt, K., Reinmuth, A., Rief, U., Diebold, J. and Brintzinger, H.-H., *Organometallics*, **16**, 1724 (1997).
84. Christopher, J. N., Diamond, G. M., Jordan, R. F. and Petersen, J. L., *Organometallics*, **15**, 4038 (1996)
85. Hlatky, G. G., Upton, D. J. and Turner, H. W., *Eur. Pat.*, 507 876, to Exxon Chemical Patents (1990).
86. Fischer, D., Langhauser, F., Kerth, J., Schweier, G., Lynch, J. and Görtz, H.-H., *Eur. Pat.*, 700 935, to BASF (1994).
87. Walzer, J. F., Jr, *Int. Pat.*, WO 96/04319, to Exxon Chemical Patents (1994).
88. Schlund, R. and Rieger, B., *Eur. Pat.*, 518 092, to BASF (1991).
89. Haylock, J. C. and Galli, P., *Proc. Metallocenes Asia '97*, 1997, p. 39 (for availability, see Ref. 30).
90. Kaminsky, W. and Renner, F., *Eur. Pat.*, 523 416, to Hoechst (1991).
91. Mc Alpin, J. J., Chen, M. C. and Mehta, A. K., *Proc. SPO '96*, 1996, p. 429 (for availability, see Ref. 30).
92. Spaleck, W., unpublished work.
93. Spaleck, W., Aulbach, M., Bachmann, B., Küber, F. and Winter, A., *Macromol. Symp.*, **89**, 237 (1995).

19

Properties of Metallocene-catalyzed Isotactic Polypropylene

WALTER SPALECK

Hoechst AG, Frankfurt am Main, Germany[†]

1 INTRODUCTION

In contrast to ethylene, linear chains of propylene molecules theoretically may have very different microstructures between complete irregularity of enchainment and several types of regularity such as isotacticity or syndiotacticity, which can be more or less perfect [1]. Several of them were first detected as side products in polymers made using conventional catalysts. With the introduction of metallocene catalysts to the linear polymerization of propylene, the variety of chain microstructures which can be synthesized in pure form has been considerably enlarged. They include hybrids of isotactic, syndiotactic and atactic structures and different patterns of chain irregularities [2]. Hence it is first necessary to define what the term 'metallocene-catalyzed isotactic polypropylene' means here when applied to a propylene homopolymer. Figure 1 shows five microstructures of isotactic polypropylene chain segments to demonstrate the relevant points. Chain segment (a) shows the perfect isotactic chain: the monomer enchainment is both isotactic (all methyl-bearing tertiary carbons having the same configuration) and '1,2-regioregular' (all units enchained in a 1,2-head-to-tail fashion). Chain segment (b) shows a single stereo-error, with the chain then going on as before; (c) is the same, but with the chain continued with the opposite configuration of tertiary carbons as before; (d) shows an

---

[†] Current address: Spaleck Research, Elsa-Brändström-Str. 32, D-28359 Bremen, Germany

Metallocene-based Polyolefins, Edited by J. Scheirs and W. Kaminsky
© 2000 John Wiley & Sons Ltd

isolated regioerror and (e) an isolated '1.3-enchainment', most probably formed by rearrangement of a type (d) regioerror. The isotactic homopolymers discussed here are isotactic chains in the sense of (a) with irregularities of type (b), (d) and (e) in a limited amount, so that their content of isotactic pentads according to ^{13}C NMR spectroscopy is above 70 %. The molecular weight (M_w) range commercially relevant is between 80 000 and 600 000 g/mol. They can be made by metallocene catalysts described in Chapter 18 and will be discussed in Section 2. Selected metallocene catalysts can also produce polypropylene homopolymers with type (c) microstructures, (partly combined with atactic chain segments) [3, 4]. These 'stereoblock' polypropylenes have elastomeric properties; they will *not* be discussed here.

The term 'isotactic polypropylene' is conventionally used not only for propylene homopolymers but also for random copolymers and reactor blends (so-called 'block copolymers') [1]. Such random copolymers are copolymers of propylene with one or two 1-olefins (typically ethylene or ethylene–1-butene) with up to 10 mol% comonomer content. The 'block copolymers' mentioned are blends of an isotactic homopolymer matrix (at least 35 wt %, typically above 50 wt %) with a propylene–ethylene rubber, made in the reactor by consecutive homo- and copolymerization (either two-stage in one reactor or two reactors in-line). Where such random copolymers and 'block copolymers' have been made with metallocene catalysts also, their properties are discussed in Section 3. With both homopolymers and copolymers, the difference between metallocene and conventional products is characterized to make understandable what metallocene iPPs might add to the polypropylene property portfolio. The discussion is limited to basic properties,

Figure 1 Microstructures of isotactic propylene chains: (a) perfect isotactic chain, (b) isolated 1,2-stereoerror; (c) stereoblock structure, (d) isolated 2,1-regioerror, (c) isolated 1,3-enchainment

their consequences for polypropylene applications are topics of other chapters (in Volume 2).

Section 4 comments on iPP polymers which are beyond the range of conventional catalysts and have been made with metallocenes for the first time. They include copolymers of propylene with unusual monomers and post-polymerization reaction products of isotactic homopolymers.

Metallocene iPP development is still at a very early stage, with about 10–15 grades offered commercially (compared with hundreds of conventional grades) and a production volume of less than 1 % of the world's polypropylene production. Parallel to this, the number of published articles, patents, etc., on metallocene iPP is very limited, constituting less than 10 % of the literature on metallocene iPP catalysts. The following discussion will therefore be rather incomplete and will need supplementing or rewriting very soon.

2 METALLOCENE IPP HOMOPOLYMERS

Conventional iPP homopolymer has a melting-point of 160–165 °C and a molecular weight distribution of $M_w/M_n = 5 - 8$. When it is fractionated into a large number fractions by a suitable elution fractionation, two groups of fractions are obtained [5, 6]. The first group is about 6–10 wt% of the material; these fractions contain polymer with reduced melting-point (140–155 °C) and molecular weight; along with this the crystallinity (as measured by DSC) and isotacticity (as measured by IR and ^{13}C NMR spectroscopy) are also reduced, and part of the material may be completely amorphous. The second group, about 90 %, just contains fractions of increasing molecular weight with fairly constant melting-points and crystallinities in the range of the unfractionated material. From these observations it can be concluded that the conventional Ziegler–Natta catalyst has several types of active centers with different stereospecificities, the lower stereospecificity centers also making a lower polymer weight.

State-of-the-art metallocene iPP homopolymers have melting-points between 140 and 162 °C† and molecular weight distributions M_w/M_n of 2.5–3.5. A fractionation of the kind cited results in fractions of increasing molecular weight, but constant melting-point and crystallinity. The xylene-soluble fraction, referred to as atactic PP (aPP) in the older literature, is clearly below 0.3 wt% (conventional iPP: 1.5–5 wt%). Obviously the metallocene catalyst has only one type of active center, hence the product is more uniform than the conventional one.‡ This difference in

---

†Including developmental products; the range commercially available is narrower, 146–152 °C.

‡This is only true for the first-generation metallocene catalysts as described in section 4 in chapter 18. Further developments of metallocene iPP catalysts will include multicenter catalysts with several metallocenes on one support, theoretically, such a metallocene multicenter system might then perfectly imitate a conventional catalyst.

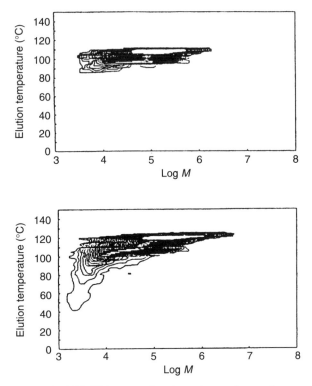

Figure 2 Graphs of TREF-SEC cross-fractionations of metallocene (above) and conventional (below) isotactic polypropylene homopolymer. Reproduced from Ref. 7 by permission of Schotland Business Research

uniformity is shown nicely in the graphs of TREF-SEC cross-fractionations depicted in Figure 2 [7].

Chain microstructure can be analyzed in detail by ^{13}C NMR spectroscopy [8]. Table 1 shows selected NMR signal intensities† of conventional and metallocene iPP samples, together with the samples' melting-points and molecular weights. Entries 1–3 in Table 1 contain data for a conventional polypropylene (gas-phase homopolymer, made with high-yield catalyst), entry 1 is the neat product, entries 2 and 3 are two of 18 elution fractions (fractions 2 and 18). Entries 4–6 are unfractionated samples of metallocene iPP, made with three different metallocene catalysts of

---

†High-resolution spectra of isotactic polypropylene are complex [8]. The occurrence of one type of misinsertion, 1,2-stereoerror, isolated or repeated, already generates a pattern of nine distinguishable methyl-carbon signals. Among them, the isotactic pentad mmmm has the highest intensity and is a measure of chain isotacticity. Obviously from the spectrum, the nine signals can be combined to three signal groups, related to the isotactic triads mm, mr and rr. 2,1- and 1,3-regiomisinsertions produce signal patterns clearly integrable and distinguishable from all other signals.

Table 1 Chain microstructure and iPP melting-point: comparison of conventional and metallocene homopolymers

Entry No.	Sample	M_w (kg/mol)	m.p. (°C)	^{13}C NMR signal intensities (% of methyl group signals integration)					Ref.
				mmmm	mm	mr	rr	2,1-m.i.	
1	Conv. iPP unfractionated	397	164	92, 1	94, 3	3, 0	2, 7	None	5 ,6
2	Conv. iPP low M_w fract.	37	146/156	83, 0	89, 7	7, 3	3 ,0	None	5, 6
3	Conv. iPP high M_w fract.	491	168	99,5	99,7	0, 2	0, 1	None	5, 6
4	Met. iPP A	35	136	86, 3	90, 1	6, 2	2, 7	1, 0	9, 10
5	Met. iPP B	190	145	88, 5	92, 6	4, 8	2, 1	0, 5	9, 10
6	Met. iPP C	740	159	95, 2	98, 9	0, 6	0, 2	0, 3	9, 10

different stereospecificities [9, 10]. Within both metallocene and conventional iPP an increase in melting-point is accompanied by an increasing share of isotactic pentads mmmm. The most obvious difference between conventional and metallocene products is the occurrence of 2,1-misinsertions in the latter; the different mm/mr/ rr ratios indicate further subtle differences in misinsertion patterns (isolated vs cumulated misinsertions along the chain) [11]. These differences influence not only the melting-points, but also all other macroscopic properties, as will be seen below. Compared with conventional iPP, metallocene iPP thus has an additional variable, constituting a wider range of possible products.

The morphologies of these products at the different scales of magnitude (crystals, lamellae, spherulites [12]) have not been examined in detail. As with the conventional product, the α-form is the primary crystalline form detected by wide-angle X-ray scattering; in samples of low stereospecifity and low molecular weight, however, γ-phase crystallinity will prevail. For different metallocene iPP homopolymers, WAXS crystallinities between 40 and 70 % were found [13–15]. Crystallization rates are also a function of chain microstructure. Only with high isotacticity product do they exceed the values of conventional material (see Figure 3) [16, 17].

Figure 4 shows rheology curves for a metallocene homopolymer (m.p. 148 °C) melt and a conventional homopolymer with equal melt flow ratio (MFR = 10 g/10 min). Only at low and medium shear rates is the melt viscosity of the metallocene product lower than that of the conventional homopolymer. This means a setback in melt flowability where high shear rates are relevant. As depicted

---

†Spiral flow test: samples are injection molded in an open-ended spiral-shaped tool under constant injection conditions. 'Easy flow' materials produce larger (longer) spirals, the length of the spiral corresponding to a measure of 'flowability' [18].

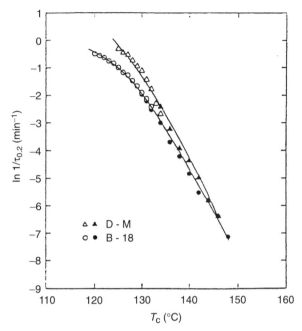

Figure 3 Plot of natural logarithum of the overall rate of crystallization, $1/\tau_{0.20}$, of a metallocene (D-M) and a conventional (B-18) iPP homopolymer (D-M: mmmm = 99.1 %, 2, 1-m.i. = 0.22 %, M_w = 575 kg/mol, M_w/M_n = 2, 4.   B-18: mmmm = 98.1 %, 2,1-m.i. none, M_w = 651 kg/mol, M_w/M_n = 3, 2) [16]

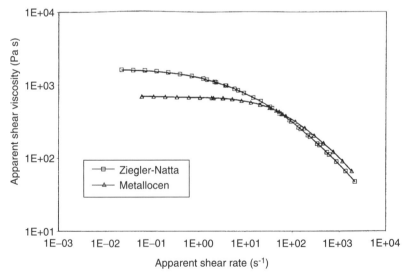

Figure 4 Rheology curves at 230 °C: metallocene vs, conventional Ziegler–Natta iPP (both samples MFR = 10) [44]

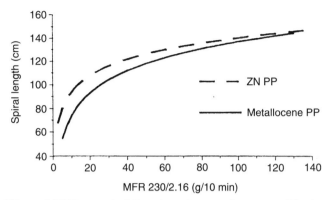

Figure 5 Effect of MFR on spiral flow length: metallocene vs Ziegler–Natta iPP homopolymer. Reproduced from Ref. 19 by permission of Schotland Business Research

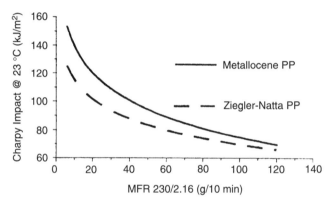

Figure 6 Effect of MFR on impact strength: metallocene vs Ziegler–Natta iPP homopolymer. Reproduced from Ref. 19 by permission of Schotland Business Research.

in Figure 5, the metallocene product has a shorter flow length in the spiral flow test,† a practical flowability test for injection molding. This can be compensated for however, by using a material with a higher MFR value due to a favorable correlation between mechanical properties and MFR (see Figure 6, as an example; impact resistance as function of MFR) [18, 19].

For the characterization of mechanical and optical properties, typical values of two metallocene iPPs with different melting-points (148 and 160 °C) will be compared with three different conventional products of equivalent melting-points: a propylene–ethylene random copolymer, a standard homopolymer and a 'high-crystallinity' homopolymer made by a conventional catalyst of optimized stereo-specificity (see Table 2 [18–24]). Concerning the low content of extractables, both

Table 2 Properties of metallocene vs Ziegler–Natta isotactic polypropylenes [18–24]^a

Property	MH1	ZN RCP	ZN H1 'standard'	MH2	ZN H2 'high cryst.'
Xylene solubles (wt%)	< 0.5	4.0	4.0	< 0.5	1.5
Melting-point, T_m (°C)	148	150	162	160	165
Crystallization temperature, T_c (°C)	110	n.m.	110	121	119
Melting enthalpy (J/g)	98	93	105	111	112
Tensile modulus (MPa)	1650	1150	1550	1900	2000
Ball indentation hardness (MPa)	76	58	77	95	90
Heat deflection temperature (°C)	98	75	95	115	110
Charpy impact strength, 23 °C(kJ/m^2)	90	180	100	105	80
Light transmission (%)	60	65	35	35	38
Haze (%)	7	7	60	n.m.	53
Gloss, 20 ° (%)	77	65	57	n.m.	41

^{a} MH metallocene homopolymer; ZN H, Ziegler–Natta homopolymer; ZN RCP, Ziegler–Natta random copolymer)

metallocene products how much lower levels than their conventional counterparts. The mechanical properties of the metallocene product with m.p. = 148 °C are equivalent to the standard conventional product with a melting-point 15 °C higher. The optical properties are clearly superior and match those of the conventional random copolymer, the stiffness and hardness of which are lower. The hitherto unknown property combination will allow the low-melting metallocene homopolymer to exceed and replace either conventional homo- or copolymers according to a critical property that may be required in a certain application. The metallocene product with m.p. = 160 °C is somewhere between standard and 'high-crystallinity' conventional iPPs. Its high heat deflection temperature is surprising; other advantages have been discovered in extrusion processes under certain process parameters (see Volume 2).

3 METALLOCENE IPP COPOLYMERS

Copolymerization of propylene with ethylene or other 1-olefins introduces short-chain branches into the polymer's linear structure, which will hamper the ability of the chain to crystallize. Macroscopic consequences of such reduced crystallinity are lower stiffness and melting-point and higher impact resistance and transparency. With metallocene catalysts this transformation of properties may start from different homopolymer property levels, according to the stereospecifity of the metallocene catalyst used (see Section 2). For example, a metallocene-made random copolymer of propylene and ethylene with a 2 wt % ethylene content may have a melting-point between 130 and 145 °C, simply as a function of the catalyst used [25]. As with homopolymers, metallocene iPP random copolymers thus exhibit a larger variety of

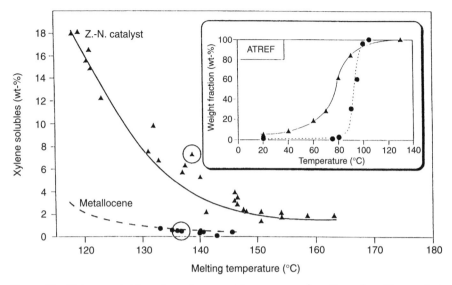

Figure 7 Xylene solubles of random copolymers as a function of melting-point: metallocene vs Ziegler–Natta products. Inset: temperature-rising elution fractionation of two samples (encircled) from the main figure. Reproduced from Ref. 19 by permission of Schotland Business Research

possible property combinations than the conventional products. Comonomer incorporation is not uniform in conventional products; their lower molecular weight fractions contain a higher percentage of comonomer than the rest, and they therefore show considerable solubility in hydrocarbons. On the other hand, metallocene copolymers have a chemical uniformity comparable to that of the homopolymers; they show low levels of solubles even at high comonomer incorporation (see Figure 7 and Table 3 [19]). This prevents particle adhesion leading to reactor fouling and lump formation in production processes and is a critical advantage in end-uses where low migration rates are demanded (e.g. medical uses, food packaging). Enhancement of optical properties by nucleation is especially effective with metallocene random copolymers. Optimized products practically match polystyrene in transparency and

Table 3 Properties of propylene–ethylene random copolymers: metallocene vs Ziegler–Natta products. Reproduced from Ref. 19 by permission of Schotland Business Research

Property	Metallocene			Ziegler–Natta	
C_2 content (mole %)	None	1.1	2.5	4.3	5.3
Melting point (°C)	145	140	132	146	140
MFR (g/10 min)	3.9	5.0	9.1	1.9	1.5
Shear modulus (N/mm^2)	670	630	530	490	370
Xylene solubles (wt %)	0.3	0.4	0.7	3.8	5.3

Table 4 Comparison of selected isotactic polypropylenes with a polystyrene grade [18, 23]

Property	Ziegler–Natta random copolymer	Metallocene homopolymer	Metallocene random copolymer	Polystyrene
Tensile modulus (MPa)	1150	1700	1200	3150
Charpy impact, 23 °C(kJ/m^2)	180	90	160	17
Heat deflection temp. (°C)	70	86	75	80
Strain at break (%)	> 100	> 200	> 200	1.5
Transparency (%)	93	93	96	98
Density (g/cm^3)	0.900	0.900	0.900	1.050

will generate considerable substitution potential owing to their better-balanced mechanical properties and their favorable volume/price ratio (Table 4) [23]. Metallocene catalysts also facilitate the incorporation of higher 1-olefins such as 1-hexene and 1-octene; but the properties of these products are still only sparsely described [26]. Long chain-branched polypropylenes might be generated by copolymerization of propylene with propylene oligomers [27], a possible route to products with high melt strength [28].,

Reactor blends of iPP homopolymer and propylene–ethylene rubber (often referred to as 'block copolymers') show a more attractive balance of stiffness and impact resistance than random copolymers and might, like these, profit from metallocene catalysts. Although the manufacture of such products has been described in several patents [29–32], no data have yet been published which allow a meaningful comparison with conventional grades. The discussion about property advantages is thus still hypothetical [22, 33].

4 FURTHER DEVELOPMENTS

A basic shortcoming of Ziegler–Natta polyolefins is the lack of polarity, which might enhance properties such as dyeability, paintability, printability, adhesion and compatibility with other functional polymers. Ziegler–Natta catalysts are poisoned by polar substances and thus cannot integrate polar monomers as comonomers into the chain. If polarity is desired, it must be added in a post-polymerization step such as irradiation or grafting. With the development of metallocene catalysts, the topic of copolymerization with polar monomers came up again. A practical result of these reinvestigations is metallocene-catalyzed copolymers of propylene with long-chain unsaturated alcohols, esters and acids [34, 35]. Functionalized polypropylenes with up to 4 wt % alcohol were prepared which showed adhesion to aluminum that was improved by a factor of > 20 compared with iPP homopolymer. The approach has also been used for *in situ* stabilization. Metallocene-catalyzed copolymerization of propylene with a 1-olefin-substituted hindered phenol yielded polypropylene with up to 5.5 wt% stabilizer incorporated and a drastically enhanced thermooxidative stability [36, 37]. It has been reported [38] that simple addition of stabilizers such as phosphites, hindered amines or phenols to the polymerization mixture also yields *in situ* stabilized metallocene iPPs without reduction of catalyst activity.

A second approach to introduce polarity into metallocene iPP is post-polymerization chemistry with chain-end double bonds [39–43]. By suitable addition reactions or reaction sequences, anhydride-, silane-, thiol-, epoxy-, borane-, carboxylic acid-, amino- and chlorine-terminated metallocene iPPs can be prepared, with up to 90% conversion of the iPP's vinyl groups. Such polypropylenes with functional end-groups may be reacted with functional end-groups of other polymers to form AB- and ABA-type block copolymers, where A represents a polypropylene segment. These block copolymers can be used as effective compatibilizers within blends of iPP and engineering resins [42].

5 REFERENCES

1. Albizatti, E., Giannini, U., Collina, G., Noristi, L. and Resconi, L., in Moore, E. P., Jr. (Ed), *Polypropylene Handbook*, Carl Hanser Munich, 1996, p. 11.
2. Brintzinger, H.-H., Fischer, D., Mülhaupt, R., Rieger, B. and Waymouth, R. M., *Angew. Chem., Int. Ed. Engl.*, **34**, 1143 (1995)
3. Llinas, G. H., Dong, S.-H., Mallin, D. T., Rausch, M. D., Lin, Y.-G., Winter, H. H. and Chien, J. C. W., *Macromolecules*, **25**, 1242 (1992).
4. Waymouth, R. M., Hauptman, E. and Coates, G. W., *Int. Pat.*, WO 95/25757, to The Board of Trustees of the Leland Stanford University (1994).
5. Paukkeri, R, Väänänen, T. and Lehtinen, A., *Polymer*, **34**, 2488 (1993).
6. Paukkeri, R. and Lehtinen, A., *Polymer*, **34**, 4075 (1993).
7. Ushioda, T., Fujita, H. and Saito, J., *Proc. SPO '97*, 1997, p. 101 (available from Schotland Business Research, 16 Duncan Lane, Skillman, NJ 08558, USA).
8. Tsutsui, T., Ishimaru, N., Mizuno, A., Toyota, A. and Kashiwa, N., *Polymer*, **30**, 1350 (1989).
9. Spaleck, W., Küber, F., Winter, A., Rohrmann, J., Bachmann, B., Antberg, A., Dolle, V. and Paulus, E. F., *Organometallics*, **13**, 954 (1994).
10. Spaleck, W., Küber, F., Bachmann, B., Fritze, C., and Winter, A., *J. Mol. Catal. A.: Chem.*, **128**, 279 (1998).
11. Randall, J. C., *Macromolecules*, **30**, 803 (1997).
12. Phillips, R. A. and Wolkowicz, M. D., in Moore, E. P., Jr, (Ed), *Polypropylene Handbook*, Carl Hanser, Munich, 1996, p. 113.
13. Antberg, M., Dolle, V., Haftka, S., Rohrmann, J., Spaleck, W., Winter, A. and Zimmermann, H. J., *Makromol. Chem., Macromol. Symp.* **48/49**, 333 (1991).
14. Haftka, S., unpublished work.
15. Fischer, D. and Mülhaupt, R., *Macromol. Chem. Phys.*, **195**, 1433 (1994).
16. Galante, M. J., Mandelkern, L., Alamo, R. G., Lehtinen, A. and Paukkeri, R., *J. Therm. Anal.*, **47**, 913 (1996).
17. Bond, E. B. and Spruiell, J. E., *Annu. Tech. Conf. (ANTEC), Soc. Plast. Eng.*, **55**, 388 (1997).
18. Grasmeder, J. R., *Proc. Metallocenes '97*, 1997, p. 323 (for availability, see Ref. 7).

19. Fischer, D., Bidell, W., Grasmeder, J., Hingmann, R., Jones, P., Kersting, M., Langhauser, F., Marczinke B., Moll, U., Rauschenberger, V., Süling, C. and Popham, N., *Proc. SPO '97*, 1997 (for availability, see Ref. 7).
20. Brekner, M. J., *Proc. Metallocenes '96*, 1996, p. 153 (for availability, see Ref. 7).
21. Kunzer, R. and Wieners, G., *Kunststoffe*, **86**, 666 (1996).
22. Galli, P., *Proc. Flexpo '97*, 1997, p. 60 (available from Chemical Market Resources Inc., 1120 Nasa Road 1, Suite 340, Houston, TX 77058-3320, USA).
23. Grasmeder, R., *Proc. PP '97, Session VII/2-1*, 1997 (available from Maack Business Services, Moosacherstr. 14, CH-8804 Au, Switzerland).
24. McAlpin, J. J., Stahl, G. A., *Proc. Metcon '94*, 1994 (available from Catalyst Consultants, P.O. Box 637, Spring House, PA 19477, USA).
25. Winter, A., *Proc. Metcon '97*, 1997 (for availability, see Ref. 7).
26. McAlpin, J. J., Mchta, J. J., Stahl, G. A. and Autran, J. P., *Int. Pat.*, WO 95/32242, to Exxon Chemical Patents (1994).
27. Seelert, S., Fischer, D., Marczinke, B. L., Langhauser, F., Kerth, J. and Müller, H. J., *DE-OS*, 4 425 787, to BASF (1994).
28. De Maio, V. V. and Dong, D., *Annu. Tech. Conf. (ANTEC), Soc. Plast. Eng.*, **55**, 1512 (1997).
29. Schreck, M., Winter, A., Spaleck, W., Kondoch, H. and Rohrmann J., *US Pat.*, 5 280 074, to Hoechst (1989).
30. Schlund, R., Müller, P., Kerth, J. and Hungenberg, K. D., *DE-OS*, 4 130 429, to BASF (1991).
31. Ueda, T., Hashimoto, M., Kawasaki, M., Fukuoka, D. and Imuta, J., *Eur Pat.*, 704 462, to Mitsui Petrochemical Industries (1994).
32. Collina, G., Dall'Occo, T., Galimberti, M., Albizatti, E. and Noristi, L., *Int. Pat.*, WO 96/11218, to Montell Technology (1994).
33. Van Os, G. and Pentz, H., *Proc. Metallocenes '95*, 1995, p. 521 (for availability, see Ref. 7).
34. Löfgren, B. and Seppälä, J., *Proc. Metcon '97* 1997 (for availability, see Ref. 24).
35. Aaltonen, P., Fink, G., Löfgren, B. and Seppälä, J., *Macromolecules*, **29**, 5255 (1996).
36. Wilén, C.-E. and Näsman, J. H., *Macromolecules*, **27**, 4051 (1994).
37. Wilén, C.-E. and Näsman, J. H., *Int. Pat.*, WO 95/27744, to Borealis Holding (1994).
38. Rotzinger, B., Schmutz, T., Brunner, M. and Stauffer, W., *Eur Pat.*, 755 948, to Ciba SC Holding (1995).
39. Shiono, T., Kurosawa, H., Ishida, O. and Soga, K., *Macromolecules*, **26**, 2085 (1993).
40. Shiono, T. and Soga, K., *Makromol. Chem., Rapid Commun.* **13**, 371 (1992).
41. Chung, T. C., Lu, H. L. and Janvikul, W., *Polymer*, **38**, 1495 (1997).
42. Mülhaupt, R., Duschek, T. and Rieger, B., *Macromol. Chem., Macromol. Symp.*, **48/49**, 317 (1991).
43. Mülhaupt, R., Duschek, T., Fischer, D. and Setz, S., *Polyme. Adv. Technol.*, **4**, 439 (1993).
44. Rauschenberger, V. and Fischer, D., unpublished work.

20

Metallocene-Catalyzed Syndiotactic Polypropylene: Preparation and Properties

TETSUNOSUKE SHIOMURA, NOBUTAKA UCHIKAWA,
TADASHI ASANUMA, RYUICHI SUGIMOTO, ICHIRO FUJIO,
SHIGERU KIMURA, SHIGERU HARIMA, MASAYA AKIYAMA,
MASAHIRO KOHNO AND NORIHIDE INOUE
Mitsui Toatsu Chemicals, Inc., Yokohama, Japan.

1 INTRODUCTION

As early as 1960, Natta *et al.* [1] succeeded in the isolation of the syndiotactic polypropylene (SPP) portion from the prevalently isotactic polypropylene (IPP), utilizing the differece in the affinity with an adsorption bed consisting of highly crystalline IPP. The SPP portion amounted to ca 1% with $TiCl_3$ catalyst and 5–10 % with $TiCl_4$ catalyst. Polymerization at low temperature and the use of a vanadium catalyst improved the yield of SPP [2]. Zambelli and co-workers [2] showed that the syndiotactic chain propagation with a V catalyst occurs via secondary (2–1) insertion, instead of primary (1–2) insertion for isotactic propagation with a Ti catalyst:

$$Ti—R + C_3H_6 \longrightarrow Ti—CH_2CH(CH_3)R$$
$$V—R + C_3H_6 \longrightarrow V—CH(CH_3)CH_2R$$

Metallocene-based Polyolefins   Edited by J. Scheirs and W. Kaminsky
© 2000 John Wiley & Sons Ltd

The stereoregulation energies for both steps were evaluated from the dyad composition of polymers produced at various polymerization temperatures (-78 to $-18.5°C$), assuming the following equations [3].

$$\ln(P_{DD}/P_{DL}) = \ln(2[m]/[r]) = \ln(a_{DD}/a_{DL}) + \Delta E_{DL-DD}/RT$$
$$\ln(P_m/P_r) = \ln([m]/[r]) = \ln(a_m/a_r) + \Delta E_{r-m}/RT$$

where [m] and [r] are the mole fractions of isotactic and syndiotactic dyads, respectively, and

$$\Delta E_{DL-DD} = 4.8 \pm 0.5 \text{ kcal/mol}$$
$$\Delta E_{r-m} = -2 \pm 0.4 \text{ kcal/mol}$$

ΔE_{DL-DD} is the difference in the activation energies for the formation of L and D placements at the D-preferring sites in the 'enantiomorphic sites' propagation process for isotactic control and ΔE_{r-m} is that for the formation of syndiotactic and isotactic dyads in the secondary insertion process. The sign of ΔE shows the type of stereoregulation ($+$ for isotactic, $-$ for syndiotactic) and the absolute values give an estimation of the stereocontrol force. Syndiotactic propagation has been attributed to the asymmetry of the last inserted monomer unit controlling the stereochemistry of polymerization. Analyzed with advanced ^{13}C NMR technology, typical SPP exhibited the following microtacticity: mmmm 0.011, mmmr 0.033, rmmr 0.031, mmrr 0.037, mmrm + rmrr 0.239, rmrm 0.109, rrrr 0.254, mrrr 0.217 and mrrm 0.069 [3c].

In 1984–85, Ewen, closely followed by Brintzinger, Kaminsky and co-workers, reported success in producing IPP with soluble metallocenes, confirming the assumption that the chiral, stereorigid (bridged) metallocene having C_2 symmetry gives an isotactic polymer complying with the Cossee–Arlman mechanism, where the metallocene has four coordination sites [two being Cp groups, one attached alkyl (or growing polymer chain) and one vacant site available for olefin coordination]. *Cis*-opening and chain migratory insertions are assumed. Two stereochemical control modes, control by chiral growing chain ends and by enantiomorphic sites, have been formulated [4]. In 1988, Ewen *et al.* designed metallocenes with C_s symmetry, e.g. isopropylidene(η^5-cyclopentadienyl) (9-η^5-fluorenyl)zirconium dichloride [$Me_2C(Cp)(Flu)ZrCl_2$] and its hafnium analog. Combined with methylalumoxane (MAO), these catalysts gave highly syndiotactic PP in high yield under conventional polymerization conditions [5a]. The microstructure of the new SPP (mainly containing ... rrmmrr ... type defects instead of ... rrmrr ... type defects) agreed well with the enantiomorphic site control model.

The origin of the stereo-control was explained with molecular mechanics (MM) calculations by Guerra and co-workers [6] and with *ab initio* molecular orbital plus MM calculations by Morokuma and co-workers [7]. According to those studies, the fluorenyl ligand determines the conformations of the polymer chain end and the fixed polymer chain end conformation, in turn, determines the sterochemistry of

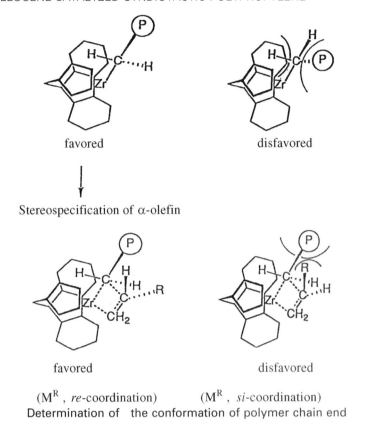

favored   disfavored

Stereospecification of α-olefin

favored   disfavored

(M^R, *re*-coordination)   (M^R, *si*-coordination)
Determination of   the conformation of polymer chain end

Scheme 1   Model mechanism of syndiotactic propagation [7b]

olefin insertion at the transition state (Scheme 1), where the relative energy for the transition states has been estimated as shown in Figure 1 [7c].

2   CATALYSTS AND POLYMERIZATION

Many studies have been dedicated to improving syndiospecificity. They can be classified as (i) modification of bridging structure, (ii) substitution at the Flu ring, (iii) substitution at Cp ring, (iv) alteration of transition metal, (v) cocatalyst component and (vi) novel-type metallocene or non-metallocene. The results are summarized in Tables 1–7.

As the polymerization conditions adopted by several authors differ in many details (source of MAO, solvent, admixing of catalyst components, monomer concentration, etc.), some variations in the catalyst performance are inevitable, as can be seen in the

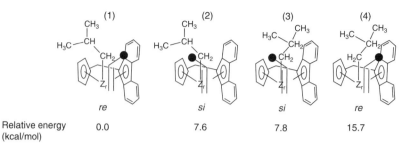

	(1)	(2)	(3)	(4)
	re	*si*	*si*	*re*
Relative energy (kcal/mol)	0.0	7.6	7.8	15.7

Figure 1 Transition state model for propylene polymerization. Reproduced by permission of Hüthig & Wepf Verlag, Zug, Switzerland from M. Kakugo, *Macromol. Symp.*, **89**, 545 (1995)

Table 1 Modification in the bridging unit: $X(Cp)(Flu)ZrCl_2$ [a]

X	Activity (kg/g h)	$M_w \times 10^3$	T_m °C	rrrr
Me_2C	62	150	148	0.92
$H(^tBu)C$	1	180	144	
Ph_2C	(17)	780	143	
$(H)(Ph)C$	61	170	147	
$(Me)(Ph)C$	64	230	150	
$(CH_2)_5C$	62	120	145	
$S(C_2H_4)_2C$	2	90	147	
$O(C_2H_4)_2C$	22	86	148	0.90
$(Me)[(p\text{-}MeOC_6H_4)CH_2]C$	24	120	145	
$Cl_2Zr(Cp)(Flu)C(C_2H_4)_2C(Cp)$ (Flu) $ZrCl_2$	33	55	146	
Me_2Si	63	100	None	
$Me_2C(Cp)(Phen)ZrCl_2$ [b] [5d]	23	154	119	
$Me_2Si(Cp)(Phen)ZrCl_2$ [c] [5d]	2.2	63	oil	

[a] $T_p = 20°C$ (30 °C for Me_2Si; 60 °C for the last two entries); solvent, toluene; C_3 pressure, $3kg/cm^2$-G; time, 1h; MAO/Zr molar ratio4, 2000.
[b]

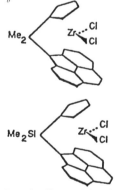

Reproduced by permission of Hüthig & Wepf Verlag, Zug, Switzerland from T. Shiomura *et al.*, *Macromol. Symp.*, **101**, 289 (1996)

Table 2 Modification in the bridging unit: X(Cp) (Flu)ZrCl$_2$, reported by other authors[a]

X	MAO/Zr	T_p (°C)	P (kg/ cm^2-G)	Solvent	Activity (kg/gh)	T_m (°C)	rrrr	M_w ($\times 10^3$)	Ref.
Me$_2$C	15000	70	(32)	Liq. C$_3$	169	128	–	90	13b
Ph$_2$C	15000	70	(32)	Liq. C$_3$	186	130	–	480	13b
Me$_2$Si	15000	70	(32)	Liq. C$_3$	55	100	–	135	13b
Ph$_2$Si	15000	70	(32)	Liq. C$_3$	30	105	–	600	13b
H$_2$C	(2000)	60	(23)	Liq. C$_3$	102	–	r = 0.935	82	14k
Me$_2$C	(2000)	60	(23)	Liq. C$_3$	75	135	r = 0.946	82	14k
(H) (Ph) C	(2000)	60	(23)	Liq. C$_3$	115	138	–	85	14k
(Me) (Ph) C	(2000)	60	(23)	Liq. C$_3$	68	–	r = 0.970	490	14k
Ph$_2$C	(2000)	60	(23)	Liq. C$_3$	155	–	r = 0.965	547	14k
(CH$_2$)$_4$C	(2000)	60	(23)	Liq. C$_3$	77	132	r = 0.951	110	14k
(CH$_2$)$_5$C	(2000)	60	(23)	Liq. C$_3$	167	133	–	–	14k
Me$_2$C	2500	30	1.1	Toluene	(1.2)	143	rr = 0.92	94	15a
(H) (^{t}Bu) C	2500	30	1.1	Toluene	(2.1)	140	rr = 0.92	91	15a
Me$_2$Si	2400	25	0.7	Toluene	0.8	none	rr = 0.12	46	15d
Me$_2$C		60	(23)	Liq. C$_3$	320	138	–	100	12b
Et$_2$Si		60	(23)	Liq. C$_3$	8.3	107	–	286	12b
Me$_2$C		40	(15)	Liq. C$_3$	120	144	0.86	138	10f,j
CH$_2$CH$_2$	3300	40	(15)	Liq. C$_3$	35	125	0.83	284	10f,j
CH$_2$CH(Ph)	2000	30	–	Toluene	–	–	0.78	160	16

[a] Liq. C$_3$ signifies that propylene polymerization was conducted in liquid propylene (bulk polymerization), where a small amount of toluene was used in order to dissolve the metallocene together with MAO. Pressure values in parentheses provide a rough measure and activity figures are expressed in kg PP/(g metallocene/h).

tables. In many cases, no data on rrrr value are available, although the melting-point of a polymer (T_m) is roughly parallel to rrrr.

In Tables 1 and 2, the effect of a Ph$_2$C bridge on polymer molecular weight (discovered at Hoechst) is remarkable, whereas an Me$_2$Si or ethylene bridge is detrimental to syndiospecificity, in sharp contrast with isospecific catalysts. Asymmetry in the bridging structure, H(^{t}Bu)C, H(Ph)C or (Me)(Ph)C, does not cause a significant change in stereospecificity.

So far, substitutions in the fluorene ring have not been very effective in improving rrrr. Unexpectedly, the octahydrofluorenyl ligand [Ph$_2$C(Cp)(H8-Flu)] gives a deteriorated rrrr (Table 3). Asymmetrically substituted (lop-sided) fluorenes show fair to good syndiospecificity. On the other hand, substitution of the Cp ligand causes drastic changes in stereoselectivity (Table 4). 3-Me-Cp gives a so-called hemi-isotactic polymer, while 3-^{t}Bu-Cp gives a highly isotactic polymer. Such a dramatic change in stereospecificity can be interpreted by assuming skipped insertions (see the next section).,

The influence of transition metal species on the syndioselectivity can be seen in Table 5. Hf provides a less syndiospecific polymer. Little is known about Ti-based metallocenes giving highly syndiotactic polypropylenes.

Table 3 Substitution at the fluorene ligand: R$_2$C(Cp) (substituted-Flu) ZrCl$_2$[a]

Ligand	T_p (°C)	MAO/Zr	Solvent	Activity (kg/gh)	M_w (×10³)	T_m (°C)	rrrr	Ref.
Me$_2$C(Cp) (Flu)	60	2000	Liq. C$_3$	188	92	137	0.82	11a-c
Me$_2$C(Cp) [2,7-di (^{t}Bu) Flu]	60	2000	Liq. C$_3$	116	75	141	–	11a-c
Me$_2$C(Cp) [2,7-di (Me$_3$Si) Flu]	40	2000	Liq. C$_3$	79	33	147	0.94	11a-c
Ph$_2$C(Cp) (Flu)	40	2000	Liq. C$_3$	39	33	142	0.92	11a-c
Ph$_2$C(Cp) (H8-Flu)	60	2000	Liq. C$_3$	101	77	None	0.58	11a-c
Me$_2$C(Cp) (Flu)	60	5500	Liq. C$_3$	280	115	137	–	5d, g, 12b
Me$_2$C(Cp) (2,7-diMeFlu)	60	6000	Liq. C$_3$	112	126	135	–	5d, g, 12b
Me$_2$C(Cp) [2,7-di (^{t}Bu) Flu]	60	4800	Liq. C$_3$	116	75	141	–	5d, g, 12b
Me$_2$C(Cp) (2,7-diFFlu)	60	3700	Liq. C$_3$	132	83	132	–	5d, g, 12b
Me$_2$C(Cp) (2,7-diClFlu)	60	600	Liq. C$_3$	8	92	136	–	5d, g, 12b
Me$_2$C(Cp) (Phen)	60	760	Liq. C$_3$	23	154	119	0.72	5d, g, 12b
Me$_2$C(Cp) (Flu)	60	2000	Liq. C$_3$	87	82	135	r = 0.95	14g
Me$_2$C(Cp) (2,7-diMeFlu)	60	2100	Liq. C$_3$	27	80	131	r = 0.95	14g
Me$_2$C(Cp) (2,7-diPhFlu)	60	2800	Liq. C$_3$	33	65	132	r = 0.93	14g
Me$_2$C(Cp) [2,7-di (^{t}Bu) Flu]	70	2600	Liq. C$_3$	110	74	142	r = 0.94	14g
Me$_2$C(Cp) [2,7-di (MeO) Flu]	60	2300	Liq. C$_3$	0.7	20	96	–	14g
Me$_2$C(Cp) (2,7-diClFlu)	60	2300	Liq. C$_3$	46	–	–	r = 0.93	14g
Me$_2$C(Cp) (2,7-diBrFlu)	60	2800	Liq. C$_3$	49	60	131	r = 0.91	14g
Me$_2$C(Cp) (4-Me-Flu)	60	1500	Liq. C$_3$	30	42	112		5g
Me$_2$C(Cp) (4-MeO-Flu)	60	5400	Liq. C$_3$	245	189	136		5g
Me$_2$C(Cp) (4-MeOCH$_2$-Flu)	60	1100	Liq. C$_3$	12	94	79		5g
Me$_2$C(Cp) (4-Me$_2$N-Flu)	60	700	Liq. C$_3$	9	150	Atactic	0.14	5g
Me$_2$C(Cp) (3-MeO-Flu)	60	1600	Liq. C$_3$	4	72	131		5g
Me$_2$C(Cp) (2-MeO-Flu)	60	3200	Liq. C$_3$	3	50	129		5g
Me$_2$C(Cp) (1-MeO-Flu)	60	3200	Liq. C$_3$	21	77	136		5g
Me$_2$C(Cp) (2,3-diMeInd)	30	1000	–	–	–	–	0.45	17
Me$_2$C(Cp) (3,4-diMeCp)	30	1000	–	–	–	–	0.14	17

[a]H8-Flu denotes 1, 2, 3, 4, 5, 6, 7, 8-octahydrofluorene-9-yl and 3, 4-diMeInd denotes 3, 4-dimethylindene-1-yl.

Table 4 Substituents in the cyclopentadiene ligand: Me_2C (3-R-Cp) (Flu) $ZrCl_2$

R	MAO/Zr	Solvent	T_p (°C)	Activity (kg/gh)	M_w ($\times 10^3$)	T_m (°C)	rrrr	Ref.
H	130	Liq. C_3	70	198	108	134	0.82	5b, d, e, h
Me	700	Liq. C_3	65	59	58	–	0.23	5b, d, e, h
^tBu	3400	Liq. C_3	60	40	81	125	Isotactic	5b, d, e, h
Me	–	Liq. C_3	60	–	42	126	0.23	10d–f
^tBu	3400	Liq. C_3	60	48	62	127	mmmm = 0.77	10d–f
Me	1800	Toluene	25	–	38	–	0.21 (mmmm = 0.14)	18b, c
Me	1800	CH_2Cl_2	25	–	43	–	0.06 (mmmm = 0.34)	18b, c

Bridge	R							
Me_2C	Me	3300	Liq. C_3	65	88	40	–	0.19 (mmmm = 0.15)	13b, c
Me_2Si	Me	3600	Liq. C_3	60	59	320	–	0.53 (mmmm = 0.14)	13b, c
Me_2Si	^iPr	1500	Liq. C_3	60	25	62	–	0.01 (mmmm = 0.64)	13b, c
Me_2Si	^tBu	15000	Liq. C_3	70	137	59	145	Isotactic	13b, c

Table 5 Zr vs Hf

Ligand	Metal	MAO/M	Solvent	T_p (°C)	Activity (kg/gh)	M_w ($\times 10^3$)	T_m (°C)	rrrr	Ref.
$Me_2C(Cp)$ (Flu)	Zr	12000	Liq. C_3	50	694	69		0.81	5a
		13000	Liq. C_3	70	304	55		0.76	5a
	Hf	800	Liq. C_3	50	3	777		0.74	5a
$Me_2C(Cp)$ (Flu)	Zr	12000	Liq. C_3	50	194	133	138	r = 0.96	10a, b, j
	Hf	1800	Liq. C_3	50	54	777	118	r = 0.74	10a, b, j
$Ph_2C(Cp)$ (Flu)	Zr	5900	Liq. C_3	50	68	560	139	r = 0.97	10a, b, j
	Hf	1600	Liq. C_3	50	12	1950	101	r = 0.90	10a, b, j

Various anions so far proposed have not surpassed MAO in terms of syndiospecificity (Table 6).

Apart from Ewen-type metallocenes, novel types of catalysts have been reported recently. One was described by Bercaw and co-workers [19] (Table 7): although it retains the main features of the original metallocene, viz (a) C_s symmetry, (b) cyclopentadienyls of differing size and (c) steric bulk flanking the metallocene wedge with an open region in the center, its performance seems to be exceptionally affected by the polymerization conditions.

Another belongs to the so-called constrained geometry catalysts. There is controversy about the stereospecificity of $Me_2Si(Flu)(N - {}^tBu)ZrX_2$ (X = Me,

Table 6 Influence of cocatalysts [11c][a]

Metallocene Me$_2$C(Cp)(Flu)ZrCl$_2$ (mg)	Cocatalyst					Yield (g)	M_w (×10^3)	T_m (°C)
	MAO (mg)	Ph$_3$CB(C$_6$F$_5$)$_4$ (mg)	Ph$_3$CAl(C$_6$F$_5$)$_4$ (mg)	Ph$_3$CGa(C$_6$F$_5$)$_4$ (mg)	AlEt$_3$ (mg)			
2.5	670	-	-	-	-	130	150	148
2.0	-	9.5	-	-	40	150	120	142
2.0	-	-	9.7	-	40	160	140	144
2.0	-	-	-	10.8	40	53	46	147

[a]Solvent, toluene; propylene pressure, 2 kg/cm^2; duration, 2 h.

Table 7   Bercaw catalyst: $(Me_2Si)_2$ (4-R-Cp$'$) (3,5-di-i Pr-Cp$'$) $ZrCl_2{}^a$

Ligand (R)	MAO/Zr	T_p (°C)	Solvent	Pressure (kg/cm^2)	Activity (kg/g/h)	rrrr	r
Et$_2$C(Cp)(Flu)	300	60	Liq. C$_3$	(32)	9.3	–	0.926
(^{i}Pr)$_2$Si(Cp) (3, 4-diMe$_3$Si-Cp)	1000	0	Liq. C$_3$	6	–	–	0.55
H	2000	0	Liq. C$_3$	(6)	2.1	0.837	0.940
H	430	25	toluene	2.7	0.3	0.273	0.741
H	300	60	Liq. C$_3$	(32)	74.2	0.760	0.926
^{i}Pr	2000	0	Liq. C$_3$	(6)	1.7	0.989	0.996
^{i}Pr	430	25	Toluene	2.7	0.23	0.388	0.756
Me$_3$Si	2000	0	Liq. C$_3$	(6)	1.7	0.959	0.990
Me$_3$Si	430	25	Toluene	2.7	0.23	0.339	0.752
^{t}Bu	2000	0	Liq. C$_3$	(6)	0.72	0.905	0.969
^{t}Bu	430	25	Toluene	2.7	0.20	0.296	0.762
MeEtCH	2000	0	Liq. C$_3$	(6)	1.5	0.831	0.944
MeEtCH	430	25	Toluene	2.7	0.20	0.200	0.620
Me$_3$C(Me)CH	2000	0	Liq. C$_3$	(6)	0.93	0.418	0.146
Me$_3$C(Me)CH	430	25	Toluene	2.7	0.16	0.0 (mmmm = 0.612)	0.146
Me$_3$C(Me)CH	430	25	Toluene	0.7	0.11	0.0 (mmmm = 0.585)	0.176

^{a}

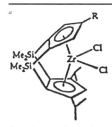

Reproduced with permission from T. A. Herzog, D. L. Zubris and J. E. Bercaw, *J. Am. Chem. Soc.*, **118**, 11988 (1996). Copyright 1996 American Chemical Society.

Cl or Me$_2$N) whether it is combined with MAO or B(C$_6$F$_5$)$_4$ anion, supposedly owing to its openness on the Zr flanks which may develop several modes in accommodating counteranions [a smaller anion, B(C$_6$F$_5$)$_4$, may result in more compact coordination than more bulky MAO]. Apart from small quantities of toluene-insoluble fraction [23a], both anions give mainly syndiotactic polymer; however B(C$_6$F$_5$)$_4$ anion brings out very diversified products (in tacticity and molecular weight), contrasting with MAO producing more homogeneous polymers [23b] (Table 8). On the other hand, Waymouth and co-workers noticed that, in combination with MAO (Al/Zr - 300). Me$_2$Si(Flu)(N-^{t}Bu)ZrCl$_2$ gives highly regioregular, high molecular weight atactic polymer with a slight syndiotactic enrichment [23f]. Starting from Me$_2$Si(Me$_4$Cp)(N-^{t}Bu)Ti$^+$ and B(C$_6$F$_5$)$_3$ or Ph$_3$CB(C$_6$F$_5$)$_4$, Chen and Marks obtained syndiotactic-enriched atactic polypropy-

Table 8 Performance of constrained geometry catalysts[a]

Metallocene	Amount (g)	Al$_i$Bu$_3$ (g)	MAO (g)	Ph$_3$CB(C$_6$F$_5$)$_4$ (g)	Solvent	T_p (°C)	P (atm)	Yield (g)	rrrr	Fr-1 (g)	Fr-2 (g)	Fr-3 (g)	Ref.
Me$_2$Si(Flu)(N$-^t$Bu)ZrNMe$_2$	0.15	4.9	3.7	–	Toluene	20	1	12.4	–	11.7	0.7	0	23 b
	0.15	4.9	3.7	–	Toluene	45	1	11.0	–	11.0	0	0	23 b
	0.15	4.9	–	0.57	Toluene	25	1	49.0	–	24.0	8.5	16.5	23 b
	0.15	4.9	–	0.57	Toluene	45	1	17.1	–	11.7	5.4	0	23 b
Me$_2$Si(Flu)(N$-^t$Bu)Cl$_2$	0.15	4.9	8.4	–	Toluene	25	1	2.2	–	2.2	0	0	23 b
	0.15	4.9	8.4	–	Toluene	40	1	3.8	–	3.8	0	0	23 b
	0.15	4.9	–	0.57	Toluene	25	1	50.4	–	36.9	5.7	7.8	23 b
	0.15	4.9	–	0.57	Toluene	60	1	11.0	–	10.5	0.5	0	23 b
									Π	**mm**	**M_w**	**M_w/M_r**	
Me$_2$Si(Flu)(N$-^t$Bu)ZrCl$_2$	0.10	–	4.5	–	Toluene	20	1	0.0	–	–	–	–	23 a
	0.20	6.7	9.5	–	Toluene	20	1	4.8	0.875	0.038	105 × 10^3	2.2	23 a
	0.20	6.7	–	[Me$_2$NHPh][B(C$_6$F$_5$)$_4$] 0.83	Toluene	20	1	0.6	0.951	0.025	60 × 10^3	5.1	23 a
	0.20	6.7	–	0.70	Toluene	20	1	0.0	–	–	–	–	23 a
									rrrr		**M_w**	**M_w/M_r**	
Me$_2$Si(Flu)(N$-^t$Bu)ZrCl$_2$	0.019	–	0.16	–	Toluene	30	4.5	0.14	0.145		652 × 10^3	1.8	23 f
	0.019	–	0.16	–	Toluene	30	9.4	0.11	0.178		779 × 10^3	1.9	23 f
Me$_2$Si(Me$_4$Cp)(N$-^t$Bu)TiCl$_2$	0.014	–	0.19	–	Toluene	30	4.5	3.10	0.177		583 × 10^3	2.0	23 f
	0.014	–	0.19	–	Toluene	30	8	*2.89	0.233		400 × 10^3	1.7	23 f
Me$_2$Si(Me$_4$Cp)(N$-^t$Bu)Ti(CH$_2$Ph)$_2$	0.072	–	–	0.014^b B(C$_6$F$_5$)$_3$	Toluene	25	1	**2.65	rr = 0.35		142 × 10^3	3.8	22 g
	0.072	–	–	0.0078^c	Toluene	25	1	0.65	rr = 0.41		169 × 10^3	2.8	22 g

[a] Duration, 1 h. Fr-1: 20 °C acetone-soluble (oil). Fr-2: hot acetone-soluble. Fr-3: hot acetone-insoluble (sticky solid). *5 min,**10 min

b

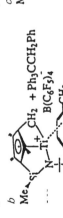

c

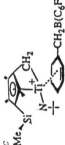

lenes having appreciable molecular weights [22g], Varying the counter-anion incurs a remarkable activity change which has been ascribed to the formation of various cationic structures having different relative coordinating abilities, which may entail different ease of displacement with an olefin. Supposedly, MAO may enforce greater cation–anion separation than $B(C_6F_5)_4$ analogs.

A remakable effect of anions on an isospecific metallocene, rac-$Me_2Si(Ind)_2ZrMe_2$, has been reported. While $Ph_3CB(C_6F_5)_4$ gives low isotactic products ($[mm] = 0.47, [mr] = 0.33, [rr] = 0.20$). a strongly ion-paired analog generated with $Ph_3CAlF(o\text{-}C_6F_5C_6F_4)_3$ affords highly isotactic polypropylene ($[mmmm] = 0.98$) [22h]. By direct support on SiO_2 of the Ewen-type metallocenes, instead of support on MAO-treated SiO_2, the catalysts have been reported to change into isospecific catalysts with a major loss in activity [43a, b]. The complete blocking on one coordinating side of the metal hinders the chain-migratory insertions and leads to isospecific catalysts.

3 MECHANISM RECONSIDERED

A marked decrease in syndiospecificity (decrease in T_m) and a complementary increase in the xrmx value have been reported by Ewen $et\ al.$ and ascribed to the skipped insertion [5c].

Buscico and Cipullo explained that the stereoerrors in producing isotactic polymer with C_2 symmetry at very low monomer concentration arise predominantly from chain-end isomerization rather than from errors in the enantiomorphic orientation of the inserting olefin [20a–c]. Resconi $et\ al.$ made a similar observation [25]. Leclerc and Brintzinger corroborated that an α-agostic interaction is operative in isotactic propylene polymerization and the same interaction leads to stereoerror formation by isomerization of the last-inserted unit [21a] (Scheme 2).

Herfert and Fink [8a] reported a marked solvent effect in syndiospecific polymerization. Using $Me_2C(Cp)(Flu)ZrCl_2/MAO$ as catalyst at $25\,°C$ and $2kg/cm^2$-G in a mixed solvent (toluene–CH_2Cl_2), the rrrr value decreases linearly from 0.89 (toluene) to 0.42 (93.8 vol. % CH_2Cl_2) ($T_m = 143–47°C$), while rmrr increases from 0.02 to 0.20, rmmr remaining unaffected (0.02) with increasing

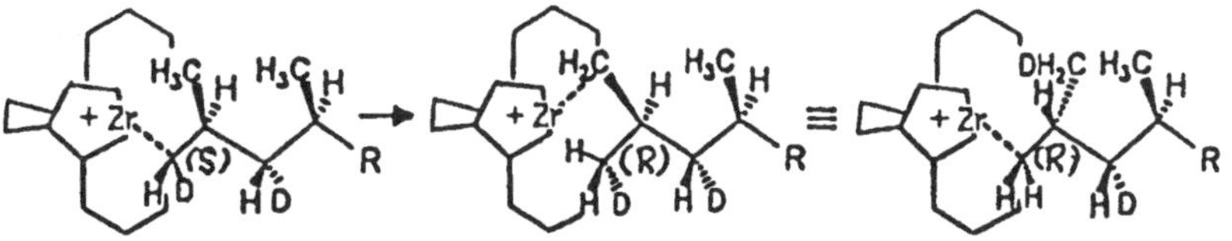

Scheme 2 Stereoerror formation by isomerization of the last-inserted unit. Reproduced with permission from M. K. Leclerc and H. H. Brintzinger, *J. Am. Chem. Soc.*, **117**, 1651 (1995). Copyright 1995 American Chemical Society

dielectric constant of the solvent mixture. Raising the polymerization temperature has a similar effect on the microstructure of the polymers. In both cases, an increase in mmmm and a decrease in rrrr and rrrm are observed. The loss in syndiospecificity must have been caused by an isomerization of the solvent-separated free zirconocene species via migration of the growing polymer chain before the next monomer insertion ('growing chain epimerization').

Buscico *et al.* recently made a surprising report that the usually syndiospecific catalyst (Me)(Ph)C(Cp)(Flu)ZrCl$_2$/MAO-AlMe$_3$ (MAO/Zr = 300, T_p = 10–60°C) polymerizes bulky 3-methyl-1-butene into isotactic products with reasonable activity (9–140 kg/g/h) [20d].

Asanuma *et al.* tried to take up the skipped insertion in C$_s$-symmetrical catalysis [24d, e] and formulated a four-parameter model to describe the pentad distribution (Scheme 3). Farina *et al.* pointed out some miscalculations and gave the correct equations [24a]. Furthermore, the latter indicated the original four-parameter model (K_1–K_4) of Asanuma *et al.* can be reduced to a simple two-parameter model ('*a–c* scheme') by presuming $K_2/K_1 = K_4/K_3$ without significant losses in accuracy. Under this condition, *a* and *c* (and also *b–h*) are equated with K_1–K_4 according to the following equations (Table 9) [24a]

$$K_1 = a(1-a), \ K_2 = (1-a)(1-c), \ K_3 = ac, \ K_4 = (1-a)c; \ a = K_1 + K_3,$$
$$c = K_2 + K_4$$

or

$$K_1 = (1-a)(1-c), \ K_2 = a(1-c), \ K_3 = (1-a)c, \ K_4 = ac; \ a = K_2 + K_4,$$
$$c = K_3 + K_4$$

The coordinating anion, situated on one side of the Zr cations with varying ease of dislocation by the incoming olefin, makes even a C$_s$-symmetrical metallocene cation a somewhat lop-sided ion pair, which may entail some loss in the syndiospecificity. More weakly coordinating anions are expected to bring better performance to the syndiospecific polymerization.

Here we recall the observation by Lovinger and co-workers in their study of the crystal structure of SPP (V catalyst based, $r = 0.77$) [32b, c]. On all occasions, they identified the presence of separate IPP crystals together with SPP single crystals, and confirmed that IPP does not constitute stereoblock defects within the SPP population, although for highly syndiotactic ($r > 0.99$) samples they could not detect the presence of a separate IPP [32d].

4 PROPERTIES

In the following sections, some physical properties of SPP are discussed to stress the effect of tacticity, which has often been neglected in earlier works.

Table 9 Pentad distribution taking into account of skipped insertion for C_s-symmetric catalyst (a–c scheme)[a]

Pentad		$a = 0.95$, $c = 0.00$, $e = 0.05$, $g = 1.00$	$a = 0.95$, $c = 0.95$, $e = 0.05$, $g = 0.05$
mmmm	$a^2(ac + eg)^2(ac^2 + eg^2) + e^2(ag + ec)^2(ag^2 + ec^2) + 2aecg(ac + eg)(ag + ec)$	0.0023	0.6342
mmmr	$2(ac + eg)(ag + ec)(a^2(ac^2 + eg^2) + e^2(ag^2 + ec^2)) + 2aecg((ac + eg)^2 + (ag + ec)^2)$	0.0045	0.1368
rmmr	$a^2(ag + ec)^2(ac^2 + eg^2) + e^2(ac + eg)^2(ag^2 + ec^2) + 2aecg(ac + eg)(ag + ec)$	0.0407	0.0074
mmrm	$2ae(ac + eg)(ag + ec)(c^2 + g^2) + 2cg(a^2(ac + eg)^2 + e^2(ag + eg)^2)$	0.0045	0.0702
mmrr	$2ae((ac^2 + eg^2)(ac + eg)^2 + (ag^2 + ec^2)(ag + ec)^2) + 2cg(a^2 + e^2)(ac + eg)(ag + ec)$	0.0857	0.0742
rmrr	$2ae((ac^2 + eg^2)(ag + ec)^2 + (ag^2 + ec^2)(ac + eg)^2) + 2cg(a^2 + e^2)(ac + eg)(ag + ec)$	0.0045	0.0118
rmrm	$2ae(ac + eg)(ag + ec)(c^2 + g^2) + 2cg(a^2(ag + ec)^2 + e^2(ac + eg)^2)$	0.0045	0.0084
mrrm	$a^2(ac + eg)^2(ag^2 + ec^2) + e^2(ag + ec)^2(ac^2 + eg^2) + 2aecg(ac + eg)(ag + ec)$	0.0023	0.0355
mrrr	$2(ac + eg)(ag + ec)(a^2(ag^2 + ec^2) + e^2(ac^2 + eg^2)) + 2aecg((ac + eg)^2 + (ag + ec)^2)$	0.0815	0.0115
rrrr	$a^2(ag + ec)^2(ag^2 + ec^2) + e^2(ac + eg)^2(ac^2 + eg^2) + 2aecg(ac + eg)(ag + ec)$	0.7738	0.0025

[a] a, b: Bernoullian parameter, probability of forming re (or si) configuration in either catalytic site (for C_s-symmetric catalyst, b = 1 − a). c: Probability of successive additions at the catalytic site related to parameter b. d: Analogous probability for the catalytic site related to parameter a. e = 1 − a, f = 1 − b, g = 1 − c, h = 1 − d. Reproduced by permission of Hüthig & Wepf Verlag, Zug, Switzerland, from M. Farina and A. Terragni, Makromol. Chem. Rapid Commun., **14**, 791 (1993).

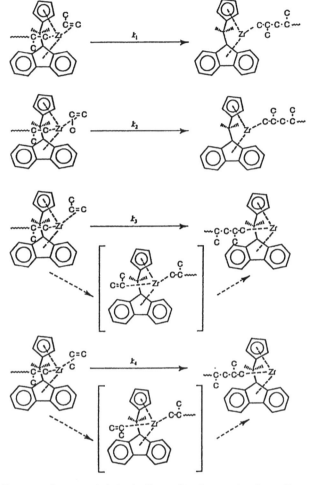

Scheme 3 Propagation model including site isomerization. Reproduced by permission of Hüthig & Wepf Verlag, Zug, Switzerland, from T. Asanuma *et al., Makromol. Chem. Rapid Commun.,* **14**, 315 (1993)

$$M^S - X + re \xrightarrow{k_1} M^R - S - X$$
$$M^S - X + si \xrightarrow{k_2} M^R - R - X$$
$$M^S - X + re \xrightarrow{k_3} M^S - R - X$$
$$M^S - X + si \xrightarrow{k_4} M^S - S - X$$
$$M^R - X + re \xrightarrow{k_3} M^S - S - X$$
$$M^R - X + si \xrightarrow{k_1} M^S - R - X$$
$$M^R - X + re \xrightarrow{k_4} M^R - R - X$$
$$M^R - X + si \xrightarrow{k_3} M^R - S - X$$

$$K_1 = \left(\frac{k_1 \cdot |re|}{k_1 \cdot |re| + k_2 \cdot |si| + k_3 \cdot |re| + k_4 \cdot |si|} \right)^S$$
$$= \left(\frac{k_1 \cdot |si|}{k_1 \cdot |si| + k_2 \cdot |re| + k_3 \cdot |si| + k_4 \cdot |re|} \right)^R$$
$$K_1 + K_2 + K_3 + K_4 = 1$$

4.1 CRYSTAL STRUCTURE

SPP has been known to be very slow in crystallization from the melt [11a, b]. Figure 2 compares the morphology of SPP with IPP, when crystallized at 50 °C from the melt (observed with a phase contrast microscope) [36d]. The morphology at low crystallization temperatures ($< 50°C$) shows no spherulite growth for SPP but accumulations of branched lamellar crystals, in contrast with IPP.

Natta *et al.* reported that the most stable structure for SPP is the $(TTGG)_2$ helix (T= *trans*, G = *gauche*) in C-centered lattices, where all the helical molecules have the same hand or are iso-chiral (cell I in Figure 3) [1]. Chatani and co-workers determined the planar zig-zag structure and a new structure assuming a (TTTTTGGTTGG) conformation, which was obtained from the all-*trans* specimen upon prolonged exposure to benzene vapor below 50 °C [33a, b].

Lovinger and co-workers elucidated that the polymorphism in SPP occurs due to the chain packing of iso- or anti-chiral helices and gave as the limiting structures cells II and III shown in Figure 3. They pointed out the existence of a pseudo *n*-pentane enchainment composed of two CH_3 groups and one CH_2 protruding in the

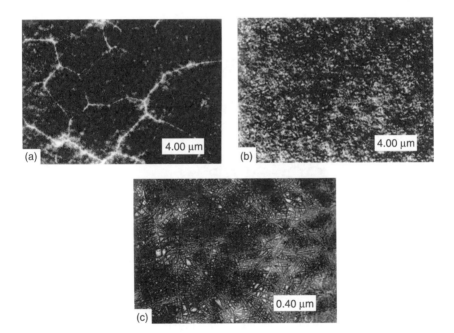

Figure 2 Morphology of IPP and SPP isothermally crystallized at 50 °C: (a) IPP at low magnification (spherulitic); (b) SPP at low magnification; (c) SPP at high magnification (lamellae arrangements). Reproduced from J. Loos and J. Petermann, *Polymer,* **37**, 4417 (1996), copyright 1996, with kind permission from Elsevier Science Ltd, The Boulevard, Langford Lane, Kidlington OX5 1GB, UK

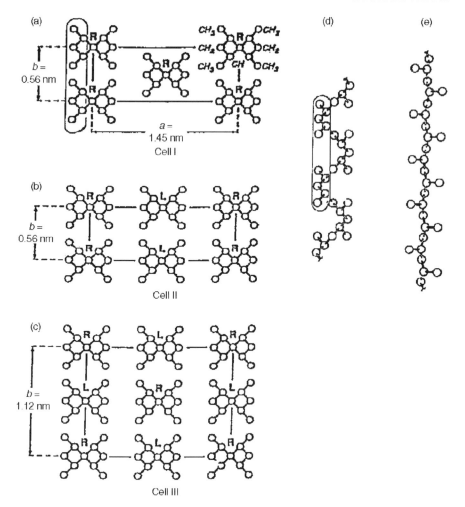

Figure 3 (a–d): Unit cells of the stable $(TTGG)_2$ form of SPP (a–c): shown in the *c*-axis projection and (d) view along the chain axis. Reproduced with permission from A. J. Lovinger *et al., Macromolecules,* **26**, 3494 (1993). Copyright 1993 American Chemical Society. (a) Cell I, fully isochiral C-centered packing; (b) Cell II, antichiral packing of chains along the *a*-axis; (c) Cell III, fully antichiral, body-centered cell. Right- and left-handed helical molecules are identified by R and L. respectively. (e) All- *trans* chain. (a–c) Reproduced with permission from A. J. Lovinger *et al., Macromolecules,* **26**, 3494 (1993). Copyright 1993 American Chemical Society. (d–e) Reproduced by permission of Hüthig & Wepf Verlag, Zug, Switzerland from P. Sozzani *et al., Makromol. Chem. Rapid Commun.,* **13**, 305 (1992)

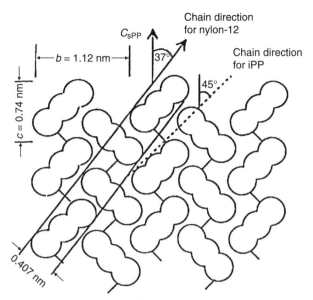

Figure 4 Epitaxial crystallization of SPP with nylon 12 or IPP. Reproduced from S. Yang, J. Petermann and D. Yang, *Polymer*, **37**, 2681 (1996), copyright 1996, with kind permission from Elsevier Science Ltd, The Boulevard, Langford Lane, Kidlington OX5 1GB, UK

b–c plane at 45 ° from the *c*-axis direction, which they confirmed with atomic force microscopy (AFM) [32f, j]. These rows of aligned *n*-pentane units afford sites for epitaxial growth when contacted with foreign substrates. IPP or PE quadrites grows in an epitaxial arrangement with flat-on SPP lamellae or, conversely, SPP crystallizes on PE orienting the protruded short pseudo *n*-pentane segments parallel to the PE chain (at 37 ° to the SPP *c*-axis) (Figure 4). Similar situations have been reported with 12-nylon and with oligophenyls [32c–j, 35a, c]. Taking full advantage of this phenomenon, we have worked out countermeasures to overcome the very sluggish rate of crystallization of SPP [11a, c].

Recent advances in solid-state NMR have spotlighted the unusaully rich T populations in both amorphous and crystalline regions of SPP [34]. The sensitivity of the methylene ^{13}C chemical shift to the main chain conformation is remarkable, since changing a *gauche* conformation into a *trans* conformation is accompanied by an increased distance of the carbon atoms three bonds apart. Thus, Sozzani *et al.* argued that the probability of occurrence of GG is 0.28 and the probability of GG following a TT arrangement is 0.37 in the unperturbed chain, that TTTT or TTTG sequences persist at high temperature well above the T_g [34d], that TTTT is abundant in both the amorphous and crystalline phases [34g], that SPP in the amorphous phase is better described as in Figure 5(d) and that the crystalline structure proposed by Chatani *et al.* may be accounted for as the defective portion in cell I [34e].

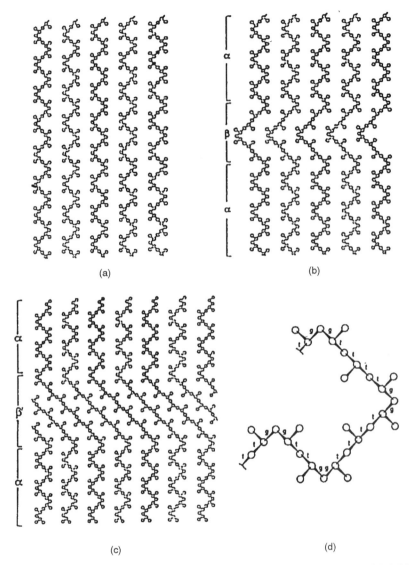

Figure 5 (a–c) Projections of C-centered structure. (a) Ordered case. (b). (c) Two examples of disordered cases: the defective portions β of the chains corresponds to a conformation $G_2T_6G_2T_6$ in (b) and G_2T_{18} in (c). (d) Chain segment of SPP describing the high content of *trans* conformation in the amorphous phase. consistent with the ^{13}C NMR chemical shifts. (a–c) Reproduced with permission from F. Auriemma *et al.*, *Macromolecules*, **28**, 6902 (1995). Copyright 1995 American Chemical Society. (d) Reproduced with permission from P. Sozzani, R. Simonutti and M. Galimbetti, *Macromolecules*, **26**, 5782 (1993). Copyright 1993 American Chemical Society

4.2 VISCOSITY (SOLUTION AND MELT), CHARACTERISTIC RATIO AND ENTANGLEMENT

The characteristic ratio. C_∞, is defined as

$$C_\infty = \langle r_0^2/nl^2 \rangle$$

where r_0^2, n and l are the mean-square end-to-end distance of an unperturbed coiled real chain, the number of bonds and the bond length, respectively. Inagaki *et al.* [26c] in 1966 reported the ratios corresponding to IPP:atactic PP:SPP = 1.00:1. 16:1.11 (in decalin at 135 °C), while the predicted values vary from 1.00:1.31:2.61 [26a, b] to 1.00:0.87:1.17 [26d] at 140 °C (Figure 6). Now that highly syndiotactic PP with a narrow molecular weight distribution ($M_w/M_n = 2$) and widely varying molecular weight has become available for the first time, we have tried to resolve this discrepancy [11b]. The intrinsic viscosity, $[\eta]$ (in tetralin at 135 °C), and molecular weight, M_w, measured with GPC (in 1,2,4-trichlorobenzene at 135 °C; expressed as the extended chain length calibrated with PS) for SPP and IPP are plotted in Figure 7. Clearly, separate groups of the correlation are discernible. If $[\eta]$ is presumed to be roughly proportional to the 3/2 power of C_∞, C $(SPP)_\infty/C(IPP)_\infty$ lies between 1.1 and 1.2.

The melt behaviors of SPP differ markedly from those of IPP, as can be seen in Figure 8 [11b].

The idea of the entanglement molecular weight (M_e) has been proposed and correlated with C_∞. Wu presented equations expressing both M_e and C_∞, where M_e is assumed to be proportional to the second power of C_∞, and C_∞, in turn, is

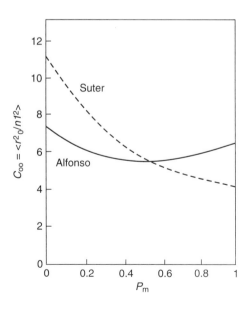

Figure 6 Characteristic ratio as the function of *meso* dyad. Reproduced with permission from U. W. Suter and P. J. Flory, *Macromolecules*, **8**, 765 (1975). Copyright 1975 American Chemical Society. Also reproduced from G. C. Alfonso, D. Yan and Z. Zhou, *Polymer*, **34**, 2830 (1993), with kind permission from Elsevier Science Ltd, The Boulevard, Langford Lane, Kidlington OX5 1GB, UK

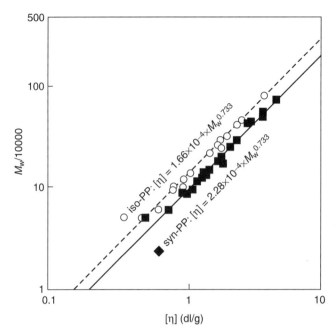

Figure 7 Tacticity (η)–molecular weight relationship. Reproduced by permission of Hüthig & Wepf Verlag, Zug, Switzerland from T. Shiomura *et al., Macromol. Symp.*, **101**, 289 (1996)

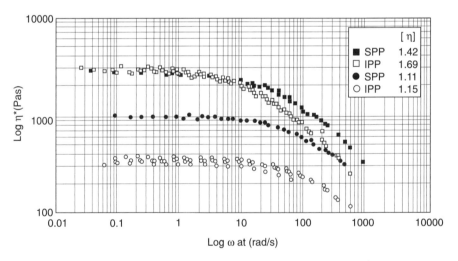

Figure 8 Melt viscosity (η^*)–shear rate (ω) relationship for SPP and IPP at 230 °C. Reproduced by permission of Hüthig & Wepf Verlag, Zug, Switzerland from T. Shiomura *et al., Macromol. Symp.*, **101**, 289 (1996)

Table 10 Entanglement molecular weight (M_e) for SPP and IPP

Sample	$[\eta]$ (dl/g)	T_m (°C)	M_w/M_n	mmmm	rrrr	Procedure 1		T_m (°C)	Procedure 2			Ref.
						G_N^0 (dyn/cm^2)	M_e		G_c (dyn/cm^2)	G_N^0 (dyn/cm^2)	M_e	
IPP	1.69	160	2.2	0.96	—	2.45×10^6	11800	230	4.00×10^5	2.96×10^6	9300	11b
IPP	1.96	158	3.0	0.96	—	2.48×10^6	11600	230	3.60×10^5	2.48×10^6	8800	11b
SPP	1.23	130	2.3	—	0.81	3.28×10^6	8700	230	7.20×10^5	5.52×10^6	5500	11b
SPP	1.68	154	2.3	—	0.94	3.28×10^6	8700	230	7.75×10^5	6.51×10^6	4400	11b
SPP	1.42	148	2.2	—	0.93	3.33×10^6	8700	230	8.00×10^5	5.92×10^6	4800	11b
SPP	1.38	130	2.6	—	0.78	3.35×10^6	8600	230	7.10×10^5	5.40×10^6	5300	11b
IPP												
PP123	0.85^a	150	1.8	0.98	—			170	5.5×10^{5a}	3.4×10^{6a}	8200^a	31
JP082	1.55	155	1.7	0.97	—			170	5.5×10^5	3.2×10^6	8800	31
JP120	2.05	156	1.9	0.96	—			170	5.0×10^5	3.3×10^6	8600	31
JP098	2.35	165	2.3	0.97	—			170	5.0×10^5	3.8×10^6	7400	31
SPP												
PP118	0.91	139	1.9	—	0.95			170	1.1×10^6	7.2×10^6	3700	31
Pp117	1.10	147	1.7	—	0.97			170	1.0×10^6	5.8×10^6	4600	31
SPP1	1.30	154	2.2	—	0.80			170	1.0×10^6	7.4×10^6	3700	31
SPH40	1.50	145	1.9	—	0.92			170	1.1×10^6	6.7×10^6	4100	31

^{a}Figures estimated by the present authors.

proportional to the fraction of *meso* diad [27]; thus, wheat obtained $C(SPP)_{\infty}/C(IPP)_{\infty} = 9.0/5.2$, M_e 1730 for SPP and 5100 for IPP [40a]. Taking into account of the lattice constants for the stable crystal structure. On the other hand, Wool [28] derived the equation

$$M_e = 30.89 C_{\infty} M_0 j (bz/C)^2$$

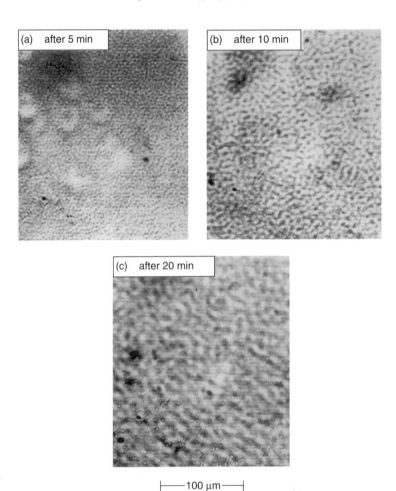

|————100 μm————|

Figure 9 Progression of spinodal decomposition for SPP/IPP(MFL) = 50/50 at 240°C. SPP: rrrr = 0.80, MFR = 10; IPP: MFL-grade (random copolymer contg. 4.9 wt-% ethylene), MFR = 8. Reproduced from T. Shiomura *et al.*, in K. Soga and M. Terano (Eds), *Catalyst Design for Tailor-made Polyolefins*, 1995, with kind permission of Elsevier Science-NL, Sara Burgerhartstraat 25, 1055 KV Amsterdam, The Netherlands

where M_0, C, b, z and j are monomer molecular weight ($= 42$). c-axis dimension ($= 7.40\,\text{Å}$ for TTGG, $5.06\,\text{Å}$ for TT and $6.50\,\text{Å}$ for TG), bond length ($= 1.54\,\text{Å}$), number of monomers per c-axis ($= 4$ for TTGG, 2 for TT and 3 for TG) and number of backbone bonds per monomer ($= 2$), respectively. This leads to $M_e = 10400$ for TTGG, 6900 for TT (for SPP) and 7600 for TG (for IPP), assuming an identical $C_\infty = 5.8$ value for all the conformations.

We have tried to estimate M_e with thermo-rheological measurements [11b]. According to a first procedure, the plateau modulus $G_N{}^0$, dynamic storage modulus G', dynamic loss modulus G'', angular frequency ω and density of melt polymer ρ at temperature T are related by [29].

$$M_e = \rho R T / G_N^0$$

$$G_N^0 = (4/\pi) \int_{-\infty}^{\omega_{max}} G''(\omega) \mathrm{d}\ln\omega$$

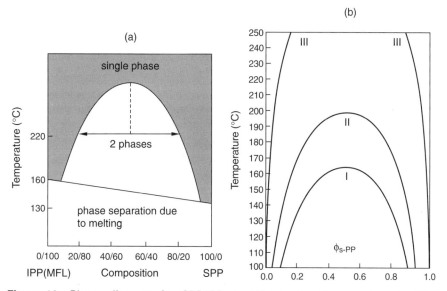

Figure 10 Phase diagram for SPP/IPP. (a) Obtained with experiments. SPP: rrrr $= 0.80$, MFR $= 10$. IPP: MFL-grade (random copolymer containing 4.9 wt % (ethylene). MFR $= 8$. (b) Computed by Kressler and co-workers [37d] for different molecular weights, (I) 25 000, (II) 30 000 and (III) 50 000. Below the respective binodals, the systems consist of two phases; above the binodals, there is a one phase-region. (a) Reproduced by permission of the Society of Polymer Science, Japan, from M. Akiyama, T. Shiomura and I. Fujio, *Polym. Prepr. Jpn*, **43**, 1232 (1994). (b) Reproduced from R. D. Maier *et al., J. Polym. Sci. Polym. Phys.*, **35**, 1135 (1997), copyright © 1997 John Wiley & Sons, Inc., by permission of John Wiley & Sons, Inc.

Our results are shown in the top part of Table 10. A second procedure calls for the crossover value G_c in G', G'' plotted against ω, correlating G_c with $G_N{}^0$ by [30].

$$\log(G_N^0/G_c) = 0.380 + 2.63 \log U/(1 + 2.45 \log U)$$

where $U = M_w/M_n$. Since Friedlich and co-workers have recently published G' and G'' data for SPP and IPP [31], calculations were made based on their data and the results are given in Table 10 together with our own results [10b]. It is clear that SPP and IPP afford different M_e values, although the two procedures above yield different numerical values. Anyway, SPP is more densely entangled than IPP in the melt.

4.3   Compatibility of SPP with other polymers

One of the most surprising aspects is the incompatibility of SPP with IPP in the melt [11a, c, 38a, b]. The phenomenon known as spinodal decomposition was investigated by (i) direct observation under a phase contrast microscope (Figure 9) and (ii) DSC measurements of the crystallization thermogram (in a particular composition range, only one T_c develops and does not change after prolonged heating and outside this composition range two T_c's appear and tend to move away from each other after prolonged heating). Thus we determined the phase diagram between the two polymers [Figure 10(a)]. Kressler and co-workers conducted thermodynamic calculations utilizing solubility parameters derived from P–V–T experiments and showed

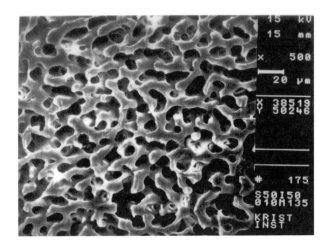

Figure 11 Scanning electron micrograph showing the interpenetrating network: SPP: IPP = 50 : 50 blend annealed for 10 min at 180 °C, then crystallized for 10 min at 135 °C, quenched in ice–water and etched in toluene at 70 °C (toluene removes the SPP phase selectively). Reproduced from R. Thomann *et al., Polymer,* **37,** 2627 (1996), with kind permission from Elsevier Science Ltd, The Boulevard, Langford Lane, Kidlington OX5 1GB, UK

that both polymers are immiscible when their molecular weights both exceed 50 000 f resp [37b] [Figure 10(b)], and attributed the incompatibility to the different conformations of the respective polymers in the melt (helical for IPP, prevalently *trans* for SPP) [37b, d].

The formation of an interpenetrating network with SPP:IPP = 50:50 at 180 °C has been demonstrated most convincingly by Kressler and co-workers [37c] (Figure 11) In this regard, SPP, in contrast with IPP, affords highly ductile (although opaque) blends with many olefin polymers, such as HDPE, LDPE, LLDPE, EPR, EVA and SBR [11b, c] (Figure 12), which may open the way to the broad utitilization of SPP. Many fields of practical applications of SPP have been referred to in the open literature from industrial sources [11, 12, 39–42].

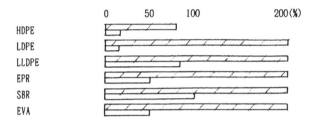

Figure 12 Elongation at break of blended systems (upper: SPP; lower: IPP) (composition ratio: PP/other polymer = 70/30)

5 REFERENCES

1. Natta, G., Pasquon, I., Corradini, P., Peraldo M. and Zambelli, A., *Rend. Acc. Naz. Lincei*, **28**, 539 (1960).
2. Natta, G., Pasquon I. and Zambelli, R. *J. Am. Chem. Soc.*, **84**, 1488 (1962).
3. (a) Zambelli, A., Locatelli, P., Zannoni G. and Bovey, F. A., *Macromolecules*, **11**, 923 (1978); (b) Zambelli A. and Allegra, G., *Macromolecules*, **13**, 42 (1980); (c) Zambelli, A., Locatelli, P., Provasoli A. and Ferro, D. R., *Macromolecules*, **13**, 267 (1980).
4. Ewen, J. A., *J. Am. Chem. Soc.*, **106**, 6355 (1984).
5. (a) Ewen, J. A., Jones, R. L., Razavi A. and Ferrara, J. D., *J. Am. Chem. Soc.*, **110**, 6255 (1988); (b) Ewen, J. A., Elder, M. J., Harlan, C. J., Jones R. L. and Atwood, J. L., *ACS Polym. Prepr.*, **32**, 469 (1991); (c) Ewen, J. A., Elder, M. J., Jones, R. L., Curtis S. and Cheng, H. N., in Keii T. and Soga, K., (Eds), *Catalytic Olefin Polymerization*, Kodansha, Tokyo. 1990, p. 439; (d) Ewen, J. A., Elder, M. J., Jones, R. L., Haspeslagh, L., Atwood, J. L., Bott S. G. and Robinson, K., *Makromol. Chem. Macromol. Symp.*, **48/49**, 253 (1991); (e) Ewen J. A., and Elder, M. J., *Makromol. Chem. Macromol. Symp.* **66**, 179 (1993); (f) Ewen, J. A., in Soga K. and Terano, M., (Eds), *Catalyst Design for Tailor-made Polyolefins*, Kodansha, Tokyo. 1994, p. 405; (g) Ewen, J. A., *Macromol. Symp.*, **89**, 181 (1995); (h) Ewen J. A. and Elder, M. J., in Fink, G., Muelhaupt R. and Brintzinger, H. H., (Eds), *Ziegler Catalysts*, Springer, Berlin. 1995, p. 99.

6. (a) Cavallo, L., Guerra, G., Vacatello M. and Corradini, P., *Macromolecules*, **24**, 1784 (1991); (b) Guerra, G., Longo, P., Cavallo, L., Corradini P. and Resconi, L., *J. Am. Chem. Soc.*, **119**, 4394 (1997).

7. (a) Kawamura-Kuribayashi, H., Koga N. and Morokuma, K., *J. Am. Chem. Soc.*, **114**, 8687 (1992); (b) Koga, N., Yoshida T. and Morokuma, K., presented at the Symposium on 40 Years of Ziegler Catalysts, Freiburg 3 September 1993; (c) Kakugo, M., *Macromol. Symp.*, **89**, 545 (1995); (d) Yoshida, T., Koga N. and Morokuma, K., *Organometallics*, **15**, 766 (1996).

8. Prosenc, M. H., Janiak C. and Brintzinger, H. H., *Organometalllics*, **11**, 4036 (1992).

9. Castonguay L. A. and Rappe, A. K., *J. Am. Chem. Soc.*, **114**, 5832 (1992).

10. (a) Razavi A. and Ferrara, J., *J. Organomet. Chem.*, **435**, 299 (1992); (b) Razavi A. and Atwood, J. L., *J. Organomet. Chem.*, **459**, 117 (1993); (c) Razavi, A., Vereecke, D., Peters, L., Hessche, D. V., DenDaw, K., Nafpliotis L. and deFroimont, Y., in *SPO '93*, Schotland Business Research, Skillman, NJ, 1993, p. 105; (d) Razavi A. and Atwood, J. L., *J. Organomet. Chem.*, **497**, 105 (1995); (e) Razavi A. and Atwood, J. L., *J. Organomet. Chem.*, **520**, 115 (1996); (f) Razavi, A., Vereecke, D., Peters, L., DenDaw, K., Nafpliotis L. and Atwood, J. L., in Fink, G., Muelhaupt R. and Brintzinger, H. H., (Eds), *Ziegler Catalysts*, Springer, Berlin. 1995, p. 111; (g) Razavi, A., Peters, L., Nafpliotis, L., Vereecke, D., DenDauw, D., Atwood J. L. and Thewald, U., *Macromol. Symp.*, **89**, 345 (1995); (h) Razavi, A., in *Metallocenes '95 (Brussels)*, Schotland Business Research, Skillman, NJ, 1995, p. 477; (j) Razavi, A., Peters L. and Nafpliotis, L., *J. Mol. Catal. A: Chem.*, **115**, 129 (1997).

11. (a) Shiomura, T., Kohno, M., Inoue, N., Yokote, Y., Akiyama, M., Asanuma, T., Sugimoto, R., Kimura S. and Abe, M., in Soga, and Terano, M., (Eds), *Catalyst Design for Tailor-made Polyolefins*, Elsevier, Amsterdam, 1994, p. 327; (b) Shiomura, T., Kohno, M., Inoue, N., Asanuma, T., Sugimoto, R., Iwatani, T., Uchida, O., Kimura, S., Harima, S., Zenkoh H. and Tanaka, E., *Macromol. Symp.*, **101**, 289 (1996); (c) Shiomura, T., Kohno, M., Akiyama, M., Asanuma, T., Sugimoto, R., Zenkoh, H., Kimura, S., Uchikawa, N., Inoue, N., Shamshoum E. S. and Wheat, W. R., in *SPO '96*, Schotland Business Research, Skillman, NJ 1996, p. 53.

12. (a) Shamshoum, E. S., in *SPO '92*, Schotland Business Research, Skillman, NJ, 1992, p. 197; (b) Shamshoum E. S. and Rauscher, D., in *MetCon '93*, Catalyst Group, Houston, 1993, p. 173; (c) Shamshoum, E. S., Kim, S., Paiz, R., Goins M. and Bartol, D., in *SPO '93*, Schotland Business Research, Skillman, NJ, 1993, p. 205; (d) Shamshoum, E. S., Sun, L., Reddy B. R. and Turner, D., in *MetCon '94*, Catalyst Group, Houston, 1994; (e) Shamshoum, E. S., Sun, L., Schardl, J., DeKunder G. and Boyle, K., in *SPO '94*, Schotland Business Research, Skillman NJ, 1994; (f) Shamshoum, E. S. and Rauscher, D., in *Polyolefins IX*, SPE, Texas, 1995, p. 143; (g) Shamshoum, E. S., Sun L. and Kim, S., in *SPO '95*, Schotland Business Research, Skillman, NJ, 1995, p. 299; (h) Shamshoum, E. S., Kim, S., Hanyu A. and Reddy, B. R., in *Metallocenes '96 (Düsseldorf)*, Schotland Business Research, Skillman, NJ. 1996, p. 259.

13. (a) Antberg, M., Dolle, V., Klein, R., Rohrmann, J., Spaleck W. and Winter, A., *Catalytic Olefin Polymerization*, 1990, p. 501; (b) Spaleck, W., Aulbach, M., Bachmann, B., Kueber F. and Winter, A., *Macromol. Symp.*, **89**, 237 (1995); (c) Spaleck, W., Antberg, M., Aulbach, M., Bachmann, B., Dolle, V., Haftka, S., Kueber, F., Rohrmann J. and Winter, A., in Fink, G., Muelhaupt R. and Brintzinger, H. H., (Eds), *Ziegler Catalysts*, Springer, Berlin, 1995, p. 83.

14. (a) Hawley, G. R., Hill, T. G., Chu, P. P., Geerts, R. L., Palackal S. J. and Alt, H. G., in *SPO '93*, Schotland Business Research, Skillman, NJ, 1993, p. 91; (b) Alt, H. G., Milius W. and Palackal, S. J., *J. Organomet. Chem.*, **472**, 113 (1994); (c) Schmid, M. A., Alt H. G. and Milius, W., *J. Organomet. Chem.*, **501**, 101 (1995); (d) Alt H. G. and Zenk, R., *J.*

Organomet. Chem., **512**, 51 (1996); (e) Alt, H. G., Zenk R. and Milius, W., *J. Organomet. Chem.*, **514**, 257 (1996); (f) Alt H. G. and Zenk, R., *J. Organomet. Chem.*, **518**, 7 (1996); (g) Alt, H. G., and Zenk, R., *J. Organomet. Chem.*, **522**, 39 (1996); (h) Alt H. G. and Zenk, R., *J. Organomet. Chem.*, **522**, 177 (1996); (i) Schmid, M. A., Alt H. G. and Milius, W., *J. Organomet. Chem.*, **525**, 9 (1996); (j) Scmidt, M. A., Alt H. G. and Milius, W., *J. Organomet. Chem.*, **525**, 15 (1996); (k) Alt H. G. and Zenk, R., *J. Organomet. Chem.*, **526**, 295 (1996).

15. (a) Fierro, R., Yu, Z., Rausch, M. D., Dong, S., Alvares D. and Chien, J. C. W., *J. Polym. Sci. Part A. Polym. Chem.*, **32**, 661 (1994); (b) Yu, Z., Chien, J. C. W., *J. Polym. Sci. Part A, Polym. Chem.*, **33**, 1085 (1995); (c) Song, W., Rausch D. and Chien, J. C. W., *J. Polym. Sci. Part A, Polym. Chem.*, **34**, 2945 (1996); (d) Chen, Y. X., Rausch M. D. and Chien, J. C. W., *J. Organomet. Chem.*, **497**, 1 (1995).

16. Rieger, B., Repo T. and Jany, G., *Polym. Bull.*, **35**, 87 (1995).

17. Kageyama, A., Mise, T., Miya, S., Yamazaki H. and Aoki, J., *Prepr. Chem. Soc. Jpn. (63rd)*, *3C*, 123 (1992).

18. (a) Herfert N., and Fink, G., *Makromol. Chem.*, **193**, 773 (1992); (b) Herfert N. and Fink, G., *Makromol. Chem. Macromol. Symp.*, **66**, 157 (1993); (c) Montag, P., Leek, Y. u. d., Angemund K. and Fink, G., *J. Organomet. Chem.*, **497**, 201 (1995).

19. Herzog, T. A., Zubris D. L. and Bercaw, J. E., *J. Am. Chem. Soc.*, **118**, 11988 (1996).

20. (a) Buscico V. and Cipullo, R., *J. Am. Chem. Soc.*, **116**, 9329 (1994); (b) Buscico V. and Cipullo, R., *Macromol. Symp.*, **89**, 277 (1995); (c) Buscico V. and Cipullo, R., *J. Organomet. Chem.*, **497**, 113 (1995); (d) Borriello, A., Buscico, V., Cipullo, R., Chadwick J. C. and Sudmeijer, O., *Macromolecular Rapid Commun.*, **17**, 589 (1996); (e) Buscico, V. Brita, D., Caporaso, L., Cipullo R. and Vacatello, M., *Macromolecules*, **30**, 3971 3971 (1997).

21. (a) Leclerc M. K. and Brintzinger, H. H., *J. Am. Chem. Soc.*, **117**, 1651 (1995); (b) Brintzinger, H. H., Fischer, D., Muelhaupt, R., Rieger B. and Waymouth, R., *Angew. Chem., Int. Ed. Engl.*, **34**, 1143 (1995); (c) Leclerc M. K. and Brintzinger, H. H., *J. Am. Chem. Soc.*, **118**, 9024 (1996).

22. (a) Yang, X., Stern C. L. and Marks, T. J., *J. Am. Chem. Soc.*, **113**, 3623 (1991); (a) 'Yang, X., Stern C. L. and Marks, T. J., *Angew. Chem., Int. Ed. Engl.*, **31**, 1375 (1992); (b) Sishta, C. Hathorn R. H. and Marks, T. J., *J. Am. Chem. Soc.*, **114**, 1112 (1992); (c) Deck P. J. and Marks, T. J., *J. Am. Chem. Soc.*, **117**, 6128 (1995); (d) Jia, L., Yang, X., Ishihara A. and Marks, T. J., *Organometallics*, **14**, 3135 (1995); (e) Jia, L., Yang, X., Stern C. L. and Marks, T. J., *Organometallics*, **16**, 842 (1997); (f) Obora, Y., Stern C. L. and Marks, T. J., *Organometallics*, **16**, 2503 (1997); (g) Chen Y. X. and Marks, T. J., *Organometallics*, **16**, 3649 (1997); (h) Chen, Y. X., Stern C. L. and Marks, T. J., *J. Am. Chem. Soc.*, **119**, 2582 (1997).

23. (a) Shiomura, T., Asanuma T. and Inoue, N., *Macromolecular Rapid Commun.*, **17**, 9 (1996); (b) Shiomura, T., Asanuma A. and Sunaga, T., *Macromolecular Rapid Commun.*, **18**, 169 (1997); (c) Canich, J. A. M., *US Pat.*, 5 055 438; *Chem. Abstr.*, **118**, 60283 (1993); (d) Turner, H. W., Hlatky G. G. and Canich, J. A. M., *PCT Int. Appl.*, WO 93/19, 103; *Chem. Abstr.*, **20**, 271442 (1994); (e) Stevens, J. C., in *MetCon '93*, Catalyst Group, Houston, 1993, p. 157; (f) McKnight, A. L., Masood Md. A. and Waymouth, R. M., *Organometallics*, **16** 2879 (1997).

24. (a) Farina M. and Terragni, A., *Makromol. Chem. Rapid Commun.*, **14**, 791 (1993); (b) Farina, M., DiSilvestro G. and Terragni, A., *Macromol. Chem. Phys.*, **196**, 353 (1995); (c) DiSilvestro, G., Sozzani P. and Terragni, A., *Macromol. Chem. Phys.*, **197**, 3209 (1996); (d) Asanuma, T., Nishimori, Y., Ito M. and Shiomura, T., *Makromol. Chem. Rapid*

Commun., **14**, 315 (1993); (e) Asanuma, T., Nishimori, Y., Ito M. and Shiomura, T., *Makromol. Chem. Rapid Commun.*, **15**, 723 (1994).

25. Resconi, L., Fait, A., Piemontesi, F., Colonnesi, M., Rychlicki H. and Ziegler, R., *Macromolecules*, **28**, 6667 (1995).

26. (a) Flory, P. J., *Statistical Mechanics of Chain Molecules*, Interscience, New York. 1969, p. 236; (b) Suter U. W. and Flory, P. J., *Macromolecules*, **8**, 765 (1975); (c) Inagaki, H., Miyamoto T. and Ohta, S., *J. Phys. Chem.*, **70**, 3420 (1966); (d) Alfonso, G. C., Yan D. and Zhou, Z., *Polymer*, **34**, 2830 (1993); (e) Matsumura, S., Nakaoki T. and Hayashi, H., *Polym. Prepr. Jpn.*, **44**, 838 (1995).

27. Wu, S., *Polym. Eng. Sci.*, **32**, 823 (1992).

28. Wool, R. P., *Macromolecules*, **26**, 1564 (1993).

29. Ferry, J. D., *Viscoelastic Properties of Polymers*, Wiley, New York, 3rd edn, 1980, Chapt. 13.

30. Wu, S., *J. Polym. Sci., Part B*, **27**, 723 (1989).

31. Eckstein, A., Friedrich, C., Lobbrecht, A., Spitz R. and Muelhaupt, R., *Acta Polym.*, **48**, 41 (1997).

32. (a) Lotz, B. Lovinger A. J. and Cais, R. E., *Macromolecules*, **21**, 2375 (1988); (b) Lovinger, A. J., Lotz B. and Davis, D. D., *Polymer*, **31**, 2253 (1990); (c) Lovinger, A. J., Davis D. D. and Lotz, B., *Macromolecules*, **24**, 552 (1991); (d) Lovinger, A. J., Lotz, B., Davis D. D. and Padden, F. J., Jr. *Macromolecules*, **26**, 3494 (1993); (e) Lovinger, A. J., Lotz, B., Davis D. D. and Schumacher, M., *Macromolecules*, **27**, 6603 (1994); (f) Stocker, W., Schumacher, Graff, S., Lang, J., Wittmann, J. C., Lovinger A. J. and Lotz, B., *Macromolecules*, **27**, 6948 (1994); (g) Schumacher, M., Lovinger, A. J. Agarwal, P., Wittmann J. C. and Lotz, B., *Macromolecules*, **27**, 6954 (1994); (h) Kopp, S., Wittman J. C. and Lotz, B., *Macromol. Symp.*, **98**, 917 (1995); (i) Lotz B. and Wittmann, J. C., *Macromol. Symp.*, **101**, 91 (1996); (j) Lotz, B., Wittmann J. C. and Lovinger, A. J., *Polymer*, **37**, 4979 (1996).

33. (a) Chatani, Y., Maruyama, H., Noguchi, K., Asanuma T. and Shiomura, T., *J. Polym. Sci. Polym. Lett.*, **28**, 393 (1990); (b) Chatani, Y., Maruyama, H., Asanuma T. and Shiomura, T., *J. Polym. Sci. Polym. Phys*, **29**, 1649 (1991).

34. (a) Sozzani, P., Galimberti M. and Balbontin, G., *Makromol. Chem. Rapid Commun.*, **13**, 305 (1992); (b) DeRosa C. and Corradini, P., *Macromolecules*, **26**, 5711 (1993); (c) Auriemma, F., DeRosa C. and Corradini, P., *Macromolecules*, **26**, 5719 (1993); (d) Sozzani, P., Simonutti R. and Galimberti, M., *Macromolecules*, **26**, 5782 (1993); (e) Auriemma, F., Born, R., Spiess, H. W., DeRosa C. and Corradini, P., *Macromolecules*, **28**, 6902 (1995); (f) Auriemma, F., Lewis, R. H., Spiess H. W. and DeRosa, C., *Macromol. Chem. Phys.*, **196**, 4011 (1995); (g) Sozzani, P., Simonutti R. and Comotti, A., *Macromol. Symp.*, **89**, 513 (1995); (h) DeRosa, C., Auriemma F. and Vinti, V., *Macromolecules*, **30**, 4137 (1997).

35. (a) Rodriguez-Arnold, J., Zhang, A., Cheng, S. Z. D., Lovinger, A. L., Hsieh, E. T., Chu, P., Johnson, T. W., Honnell, K. G., Geerts, R. E., Palackal, S. J., Hawley G. R. and Welch, M. B., *Polymer*, **35**, 1884 (1994); (b) Rodriguez-Arnold, J., Bu, Z., Cheng, S. Z. D., Hsieh, E. T., Johnson, T. W., Geerts, R. G., Palackal, S. J., Hawley G. R. and Welch, M. B., *Polymer*, **35**, 5194 (1994); (c) Rodriguez-Arnold J. and Cheng, S. Z. D., *J. Macromol. Sci. Rev. Macromol. Chem. Phys. C*, **35**, 117 (1995); (d) Bu, Z., Yoon, Y., Ho, R. M., Zhou, W., I.Jangchud, Eby, R. K., Cheng, S. Z. D., Hsieh, E. T., Johnson, T. W., Geerts, R. G., Palackal, S. J., Hawley, C. W. and Welch, M. B., *Macromolecules*, **29**, 6575 (1996).

36. (a) Petermann, J., Xu, Y., Loos, J. and Yang, D., *Polymer*, **33**, 1095 (1992); (b) Loos, J., Buhk, M., Petermann, J., Zoumis, K. and Kaminsky, W., *Polymer*, **37**, 387 (1996); (c) Yang, S., Petermann, J. and Yang, D., *Polymer*, **37**, 2681 (1996); (d) Loos, J. and

Petermann, J., *Polymer*, **37**, 4417 (1996); (e) Loos, J., Schauwienold, A. M., Yan, S., Petermann, J. and Kaminsky, W., *Polym. Bull.*, **38**, 185 (1997).

37. (a) Thomann, R., Wang, C., Kressler, J., Juengling, S. and Muelhaupt, R., *Polymer*, **36**, 3795 (1995); (b) Thomann, R., Kressler, J., Setz, S., Wang, C. and Muelhaupt, R., *Polymer*, **37**, 2627 (1996); (c) Thomann, R., Kressler, J., Rudolf, B. and Muelhaupt, R., *Polymer*, **37**, 2635 (1996); (d) Maier, R. D., Thomann, R., Kressler, J., Muelhaupt, R. and Rudolf, B., *J. Polym. Sci. Polym. Phys.*, **35**, 1135 (1997).

38. (a) Akiyama, M., Shiomura, T. and Fujio, I., *Polym. Prepr. Jpn.*, **43**, 1232 (1994); (b) Shiomura, T. and Sugimoto, R., *Seikeikako '94 Jpn. Prepr.*, 202 (1994).

39. (a) Sugimoto, R., Uchikawa, N. and Yoshino, K., *J. Soc. Elec. Mater. Eng.*, **3**, 3 (1994); (b) Yoshino, K., Yin, X. H., Tada, K., Kawai, T., Hamaguchi, M., Araki, H., Sugimoto, R., Uchikawa, N., Asanuma, T., Kawahigashi, M. and Kato, H., *IEEE Trans. Diel. Elec. Insul.*, **3**, 331 (1996); (c) Fujio, I. and Sugimoto, R., *Polym. Prepr. Jpn.*, **46**, 854 (1997).

40. (a) Wheat, W. R., in *ANTEC SPE. 53rd*, 1995, Vol. 2, p. 2275; (b) Wheat, W. R., in *ANTEC SPE*, 5th, 1997, Vol. 2, p 1968; (c) Schardl, J., Sun, L., Sugimoto, R. and Kimura, M., in *ANTEC SPE (May)*, 1995; (d) Buras, P., Sun, L. and Baumgartner, A., in *RETEC SPE*, 1995; (e) Donohue, J., in *SPO '95*, Schotland Business Research, Skillman, NJ, 1995.

41. (a) Antberg, M., Dolle, V., Haftka, S., Rohrmann, J., Spaleck, W., Winter, A. and Zimmermann, H. J., *Makromol. Chem., Macromol. Symp.*, **48/49.**, 333 (1991); (b) Haftka, S. and Koennecke, K., *J. Macromol. Sci. Phys. B*, **30**, 319 (1991).

42. Hawley, G. R., Hill, T. G., Chiu, P. P., Geerts, R. L., Palackal, S. J. and Alt, H. G., in *SPO '93*, Schotland Business Research, Skillman, N. J., 1993, p. 91.

43. (a) Kaminsky, W., *Macromol. Symp.*, **89** 203 (1995); (b) Mitsui Toatsu Chemicals, *Jpn. Kokai*, 7–165812 (Inoue, N., Kohno, M. and Shiomura, T.), *Chem. Abstr.*, **115**, 184145 (1995).

21

Synthesis of Atactic Polypropylene Using Metallocene Catalysts

LUIGI RESCONI
Montell Italia, G. Natta Research Center, Ferrara, Italy

1   INTRODUCTION

The steady appearance of patents and publications on the use and properties of amorphous polypropylene testifies to the practical interest in this material, which finds applications as a component in hot melt adhesives, paper coatings and as a bitumen additive [1] Amorphous polypropylene [2] can be obtained from hetero-geneous, Ti-based catalysts [3], but this product has moderate molecular weight and a broad molecular weight distribution and contains some residual crystallinity [1, 4].

The recent development of fully amorphous (atactic) polypropylene of higher molecular weight should lead to more applications, notably in blends with crystalline polypropylene for PVC replacement [5].

Atactic polypropylene, lacking any long-range stereochemical order in the polymer chain [2a–c, 6], is fully amorphous, although other types of low-order microstructures (e.g. hemi-isotactic) can lead to a fully amorphous polypropylene. Original studies of purely atactic polypropylene (a-PP) were performed on mono-disperse samples obtained by hydrogenation of poly(1,3-dimethyl-1-butenylene) [2f,g]

Homogeneous, single-center catalyst systems are more suitable for the production of a-PP, as they can be designed to lack any monomer enantioface selectivity and produce homopolymers over the whole range of molecular weights. Depending on

Metallocene-based Polyolefins, Edited by J. Scheirs and W. Kaminsky
© 2000 John Wiley & Sons Ltd

its molecular weight, a-PP has a physical appearance ranging from an oil to a soft, transparent rubber.

A wide variety of soluble catalysts to produce atactic PP have been reported. Apart from metallocenes, examples include MAO-activated $VOCl_3$ [7], Ti and Zr alkoxides and alkoxychlorides [8], monocyclopentadienyl complexes [9, 10], Cp-Ti-amides [11] and Cp-Ti-amines [12] and the Ni- and Pd-diimine complexes described by Brookhart and co-workers [13]. Most of these catalysts produce a-PP with low regioregularity: for example, most 'constrained geometry' catalysts produce a-PP with 2–15 % of 2,1 units in the chain [11]. The presence in the polymer of 3,1 units formed by the isomerization of secondary units, as is the case with the Ni-α-diimine catalysts, makes a-PP akin to an ethylene–propylene rubber, with a T_g as low as $-55\,°C$ [13].

At the moment, metallocenes are by far the most active and most versatile systems for the production of amorphous polypropylene. While a-PP from metallocenes is almost invariably highly regioregular, in terms of tacticity, three types of propylene polymers can be made with metallocene catalysts which are fully, or mostly, amorphous. These are atactic, hemi-isotactic and isotactic–atactic stereoblock polypropylenes, with a continuum of tacticity distributions covering the whole range between these limiting structures.

Isotactic–atactic stereoblock polypropylenes range in properties from fully amorphous to thermoplastic–elastomeric, depending on the length and length distribution of the isotactic sequences. These materials have been produced with heterogeneous, Zr-based catalysts [14] and metallocene catalysts [15]. Waymouth and co-workers' catalysts are especially interesting as they produce isotactic–atactic stereoblock polypropylenes in which the length and length distribution of the isotactic sequences can be varied by ligand design, so that fully amorphous polymers can also be produced [15d–f].

Hemi-isotactic PP [6, 16] is another type of amorphous polymer that can be produced with metallocene catalysts [17]. This polymer has been important in confirming the polymerization mechanism of metallocene catalysts [17a, 18]. The presence on the same metal center of one isospecific and one aspecific site, together with the requirement of chain migratory insertion with site switching at (almost) every insertion, generate a unique polymer structure, in which every other methyne is in the same (isotactic) configuration, while the remaining alternating methynes are in a random configuration. This results in a well defined pentad distribution (ideally 3:2:1:4:0:0:3:2:1) [16]. The best catalysts are $Me_2C(3\text{-Me-Cp})(9\text{-Flu})_2ZrCl_2$ and its Hf analog [17b]. The highest molecular weights ($\bar{M}_w\,200\,000 - 300\,000$) are obtained with the hafnium derivative [17b], while the zirconium complex gives low molecular weights ($\bar{M}_w \approx 50\,000$). Both catalysts have a fairly low activity compared with other metallocene catalysts. No physical properties of this polymer have been reported, but it is expected to be similar to atactic polypropylene, with possibly better elastomeric properties provided that its molecular weight is high enough.

Other metallocenes with the proper ligand symmetry for hemi-isospecific polymerization have been reported [19], but they produced low molecular weight materials.

For the purpose of this chapter, a polypropylene is defined as 'atactic' if, in addition to being fully amorphous, the propylene units in the polymer chain are arranged in a random sequence of r and m dyads obeying the relationship $B = 4[\text{mm}][\text{rr}]/[\text{mr}]^2 \approx 1$. In this chapter, the synthesis of atactic polypropylene with metallocene catalysts is reviewed, with emphasis on molecular weight control.

Atactic polypropylene is produced by two types of metallocenes: achiral, C_{2v}-symmetric unbridged (e.g. Cp_2ZrCl_2) and by extension any alkylcyclopentadienyl or indenyl metallocene lacking stereorigidity [e.g. $(\text{MeCp})_2ZrCl_2$, Ind_2ZrCl_2] [20, 21] as well as bridged, stereorigid C_{2v}-symmetric metallocenes {e.g. $\text{Me}_2\text{Si}(Cp)_2ZrCl_2$ or $\text{Me}_2\text{Si}(\text{Me}_4Cp)_2ZrCl_2$ [21] and the *meso* isomers of *ansa*-metallocenes [e.g. *meso*-ethylene(1-Ind)$_2$TiCl$_2$, *meso*-ethylene(1-Ind)$_2$ZrCl$_2$, *meso*-ethylene (H$_4$Ind)$_2$-ZrCl$_2$, and the like] [22].

Medium to high molecular weight a-PP has been obtained with a series of *meso*-Me$_2$Si(4-R-2-Me-Ind)$_2$ZrCl$_2$/MAO catalysts [22c].

Recently, we found that some C_{2v} symmetric *ansa*-bisfluorenylzirconium complexes [23] are highly efficient in the synthesis of high molecular weight atactic polypropylene (HMW-aPP) [24], the first example of a truly atactic PP with a molecular weight high enough to prevent cold-flow and to induce elastomeric properties. One of the catalysts suitable for the production of HMW-a-PP is Me$_2$Si(9-Flu)$_2$ZrCl$_2$, a synthetically simple zirconocene [24] able to produce HMW-a-PP with molecular weights ($\bar{M}_w$) in the 100 000–500 000 range at practical polymerization temperatures (50–70 °C).

The molecular structure of the related, more soluble Me$_2$Si(9-Flu)$_2$ZrMe$_2$ [24d] is shown in Figure 1 with its relevant geometrical parameters.

Lower molecular weights are obtained with the related C$_2$H$_4$(9-Flu)$_2$MCl$_2$/MAO catalysts (M = Zr, Hf) [23–25].

Key properties of HMW-a-PP include high solubility in hydrocarbons (e.g. ca 10 g/l in hexane at room temperature) and ethers, but insolubility in liquid propene (50 °C), good transparency, low hardness, a lower density than i-PP and elastic properties, as shown by its high elongation and elastic recovery. Its properties [26] and some potential applications [5a–f] have been described.

2   MOLECULAR WEIGHT CONTROL BY LIGAND SELECTION

The number, position and size of the alkyl substituents on the two π-Cp ligands control both the propagation rate and type and extent of chain transfer reactions [27]. The ratio between these rates results in the molecular weight of the polymer.

Two major chain transfer reactions have been observed in low molecular weight atactic polypropylene: bimolecular β-hydrogen transfer [28] and β-methyl transfer

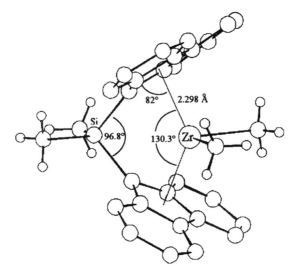

Figure 1   Selected bond distances and angles in $Me_2Si(9\text{-}Flu)_2ZrMe_2$ (from Ref. 24d)

(Scheme 1) [21, 29]. Bimolecular β-hydrogen transfer after a secondary insertion to give a 2-butenyl unsaturation and chain transfer to the Al cocatalyst to give a saturated chain-end are only minor reactions.

Typical examples are Cp_2ZrCl_2/MAO, which produces vinylidene-terminated oligomers with $\bar{P}_n \approx 20$ as a viscous liquid, and $Cp_2^*ZrCl_2/MAO$, which produces allyl-terminated, liquid oligomers of $\bar{P}_n \approx 4$ [21]. Both products are of potential interest: the vinylidene-terminated oligomers as functionalizable substrates [30] and

Scheme 1

the allyl-terminated oligomers as a possible source of 4-methyl-1-pentene monomer and polymerizable macromonomers [31].

We have extended this study to a larger set of aspecific metallocenes: the results obtained in liquid monomer at 50 °C are reported in Table 1.

Indenyl complexes produce higher molecular weights than biscyclopentadienyl compounds, with a-PP from substituted indenyl complexes having a slightly higher molecular weight than a-PP from Ind_2ZrCl_2. Also, the activities of indenyl-based zirconocenes are heavily dependent on the indenyl substitution: Ind_2ZrCl_2/MAO is among the most active aspecific zirconocenes, whereas $(4,7-Me_2Ind)_2ZrCl_2$/MAO is almost inactive. *ansa*-Bisfluorenyl complexes produce by far the highest molecular weights, while keeping good activity, hence representing the best catalysts in terms

Table 1 Propylene polymerization with aspecific metallocene/MAO and related catalyst systems: influence of the ligand structure[a]

Metallocene	Al/Zr molar ratio	Activity $[kg_{pp}/(mmol_M \, h)]$	$[\eta]^b$ (dl/g)	$\overline{M}_w{}^c$	$\bar{P}_n$
Cp_2ZrCl_2	1500	5.7			19^d
Cp_2HfCl_2	1400	4.6			137^d
$(MeCp)_2ZrCl_2$	1400	24.5			35^d
$(Me_5Cp)_2ZrCl_2$	1500	3.6			4.5^d
$(Me_5Cp)_2HfCl_2$	1500	12.6			3.4^d
$Me_2SiCp_2ZrCl_2$	1500	9.1			17^d
$Me_2Si(Me_4Cp)_2ZrCl_2$	1500	3.7	0.23	15 800	316^d
Ind_2ZrCl_2	1500	18.0			110^d
$(2-Me-Ind)_2ZrCl_2$	1000	10.0	0.29	21 600	220^d
$(2-Me-H_4Ind)_2ZrCl_2$	3000	9.2	0.23	15 800	188^e
$(4,7-Me_2-Ind)_2ZrCl_2$	1000	0.2			376^d
$C_2H_4(9-Flu)_2ZrCl_2$	1000	17.7	1.15	140 400	1670^e
$C_2H_4(9-Flu)_2HfCl_2$	1900	8.8	1.14	138 700	1650^e
$Me_2Si(9-Flu)_2ZrCl_2$	1000	16.5	2.56	415 700	4900^e
$Me_2Si(9-Flu)_2ZrMe_2$	2800	29.0	2.10	317 800	3800^e
$(2-Ph-Ind)_2ZrCl_2{}^f$	3000	2.5	0.7	64 600	850^e
$meso\text{-}C_2H_4(Ind)_2ZrCl_2$	3000	n.d.^{g}			75^d
$meso\text{-}C_2H_4(H_4Ind)_2ZrCl_2$	3000	n.d.^{g}			104^d
$meso\text{-}Me_2Si(2-Me-Ind)_2ZrCl_2$	3000	n.d.^{g}			309^d
$Me_2Si(Me_4Cp)(N\text{-}^tBu)ZrCl_2$	3000	22.0	1.56	212 300	2500^e

[a] In liquid propylene at 50 °C, 1 h. Metallocene/MAO premixed in toluene (10–20 ml) for 5–10 min at room temperature.
[b] $[\eta]$ = Intrinsic viscosity in THN at 135°C.
[c] Calculated from the $[\eta] = 1.85 \times 10^{-4}(\overline{M}_w)^{0.737}$.
[d] Number-average polymerization degree from ^{1}H NMR, assuming one double bond per polymer chain.
[e] Number-average polymerization degree from viscosity, assuming $\overline{M}_w/\overline{M}_n = 2$.
[f] Waymouth's catalyst produces a partially crystalline, elastomeric polypropylene.
[g] The three samples from *meso/rac*-zirconocene mixtures have been exhaustively extracted with Et_2O in a Kumagawa extractor. The fraction soluble in Et_2O at room temperature is only the atactic part of the polymer mixture, no isotactic polymer being soluble.

of both performance and availability [24]. The rationale of the choice of the Me$_2$SiFlu$_2$ ligand has been discussed in detail [24c].

3   INFLUENCE OF THE POLYMERIZATION CONDITIONS

As generally observed for metallocene-catalyzed olefin polymerization, both poly-merization temperature and concentration of propylene heavily affect the molecular weight, while the Al/Zr ratio (for MAO activated metallocene catalysts) has in most cases only a minor influence [32].

HMW-a-PP with average viscosimetric molecular weights in the range $10^5 - 10^6$ is obtained with the Me$_2$Si(9-Flu)$_2$ZrCl$_2$/MAO (**1**/MAO) catalyst by changing the polymerization conditions (e.g. polymerization temperature, Al/Zr molar ratio and monomer concentration). In the following, the influence of the polymerization parameters on the behavior of **1**/MAO in the synthesis of HMW-a-PP is reviewed, and in some cases compared with other two systems: C$_2$H$_4$(9-Flu)$_2$ZrCl$_2$/MAO (**2**/MAO) and (2-Me-Ind)$_2$ZrCl$_2$/MAO (**3**/MAO) [24f, 33].

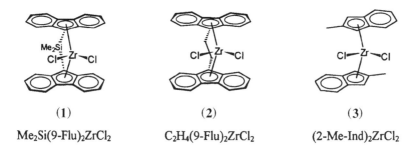

(**1**)	(**2**)	(**3**)
Me$_2$Si(9-Flu)$_2$ZrCl$_2$	C$_2$H$_4$(9-Flu)$_2$ZrCl$_2$	(2-Me-Ind)$_2$ZrCl$_2$

1/MAO, as typical of zirconocene catalysts, is long lived: no relevant catalyst decay was observed in liquid propylene at 50 °C for up to 4 h. The influence of monomer concentration on a-PP molecular weight with **1**/MAO has been reported [24d]: the catalyst activity increases linearly with [propylene], and the molecular weights show the typical dependence due to the presence of (at least) two chain transfer reactions, one unimolecular (chain transfer to the metal) and the other bimolecular (chain transfer to the monomer), as reported also for isospecific zirconocenes [34]. This is shown in Figure 2.

3.1   EFFECT OF HYDROGEN

Most metallocenes are highly sensitive towards the addition of hydrogen: in general, an activating effect is observed at low levels of hydrogen addition for isospecific

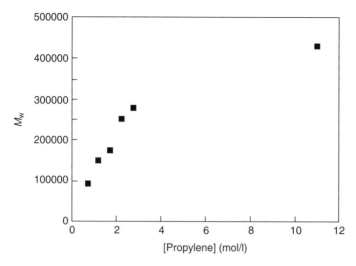

Figure 2   Propylene polymerization with $Me_2Si(9\text{-}Flu)_2ZrCl_2$/MAO: influence of propylene concentration on a-PP molecular weight (hexane, 50 °C, Al/Zr = 1200)

zirconocenes [35]. In the case of **1**/MAO, hydrogen has an adverse effect on catalyst activity: at 50 °C in 0.5 l of liquid monomer neither the activity nor the molecular weight is affected by the addition of 50 ml of H_2 {Activity = 21 400 g_{PP}/($mmol_M$ h), $[\eta] = 2.09$, $\overline{M}_w = 315\,700$ corresponding to $P_n \approx 3700$}, in the presence of 200 ml of H_2, however, the activity drops by 75 % and the resulting a-PP has a lower molecular weight {Activity = 5400g_{PP}/($mmol_M$h), $[\eta] = 1.52$, $\overline{M}_w = 204900$ corresponding to $\overline{P}_n \approx 2400$}. Addition of hydrogen has an adverse effect on catalyst activity also in the case of ethylene polymerization: addition of hydrogen to a slurry polymerization of ethylene (hexane, 4 bar, 50 °C) totally suppresses polymer formation. This effect is reversible: venting the hydrogen–ethylene gas mixture and replacing it with pure ethylene restarts the polymerization activity.

3.2   INFLUENCE OF THE COCATALYST

Compound **1** is equally activated by MAO and $B(C_6F_5)_3$, giving a-PP with the same molecular weights and the same activity, at least in hexane solution. A broader investigation on different cocatalysts under different polymerization conditions would be required to assess better if there is a cocatalyst effect on chain transfer. However, different types of MAO and different MAO/Zr ratios have an effect on both catalyst activity and a-PP molecular weight. The results obtained for solution polymerizations in hexane at 4 bar propylene and 50 °C are shown in Figures 3 and

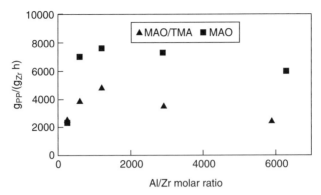

Figure 3 Propylene polymerization with Me$_2$Si(9-Flu)$_2$ZrCl$_2$/MAO: influence of the Al/Zr ratio on activity (hexane, 50 °C, 4 bar)

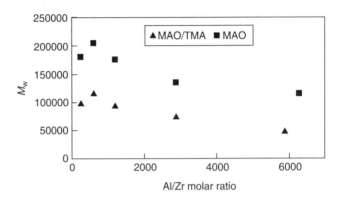

Figure 4 Propylene polymerization with Me$_2$Si(9-Flu)$_2$ZrCl$_2$/MAO: influence of the Al/Zr ratio on a-PP molecular weight (hexane, 50 °C, 4 bar)

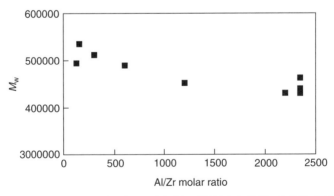

Figure 5 Propylene polymerization with Me$_2$Si(9-Flu)$_2$ZrCl$_2$/MAO: influence of the Al/Zr ratio on a-PP molecular weight (liquid monomer, 50 °C)

Table 2 Influence of Al/Zr ratio on molecular
weight for catalysts **1**, **2** and **Hf-2**^{a}

Catalyst	Al/Zr molar ratio	$[\eta]^b$ (dl/g)	$\overline{M}_w{}^c$
1/MAO	150	2.94	502 000
1/MAO	300	3.26	577 000
1/MAO	600	2.83	476 000
1/MAO	1000	2.62	429 000
1/MAO	2300	2.58	420 900
2/MAO	1000	1.15	140 300
2/MAO	100	1.47	195 800
2/MAO/TMA	100	1.35	174 500
2/M-MAO	1000	0.87	96 100
Hf-2/MAO	1900	1.14	138 700
Hf-2/MAO	100	1.06	125 700

^{a} Polymerization conditions: $T_p = 50\,°C$, 1 l or 2 l stainless-steel
autoclave, 0.5 or 1 l liquid propylene, 1–2 h, metallocene/MAO
solutions aged in toluene (10–20 ml) for 10 min at room temperature.
^{b} $[\eta]$ = Intrinsic viscosity in THN at 135 °C.
^{c} Estimated from $[\eta] = 1.85 \times 10^{-4}(\overline{M}_w)^{0.737}$.

4: increasing the Al/Zr ratio leads to an increase in catalyst activities up to about
Al/Zr = 1000 − 2000, and then a decrease at higher ratios, while molecular weights
decrease from Al/Zr = 500 upward; MAO solutions containing free $AlMe_3$ (ca
25 %) consistently give lower activity and lower molecular weights than pre-dried
MAO. The results obtained in liquid propylene at 50 °C are shown in Figure 5 for
catalyst **1**/MAO and collected in Table 2 for the three catalysts **1**/MAO, **2**/MAO and
Hf-2/MAO.

3.3 INFLUENCE OF POLYMERIZATION TEMPERATURE

The influence of polymerization temperature on catalyst activity and a-PP molecular
weight is shown in Table 3 for catalysts **1** and **2** and in Table 4 for catalysts
Ind_2ZrCl_2 and **3** [24f]. The molecular weight dependence on T_p for the four catalysts
is shown in Figure 6.

The polymerization results show an increased activity for the **2**/MAO over the **1**/
MAO system. As already reported, the insolubility of the $Me_2Si(9\text{-Flu})_2ZrCl_2$
precatalyst does not allow its isolation in pure form. Compound **1** was obtained in
85 % purity, with $LiCl(Et_2O)_x$ as the by-product [24d].

As the molecular weight of the polymer is function of both the rate of chain
transfer and the rate of propagation, the influence of polymerization temperature (T_p)
was investigated to obtain the energy of activation for propagation versus transfer.

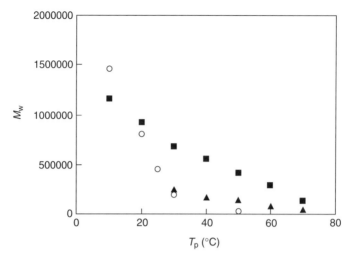

Figure 6 Influence of the polymerization temperature on a-PP molecular weight (liquid monomer). ■, $Me_2Si(9\text{-Flu})_2ZrCl_2/MAO$;   ○, $(2\text{-MeInd})_2ZrCl_2/MAO$; ▲, $C_2H_4(9\text{-Flu})_2ZrCl_2/MAO$

Table 3 Influence of T_p on molecular weight for catalysts **1** and **2**[a]

Catalyst	T_p (°C)	Al/Zr molar ratio	Viscosity[b]		SEC[d]		
			$[\eta]$ (dl/g)	$\overline{M}_w{}^c$	$\overline{M}_w$	$\overline{M}_n$	$\overline{M}_w/\overline{M}_n$
1/M-MAO	10	1000	5.46	1 162 000	1 253 000	699 000	1.8
1/M-MAO	20	1000	4.63	929 000	1 070 000	584 000	1.8
1/M-MAO	30	1000	3.70	685 000	852 000	460 000	1.9
1/M-MAO	40	1000	3.19	560 000	625 000	271 000	2.3
1/M-MAO	50	1000	2.56	415 700	595 000	290 000	2.1
1/M-MAO	60	1000	1.99	295 000	419 000	171 000	2.5
1/MAO	70	2000	1.09	130 000			
2/M-MAO	30	1000	1.78	253 900	257 600	124 800	2.1
2/M-MAO	40	1000	1.32	169 200	171 300	77 900	2.2
2/MAO	50	1000	1.15	140 300			
2/M-MAO	50	1000	0.87	96 100	115 800	47 300	2.5
2/M-MAO	60	1000	0.75	78 600	92 200	36 600	2.5
2/MAO	70	1000	0.47	41 700			
2/M-MAO	70	1000	0.39	32 400			

[a]Polymerization temperature typically ± 2 °C. Polymerization conditions: 1 l or 2 l stainless-steel autoclave, 0.5 or 1 l liquid propylene, 1–2 h, metallocene/MAO solutions aged in toluene (10–20 ml) for 10 min at room temperature.
[b]$[\eta]$ = Intrinsic viscosity in THN at 135 °C.
[c]Estimated from $[\eta] = 1.85 \times 10^{-4}(\overline{M}_w)^{0.737}$.
[d]SEC values (o-DCB, 135 °C).

Table 4   Propylene polymerization with Ind_2ZrCl_2/ MAO and **3**/MAO catalysts[a]

Catalyst	T_p (°C)	Al/Zr molar ratio	$[\eta]^b$ (dl/g)	$\overline{M}_w^{\,c}$
Ind_2ZrCl_2	50	1500	0.15	9 600
Ind_2ZrCl_2	20	3000	0.63	62 000
Ind_2ZrCl_2	0	3000	1.82	261 700
3	50	1000	0.29	21 700
3	30	3000	1.45	192 200
3	20	3000	4.18	808 600
3	10	3000	6.46	1 460 000
3	0	3000	11.43	3 166 000

[a]Polymerization conditions: 1 l stainless-steel autoclave, liquid propylene. Metallocene/MAO solutions aged in toluene (10–20 ml) for 5–10 min at room temperature.
[b]$[\eta]$ = Intrinsic viscosity (in THN at 135 °C).
[c]Estimated from $[\eta] = 1.85 \times 10^{-4}(\overline{M}_w)^{0.737}$.

Ethylenebis(9-fluorenyl)zirconium dichloride (**2**/MAO), even at low Al/Zr ratios, always gives lower molecular weight polypropylenes than **1**/MAO. The lower molecular weights modify the physical properties of the material: whereas a-PP from **1** and **2** are solid, non-sticky, elastomeric materials, samples from **3** show enhanced cold-flow or (for polymer samples prepared at $T_p = 70$ °C) are very viscous and sticky. The Arrhenius plots ($\ln \overline{P}_n$ vs $1/T_p$) for **1**/MAO and **2**/MAO give $\Delta\Delta E^{\ddagger}$ values of 4.9 ($T_p = 10 - 60$ °C, $R = 0.994$) and 10.1 kcal/mol ($T_p = 30 - 70$ °C, $R = 0.956$), respectively. At 70°C, we observe a lowering of catalyst activity for both catalysts. As for the molecular weights, we observe both a deviation from the linear behavior in the Arrhenius plot (molecular weights lower than expected) and a broadening of the molecular weight distributions for both **1**/MAO and **2**/MAO. These observations indicate some modification in the structure of the active centers induced by the temperature.

Finally, we have found that the non-bridged aspecific catalyst (2-methylindenyl)$_2$ZrCl$_2$ (**3**/MAO) shows a much stronger dependence of molecular weight on T_p. The polymerization results for propylene polymerizations in liquid monomer with **3**/MAO are compared with the simpler (Ind)$_2$ZrCl$_2$/MAO in Table 4.

4   NMR ANALYSIS

^{13}C NMR analysis confirms the atactic nature of the a-PP samples discussed above. All samples are highly regioregular with the exception of a-PP from Ind$_2$ZrCl$_2$ (which contains ca1 % of 3,1 units) and a-PP from Me$_2$Si(Me$_4$Cp)(N-^{t}Bu)TiCl$_2$ (CGC-Ti, which contains ca 2 % of 2,1 units). The pentad region of the a-PP

samples from Ind_2ZrCl_2, CGC-Ti and **1** are compared in Figure 7. In the range of polymerization temperatures investigated, both the C_{2v}-symmetric metallocenes and the freely rotating $(2\text{-Me-Ind})_2ZrCl_2$ produce atactic or nearly atactic polypropylenes, and the polymerization mechanism is in all cases Bernoullian (at the dyad level). It can be observed that a-PP from **1** (and also a-PP from **2**) is slightly

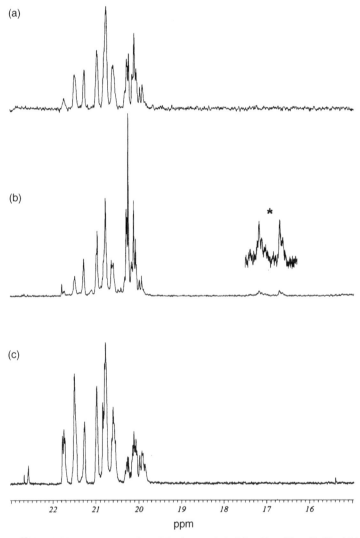

Figure 7 ^{13}C NMR spectra of a-PP from (a) $Me_2Si(9\text{-Flu})_2ZrCl_2/MAO$, (b) $Me_2Si(Me_4Cp)(N\text{-}^tBu)TiCl_2/MAO$ (The asterisk indicates peaks due to 2,1 insertions) and (c) Ind_2ZrCl_2/MAO. All samples produced at 50 °C in liquid propylene

syndiotactoid, whereas a-PP from Ind_2ZrCl_2 (as in the case of **3**) has an excess of m dyads. Deviation from a statistically atactic dyad distribution (m = r = 0.5) seems to be induced by a slight chain-end control. In any case, the degree of chain-end control is very low and is the same for both catalysts **2** and **3** ($\Delta\Delta E^{\ddagger} = 0.17$ kcal/mol) [24f].

5 INFLUENCE OF COMONOMERS

One important property of a-PP is its glass transition temperature (T_g), which defines the lower operating temperature. The T_g is somewhat dependent on the molecular weight [2d], as shown in Figure 8. One drawback of HMW-aPP is its relatively high glass transition temperature (ca 0 °C). Copolymerization of propylene with ethylene or higher α-olefins produces copolymers with lower T_g; high comonomer contents are needed when using higher α-olefins to obtain an appreciable effect, while ethylene is more efficient (Table 5). The decrease of T_g is linear with the ethylene content.

Ethylene increases the polymerization activity for both **1**/MAO and **3**/MAO catalytic systems while the copolymer molecular weight is increased only in the case of **3**. In the case of **1**/MAO, ethylene is a very effective chain transfer agent, and can be used as a molecular weight regulator in place of hydrogen.

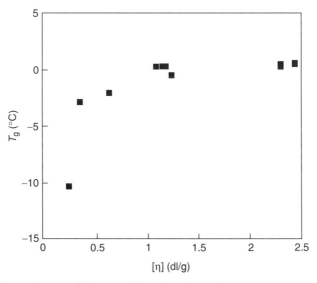

Figure 8 Dependence of T_g on a-PP molecular weight

Table 5 Influence of comonomer on T_g

Comonomer	Concentration (mol%)	$T_g(^{\circ}C)$
1-Butene	33	−9
1-Hexene	39	−16
1-Octene	3	−2
1-Octene	20	−16
1-Octene	42	−32
Ethylene	5	−13
Ethylene	12	−17
Ethylene	40	−50

6 CONCLUSIONS

Metallocene catalysts produce atactic polypropylene in the whole range of molecular weights, from oligomers (with different end groups) to very high molecular weight polymers. The degree of polymerization increases in the order $Cp^* < Cp < Ind < Flu$. The C_{2v}-symmetric $Me_2Si(9\text{-Flu})_2ZrCl_2$ produces a-PP with the highest molecular weight at the highest polymerization temperature.

The polymerization temperature, the Zr/Al ratio and the monomer concentration have been found to affect the molecular weight of HMW-a-PP.

The influence of the addition of hydrogen or ethylene on the polymerization of propylene with $Me_2Si(9\text{-Flu})_2ZrCl_2/MAO$ has also been investigated. Whereas hydrogen lowers the catalyst activity without being a good chain transfer agent, ethylene proved very effective in lowering the molecular weight of a-PP even at very low ethylene/propylene ratios. At the same time, ethylene addition increases the catalyst activity and produces amorphous propylene–ethylene copolymers with lower T_g [24f].

7 EXPERIMENTAL

7.1 GENERAL PROCEDURES

All catalyst manipulations were performed under nitrogen and using conventional Schlenk-line techniques. Toluene was distilled over $Al(^iBu_3)$ and stored under nitrogen. Hexane was purified over alumina. The typical residual water content was 2 ppm. Polymerization-grade propylene was obtained directly from the Montell Ferrara plant. Commercial MAO (Witco, 30 % w/w in toluene) was either used as received or brought to dryness and further treated under vacuum (2 h, 60 °C) in order to remove most of the unreacted TMA. Commercial M-MAO (Albemarle, 2.3 M solution in Isopar C) was used as received. $Me_2Si(Flu)_2ZrCl_2$ (**1**) and $Et(Flu)_2ZrCl_2$ (**2**) were prepared as described previously [24].

7.2 POLYMERIZATIONS

Polymerizations were carried out in liquid propylene or hexane in either a 1 l (0.5 l propylene) or 2 l (1 l propylene) stainless-steel autoclave at constant pressure for 1 h. Temperature control was within $\pm 2\,^{\circ}C$. Stirring was maintained at 800 rpm by means of a three-blade propeller. Metallocene and MAO were precontacted for 10 min in toluene solution (10 ml), then added to the monomer–solvent mixture at the polymerization temperature. The polymerizations were quenched with CO; the polymers were isolated by distilling off the solvents under reduced pressure, then dried at $50\,^{\circ}C$ *in vacuo*.

7.3 POLYMER ANALYSIS

Intrinsic viscosities were measured in tetrahydronaphthalene (THN) at $135\,^{\circ}C$.

The weight average molecular weights of a-PP were obtained from its intrinsic viscosity and the Mark–Houwink–Sakurada parameters derived by Pearson *et al.* [28].

GPC measurements were carried out on a Waters 150-C GPC system at $135\,^{\circ}C$ in *o*-dichlorobenzene. Monodisperse fractions of polystyrene were used as a standard. Solution ^{13}C NMR spectra were run at 100 MHz on a Bruker DPX-400 NMR spectrometer in $C_2D_2Cl_4$ at $130\,^{\circ}C$.

8 ACKNOWLEDGEMENTS

Thanks are due to M. Colonnesi for the polymerization experiments, R. L. Jones and D. Balboni for the synthesis of the catalysts and I. Mingozzi, A. Marzo, C. Baraldi, H. Rychlicki, I. Camurati, F. Piemontesi, M. Cappati, A. Celli and I. Giulianelli for the polymer analysis.

9 REFERENCES

1. See, for example, Foster, B. W. and Hilscher, L. W., *US Pat.*, 4 749 739, to Eastman Kodak (1988); Pellon, B. J. and Allen, G. C., *Eur. Pat. Appl.*, 475 306, to Rexene (1992); Lakshmanan, P. R. and Tayebianpour, A., *US Pat.*, 5 397 843, to IGI Baychem (1995); Robe, G. R. *Adhe. Age*, February, 26 (1993); Pellon, B. J. in *Proceedings of SPO '93*, Schotland Business Research, Skillman, NJ, 1993, p. 401.
2. (a) See, for example, Schilling, F. and Tonelli, A., *Macromolecules*, **13**, 270 (1980); (b) Suter, U. and Neuenschwander, P., *Macromolecules*, **14**, 528 (1981); (c) Plazek, D. L. and Plazek, D. J., *Macromolecules*, **16**, 1469 (1983); (d) Burfield, D. R. and Doi, Y.,

Macromolecules, **16**, 702 (1983); (e) Cheng, H. N. and Lee, G. H., *Polym. Bull.,* **13**, 549 (1985); (f) Zhongde, X., Mays, J., Xuexin, C., Hadjichristidis, N., Schilling, F., Bair, H., Pearson, D. and Fetters, L., *Macromolecules,* **18**, 2560 (1985); (g) Pearson, D., Fetters, L., Younghouse, L. and Mays, J., *Macromolecules,* **21**, 478 (1988); Marigo, A., Marega, C., Zanetti, E., Zannetti, R. and Paganetto, G., *Makromol. Chem.,* **192**, 523 (1991); Alfonso, G., Yan, D. and Zhou, Z., *Polymer,* **34**, 2830 (1993); Sakurai, K., MacKnight, W. J., Lohse, D., Schulz, D. N. and Sissano, J. A. *Macromolecules,* **27**, 4941 (1994).

3. See, for example, Smith, T. W., Ames, W. A., Holliday, R. E. and Pearson, N., *Eur. Pat. Appl.,* 232 201, to Eastman Kodak (1987); Soga, K., Park, J., Uchino, H., Uozumi, T. and Shiono, T., *Macromolecules,* **22**, 3824 (1989); Smith, C., *Eur. Pat. Appl.,* 423 786, to Himont (1991); Pellon, B. J. and Allen, G. C., *Eur. Pat. Appl.,* 475 307, to Rexene (1992); Job, R., *US, Pat.,* 5 270 410, to Shell Oil (1993).

4. Van der Ven, S., *Polypropylene and Other Polyolefins. Studies in Polymer Science,* Vol. 7, Elsevier, Amsterdam, 1990, Chapt. 2, pp. 134–212; Kakugo, M., Miyatake, T., Naito, Y. and Mizunuma, K., *Makromol. Chem.,* **190**, 505 (1989); Busico, V., Corradini, P., De Martino, L., Graziano, F. and Iadicicco, A., *Makromol. Chem.,* **192**, 49 (1991).

5. (a) Pelliconi, A., Silvestri, R., Braga, V. and Resconi, L., *Eur. Pat. Appl.,* 666 284, to Spherilene (1995); (b) Silvestri, R., Resconi, L. and Pelliconi, A., *Eur. Pat. Appl.,* 697 436, to Montell (1995); (c) Braga, V. and Sartori, F., *PCT Int. Appl.,* WO 97/31065, to Montell (1997); (d) Braga, V. and Sartori, F., *Eur. Pat. Appl.,* 753 529, to Montell (1996); (e) Yang, H. W., Canich, J. A. M. and Licciardi, G. F., *US Pat.,* 5 516 848, to Exxon (1996); (f) Canich, J. A. M., Yang, H. W. and Licciardi, G. F., *US Pat.,* 5 539 056, to Exxon (1996).

6. Farina, M., *Top. Stereochem.,* **17**, 1 (1987).

7. Kakugo, M., Miyatake, T., Mizunuma, K. and Yagi, Y., *Eur. Pat. Appl.,* 371 411, to Sumitomo Chemical (1990).

8. Kakugo, M., Miyatake, T., Kawai, Y., Shiga, A. and Mizunuma, K., *Eur. Pat. Appl.,* 241 560, to Sumitomo Chemical (1987); Miyatake, T., Mizunuma, K., Seki, Y. and Kakugo, M., *Makromol. Chem., Rapid Commun.,* **10**, 349 (1989); Miyatake, T., Mizunuma, K. and Kakugo, M., *Makromol. Chem., Macromol. Symp.,* **66**, 203 (1993).

9. Soga, K. and Lee, D. H., *Makromol. Chem.,* **193**, 1687 (1992); Park, J. R., Shiono, T. and Soga, K., *Macromolecules,* **25**, 521 (1992).

10. Pellecchia, C., Proto, A., Longo, P. and Zambelli, A., *Makromol. Chem., Rapid Commun.,* **13**, 277 (1992); Salssmannshausen, J., Bochmann, M., Rösch, J. and Lilge, D., *J. Organomet. Chem.,* **548**, 23 (1997).

11. Canich, J. A. M., *US Pat.,* 5 504 169, to Exxon (1996); McKnight, A. L., Masood, Md. A., Waymouth, R. M. and Straus, D. A., *Organometallics,* **16**, 2879 (1997).

12. Flores, J. C., Chien, J. C. W. and Rausch, M. D., *Organometallics,* **13**, 4140 (1994); Flores, J. C., Chien, J. C. W. and Rausch, M. D., *Macromolecules,* **29**, 8030 (1996).

13. Johnson, L. K., Killian, C. M. and Brookhart, M., *J. Am. Chem. Soc.,* **117**, 6414 (1995); Killian, C. M., Tempel, D. J., Johnson, L. K. and Brookhart, M., *J. Am. Chem. Soc.,* **117**, 11664 (1996).

14. Collette, J., Tullock, C. W., MacDonald, R. N., Buck, W. H., Su, A. C. L., Harrell, J. R., Mülhaupt, R. and Anderson, B. C. *Macromolecules,* **22**, 3851 (1989); Tullock C., Tebbe, F. N., Mülhaupt, R., Ovenall, D. W., Setterquist, R. A. and Ittel, S. D. *J. Polym. Sci. Part A: Polym. Chem.,* **27**, 3063 (1989); Ittel, S., *Polym. Prepr. Am. Chem. Soc. Div. Polym. Chem.,* **35**, 665 (1994); Gahleitner, M., Ledwinka, H., Hafner, N., Heinemann, H. and Neissl, W., in *Proceedings of SPO '96,* Schotland Business Research, Skillman, NJ, 1996.

15. (a) Chien, J. C. W., Llinas, G. H., Rausch, M. D., Lin, Y.-G., Winter, H. H., Atwood, J. L. and Bott, S. G., *J. Polym. Sci. Part A: Polym. Chem.,* **30**, 2601 (1992); (b) Coates, G. W. and Waymouth, R. M., *Science,* **267**, 217 (1995); (c) Hauptman, E., Waymouth, R. M. and Ziller, J. W., *J. Am. Chem. Soc.,* **117**, 11586 (1995); (d) Maciejewski Petoff, J. L., Bruce, M.

D., Waymouth, R. M., Masood, A., Lal, T. K., Quan, R. W. and Behrend, S. J., *Organometallics*, **16**, 5909 (1997); (e) Kravchenko, R., Masood, A. and Waymouth, R. M., *Organometallics*, **16**, 3635 (1997); (f) Bruce, M. D., Coates, G. W., Hauptman, E., Waymouth, R. M. and Ziller, J. W., *J. Am. Chem. Soc.*, **119**, 11174 (1997).

16. Farina, M., Di Silvestro, G. and Sozzani, P., *Macromolecules*, **26**, 946 (1993).
17. (a) Ewen, J. A., presented the 198th ACS National Meeting, Miami Beach, Fl, September 10–15, 1989; (b) Dolle, V., Rohrmann, J., Winter, A., Antberg, M. and Klein, R. *Eur. Pat. Appl.*, 399 347, to Hoechst (1990); (c) Antberg, M., Dolle, V., Klein, R., Rohrmann, J., Spaleck, W. and Winter, A., in Keii, T. and Soga, K. (Eds), *Catalytic Olefin Polymerization. Studies in Surface. Science and Catalysis*, Vol. 56, Kodansha–Elsevier, Tokyo, 1990, p. 501; (d) Ewen, J. A., *Eur. Pat. Appl.*, 423 101, to Fina (1991); (e) Ewen, J. A., Elder, M. J., Harlan, C. J., Jones, R. L., Alwood, J. L., Bott, S. G. and Robinson, K., *Polym. Prepr. Am. Chem. Soc. Div. Polym. Chem.*, **32**, 469 (1991); Fierro, R., Chien, J. C. W. and Rausch, M. D., *J. Polym. Sci. Part A: Polym. Chem.*, **32**, 2817 (1994); Gauthier, W. J., Corrigan, J. F., Taylor, N. J. and Collins, S., *Macromolecules*, **28**, 3771 (1995); Gauthier, W. J. and Collins, S., *Macromolecules*, **28**, 3779 (1995), and references therein.
18. Ewen, J. A., Elder, M. J., Jones, R. L., Haspeslagh, L., Alwood, J. L., Bott, S. G. and Robinson, K., *Makromol. Chem., Macromol. Symp.*, **48/49**, 253 (1991); Herfert, N. and Fink, G., *Makromol. Chem., Macromol. Symp.*, **66**, 157 (1993).
19. Rohrmann, J., *Eur. Pat. Appl.*, 528 287, to Hoechst (1993); Green, M. L. H. and Ishihara, N., *J. Chem. Soc., Dalton Trans.*, 657 (1994); Collins, S. and Gauthier, W., in *Proceedings of SPO '93*, Schotland Business Research, Skillman, NJ, 1993, p. 149; Fierro, R., Chien, J. C. W. and Rausch, M. D., *J. Polym. Sci. Part A: Polym. Chem.*, **32**, 2817 (1994).
20. Kaminsky, W., Hähnsen, H., Külper, K. and Woldt, R., *US Pat.*, 4 542 199, to Hoechst (1985); Kaminsky, W., *Makromol. Chem., Macromol. Symp.*, **3**, 377 (1986); Miya, S., Harada, M., Mise, T. and Yamazaki, H., *US Pat.*, 4 874 880, to Chisso (1989).
21. Resconi, L., Piemontesi, F., Franciscono, G., Abis, L. and Fiorani, T., *J. Am. Chem. Soc.*, **114**, 1025 (1992).
22. (a) Ewen, J., *J. Am. Chem. Soc.*, **106**, 6355 (1984); (b) Collins, S., Gauthier, W., Holden, D., Kuntz, B., Taylor, N. and Ward, D., *Organometallics*, **10**, 2061 (1991); (c) Winter, A., Antberg, M., Bachmann, B., Dolle, V., Küber, F., Rohrmann, J. and Spaleck, W., *Eur. Pat. Appl.*, 584 609, to Hoechst (1994).
23. Alt, H., Milius, W. and Palackal, S., *J. Organomet. Chem.*, **472**, 113 (1994); Alt, H., Palackal, S., Patsidis, K., Welch, M., Geerts, R., Hsieh, E., McDaniel, M., Hawley, G. and Smith, P., *Eur. Pat. Appl.*, 524 624, to Phillips Petroleum (1993).
24. (a) Resconi, L. and Jones, R. L., *Eur. Pat. Appl.*, 96103118.4, to Montell (1994); (b) Resconi, L. and Albizzati, E., *Eur. Pat. Appl.*, 604 917, to Montell (1994); (c) Resconi, L., Jones, R. L., Albizzati, E., Camurati, I., Piemontesi, F., Guglielmi, F. and Balbontin, G., *Polym. Prepr. Am. Chem. Soc. Div. Polym. Chem.*, **35**, 664 (1994); (d) Resconi, L., Jones, R. L., Rheingold, A. L. and Yap, G. P. A., *Organometallics*, **15**, 998 (1996); (e) Jones, R. L., Resconi, L. and Rheingold, A. L., in *Proceedings of SPO '95*, Schotland Business Research, Skillman, NJ (1995); (f) Resconi, L., Piemontesi, F. and Jones, R.L., presented at the RETEC International Conference on Polyolefins, Houston, February 23–26 1997.
25. Chen, Y.-X., Rausch, M. D. and Chien, J. C. W., *Macromolecules*, **28**, 5399 (1995).
26. Silvestri, R., Resconi, L. and Pelliconi, A., in *Proceedings of the International Congression Metallocene Polymers*, Brussels, 1995, p. 207; Resconi, L. and Silvestri, R. in Salamone, J. C. (Ed), *The Polymeric Materials Encyclopedia*, CRC Press, Bota Raton, FL, 1996; Eckstein, A., Suhm, J., Friedrich, C., Maier, R.-D., Sassmannshausen, J., Bochmann, M. and Mülhaupt, R., *Macromolecules*, **31**, 1335 (1998).

27. Ewen, J., in Keii, T. and Soga, K. (Eds), *Catalytic Polymerization of Olefins. Studies in Surface Science and Catalysis,* Vol. 25, Elsevier, New York, 1986, p. 271; Brintzinger, H.-H., Fischer, D., Mülhaupt, R., Rieger, B. and Waymouth, R. M., *Angew. Chem., Int. Ed. Engl.,* **34**, 1143 (1995).

28. Tsutsui, T., Mizuno, A. and Kashiwa, N., *Polymer,* **30**, 428 (1989).

29. Watson, P. L. and Roe, D. C., *J. Am. Chem. Soc.,* **104**, 6471 (1982); Eshuis, J.J.W., Tan, Y. Y., Meetsma, A., Teuben, J. H., Renkema, J. and Evens, G. G., *Organometallics,* **11**, 362 (1992).

30. Shiono, T. and Soga, K., *Macromolecules,* **25**, 3356 (1992); Shiono, T., Kurosawa, H., Ishida, O. and Soga, K., *Macromolecules,* **26**, 2085 (1993); Hungenberg, K.-D., Kerth, J., Langhauser, F., Müller, H.-J. and Müller, P., *Angew. Makromol. Chem.,* **227**, 159 (1995).

31. Shiono, T., Moriki, Y., Ikeda, T. and Soga, K., *Macromol. Chem. Phys.,* **198**, 3229 (1997).

32. Our experience with chiral zirconocenes is that the aluminum concentration has a major influence on catalyst activity but not on iPP molecular weight when the polymerization is carried out in liquid propylene. At lower propylene concentration,aluminum concentration does have an effect on the molecular weight as well. See, for example, Jüngling, S. and Mülhaupt, R., *J. Organomet. Chem.,* **497**, 27 (1995).

33. Van Beek, J., De Vries, J., Persad, R. and Van Doremaele, G., *PCT Int. Appl.,* WO 94/11406, to DSM (1994); Resconi, L., Piemontesi, F. and Balboni, D., *Eur. Pat. Appl.,* 693 506, to Spherilene (1996).

34. Stehling, U., Diebold, J., Kirsten, R., Röll, W., Brintzinger, H.-H., Jüngling, S., Mülhaupt, R. and Langhauser, F., *Organometallics,* **13**, 964 (1994).

35. Tsutsui, T., Kashiwa, N. and Mizuno, A. *Makromol. Chem., Rapid Commun.,* **11**, 565 (1990); Kashiwa, N. and Kioka, M., *Polym. Mater. Sci. Eng.,* **64**, 43 (1991); Kioka, M., Mizuno, A., Tsutsui, T. and Kashiwa, N., in Vandenberg, E. J., Salamone, J. C., (Eds); *Catalysis in Polymer Synthesis. ACS Symposium Series,* Vol. 496, American Chemical Society, Washington, DC, 1992, p. 72, Busico, V., Cipullo, R. and Corradini, P., *Makromol. Chem., Rapid Commun.,* **14**, 97 (1993); Jüngling, S., Mülhaupt, R., Stehling, U., Brintzinger, H.-H., Fischer, D. and Langhauser, F., *J. Polym. Sci. Part A: Polym. Chem.,* **33**, 1305 (1995).

36. Ewen, J. A. and Elder, M. J., *Eur. Pat. Appl.,* 427 697, to Fina (1991); Yang, X., Stern, C. L. and Marks, T. J., *J. Am. Chem. Soc.,* **113**, 8570 (1991).

22

Design of Ethylene-bridged *ansa*-Zirconocene Dichlorides for a Controlled Propene Polymerization Reaction

JOHANNA VOEGELE, ULF DIETRICH, MARTIJN HACKMANN
AND BERNHARD RIEGER
University of Ulm, Ulm, Germany

1 INTRODUCTION

In the last 15 years, the field of polyolefin chemistry has gone through a renaissance with major achievements in control over the polymer microstructures and hence over the bulk properties of this important family of organic polymers. This development was triggered by the synthesis of stereorigid *ansa*-zirconocene dichlorides by Brintzinger's group at the University of Konstanz at the beginning of the 1980s [1]. Complexation of two indenyl ligands to a Zr(IV) center affords the formation of the stereoisomers **1a**, **1b** and **1c** (Figure 1). In addition to the two C_2-symmetric enantiomers (*S*)-(**1a**) and (*R*)-(**1b**), there exists *a meso* isomer (**1c**) [2], in which both indenyl fragments are directed towards the same side of the complex. In all three species, the indenyl groups define a certain stereochemistry and the ethylene bridge hinders rotation of these moieties and thus fixes their position relative to each other.

In the following years, John Ewen (1984) [3] and Walter Kaminsky (1985) [4] and their co-workers succeeded in the preparation of isotactic polypropene by applying the racemic mixture of the *S*- and *R*-enantiomers, activated by methylaluminoxane (MAO). For the first time it was possible to correlate a molecular defined active

Metallocene-based Polyolefins   Edited by J. Scheirs and W. Kaminsky
© 2000 John Wiley & Sons Ltd

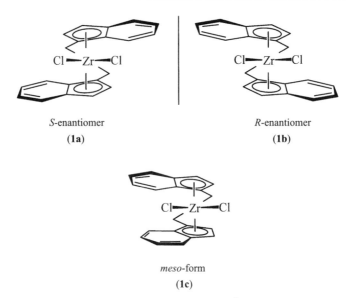

S-enantiomer   R-enantiomer
(1a)   (1b)

meso-form
(1c)

Figure 1 Stereoisomers of ethylene-bridged bis(η^5-1-indenyl)zirconium dichloride

catalyst species with the microstructure of a stereoregular polymer chain. This opened the way to a new catalyst generation, where all the active species have a uniform structure ('single-site catalysts') which can be determined by the use of well known techniques such as single-crystal X-ray structure analysis and NMR spectroscopy.

The discovery that organic and organometallic tools can be applied for a rational catalyst design laid the foundation for worldwide activity both in academia and in industry, leading to several types of metallocenes with the above-mentioned, precise correlation with the corresponding polymer microstructures (Figure 2). These catalysts not only allow the synthesis of highly isotactic or more or less atactic polypropenes, but also syndiotactic, hemi-isotactic and different kinds of stereoblock polymer structures are accessible, providing a portfolio of tailor-made microstructures for different industrial applications [5].

2 INFLUENCE OF THE POLYMERIZATION TEMPERATURE ON THE STEREOSELECTIVITY AND ACTIVITY OF THE CATALYST

Most of the conventional Ziegler–Natta catalysts show a dependence of the activity (A_p) on the polymerization temperature (T_p). In a typical run, A_p at first increases with increase in temperature, then, after passing through a maximum, a further rise

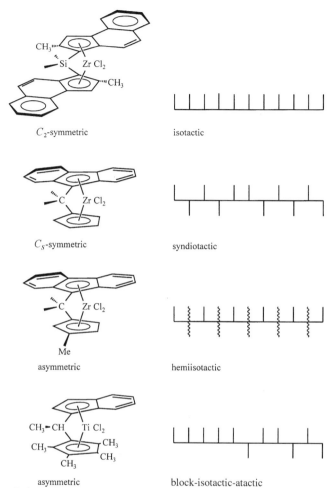

C_2-symmetric   isotactic

C_s-symmetric   syndiotactic

asymmetric   hemiisotactic

asymmetric   block-isotactic-atactic

Figure 2 Selected metallocene dichlorides and the corresponding polymer microstructures

of T_p results in a decrease in the catalyst activity. This behavior can be attributed to a slow activating reaction at low and a fast catalyst decomposition at higher temperatures [6]. The catalyst activity of a racemic mixture of **1a** and **1b** (*rac*-Et[Ind]$_2$ZrCl$_2$) increases continually to a factor of 3000 on raising T_p from -55 to $+70°$C (Figure 3) [7].

An Arrhenius plot of A_p shows a linear dependence on the polymerization temperature [8] ($A_p \sim 1/T_p$), from which the activation energy of the polymerization reaction can be estimated to be about 10 kcal/mol. Similar values were determined for catalyst systems such as Cp$_2$TiCl$_2$/(CH$_3$)$_2$AlCl [9]; δ-TiCl$_3$/(C$_2$H$_5$)$_2$AlCl [10] and MgCl$_2$-supported-titanium chlorides [11]. As the activity increases to more

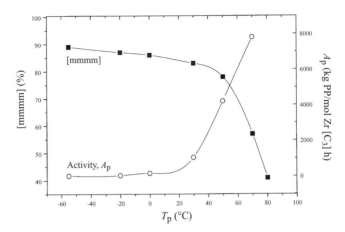

Figure 3 Catalyst activity (A_p) of $rac-$Et[Ind]$_2$ZrCl$_2$ and polymer stereoregularity ([mmmm]) versus the polymerization temperature (T_p)

technically important values above room temperature, the stereoregularity [12] of the polymer products (% [mmmm] sequences) decreases rapidly with increase in T_p. At low T_p, powdery, isotactic polypropenes are formed, which show variable but defined melting transitions, whereas oligopropenes of low isotacticity with reduced melting-points are obtained at an industrially interesting temperature of $T_p = 70\,^{\circ}$C.

This strong decline in catalyst stereoselectivity with temperature is much less pronounced for mono-atom bridged complexes [13], indicating that it should be possible to identify a structural feature, typical for ethylene-bridged complexes, being responsible for this drastic loss of stereoregularity. A closer comparison of mono-atom and ethylene-bridged zirconocene dichlorides shows that the mono-atom bridge forces the two π-ligands in an eclipsed position (Figure 4, structure **2**), whereas the sterically more demanding ethylene bridge causes a staggered arrangement of the η^5-coordinated fragments (Figure 4, structure **1a**).

A view to the front side of such C$_2$H$_4$-bridged complexes into the chiral cage defined by the position of the indenyl ligands, or more precisely by the β-substituents [14] of the cyclopentadienyl rings, shows that the twisted metallacycle

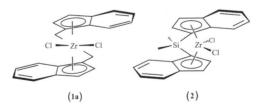

(1a)   (2)

Figure 4 Staggered (**1b**) and eclipsed (**2**) arrangements of the indenyl fragments in a C$_2$H$_4$-bridged and in a mono-atom-bridged *ansa*-metallocene dichloride

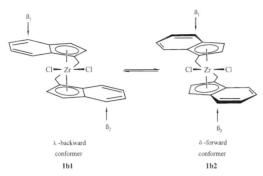

Figure 5 Bridge conformation and relative position of the β-substituents in diastereomeric forms of (*R*)-Et[Ind]$_2$ZrCl$_2$

can exist as two different stereoisomers (Figure 5) [15]. The λ-backward conformer (**1b1**) shows a 'right–left–right' arrangement of the two CH$_2$ groups of the ethylene bridge. This geometry pushes the β-substituents of the cyclopentadienyl ligands into a backward position where the two β-CH groups have a maximum distance from each other, thus defining an 'open' chiral cage. A flip of the bridge into the 'left–right–left' -arranged δ-forward conformation (**1b2**) accompanies a rotation of the η^5-coordinated indenyl ligands around the Cp-centroid—Zr bonds. This forces the β-substituents to the front side of the complex where they have a minimum distance from each other. This arrangement defines a tighter chiral cage around the Cl—Zr—Cl plane where the stereoselectivity is determined. The equilibrium between λ- and δ-conformers is fast in solution, at least for C$_2$-symmetric dichlorides, and allows the occurrence of diastereomeric species in a racemic mixture of chiral *ansa*-metallocene dichlorides.

Since the λ- and δ-chelate ring conformers are diastereoisomers, which in principle can have different stereoselectivities, it is meaningful to ask whether the equilibrium between such stereoisomeric species could be responsible for the decline in isotacticity with increasing polymerization temperature [8,16]. This hypothesis is supported by polymerization experiments with mono-atom bridged complexes where the effect of temperature on stereoregularity is much less pronounced [13]. Therefore, a new strategy has to be worked out for the synthesis of ethylene-bridged *ansa*-metallocenes bearing bulky bridge substituents, because more stereoridgid chelate ring conformers are expected from complexes of this kind.

3   PREPARATION OF CHIRAL ETHYLENE-BRIDGED *ANSA*-ZIRCONOCENE COMPLEXES BEARING DIFFERENT BACKBONE SUBSTITUENTS

Epoxides can serve as cheap and excellent starting materials for the preparation of a variety of stereorigid biscyclopentadiene ligand precursor systems bearing a variable backbone substitution pattern [17–19]. A ring opening reaction of the epoxides **3** by indenyl (Ind)- or fluorenyl (Flu)-lithium (Figure 6) affords a nearly quantitative formation of the alcohol. If epoxystyrene is used, a mixture of secondary and primary alcohols (1:3 ratio) is obtained. The primary alcohol **4** crystallizes in its pure form from toluene–hexane, leaving a 1:1 mixture of both alcohols in solution. A quantitative separation can be achieved by column chromatography over silica [17, 19]. The primary alcohol **4** can be converted into dicyclopentadiene ligand precursors by substitution of the hydroxyl function, following two different synthetic strategies (Figure 6, routes I and II). Route I uses trifluoromethane sulfonate derivatives (**5**) to introduce the indenyl fragment by substitution of the leaving group, resulting in the formation of the ethylene bridged biscyclopentadienes **9–12**. The relative position of the phenyl group remains unchanged. In route II, deprotonation of the methylsulfonate **6** with a sterically hindered base leads to the spiro compound **7** by an intramolecular substitution of the leaving group. The high ring tension of **7** combined with the ability of the indenyl group to stabilize a negative

Route I:

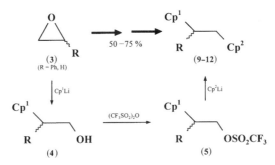

Route II:

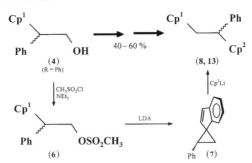

Figure 6 Two possible routes for the preparation of dicyclopentadiene ligand precursors

charge permits the introduction of a Cp fragment by a ring opening reaction of the spirocyclopropane. The ring opening of **7** is highly regiospecific and leads exclusively and in high yields to the ligand precursors **8** and **13** where the bridge substituent is located at the α-position relative to the incoming Cp group. Both reaction paths (**I, 3** $\longrightarrow$ **9–12; II, 3** $\longrightarrow$ **8, 13**) allow to control the backbone substitution pattern over a broad range of substituents.

An interesting extension of the above-mentioned reaction is the application of an enantiomerically pure epoxide. The ring opening of, e.g. (*R*)-(−)-epoxystyrene with (Flu)Li proceeds stereospecifically, leading exclusively to (*S*)-(−)-(2-fluorenyl-2-phenyl)ethanol (**4***). In order to determine the enantiospecificity of the ring opening reaction, the enantiomerically pure primary alcohol **4*** was converted to the corresponding ester **15*** by use of the chiral auxiliary (*R*)-α-aminophenylacetic acid (**14***) (Figure 7) [20]. The reaction is carried out in the melt, composed of a mixture of alcohol, amino acid and *p*-toluenesulfonic acid, at about 200 °C and affords the formation of the hydrochloride **15*** (after acidic work-up) in quantitative yield. ^{1}H NMR analysis of **15*** reveals only one set of resonances, indicating the presence of a single diastereoisomer [21]. Starting from (*S*)-(−)-(2-fluorenyl-2-phenyl)ethanol (**4***), it is possible to prepare directly optically active ligand precursors and their corresponding complexes by a chiral pool synthesis without the need for separation of the enantiomeric complexes.

Preparation of the zirconocene dichloride complexes was accomplished by transformation of the biscyclopentadienes **8–13** into the dilithio salts by reaction with *n*-butyllithium in diethyl ether. Subsequent treatment with $ZrCl_4$ in dichloromethane solution at −78 °C afforded the *ansa*-metallocene dichlorides **16–21** in isolated yields up to 93 % [17, 18]. This synthetic concept allows one to prepare a variety of ethylene-bridged complexes bearing two equivalent or two different cyclopentadienyl units and a broad range of bridge substituents. Some selected species, based on epoxystyrene, are summarized in Figure 8.

Racemic epoxystyrene affords the formation of the complexes **16** and **18** (Flu-Ph-Cp, Flu-Ph-Flu) as racemic mixtures, due to the single stereogenic backbone center. Ethylene oxide leads to complex **17** (Flu-H-Ind) having the chiral carbon atom of the indenyl moiety [22]. Each of the complexes **19** and **20** (Cp-Ph-Ind, Flu-Ph-Ind),

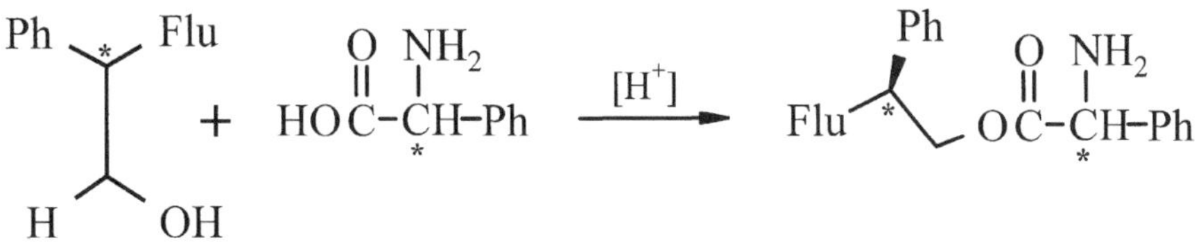

Figure 7 Enantiomerically pure ligands from (*R*)-(−)-epoxystyrene

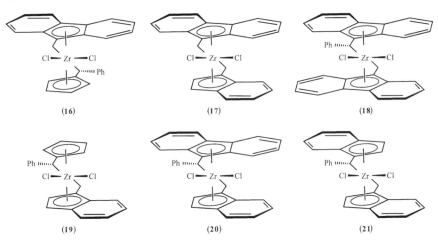

Figure 8 A selection of ethylene-bridged *ansa*-metallocene dichloride complexes

which are composed of a stereogenic carbon bridge together with one indenyl group, exists in two diastereomeric forms, each of which consists of two enantiomeric mirror images (e.g. **20a** = RR, *SS*; **20b** = RS, *SR*). The presence of a second indenyl ligand, as in **21** (Ind-Ph-Ind), leads to four racemic diastereoisomers [23].

As examples, two of the four possible forms of **20** {**20a**, [1-(η^5-9-fluorenyl)-1-(*R*)-phenyl-2-(η^5-1-(*R*)-indenyl)ethane]ZrCl$_2$; **20b**, [1-(η^5-9-fluorenyl) $-$ 1 $-$ (*S*)-phenyl-2-(η^5-1-(*R*)-indenyl)ethane]ZrCl$_2$} are sketched in Figure 9 [24]. The bulky phenyl groups occupy the energetically favored equatorial positions of the metallacycles [25] in both complexes. This leads to preferred conformations of the chelate rings, depending on the nature of the stereogenic backbone center. An *R*-configuration causes a δ-conformation (**20a**), whereas the *S*-stereocenter gives rise to the λ-conformer (**20b**), while the stereochemistry of the indenyl fragments remains unchanged (*R*-configuration in Figure 9) [15]. The different bridge twists result in staggered arrangements of the fluorenyl and indenyl units within **20a, b**. This allows control of the relative positions of the β-CH substituents to each other (Figure 9), which are assumed to play a major role in the enantiofacial discrimination of the inserting propene monomer [13]. According to a notation introduced by Brintzinger and co-workers [26], the δ-conformation places the opposite standing substituents β^1 and β^3 in a forward position, minimizing the distance between the two groups. Consequently, a backward arrangement characterizes the λ-conformation with a maximum distance between β^1 and β^3 [17].

In all these asymmetric complexes, the ethylene backbone substituent, which originates from the epoxide starting material, creates a stereogenic carbon center. In combination with the chiral coordination of the indenyl group, **19a, b** and **20a, b** exist as diastereomers, which can be separated by crystallization. Compounds **19a**

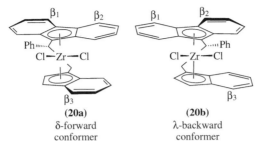

Figure 9 λ-backward and δ-forward conformers of [1-(η^5-9-fluorenyl)-1-(*R*, *S*)-phenyl-2-(η^5-1-(*R*)-indenyl)ethane]ZrCl$_2$. Reproduced by permission of the American Chemical Society, 1998

and **20a** are obtained from toluene solution at −30 °C in their pure form. Extraction of the residue with 1,2-dimethoxy ethane leaves **19b** and **20b** in roughly 80 % excess. Subsequent recrystallization from toluene results in precipitation of the pure diastereomers **19b** and **20b**. For the formation of **20a, b** a remarkable dependence of the diastereoselectivity on solvent and temperature is observed. Preparation of the complexes according to the above-mentioned procedure in CH$_2$Cl$_2$ gives a 1:3 mixture of **20a** and **20b**. Reaction of the dilithio salt of **12** with ZrCl$_4$ in hexane under reflux affords an excess of **20a** (ratio 3:1) (Figure 10). Analogous solvent influences were reported by Chien and co-workers [27] for the preparation of asymmetric diastereomers with a single carbon bridge.

The solid-state structures of **20a, b** were determined by X-ray structure investigations (Figure 11) [17]. Compound **20a** shows a δ-conformation of the ideally twisted

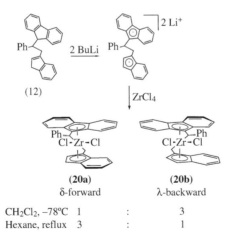

Figure 10 Preparation and separation of the diastereoisomers of the complexes **20a, b**

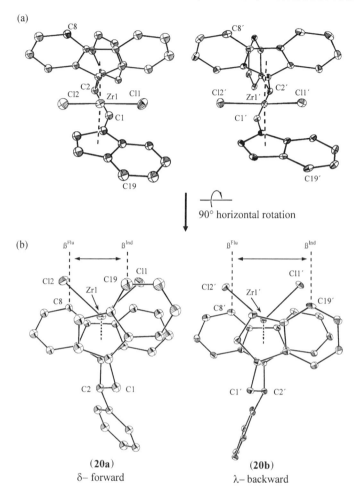

Figure 11 Solid-state structures of [1-(η^5-9-fluorenyl)-1-(R)-phenyl-2-(η^5-1-(R)-indenyl)ethane]ZrCl$_2$ (**20a**) and [1-(η^5-9-fluorenyl)-1-(S)-phenyl-2-(η^5-1-(R)-indenyl)ethane]ZrCl$_2$ (**20b**), (a) in front view and (b) after a 90° horizontal rotation (view perpendicular to the indenyl plane). Reproduced by permission of the American Chemical Society, 1998

metallacycle, as expected. The phenyl substituent is located in the energetically favored equatorial position of the cycle and the β-CH groups (C8, C19 in **20a**) are found in a forward arrangement. The molecular structure of **20b** verifies the λ-backward conformation of this complex type. Interestingly, the inversion of the backbone chirality and hence the conversion from the δ-forward to the λ-backward conformation do not lead to a distortion of the gross coordination geometry. Both

Table 1 Selected bond lengths (Å) and angles (°) for the complexes **20a** and **20b**

20a		20b	
Zr—Cl(1)	2.398(6)	Zr—Cl(1)	2.418(1)
Zr—Cl(2)	2.426(6)	Zr—Cl(2)	2.410(1)
Zr—Flu (centr.)	2.259	Zr—Cp (centr.)	2.275
Zr—Ind (centr.)	2.226	Zr—Ind (centr.)	2.217
Cl(1)—Zr—Cl(2)	96.1(2)	Cl(1)—Zr—Cl(2)	95.6(1)
Flu (centr.)—Zr—Ind (centr.)	127.3	Flu (centr.)—Zr—Ind (centr.)	127.7

Reproduced by permission of the American Chemical Society (© 1998)

bond lengths and angles are in accordance with expectation (Table 1). The effect of the different bridge arrangements is nicely demonstrated by a comparison of a front and a bottom view of the two diastereoisomers (Figure 11). The fluorenyl-ZrCl$_2$ fragments of both complexes can be superimposed, demonstrating the similarity of the coordination geometry. Whereas the forward conformer (**20a**) comprises an (indenyl)-shielded Cl1—Zr—Cl2 fragment (Figure 11(b), 'tight cage'), the bridge-twist to the backward isomer (**20b**) results in a sterically more open environment ('open cage'), indicated by an increased distance between β^{Flu} (C8) and β^{Ind} (C19) on going from **20a** compared with **20b**.,

The conformations found in the solid state are also expected to be preferred in solution, owing to the steric demand of the bulky bridge substituents. If any transition between the δ- and λ-conformations occurs, it has to be extremely fast, since variable-temperature ^{1}H NMR experiments performed on the backbone ABX spin system of **20a, b** showed no line broadening at temperatures down to $-100°$C which could be indicative of such a process [28].

4 INFLUENCE OF POLYMERIZATION TEMPERATURE AND MONOMER CONCENTRATION ON THE POLYMERIZATION PROPERTIES IN HOMOGENEOUS PROPENE POLYMERIZATION REACTIONS

The observation of the molecular structure of diastereomeric complexes such as **20a, b** reveals a clear difference of the arrangements of fluorenyl and indenyl ligands within one particular ligand system. However, this difference is expected to be negligible compared with the influence of different Cp substituents (e.g. alkyl, aryl, condensed cycloalkyl, aromatics) on the stereoselectivity of the C—C bond-forming reaction during chain growth. The following three sub-sections will show that fine tuning of a particular geometry is important for a rational design of selective polymerization catalysts.

4.1 TEMPERATURE EFFECTS

In a first series of experiments, propene polymerization was carried out at different temperatures. The catalyst systems **17** (Flu-H-Ind) and **20a, b** (Flu-Ph-Ind)/MAO were selected for propene polymerization in toluene at 50, 70 and 85 °C at a constant monomer concentration ($[C_3] = 0.71$ mol/l) [29]. A comparison of the polymerization data shows the typical activity increase with temperature (T_p) for all three complexes (Table 2). The propene consumption is constant at 50 and 70 °C over the entire polymerization period. Although the activity is highest at 85 °C, **17** and **20a, b** undergo a slow decomposition reaction at and above this temperature, which results in a steady decline of the monomer consumption with time.

The unsubstituted system **17**/MAO displays the highest polymerization activity in the present series of catalysts. The activities of the diastereomeric complexes **20a** and **20b** are comparable, although the activity of the higher selective **20a** slightly exceeds that of **20b**. The weight-average molecular weights of the polymers decrease with increasing T_p, as expected. Within the error limits of the GPC measurements, the products of all three catalysts have equivalent molecular weights. Interestingly, no regio-misinsertions, such as head-to-head combinations or 1,3-insertions, could be detected for all three catalysts, either in the total polymers or in the fractions [30].

However, the most interesting results arise from a comparison of the product stereoregularities. The δ-forward conformer **20a** is by far the most selective catalyst. The polymers produced with **20a**/MAO are crystalline materials with defined melting transitions at all applied polymerization temperatures. One of the major implications of the stabilizing bridge substituent is demonstrated by an increase in T_p from 50 to 70 °C (constant monomer concentration, 0.71 mol/l; Figure 12). In

Table 2 Polymerization data for **17**, **20a** and **20b**/MAO at different temperatures and constant monomer concentration[a]

Complex	T_p (°C)	t_p (s)	$[Zr]\,10^{-5}$ (mol/l)	Yield (g)	Activity[b]	[mmmm][c] (%)	T_m[d] (°C)	$\overline{M}_w$[e] (kg/mol)	$\overline{M}_w/\overline{M}_n$[e]
17	50	1230	1.1	7.1	2660	63.9	110	27.8	1.8
	70	1020	1.1	11.4	5150	59.3	104	19.3	1.4
	85	1070	1.1	13.2	5690	39.2	Wax	8.6	1.9
20a	50	2360	1.1	5.1	990	70.5	121	26.6	1.4
	70	2650	1.1	26.2	4560	69.8	119	20.8	1.6
	85	1690	1.1	19.8	5400	64.4	98	8.9	2.1
20b	50	2740	1.7	8.3	900	36.0	Wax	27.9	1.8
	70	2330	1.7	22.1	2830	29.7	Wax	21.9	2.0
	85	1670	1.1	18.3	5050	26.5	Oil	9.3	1.7

[a] Propene concentration, $[C_3] = 0.71$ mol/l; Al/Zr = 2000.
[b] Activity in kg PP/([Zr] $[C_3]$ h).
[c] Determined by ^{13}C NMR analysis.
[d] T_m, maximum of polymer melting transition, measured by differential scanning calorimetry.
[e] Determined by GPC in 1,2,4-trichlorobenzene at 140 °C.
Reproduced by permission of the American Chemical Society (© 1998)

contrast to *rac*-ethylenebis(indenyl)zirconium dichloride [8], there is no significant effect on the stereoregularities of the polymers. The polypropenes prepared at these temperatures have mmmm pentad contents ([mmmm]) of 70.5 and 69.8 %, respectively. Only a further increase in T_p to 85 °C reduces [mmmm] to 64.4%. Inversion of the backbone twist to the λ-backward conformer results in a nearly complete loss of stereoselectivity. Compound **20b** produces atactic oils or waxes with [mmmm] values between 26.5 % ($T_p = 85$ °C) and 36.0% ($T_p = 50$ °C). The stereoregularities of the polypropenes resulting from the unsubstituted derivative **17** are between those of **20a** and **20b**. The structure of **17** resembles that of the ethylene-bridged bisindenyl systems **1a, b**. As there, the lack of the stabilizing phenyl group leads to a rapid decline of [mmmm] from 63.9 % ($T_p = 50$ °C) to 39.2 % ($T_p = 85$ °C).

For described sets of diastereomeric catalysts, the reported differences in stereo-selectivity are explained, since they consist in most cases of a racemic C_2-symmetric complex mixture giving isotactic polypropene and an asymmetric diastereomer with reduced stereoselectivity. Well known examples are the racemic and *meso* derivatives of the ethylenebis(indenyl) system [8] or their alkyl-substituted counterparts [31]. In the present series of catalysts, both diastereomers (forward and backward conformers) are asymmetric with comparatively small differences in their ligand arrangements. At present we have no genuine explanation for the unexpected high selectivity difference between **20a** and **20b**. However, **20a** is expected to provide a tight chiral coordination cage, owing to the forward position of the β-substituents (see Figures 9 and 11). To account for the stereoregularity of the polypropenes prepared with **20a**, one could assume that the stereorigidity of this template favors

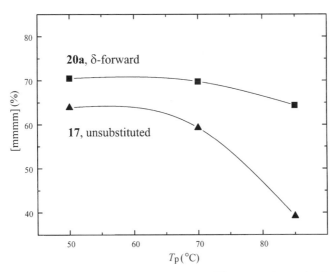

Figure 12 Stereoselectivity of **17** and **20a** at different polymerization temperatures

one particular geometry of the chain and the propene monomer. It would then be easy to visualize that the opened cage in **20b** allows a less rigid arrangement with more than one possible olefin-insertion transition state. Our reasoning is supported by the polymerization results for **17** (cf. Table 2), which carries no bulky bridge substituent and should therefore be free to adopt a sterically 'optimum' geometry during chain propagation, somewhere between the two extremes **20a** and **20b**. This results in the highest activity in the series of the complexes **17**, **20a** and **20b** but also in a significantly reduced stereoselectivity compared with the δ-forward conformer **20a**. The importance of an optimum coordination gap geometry for the design of highly selective catalysts was pointed out by Hortmann and Brintzinger [32].

4.2 Monomer Concentration

A further series of propene polymerization experiments are directed towards the influence of the monomer concentration on the stereoselectivity of asymmetric catalysts. In most of the metallocene catalysts used for propene polymerization, both coordination sides are related to each other by a C_2-symmetric axis. This leads to the same enantiofacial preference for each inserting olefin and hence to the formation of isotactic polymers. Asymmetric catalysts have two different coordination sides with different rates for the selectivity-determining step. If the rate of monomer coordination is of the same order of magnitude as the selectivity-determining step, the stereoselectivity of these complexes should depend on the monomer concentration [33]. To test this hypothesis, **17** (Flu-H-Ind), **19a, b** (Cp-Ph-Ind) and **20a, b** (Flu-Ph-Ind)/MAO were used for propene polymerization reactions at various monomer concentrations and at a constant polymerization temperature of 50 °C.

The results show that the highly substituted catalysts **17, 20a** and **20b**/MAO produce polypropenes with reduced stereoregularities on increasing the monomer concentration [34] (Table 3). Compound **20a** is the most selective species, but its stereoselectivity declines from [mmmm] = 80.4 % (0.45 mol C_3/L) to 46.5 % (3.38 mol C_3/L) with increasing propene concentration. An analogous dependence was found for the less selective derivative **17** and for the unselective λ-backward conformer **20b** (Figure 13) [35]. Complexes **19a** and **19b** are provided with only one sterically demanding β-CH substituent allowing facile monomer coordination at each catalyst side. Consequently, no influence of propene concentration on polymer stereoregularity should be observed. The [mmmm] pentad content of the polymers prepared with **19a,b**/MAO remains unchanged with variations in monomer concentration. One β-substituent is not sufficient to define a chiral cage which is tight enough to favor a single transition state geometry. Hence the polymer stereoregularities are low. However, even here the δ-forward conformer (**19a**) shows a tendency towards the formation of higher stereoregular polypropene.

A theoretical investigation of the dependence of the stereoselectivity of asymmetric complexes such as **17** and **20a, b** on the monomer concentration was recently reported by Guerra *et al.* [36]. They presented a force-field-approach which is

Table 3 Polymerization data for **17**, **19a, b** and **20a, b**/MAO at various monomer concentrations and constant temperature[a]

Complex	$[C_3]$ (mol/l)	t_p (s)	$[Zr]\,10^{-5}$ (mol/l)	Yield (g)	Activity[b]	[mmmm] (%)	$\overline{M}_w$ (kg/mol)	$\overline{M}_w/\overline{M}_n$
17	0.45	1480	1.1	5.7	1260	68.1	n.d.[c]	n.d.
	1.16	1430	1.1	9.6	2200	52.4	n.d.	n.d.
	1.76	1340	1.1	16.6	4060	42.8	n.d.	n.d.
	3.38	1620	1.1	20.3	4100	37.5	n.d.	n.d.
19a	0.45	3020	3.1	1.3	50	40.6	0.5	1.8
	1.16	3170	3.1	12.0	440	39.9	1.4	2.1
	1.76	2120	3.1	15.5	850	40.1	4.2	1.7
	3.38	2720	1.5	21.5	1900	39.3	11.0	1.5
19b	0.45	3070	1.5	0.7	60	35.4	0.8	2.0
	1.16	2850	1.5	2.3	190	35.0	1.8	1.8
	1.76	2770	1.5	6.6	570	35.7	4.5	1.7
	3.38	2630	1.5	13.0	1190	34.4	10.8	1.7
20a	0.45	1790	1.5	8.0	1070	80.4	11.5	1.6
	1.16	2010	1.5	16.1	1920	63.2	25.1	1.8
	1.76	1640	1.1	14.4	2870	53.1	36.7	1.4
	3.38	1470	1.1	16.4	3660	46.5	49.0	2.0
20b	0.45	1930	1.5	7.0	870	28.3	13.7	1.8
	1.16	1910	1.5	12.1	1520	21.3	27.0	1.5
	1.76	1920	1.5	14.5	1810	17.7	47.5	2.1
	3.38	1570	1.1	14.2	2960	15.5	62.9	1.4

[a] Polymerization temperature, $T_p = 50°C$; Al/Zr = 2000.
[b] Activity in kg PP/([Zr] h).
[c] n.d.: not determined.
Reproduced by permission of the American Chemical Society (© 1998)

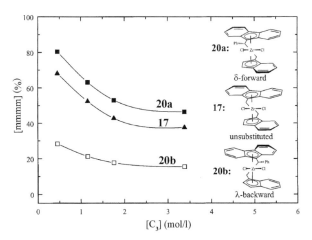

Figure 13 Polypropene stereoregularity ([mmmm]) versus monomer concentration ($[C_3]$) for different complexes at a constant polymerization temperature ($T_p = 50°C$)

applicable to diastereomeric catalytic centers with a substantially different energy of the two possible alkene-bound intermediates. Monomer coordination and insertion can occur at each of the two sides of either diastereomer, which leads to four differently ligated coordination compounds (Figure 14). The catalytically active cation exists in two arrangements depending on the coordination of the growing chain (**B** and **C**). Monomer coordination and insertion should be fast and highly selective going from state **C** through **A** to **B** (after a stereoregular insertion). A plot of the energy [36] of the two alkene-coordinated intermediates (**A** and **D**) demonstrates nicely that the coordination of the incoming monomer on the sterically more hindered side (between Flu and Ind ligands) is favored by the forward conformer, depicted as **A** in Figure 14. After each migratory insertion the growing chain moves back to the previous position (back-skip of the growing chain) so that always the same coordination side is favored by the chain in successive insertion reactions (**A** $\longrightarrow$ **B** $\longrightarrow$ **C**). The probability of this back-skipping process of the chain in the alkene-free state is dependent on the difference between the activation energies of the back-skip (**B** $\longrightarrow$ **C**) and the formation of the high-energy alkene-coordinated intermediate (**B** $\longrightarrow$ **D**). Guerra *et al.* [36] showed that the intermediate **A** leads to the stereoselective insertion (**A** $\longrightarrow$ **B**), whereas the path **D** $\longrightarrow$ **C** shows a considerably lowered stereoselectivity.

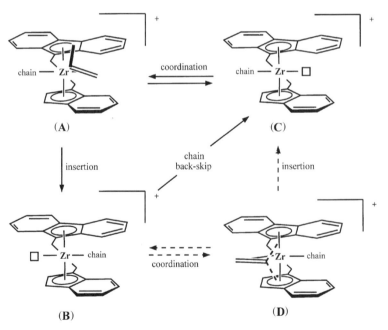

Figure 14 Monomer coordination, insertion and chain back-skip at an asymmetric fluorenyl–indenyl catalyst

By increasing the monomer concentration, the concentration of state **D** is enhanced, which leads to a higher probability of unselective insertions (**D** → **C**). This explains the decline of the stereoselectivity of these asymmetric catalytic systems at increased monomer concentrations.

Thus, the rate of the back-skipping process is in a range where a variation in monomer concentration can be used to change the probability for a stereoirregular insertion. Despite the fact that this control mechanism does not allow one to place such an error at any position of the polymer chain, it is to our knowledge the first example of controlling the amount and the distribution of single stereoerrors along one particular chain in an arbitrary manner and not just statistically by a randomly occurring fluctuation.

4.3 Catalyst Structure

In addition to the design of the stereoselectivity of a catalyst by fine tuning its coordination geometry, the position and the number of substituents, e.g. condensed aromatic units on the cyclopentadienyl rings, play a major role in catalyst performance. In order to investigate the influence of stereoridgid (substituted) bridges for fluorenyl–cyclopentadienyl and fluorenyl–fluorenyl combinations, we performed a series of experiments with the catalysts **16**/MAO (Flu-Ph-Cp, Figure 8) and **18**/MAO (Flu-Ph-Flu, Figure 8) and compared the results with those for related complexes in the literature.

The first complex employed for polymerization experiments was *rac-*{$1, 2 - bis[\eta^5 - (9 - fluorenyl)] - 1 - phenylethane$} zirconium dichloride (**18**), which contains two C_2-symmetric fluorenyl groups [37]. This complex was used together with MAO for the polymerization of propene at 30, 50 and 70 °C at constant monomer concentration (0.71 mol C_3/L). Comparison of the polymerization data (Table 4) shows that all of the polymers produced with **18**/MAO are materials of moderate isotacticity with [mmmm] pentad contents between 31.2 and 64.1 %, depending on the polymerization temperature applied as described above. Replacement of the chiral ethylene bridge in **18** by a dimethylsilane group forces both fluorenyl units back into an eclipsed orientation, giving rise to a C_{2v}-symmetric,

Table 4 Polymerization data for **18**/MAO at different temperatures and constant monomer concentration[a]

T_p (°C)	t_p (s)	[Zr] 10^{-5} (mol/l)	Yield (g)	Activity[b]	[mmmm] (%)	$\overline{M}_w$ (kg/mol)	$\overline{M}_w/\overline{M}_n$
30	2720	2.9	1.1	70	64.1	21.2	1.7
50	3140	3.1	12.6	650	49.6	18.5	1.4
70	1430	3.1	25.6	2920	31.2	1.8	1.9

[a] Monomer concentration, [C_3] = 0.71 mol/L; Al/Zr = 2000
[b] Activity in kg PP/([Zr] [C_3] h)
Reproduced by permission of Springer Verlag (© 1998)

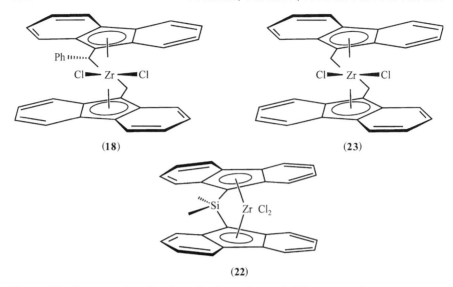

Ph
Cl
Zr
Cl

(**18**)

Cl
Zr
Cl

(**23**)

Si
Zr Cl2

(**22**)

Figure 15 Staggered and eclipsed orientations of different catalyst structures

achiral structure (**22**, Figure 15). The catalyst **22**/MAO produces atactic oligo- and polypropenes [38], as expected. The staggered arrangement of both fluorenyl groups remains on removing the bridge substituent from **18**. However, the structure becomes much more flexible (**23**, Figure 15) [39], which again produces basically atactic polymers. This demonstrates clearly that the stereoselectivity of **18** is a consequence of the staggered and hence chiral arrangement of both fluorenyl groups, which is stabilized by the bulky bridge substituent.

The major chain termination reaction in metallocene polymerization catalysts seems to be a spontaneous β-hydride elimination, leading to 2,2-substituted vinyl chain ends [Figure 16 (a)]. Interestingly, only a low degree of vinylidene end groups

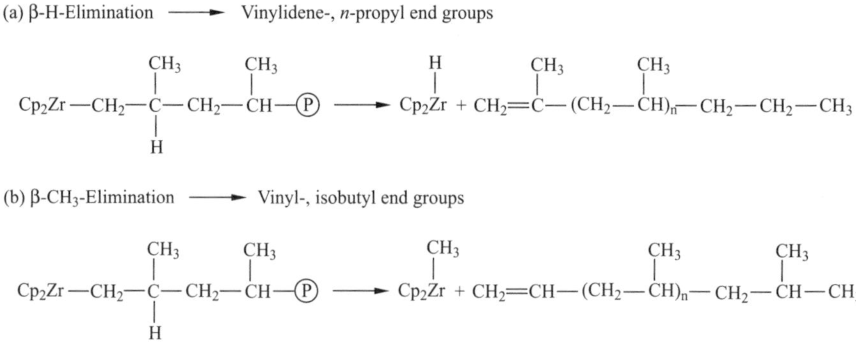

(a) β-H-Elimination → Vinylidene-, n-propyl end groups
Cp2Zr—CH2—C(CH3)(H)—CH2—CH(CH3)—(P) → Cp2Zr(H) + CH2=C(CH3)—(CH2—CH(CH3))n—CH2—CH2—CH3
(b) β-CH3-Elimination → Vinyl-, isobutyl end groups
Cp2Zr—CH2—C(CH3)(H)—CH2—CH(CH3)—(P) → Cp2Zr(CH3) + CH2=CH—(CH2—CH(CH3))n—CH2—CH(CH3)—CH3

Figure 16 β-Hydride versus β-methyl elimination and formation of vinyl-terminated macromonomers

could be detected in the ^{1}H NMR spectra of the polymers produced by the sterically crowded catalysts **18**, **22** and **23**/MAO, which indicates that such a β-hydride-elimination process occurs. In contrast, the proton spectra show a dominance of vinyl end groups formed by a $\beta - CH_3$ elimination [Figure 16 (b)]. This alternative termination mechanism has been found for bisfluorenyl catalysts, where the substitution pattern suppresses a fast β-H elimination [40]. Although the $\beta - CH_3$ elimination creates vinyl end groups, which in principle represent macromonomers, no presence of long-chain branched polypropenes could be proved so far.

The addition of an aromatic group to the fluorenyl–indenyl complexes leads to the partly isospecific polymerization catalyst **18**. The reduction of the steric demand going from **20** to the asymmetric (symmetry around the ZrCl$_2$ moiety resembles that of a C_s-symmetric catalytic system) species **16** {[1 $-$ (η^5-cyclopentadienyl)-1-(R,S)-phenyl-2-(η^5-9-fluorenyl)ethane]zirconium dichloride, Figure 17} results in a syndiospecific catalyst after activation with MAO. The system **16**/MAO contains a phenyl-substituted ethylene bridge and displays the highest activity of all mentioned polymerization catalysts. It leads to relatively high molecular weight products (ca 85 kg/mol; 0.6% of the polymer consists even of M_w > 600 kg/mol) of a partly syndiotactic microstructure. These high molecular weight products are thermoplastic elastomers after isolation. Unfortunately, they lose these properties after some weeks, probably owing to recrystallization phenomena.

Also here the stereoregularities depend on both polymerization temperature and monomer concentration (Tables 5 and 6) [41]. High syndiotacticities ([rrrr]) up to 85.3% can be achieved at low polymerization temperatures (30 °C) and, in contrast to the isotactic polypropenes resulting from **20a, b**, at increased monomer concentrations. The rrrr pentad results if each chain migration is followed by a monomer insertion into the growing polymer chain (catalytic site control). At low monomer concentrations there is also the chance of a back-skip of the chain without insertion,

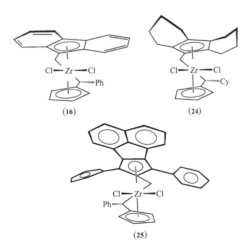

Figure 17   Comparison of three different sterically demanding complexes

Table 5 Polymerization results for **16**/MAO at different temperatures and constant monomer concentration

T_p (°C)	t_p (s)	[Zr] 10^{-5} (mol/l)	Yield (g)	Activity[c]	[rrrr] (%)	$\overline{M}_w$ (kg/mol)	$\overline{M}_w/\overline{M}_n$
30^a	1318	0.5	7.9	3320	82.3	85.0	1.8
50^b	1317	0.8	9.1	4380	36.7	28.0	1.9
70^b	505	0.8	9.8	12300	31.7	16.7	2.2
80^b	500	0.8	11.2	14200	20.3	8.3	2.0

[a] Monomer concentration, $[C_3] = 1.30$ mol/l; Al/Zr = 2000.
[b] Monomer concentration, $[C_3] = 0.71$ mol/l; Al/Zr = 2000.
[c] Activity in kg PP/([Zr] [C_3] h).
Reproduced by permission of Springer Verlag (© 1998)

Table 6 Polymerization results for **16**/MAO at various monomer concentrations[a]

T_p (°C)	$[C_3]$ (mol/l)	t_p (s)	[Zr] 10^{-5} (mol/l)	Yield (g)	Activity[b]	[rrrr] (%)	$\overline{M}_w$ (kg/mol)	$\overline{M}_w/\overline{M}_n$
50	0.71	1317	0.8	9.1	4380	36.7	28	1.9
	1.16	509	0.8	7.0	5340	43.1	48	1.6
	1.76	561	0.8	14.3	6520	46.7	79	2.1
	3.38	663	0.8	36.3	7290	53.5	110	1.8
30	3.08	1030	0.5	28.1	6380	85.3	160	2.0

[a] Al/Zr = 2000.
[b] Activity in kg PP/([Zr] [C_3] h).
Reproduced by permission of Springer Verlag (© 1998)

leading to an isolated stereoerror, indicated by the rrmr pentad. Increased propene concentrations reduce the content of rrmr pentads owing to the low probability of a chain migration without insertion.

Interestingly, hydrogenation of the fluorenyl fragment **16** to the octahydrofluorenyl complex **24** (Figure 17) leads to a drastic reduction in activity, stereoselectivity and molecular weights of the polymers (Table 7). The same effect results after replacement of the fluorenyl group by an extended aromatic π-system, as in [1 − (η^5-7,9-diphenyl-cyclopenta[a]acenaphthadienyl)-1-(R,S)-phenyl-2-(η^5-9-cyclopentadienyl)ethane]zirconium dichloride (**25**) [42]. The catalyst **25**/MAO produces only atactic polypropenes with reduced activity (Table 7). This behavior is attributed to the steric influence of the bulky 7,9-diphenyl-cyclopenta[a]acenaphthadienyl group [or cyclohexyl fragments (**24**)] which hinders an effective orientation of the incoming monomer [43].

5 TRANS-CYCLOHEXANE-BRIDGED ANSA-METALLOCENE COMPLEXES

The polymerization properties of the δ-in/λ-out conformers indicate that conformationally rigid complexes such as **20a** (Figure 10, Tables 2 and 3) can show

Table 7 Polymerization results for **24** and **25**/MAO under various polymerization conditions^a

Complex	T_p (°C)	$[C_3]$ (mol/l)	t_p (min)	Activity^b	$\overline{M}_w$ (kg/mol)
24	30	1.06	219	386	15.1
	50	0.25	103	2311	3.0
	50	0.71	106	1661	5.9
	50	1.89	53	1348	10.1
	70	0.37	176	2641	1.8
25	30	0.71	48	1150	1.8
	50	0.71	33	1440	1.7
	50	1.16	32	1720	1.7
	50	1.76	10	3650	1.9
	50	3.38	11	5700	1.8

^{a} Al/Zr = 2000; [Zr] = 10^{-5} mol/l.
^{b} Activity in kg PP/(mol Zr [C$_3$] h).

improved stereoselectivities, compared with their unsubstituted analogs. However, no highly isotactic polypropene ([mmmm] $\geqslant$ 90 %) was achieved by applying these asymmetric fluorenyl–indenyl catalysts. This might be due to the presence of two coordination sites showing different stereoselectivities. Therefore, by applying the concept of conformational rigidity to C_2-symmetric complexes, bearing two bridge substituents, an additional increase in catalyst performance is expected. We followed a synthetic route which leads to stereorigid *trans*-1,2-cycloalkane bridged complexes (Figure 18). The substrates *trans*-1,2-dihydroxycyclopentane (**26**) and *trans*-1,2-dihydroxycyclohexane (**30**), arising from oxidation of the corresponding cycloalkenes with H_2O_2 under acidic conditions, are converted into the corresponding bis(methylsulfonates) **27** and **31**. Successive treatment with indenyllithium in DMF leads to the ligand precursors **28** and **32** in high yield [44, 45]. The use of 1,2-(*R,R*)-dihydroxycyclohexane [46] leads to the formation of the single enantiomer 1,2-(*S,S*)-diindenylcyclohexane (**32**) in nearly quantitative yield and thus allows the direct preparation of enantiomerically pure *ansa*-metallocene dichlorides (**33**) without the need to separate the enantiomeric complexes.

Starting from the ligand precursor **32**, deprotonation with *n*-butyllithium and successive complexation with ZrCl$_4$ afford the formation of three diastereomeric complexes in a 1 : 1 : 2 ratio [δ-forward (**33a**): λ-backward (**33b**): asymmetric (**33c**); Figure 19]. Complex **33c** can be isolated by crystallization from toluene. A further separation of the two C_2-symmetric isomers remained unsuccessful.

However, even polymerization experiments with a 1 : 1 mixture of the C_2-symmetric diastereomers **33a, b** revealed an increase in stereoselectivity together with better thermal stability compared with the non-substituted racemic mixture of **1a, b** (Figure 20) [45].

506

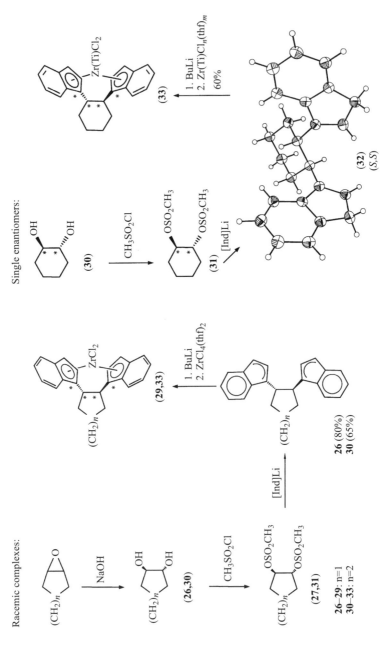

Figure 18 Preparation of racemic and enantiomerically pure *trans*-1,2-cycloalkane-bridged ligand precursors and their related Ti(IV) and Zr(IV) complexes

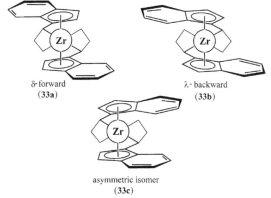

δ-forward
(33a)

λ-backward
(33b)

asymmetric isomer
(33c)

Figure 19 Diastereomers of *trans*-1,2-[bis-(1-η^5-indenyl)cyclohexane]zirconium dichloride (chlorine atoms are omitted)

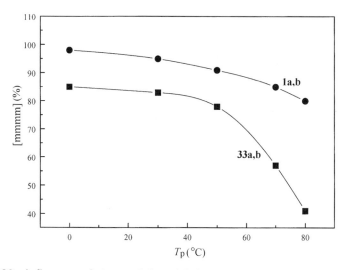

Figure 20 Influence of the cyclohexyl bridge on the stereoregularity of the corresponding polypropenes

An ultimate separation of the isomers could be achieved after hydrogenation of the complexes to their tetrahydroindenyl derivatives **34a–c** (Figure 21). After extraction of the soluble asymmetric isomer **34c** with boiling hexane, a 1 : 1 mixture of the C_2-symmetric complexes remained as an insoluble residue. Fractional crystallization from toluene allowed the separation of the C_2-symmetric isomers.

The structure of one hydrogenated species (**34b**) was determined by X-ray structure analysis. This species shows a λ-backward conformation of the metallacycle ('open chiral cage') (Figure 22).

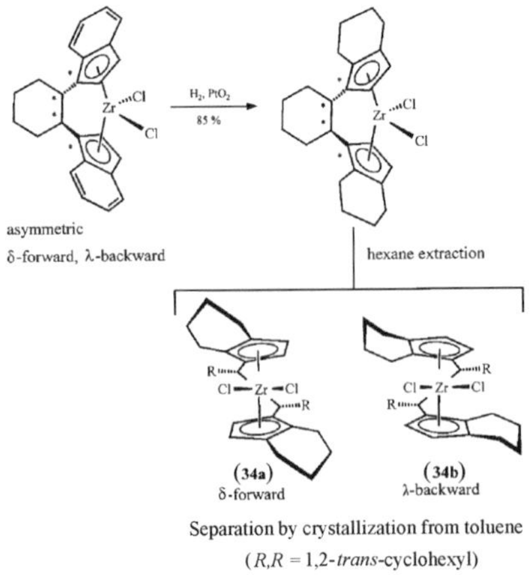

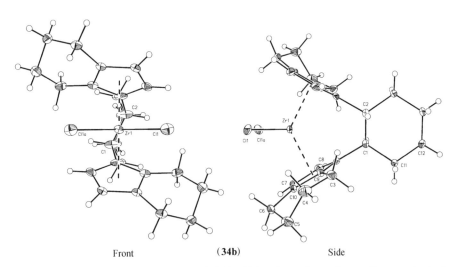

Figure 21 Hydrogenation and separation of the *trans*-cyclohexyl-bridged complexes **33a–c**

The boot conformation of the bridging *trans*-1,2-cyclohexyl unit leads to an equatorial position of the CH$_2$ substituents and effectively hinders the interconversion of δ- and λ-diastereomers.

Propene polymerization experiments were performed under identical conditions with all three hydrogenated isomers (Figure 23). The asymmetric form (**34c**) displays no significant stereoselectivity and leads to atactic waxes with low activity,

Figure 22 Solid-state structure of **34b** [λ-backward conformation; selected bonds and angle: Zr(1)—Cp-centroid, 2.248(4) Å; Zr(1)—Cl(1), 2.447(3) Å; Zr(1)—Cl(1a), 2.447(5) A ; Cl(1)—Zr(1)—Cl(1a), 98.85(8)°]

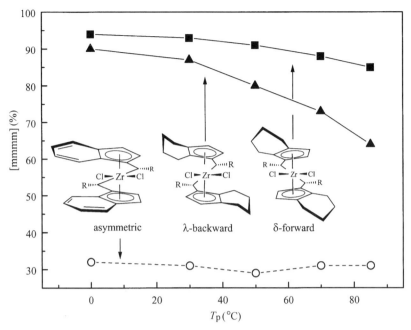

Figure 23 Dependence of the isotacticity ([mmmm]) on the polymerization temperature (T_p) of **34a–c**/MAO

as expected. Application of the λ-backward complex **34b** resulted in the formation of isotactic polypropenes, the isotacticity ([mmmm] %) of which varied from 90 to 75 % on increasing T_p from 30 to 85 °C. The δ-forward isomer **34a** leads to isotactic polypropene ($95 \geqslant [mmmm] \geqslant 85$ %, $30 \leqslant T_p \leqslant 85$ °C), where the influence of T_p on the stereoselectivity is lowest in comparison with the other cyclohexyl-bridged isomers.

The results show how sensitively the polymerization properties of both symmetric and asymmetric *ansa*-metallocene catalysts are influenced not just by the specific Cp substitution pattern applied, (alkyl, aryl; condensed aromatic, cycloalkyl), but also by fine tuning of the geometry of both Cp ligands within one particular ligand system. For the formation of highly isotactic, high molecular weight polypropenes, a portfolio of 2,4-substituted 'leading structures' (Figure 2) have been provided in recent years. Owing to the easy formation of silane-bridged, rigid structures, the application of *trans*-cyclohexyl-bridged species might be limited for that purpose. However, the availability of enantiomerically pure *trans*-dihydroxycycloalkanes might open the way to the development of other optically active C_2-symmetric complexes without separation of the enantiomers. This could help to develop the application of these systems for other enantioselective transformations. The most important results arise from the influence of the monomer concentration on the stereoselectivity of the asymmetric complexes. By using these unique species it is possible to influence the rate and the position of stereoerror formation along the chain by a second parameter, different from temperature, which can be easily controlled. Future research will show whether this concept can be used for the arbitrary control of the crystalline/non-crystalline segments within an isotactic polymer chain and hence to conduct phase separation phenomena leading to stereoregular polyolefins with new properties [47].

6 REFERENCES

1. (a) Schnutenhaus, H. and Brintzinger, H. H., *Angew. Chem.*, **91**, 837 (1979); (b) Wild, F. R. W. P., Zsolnai, L., Huttner, G. and Brintzinger, H. H., *J. Organomet. Chem.*, **232**, 233 (1982); (c) Wild, F. R. W. P., Wasiucionek, M., Huttner, G. and Brintzinger, H. H., *J. Organomet. Chem.*, **288**, 63 (1985).
2. All ethylene-bridged, twisted metallacycles can exist in two different conformers (δ or λ, see Section 2), leading to staggered arrangements of the indenyl moieties (cf. Figure 4). In this respect isomer **1c** can be considered as a *'meso'* structure only in time average; neither twisted metallacycle contains a C_s-symmetry element and are hence better described as asymmetric isomers.
3. Ewen, J. A., Jones, R. and Razavi, A., *J. Am. Chem. Soc.*, **106**, 6355 (1984).
4. Kaminsky, W., Külper, K., Brintzinger, H. H. and Wild, F. R. W. P., *Angew. Chem.*, **97**, 507 (1985).
5. Hackmann, M. and Rieger, B., *CaTTech*, **2**, 79 (1997).
6. (a) Chien, J. C. W., *J. Polym. Sci., Part A*, **425**, 1839 (1963); (b) Keii, T., *Kinetics of Ziegler–Natta Polymerization,* Kodansha, Tokyo, 1972.
7. All polymerization experiments were carried out by using an Al/Zr ratio of 2000.
8. Rieger, B., Mu, X., Mallin, D. T., Rausch, M. D. and Chien, J. C. W., *Macromolecules*, **23**, 3559 (1990).
9. Chien, J. C. W., *J. Am. Chem. Soc.*, **81**, 86 (1959).
10. (a) Reichert, K. H., *Angew. Makromol. Chem.*, **84**, 1 (1981); (b) Chien, J. C. W. and Hu, Y., *J. Polym. Sci., Part A*, **26**, 2973 (1988).
11. Chien, J. C. W., *J. Polym. Sci., Polym. Chem. Ed.*, **1**, 425 (1963).
12. The stereoregularity of the polypropenes was determined by ^{13}C NMR spectroscopy by the degree of [mmmm] sequences, which also classifies the selectivity of the catalyst; see also Frisch, H. L., Mallows, C. L. and Bovey, F. A., *J. Chem. Phys.*, **45**, (1966).
13. Rieger, B., Reinmuth, A., Röll, W. and Brintzinger, H. H., *J. Mol. Catal.*, **82**, 67 (1993).
14. The term β relates to the position of the Cp substituents relative to the bridge connection: see also Hortmann, K. and Brintzinger, H.H., *New J. Chem.*, **16**, 51 (1992).
15. For the notation of chiral metallacycles, see Corey, E. J. and Bailar J. C., Jr, *J. Am. Chem. Soc.*, **81**, 2620 (1959).
16. Mallin, D. T., Rausch, M. D., Lin, Y. G. and Chien, J. C. W. J., *J. Am. Chem. Soc.*, **12**, 2030 (1990).
17. Rieger, B., Jany, G., Fawzi, R. and Steimann, M., *Organometallics*, **13**, 64 (1994), and references therein.
18. Rieger, B., Jany, G., Steimann, M. and Fawzi, R., *Z. Naturforsch., Tal B*, **49**, 451 (1994), and references therein.
19. Rieger, B., Steimann, M. and Fawzi, R., *Chem. Ber.*, **125**, 2373 (1992).
20. Rieger, B., Jany, G., Fawzi, R. and Steimann, M., *Organometallics*, **16**, 544 (1997).

21. By the use of a racemic mixture of **4** two sets of resonance signals appear in the ^{1}H NMR spectrum. All peaks could be assigned to the corresponding protons by ^{1}H/^{1}H-COSY NMR experiments.
22. Schlögl, K., *Top. Curr. Chem.*, **6** 479 (1966).
23. All trials to achieve the formation of the complex **21** diastereoselectively or to separate one isomer have remained unsuccessful up to now.
24. The exact notation of the four theoretically possible enantiomers is as follows: **20a**, *RR* = δ-forward and *SS* = λ-forward; **20b**, *RS* = λ-backward and *SR* = δ-backward.
25. Buckingham, D. A. and Sargeson, A. M., *Top. Stereochem.*, 219 (1971).
26. (a) Schäfer, A., Karl, E., Zsolnai, L., Huttner, G. and Brintzinger, H. H., *J. Organomet. Chem.*, **328**, 87 (1987); (b) Brintzinger, H. H., in Kaminsky, W. and Sinn, H.-J., (Eds), *Transition Metals and Organometallics as Catalysts for Olefin Polymerization*, Springer, Berlin, 1988, p. 249.
27. Llinas, G. H., Day, R. O., Rausch, M. D. and Chien, J. C. W. J., *Organometallics*, **12**, 1283 (1993).
28. A flip from the preferred δ-conformation of **20a** to the λ-form would bring the phenyl bridge substituent into close vicinity to the aromatic fluorenyl protons, which, in our eyes, would create strong repulsive interactions. However, such a flip would lead to a new isomer with it own NMR pattern. Since ^{1}H NMR has a resolution of about 98% and this second pattern cannot be detected, the population of such a second state is lower than 2%.
29. Propene concentrations were measured by the use of a calibrated gas flow meter (Bronkhurst F-111-C-HD-33-V). The experimental results were checked by comparison with calculated data. For an appropriate equation of state for the propene–toluene system, see, (a) Plöcker, U., Knapp, H. and Prausnitz, J., *Ind. Eng. Chem. Process Des. Dev.*, **17**, 324 (1978); (b) Tyvina, T. N., Efremova, G. D. and Pryanikova, R. O., *Russ. J. Phys. Chem.*, **47**, 1513 (1973).
30. Selected polymer samples were fractionated according to Pasquon's method; see Ref. 8 and references therein
31. Collins, S., Gauthier, W. J., Holden, D. A., Kuntz, B. A., Taylor, N. J. and Ward, D. G., *Organometallics*, **10**, 2061 (1991).
32. Hortmann, K. and Brintzinger, H. H., *New J. Chem.*, **16**, 51 (1992).
33. An increase in polymer stereoregularity with increase in monomer concentration was found for C_s-symmetric isopropylidene[(9-fluorenyl)cyclopentadicnyl]zirconium dichloride. With the same system a decrease in the rrmr pentad at higher monomer concentration was reported. This finding was attributed to a facile chain migration at low C_3 concentration between two successive insertions (skipped insertion) as a consequence of the coordinative unsaturated zirconium cation; cf. (a) Ewen, J. A., *Stud. Surf. Sci. Catal.*, **56**, 439 (1990); (b) Herferth, N. and Fink, G., *Makromol. Chem., Makromol. Symp.*, **66**, 157 (1993).
34. Polymerization experiments with C_2-symmetric *rac*-ethylenebis(indenyl)zirconium dichloride did not show any influence of monomer concentration (in the range applied here) on the polymer stereoregularity.
35. The polypropene products of **20b** show a slight increase in the syndiotactic rrrr pentad (up to 10%; [C_3] = 3.38 mol/l) at increased monomer concentrations. This could be attributed to a structural relation of this complex to the known, highly syndioselective isopropylidene[(9-fluorenyl)cyclopentadienyl]zirconium dichloride.
36. Guerra, G., Cavallo, L., Moscardi, G., Vacatello, M. and Corradini, P., *Macromolecules*, **29**, 4834 (1996).
37. Rieger, B., *Polym. Bull.*, **32**, 41 (1994).

38. Resconi, L., Jones, R. L., Rheingold, A. L. and Yap, G. P. A., *Organometallics*, **15**, 998 (1996).
39. Alt, A., Milius, W. and Palackal, S., *J. Organomet. Chem.*, **472**, 113 (1994).
40. (a) Resconi, L., Piemontesi, F., Franciscono, G., Abis, L. and Fiorani, T., *J. Am. Chem. Soc.*, **114**, 1025 (1992); (b) Resconi, L., Fait, A., Piemontesi, F., Colonnesi, M., Rychlicki, H. and Zeigler, R., *Macromolecules*, **28**, 6667 (1995).
41. Rieger, B., Repo, T. and Jany, G., *Polym. Bull.*, **35**, 87 (1995).
42. Repo, T., Jany, G., Hakala, K., Klinga, M., Polamo, M., Leskelä, M. and Rieger, B., *J Orgmet. Chem.*, **549**, 177 (1997).
43. See also (a) Ewen, J. A., Elder, M. J., Jones, R., Curtis, S. and Cheng, H. N., *Stud. Surf. Sci. Catal.*, **56**, 439 (1990); (b) Ewen, J. A., Elder, M. J., Jones, R. L., Haspeslagh, L., Atwood, J. L., Bott, S. G. and Robinson, K., *Macromol. Chem., Macromol. Symp.*, **48/49**, 253 (1991); (c) Ewen, J. A. and Elder, M. J., *Macromol. Chem., Macromol. Symp.*, **76**, 179 (1993)
44. Rieger, B., *J. Orgmet. Chem.*, **428**, C33 (1992).
45. For an analogous approach to *trans*-cycloalkyl bridged complexes, see Erker, G., Steinhorst, A., Grehl, M. and Fröhlich, R., *J. Organomet. Chem.*, **542**, 191 (1997).
46. *Org. Synth., Coll. Vol.*, **3**, 544 (1955).
47. Hoveyda, A. H. and Morken, J. P., *Angew. Chem.*, **108**, 1378 (1996).

INDEX

Contents of Volume Two